The *Spherics* of Theodosios

This book provides the first English translation of the Greek text of the *Spherics* of Theodosios (2nd–1st century BCE), a canonical mathematical and astronomical text used from as early as the 2nd century CE until the early modern period.

Accompanied by an introduction to the life and works of Theodosios and a contextualization of his *Spherics* among other works of Greek mathematics and astronomy, the translation is followed by a detailed commentary, and an accessible English paraphrase accompanied with mathematically generated diagrams. The volume has a broad appeal to both general and specialist readers who do not read ancient Greek – allowing readers to understand the mathematical and astronomical principles and methods used by ancient and medieval readers of this important text. The paraphrase with its mathematical diagrams will be useful for readers with a scientific and mathematical background.

This study of one of the canonical mathematical and astronomical texts of the ancient Greco-Roman, classical Islamic, and medieval Christian worlds provides an invaluable resource for historians of science, astronomy, and mathematics, and scholars of the ancient and medieval periods.

Nathan Sidoli is a professor of the history and philosophy of science at Waseda University, Tokyo. He has published widely on Greek and Islamic mathematics and astronomy, including *Thābit ibn Qurra's Restoration of Euclid's Data*, with Yoichi Isahaya, and is currently co-editor of *Historia Mathematica* and *SCIAMVS*.

R.S.D. Thomas is an emeritus professor of mathematics at the University of Manitoba and fellow of St John's College in Winnipeg, Canada. His previous historical work was *Euclid's Phaenomena*, with J.L. Berggren. His main scholarly activity since 1992 has been editing the journal *Philosophia Mathematica*.

W0234967

Scientific Writings from the Ancient and Medieval World
Series editor
John Steele
Brown University, USA

Scientific texts provide our main source for understanding the history of science in the ancient and medieval world. The aim of this series is to provide clear and accurate English translations of key scientific texts accompanied by up-to-date commentaries dealing with both textual and scientific aspects of the works and accessible contextual introductions setting the works within the broader history of ancient science. In doing so, the series makes these works accessible to scholars and students in a variety of disciplines including history of science, the sciences, and history (including Classics, Assyriology, East Asian Studies, Near Eastern Studies and Indology).

Texts will be included from all branches of early science including astronomy, mathematics, medicine, biology, and physics, and which are written in a range of languages including Akkadian, Arabic, Chinese, Greek, Latin, and Sanskrit.

The Medicina Plinii
Latin Text, Translation, and Commentary
Yvette Hunt

Learning With Spheres
The *golādhyāya* of Nityānanda's *Sarvasiddhāntarāja*
Anuj Misra

The *Arithmetica* of Diophantus
A Complete Translation and Commentary
Jean Christianidis and Jeffrey Oaks

Aristarchus of Samos: *On Sizes and Distances of the Sun and Moon*
Greek Text, Translation, Analysis, and Relevant Scholia
Christián C. Carman and Rodolfo P. Buzón

The *Spherics* of Theodosios
Nathan Sidoli and R.S.D. Thomas

https://www.routledge.com/classicalstudies/series/SWAMW

The *Spherics* of Theodosios

Nathan Sidoli and R.S.D. Thomas

Routledge
Taylor & Francis Group

LONDON AND NEW YORK

First published 2023
by Routledge
4 Park Square, Milton Park, Abingdon, Oxon OX14 4RN

and by Routledge
605 Third Avenue, New York, NY 10158

Routledge is an imprint of the Taylor & Francis Group, an informa business

British Library Cataloguing-in-Publication Data
A catalogue record for this book is available from the British Library

Library of Congress Cataloging-in-Publication Data
A catalog record for this book has been requested

ISBN: 978-0-367-55730-0 (hbk)
ISBN: 978-0-367-69535-4 (pbk)
ISBN: 978-1-003-14216-4 (ebk)

DOI: 10.4324/9781003142164

Typeset in Latin Modern
by Apex CoVantage, LLC

This book has been prepared from camera-ready copy provided by the author Nathan Sidoli.

For Junpei, and Hugh and Michael

Contents

Figures

Part IV: Paraphrase   361

Preface

This book was written off and on over the course of more than ten years, with each of us putting in work as our other projects and responsibilities allowed. We came at this text with rather different backgrounds and expectations, and because our communication was almost always through email we had a number of lively disagreements about various aspects of the project. Nevertheless, we hope that we have resolved those disputes in a way that will be of some benefit to the reader.

The primary goal of this book is to put out the first English translation of the Greek text of the *Spherics* of Theodosios, which is also a scholarly translation based on the most recent work of philologists studying the manuscripts of the text and of classicists studying ancient Greek usage in mathematical texts. This is accompanied with material introducing the reader to Theodosios, his text, and the intellectual contexts in which the text was written and read, along with commentaries to help the reader follow the details of the text itself. Furthermore, since we believe that many readers will be interested in the mathematical ideas introduced in the text, but not feel the need to follow these in their historical presentation, we have included a paraphrase, which renders the text into more idiomatic English and is accompanied with perspective diagrams produced using Mathematica®.

We wish to express our gratitude to the editors at Routledge, and in particular John Steele for encouraging our work and giving us feedback on our manuscript. We would like especially to thank Claire Czinczenheim, who kindly supplied us with electronic files of her edition, and without whose excellent scholarship our work would not have been possible. We thank Michalis Sialaros for helping us secure images and reproduction permissions from the Ministry of Culture and Sports, Greece; and Elizabeth Evans for permission to print her photograph of the so-called Solunto sphere. Nathan Sidoli wishes to thank his seminar students, especially Shun Iwamitsu, Drake Mitsukawa, and Aoi Sekiyama, who read through an early draft of the translation and gave useful feedback.

This book was typeset using X∃LↃTEX. The font for the Roman script is GUST e-foundry's Latin Modern; The Greek font is Old Standard TT, designed by Alexey Kryukov; and the Arabic font is Amiri, designed by Khaled Hosny. These fonts are all available under the Open Font License.

The frontispiece image is of a ring of engraved lapis lazuli set in gold, dated to the 16th century.[1] It depicts a bearded philosopher or mathematician sitting before an engraved celestial sphere, compass in hand, looking up at the sun, moon, and stars, which are displayed in gold (Fitzwilliam Museum, Cambridge, Acc. no. CG 656. Photograph © The Fitzwilliam Museum).

Mathematical notation

In order to discuss the mathematical content of the work outside of the Translation and Paraphrase, we introduce the following notational conventions:[2]

Points: We denote points with italic type, A, B, C, and so on.

Lines: When discussing the details of the text we denote a line using two or more points that lie on the line, AB, CDE, and so on. In order to facilitate describing the mathematical ideas in the text at a more general level, without entering into the details of the text, we also use lowercase italic type to denote lines with a single letter, such as a, b, and so on. If a line with a single construction or defining characteristic takes on a number of different positions, we use subscripts, as a_1, a_2, a_n.

Figures: We denote figures with bold type, such that a triangle with vertices A, B, C is denoted as $\mathbf{T}(ABC)$, a square on AB as $\mathbf{S}(AB)$, a quadrilateral as $\mathbf{Quad}(ABCD)$, a segment as $\mathbf{Seg}(ABC)$, and so on. For general discussions, in which individual objects are named, we sometimes refer to these as *Quad*, *Seg*, and so on.

Circles: A circle is denoted as $\mathbf{C}(ABC...)$, where A, B, C, ... , are points through which it passes. Since great and small circles are key objects, we denote a great circle as $\mathbf{gC}(AB...)$, and a small circle as $\mathbf{sC}(CD...)$. For general discussions, we often refer to these as C, gC and sC, with subscripts to differentiate the various circles.

Starting from *Spherics* II, we are introduced to bundles of parallel circles as essential objects of the theory. We denote members of such a bundle with $\mathbf{pC}(AB...)$ and we call the greatest of these circles $\mathbf{gpC}(CD...)$. Again, we also use pC, with subscripts, and gpC.

---

[1] This ring was previously considered to be a Roman item (Richter 1971, no. 436; Arnaud 1984, 91, fig. 35; Gundel 1992, no. 329), but this dating has been revised by curators at the Fitzwilliam Museum, which was personally confirmed by Victoria Avery, Keeper of Applied Arts.

[2] Our notation is modeled on that introduced by Dijksterhuis (1987, 51–52) in the 1938 Dutch original of his work on Archimedes.

Angles and Arcs: When discussing angles or arcs with their textual letter-name, we use **Ang**(*ABC*) or **Arc**(*ABC*), following the naming convention of the text. For more general discussions, we also use lowercase Greek letters to denote angles and arcs, such as ϑ, α, β, and so on. Again, if an angle or arc with a certain construction or definition is found in different positions we distinguish these with subscripts. When angles are equal we use **Ang**(*ABC*) = **Ang**(*DEF*) or $\alpha_1 = \alpha_2$, and when arcs of different circles are similar we use **Arc**(*ABC*) $\sim$ **Arc**(*DEF*) or $\beta_1 \sim \beta_2$, and for the relation *greater than similar to*, used in *Spherics* II and III, we use **Arc**(*ABC*) $\succ$ **Arc**(*DEF*) or $\gamma_1 \succ \gamma_2$.

Symbols: We use a number of standard mathematical symbols. Perhaps less standard are $\cong$, which means "congruent to"; $:=$, which means "set equal to," or "assigned as"; and $\mapsto$ which means "goes to," or "is represented by." A group of objects that is either a modern set or what Greek mathematicians called a multitude ($\pi\lambda\tilde{\eta}\vartheta\text{o}\varsigma$) are inclosed in {curly braces}.

Astronomical terminology and notation

Although never discussed explicitly, one of the core goals of this text is to set a foundation for spherical astronomy. The basic concepts of spherical astronomy are explained below (see p. 29), but here we set out the following terminology that will be used to explicate the astronomical context of the treatise:[3]

The celestial sphere: The celestial sphere is considered to rotate regularly once a day about an axis joining the *visible pole*, P_v, with the non-visible, or *invisible pole*, P_i. Points fixed on this sphere will sweep out parallel circles, which we call *δ-circles*. The greatest of the δ-circles is the *celestial equator*. Great circles that are perpendicular to the equator are known as *colures*. Arcs of the equator, known as arcs of *right ascension*, will often be denoted with α, while arcs of a colure, known as arcs of *declination*, will be denoted with δ.

The ecliptic: The *ecliptic* is a great circle fixed in the celestial sphere and rotating with it, but oblique to the equator. There are four cardinal points of the ecliptic determined by its intersections with the equator: the *vernal equinox*, *VE*, the *summer solstice*, *SS*, the *autumnal equinox*, *AE*, and the *winter solstice*, *WS*. The *equinoctial points*

---

[3] Most of this notation follows that used by Neugebauer (1975, 1204–1206).

are the intersections of the equator and the ecliptic, and the *solstitial points* are taken 90° along the ecliptic from these. The colures through the cardinal points are known as the *equinoctial* and *solstitial colures*. The angle between the equator and the ecliptic, known as its *obliquity*, is denoted with ε. This can be considered either as the angle between the planes of the great circles, or as the great-arc distance between the two circles measured along the solstitial colure. Arcs of the ecliptic, known as arcs of celestial longitude, will be denoted with λ, sometimes with subscripts.

The horizon: The local *horizon* is a great circle projected into the celestial sphere with respect to which the celestial sphere and the ecliptic can be considered to rotate. Its angle with the equator determines the local configuration. This angle, φ – which is determined by either the great-arc height of the pole above the horizon or the great-arc distance of the terrestrial latitude from the terrestrial equator – lies in the closed interval $0° \leq \varphi \leq 90°$. The configuration in which $\varphi = 0°$ is known as the *upright sphere*, while that in which $0° < \varphi < 90°$ is called the *inclined sphere*. Arcs of the horizon will be denoted with η, sometimes with subscripts.

Rising arcs and times: Any given arc of the ecliptic, λ, rises, and also sets, during the same time as a certain arc of the equator, known as its *rising* or *setting time*, which is a function of both the arc of the ecliptic and the terrestrial latitude, φ; and it rises and sets over diametrically opposite arcs of the horizon, whose great-arc measure is known as its *ortive amplitude*, η. In general, rising times of a given arc of the ecliptic, λ, are denoted with $\varrho(\lambda, \varphi)$, and setting times with $\bar{\varrho}(\lambda, \varphi)$. On the upright sphere, the ortive amplitude of an arc of the ecliptic is equal to its arc of declination, $\eta(\lambda, 0°) = \delta(\lambda)$, and its rising and setting times are both equal to its right ascension, $\varrho(\lambda, 0°) = \bar{\varrho}(\lambda, 0°) = \alpha(\lambda)$.

In discussing the rising times of the arcs of the ecliptic, it is often useful to refer to a specific segment of the ecliptic, starting from the first point of a particular sign of the zodiac and ending at the last point of another. For such occasions we will use **Quad**(Ari → Gem) to mean the quadrant going from Aries, to Taurus and then Gemini, **semiC**(Can → Sag) to mean the semicircle that consists of Cancer, Leo, Virgo, Libra, Scorpio, and Sagittarius, in that order, and so on. The rising time of a specific sign is denoted with $\varrho(\text{Sign}, \varphi)$, and the setting time with $\bar{\varrho}(\text{Sign}, \varphi)$.

Transcription of geometric letter-names

Greek	Roman	Arabic
A	A	ا
B	B	ب
Γ	G	ج
Δ	D	د
E	E	ه
Z	Z	ز
H	H	ح
Θ	Q	ط
I	Ø	ى
K	K	ک
Λ	L	ل
M	M	م
N	N	ن
Ξ	X	
O	O	
Π	P	
P	R	
Σ	S	
T	T	ش
Υ	U	ت
Φ	F	ث
X	C	
Ψ	Y	
Ω	W	
Ϛ	I	
ϙ	J	
ϡ	V	

Part I

Introduction

Introduction

This book is divided into four parts. The Introduction sets out what is known about Theodosios and his influence, background information that will be useful for making an informed reading of the *Spherics*, and discussions of the editions that we used and our practices in making the translation. This is followed by our Translation of the *Spherics*. The Commentary sets out summaries of the mathematical and astronomical content of the treatise from a historical perspective and provides some philological discussions where we thought necessary. Finally, the Paraphrase provides a more modernized, or idiomatic, version of the treatise, accompanied by diagrams that many modern readers will find more intelligible than those in the medieval manuscripts.

The division of labor in producing this book has been as follows. The Introduction and Commentary were written by Sidoli, the Paraphrase by Thomas, and the Translation was produced by both of us. We both read each other's material and made comments, and also some ideas that originated with one of us have been worked into a section written by the other – in particular, a number of Thomas's insights and comments have been placed in the Commentary.

Theodosios's time and works

In his *Geography*, finished in the 20s CE, Strabon gives a list of the names of famous men from Bithynia, followed by their vocations, which includes "Hipparchos, Theodosios and his sons, mathematicians" (*Geog.* XII.4.9).[1] This Theodosios is understood to be the author of the *Spherics*. In some of the manuscripts containing the text, he is called Theodosios of Tripoli, which might be supported by an entry in the *Suda*, if this entry did not clearly mix together material concerning our Theodosios with that of another (*Suda*, Theta–142; Czinczenheim 2000, 11). Although most scholars now consider the correct name to be Theodosios of Bithynia, Tripoli is common in older literature and some scholars still prefer it.[2]

---

[1] In Greco-Roman texts those who were designated as mathematicians were the practitioners of disciplines that used mathematics or studied ideal or actual mathematical objects, such as number theory (arithmetic), geometry, computation, astronomy and astrology, music (harmonics), mechanics, optics, spherics, sphere-making, sundial theory (gnomonics), and so on.

[2] Heiberg (1927, xvi) chose to cut the word "Tripolites" from the title of his edition between the typesetting and printing. Ver Eecke (1927, viii) gives an argument for tak-

All of Theodosios's surviving work deals with spherical geometry and various geometrical properties of the celestial and terrestrial spheres. Two other surviving mathematical treatises by Theodosios are often transmitted in the same manuscripts as the *Spherics* – namely *Habitations* and *Days and Nights* (Fecht 1927; Kunitzsch and Lorch 2011a;b). These minor texts treat subject matter related to that in the *Spherics*, and, although they contain astronomical terminology, which is virtually absent from the *Spherics*, they are no less works of pure theory, having little to do with observation.

Vitruvius, writing in the 1st century BCE, says that Theodosios and Andrias made a sundial for all climes, or latitudes, referring to this with what was probably the technical name for this type of sundial, *pros pan clima* (*Arch.* IX.8.1). The mathematics of such a dial would have been developed as part of gnomonics, the study of sundials, which was an area of the mathematical sciences closely related to spherics.

The *Suda* (Theta–142) credits Theodosios with a commentary on Archimedes' *Method*, a text of some fifteen propositions, written in the form of an open letter to Eratosthenes, that develops a heuristic technique involving a virtual balance that can be used to investigate the areas and volumes of certain curvilinear figures. Although the mathematical approach of the *Method* is unrelated to what we read in Theodosios's surviving work, the most advanced part of the *Spherics* makes use of techniques that were used by Archimedes in his *Sand Reckoner* and *Equilibrium of Planes*. Hence, it is clear that Theodosios was familiar with Archimedes' work, and there is no reason to suspect that the commentary to the *Method* is attributed to him in error.

Based on this, Theodosios must have lived sometime between the time of Archimedes and that of Vitruvius or his younger contemporary Strabon – that is, sometime in the interval of, roughly, 200 to 50 BCE. Strabon, however, has a tendency to give lists of notable individuals in chronological order, so it is tempting to set Theodosios's time after that of Hipparchos (Ver Eecke 1927, viii; Neugebauer 1975, 749). To this it might be objected that the character of the *Spherics* suggests that it should have been written before Hipparchos (Bjørnbo 1902, 65; Heath 1921, 2.246), whose work with chord-table trigonometry has been held to supersede Theodosios's purely geometric approach. But such an objection should be rejected on two counts. In the first place, there are many examples in the ancient and medieval exact sciences of an earlier, supposedly inferior, method being practiced, indeed further developed, for centuries after a putatively superior method had been introduced. Secondly, Theodosios's theorems, especially those in *Spherics* II and III, can be read as providing geometrical explanations for certain properties that obtain for the instantaneous positions, or ris-

ing the association with Tripoli as appropriate to a later stage of this Theodosios's life. Nikolantonakis (2016) prefers to associate Theodosios with Tripoli.

ing times, of the ecliptic, which are simply exhibited, but not explained, in the approach using chord-table trigonometry – at least as it was later developed in Ptolemy's *Almagest*. Furthermore, it is certainly the case that mathematicians remained interested in Theodosios's *Spherics*, and worked on it carefully, long after the chord-table trigonometric methods were well known, so it is also possible that Theodosios himself decided to take up this project after chord-table trigonometric methods were already in circulation. For example, Menelaos made explicit and extensive improvements to Theodosios's work and Pappos wrote commentaries on both the *Spherics* and the *Almagest*. Hence, there should be no reservation in following Strabon's lead and situating Theodosios after Hipparchos. This would mean that Theodosios's dates should be placed toward the end of the Hellenistic period – somewhere between, say, 120 and 50 BCE.[3]

The significance of Theodosios's *Spherics*

The *Spherics*, τὰ σφαιρικά – meaning spherical things, or perhaps theorems – almost certainly included and reworked some earlier material dealing with similar topics. There are many signs of editorial processes in the surviving text itself, and a number of geometrical facts that are established in Theodosios's *Spherics* had been used in the earlier *Moving Sphere* by Autolykos and *Phenomena* by Euclid. Hence, it has long been assumed that there must have been earlier work that included some of this material before Autolykos and Euclid produced their own texts. In fact, however, we have no reference to such work, or even to the field of spherics itself, prior to the time of Theodosios,[4] and we must accept that we do not know what format such earlier material might have had. In particular, we do not know if this material was more geometrical, in the sense of Theodosios's *Spherics*, more astronomical, in the sense of Euclid's *Phenomena*, or some combination of both. Furthermore, the ancient sources transmit no names of individuals or treatises associated with this supposed early work in spherics, as will be discussed below.

Whatever the earlier material may have been, Theodosios reworked it into a new treatment, probably in the 1st century BCE, when numerical methods based on the analemma diagram and chord-table trigonometry

---

[3] Czinczenheim (2000, 12–17) provides a discussion of the scholarly debate concerning Theodosios's time and works.

[4] The use of the word σφαιρική, which had been associated with the Pythagoreans (Heath 1921, 2.245; Bulmer-Thomas 1970, 13.319 col.2), is now recognized to originate in later material. The quotes attributed to Pythagoras, in which this word is found, are understood to be spurious, and the attribution to Archytas of this term by Nikomachos has been removed by the most recent editor, independently of considerations from the history of mathematics (Huffman 2005, 103).

were already in circulation. Because we cannot make a direct comparison, it is not certain to what extent Theodosios included his own work in this process of synthesis and compilation. Opinions on this might vary from the belief that Theodosios added nothing new and simply rearranged things that were already fully established, to the position that he took an active role in formulating the theory and added his own results, or even go so far as to say that he produced proofs for claims that were simply asserted by his predecessors. Certainly, in its current state, the structural form and mathematical approach of the *Spherics* seems to be based on that of Euclid's *Elements*, and it is tempting to read this as deliberate. Indeed, the *Elements* itself is often described as a work of compilation and Euclid as an editor (Knorr 1975, 303–306; Fowler 1999, 205), but all of the earlier work, on the basis of which it is believed to have been composed, is only known to us through the *Elements* itself and our understanding of this earlier material depends on reconstructions made by modern scholars, with all of the assumptions and guesses that these involve. Furthermore, it is also recognized that some material now found in the *Elements* is also due to later editors (Vitrac 2012), and this is certainly true of the *Spherics* as well. Theodosios probably reworked the order and presentation of material in earlier sources, modeling the presentation on that of the *Elements*, formulating definitions and adding propositions where necessary. Furthermore, the most important results of *Spherics* II and III – namely, *Sph.* II.21–II.23, III.5–III.12 – are either not found in, or are clear improvements on, the earlier spherics, and, as they now stand, they are likely due to Theodosios (see p. 47, below). As it turned out, Theodosios's treatment of the material of the early spherics was so complete as to render this previous work superfluous, and it was probably Theodosios's intention that it should have been made so. Hence, the names of any earlier texts and authors were forgotten, whereas Theodosios's work was studied and improved upon from the Roman Imperial period until early modern times.

In claiming that the *Spherics* was a work of compilation, in the same genre as the *Elements*, however, we must pause to consider the role of genre in the production of mathematical texts in antiquity. Because we are accustomed to scientific communication being carried out through specialized publications for research results, *vademecum* books and articles for synthesis and organization, textbooks for education, and popular publications for general consumption (Fleck 1935/1979, 111–125), it is easy for us to imagine that this must have everywhere and always been the case. The social structures that make modern scientific communication possible, however, did not exist in Theodosios's milieu. In particular, there was not a large community of professional researchers,[5] and there were no intergenerational

---

[5] Netz (1997) sets the number of named mathematicians in antiquity at 144, while Zhmud and Kouprianov (2018) are more generous with 407. Looking into the names

teaching institutions that taught mathematics to great numbers of students and required textbooks (Bernard et al. 2014, 38–51; Sidoli 2015). Hence, we should look at the varieties of texts that ancient mathematicians produced to understand the ways in which they sought to establish their names and reputations as mathematicians.

The bulk of the texts of theoretical mathematics that have come down to us are in the form of individual letters, synthetic treatises, sometimes accompanied by an introductory letter or lecture notes, question-and-answer sets, poems, introductions, and commentaries (Taub 2013). While it is tempting to understand letters as a format for introducing new results and introductions and commentaries as formats for producing educational material (Mansfeld 1998), the situation is often not so clear cut. For example, Hipparchos begins his *Commentary on the Phenomena of Aratos and Eudoxos* in the form of a letter, then points out that many others have written an exegesis on Aratos's *Phenomena*, and notes that he will make special use of that of his contemporary Attalos, the mathematician (*In Arat.* I.1). It is clear that this project is meant to be a contribution to knowledge (Mastorakou 2020, 387–391), not a teaching text – although Hipparchos probably did hope that teachers of elementary astronomy would learn something from his work and stop teaching the errors of his predecessors. On the other hand, one of the most explicitly didactic texts in the mathematical sciences, namely Archimedes' *Method*, was written as a letter to a colleague. Of course, the *Method* is not a textbook, but neither does it announce new results. Rather, it sets out a mechanical way of thinking about mathematical problems, which Archimedes explicitly says will be useful in finding new results, as a way of teaching this approach to other mathematicians. In fact, in the Hellenistic and Imperial periods, there was no clear division among treatises that grouped together texts in the same genre with the intended level of the audience.

With regard to Euclid's *Elements* itself, while some scholars have referred to it as a textbook (Knorr 1986a, 101), others have denied that this could have been the case (Fowler 1999, 286). In thinking about whether or not the *Elements*, the *Spherics*, and other texts in early spherics were

in both of these data sets, however, one sees that many individuals who were not doing research are included. Moreover, when one considers that these are distributed unevenly over the whole of the ancient period, one appreciates that the number of people doing research in most periods was quite small. From a range of considerations, Netz (2002) estimates that there were probably 1000 mathematicians in the ancient Greco-Roman sphere, with maybe 100 living at any one time. Such an impression is supported by discussions of the research community in the introductions of the works of mathematical authors like Archimedes, Apollonios and Hypsikles (Netz 2002, 203–204). Furthermore, in referring to others working in their field, mathematicians often mention individuals who lived generations, even centuries, earlier, but less often their own contemporaries.

written to be textbooks,[6] we should take into consideration the evidence
for education in the Greco-Roman world. Although there are sources of
material composed for the classroom, such as lists of vocabulary, glosses
on difficult passages, and problem sets (Cribiore 2009, 323; Sidoli 2015,
390–391), the *Elements* and the *Spherics* are clearly not texts of this kind.
Nevertheless, it is certain that from some point on they were both used
in education, and teachers probably did compose material to facilitate the
process of studying them – such as the list of definitions from *Elements* I in
a good hand found in a 3rd-century documentary papyrus (P.Mich. III.143;
Turner et al. 1985). We do not know, however, when each of these texts began to be used in teaching the mathematical sciences, and since mathematics
education was essentially private throughout the whole of the Greco-Roman
period, this probably took place at different times in different places. Moreover, the fact that these treatises were eventually used as school texts, does
not mean that they were originally composed for this purpose. Indeed, when
we consider that the works most commonly used in advanced literary education were the poems of Homer or the orations of Isocrates and Demonthenes
(Cribiore 2001, 220–244; 2009, 329–330), we see that teachers chose texts
for instruction not because they were expository and easy to understand,
but rather because they were generally acknowledged to be excellent. We
should think of the choice of the *Elements* or the *Spherics* as a teaching
text in such terms – they were chosen not because they were easy or had
been written for students, but rather because there was a gradual agreement
among mathematicians that they were the best, or most suitable, texts available on their subjects. Hence, it is clear that the production of systematic
works such as the *Elements* was one way in which a mathematician could
make a contribution to knowledge. For example, Apollonios rewrote the
material of *Elements* I in the 2nd century BCE, and Menelaos again in the
1st century CE, but they did not, as far as we know, do this to present new
results. They gave new organizations and used different principles, such as
alternative construction procedures or adherence to direct argument. That
is, they were engaged in the same sort of project as Euclid in writing the
Elements itself – a rearrangement and reorganization of largely known results on a new foundation and with a new orientation. This is probably the
genre in which we should understand Theodosios's *Spherics* to have been
composed.

Theodosios begins his text with a treatment of some basic topics in the
geometry of the sphere and some constructions that will be of use in the rest
of the treatise. The guiding idea seems to be to establish analogies between
the geometry of the sphere and the geometry of the circle, and specifically
to find analogs for various propositions in *Elements* III, Euclid's treatment

---

[6] Neugebauer (1975, 749) and Evans (1998, 91) claim that the treatises of the early
spherics must have been textbooks because they are elementary and pedantic.

of the circle. Theodosios does not attempt to develop spherical geometry in the sense of a study of figures on the surface of the sphere. That is, he does not attempt to develop analogies to the geometry of *Elements* I, Euclid's geometry of the plane, although there are some adumbrations in this direction. In *Spherics* II, Theodosios treats the concepts of parallelism and tangency for circles in a sphere, and then combines these to explore the relationships between a bundle of parallel circles and one or two great circles that intersect this bundle. This culminates in some propositions that can be understood to describe the instantaneous positions of the ecliptic relative to the horizon, although Theodosios uses a purely geometrical articulation, which might obscure the astronomical interpretation. *Spherics* III begins with some lemmas of solid geometry for establishing relations between arcs of great circles on a sphere, and then culminates with a series of propositions that can be interpreted as describing the relationships between various arcs of the ecliptic and arcs of the equator with which, or arcs of the horizon over which, they rise – again with no mention of any astronomical concepts, nor any attempt to help the reader comprehend the underlying subject matter. Although the material in the *Spherics* can be understood as providing geometrical explanations for a number of important facts about the celestial sphere, the content of this explanation is entirely geometric and there is no material that can be regarded as explicitly didactic. In fact, one might develop the impression that Theodosios expects his readers to already know what is being discussed.

There is little clear evidence to indicate that Theodosios's *Spherics* was read around the end of the Hellenistic period. For example, one looks in vain through the *Introduction to the Phenomena* by Geminos, written within a century of the turn of the era, for any material that certainly derives from Theodosios. Although Geminos describes the principal characteristics of the celestial sphere with reference to the use of solid and armillary spheres, and discusses material connected to Autolykos's *Moving Sphere* and *Risings and Settings* and early propositions of Euclid's *Phenomena* (Evans and Berggren 2006, 203 nn. 10–12, 205 nn. 2,3), he does not mention any of the more advanced material of the early spherics, nor any purely geometric properties, such as are demonstrated in Theodosios's *Spherics*. In particular, Geminos's treatment of the rising times of arcs of the ecliptic is derived from Babylonian methods, probably as expounded by Hypsikles in his *Ascensions* (see p. 53, below), and does not make any reference to the geometrical methods developed by Euclid and Theodosios.

Of course, it may be that this is because Geminos's discussion is meant to be pitched at an elementary level, and such an impression is supported by a passage from Strabon's *Geography*. In describing the prerequisites necessary for understanding his work, Strabon makes the following remark (*Geog.* I.1.21):[7]

> Now, someone encountering this book must neither be so simple
> nor lazy as to have neither seen a sphere nor the circles in it –
> some parallel and others upright on these, and others oblique
> – nor the position of the tropics, the equator and the zodiacal
> ⟨circle⟩, through which the sun, carried, indicates the differences
> of the climes and of the winds. For, anyone having somehow
> understood these things – the horizons, the arctic ⟨circles⟩, and
> all of the others relating to what is taught in the first course of
> the mathematical sciences (τὰ μαθήματα) – is able to follow the
> statements herein.

The impression that one draws from both of these authors is that in the late Hellenistic and early Imperial periods, the introductory level of instruction in the mathematical discipline of astronomy was largely descriptive, based on the use of an inscribed or armillary sphere, and involved little or no study of the sorts of purely geometric theorems that we read in the *Spherics*. Hence, if Theodosios's *Spherics* was studied during this period, this would have been done at a more advanced level, which would have depended on the choices of the individual teachers, and was probably only undertaken by a fairly small number of students. This would be in line with the little that we know about instruction in other texts of more advanced mathematical authors at this time (Jones 2009, 343; Sidoli 2015). The fact that there is no evidence for the use of the *Spherics* in mathematics instruction in the period immediately after its composition makes it even less likely that it was originally composed for such a purpose.

The first time that Theodosios is explicitly referred to in our sources is by Menelaos, in his own *Spherics*, written around the turn of the 2nd century CE. From a mathematical perspective, Menelaos's *Spherics* was the most important text in the genre composed in the ancient period, and was the first to develop an intrinsic geometry of figures defined and produced on the surface of the sphere. That is, Menelaos considers the spherical surface as analogous to the plane and begins with a theory of the congruence of spherical triangles, on analogy with the first part of *Elements* I. In two passages, Menelaos explicitly refers to Theodosios – in both places to underscore his own improvements, but the second time also to point out that the work by both Theodosios and himself provides a basis for something that Apollonios had done (Krause 1936, 239 n.1; Sidoli and Kusuba 2014, 167, 172; Rashed and Papadopoulos 2017, 697, 769). The fact that Menelaos refers to Theodosios by name, however, and once in the same breadth as

---

[7] See also the translations of Jones (1949, I.45) and Roller (2014, 46).

Apollonios, indicates that Menelaos understood Theodosios to be his only significant predecessor in this work and believed that together they were engaged in something important, because it could be used to demonstrate a claim assumed by Apollonios (Sidoli and Kusuba 2014, 172–173). Both of Menelaos's explicit criticisms of Theodosios's *Spherics* deal with the material that Theodosios developed in *Spherics* III. There is also, however, an implicit criticism implied by the rest of the work as a whole, because Menelaos's use of figures bounded by great circles is intended to supersede Theodosios's approach of focusing on the relationship between a bundle of parallel circles and intersecting great circles. Nevertheless, despite the fact that he meant to improve on Theodosios's work, Menelaos apparently did not intend to make it irrelevant and he applied results from the former *Spherics* that he does not establish in his own, so that he must have expected his *Spherics* to be read by someone who had already mastered that of Theodosios.

Although he does not mention the name explicitly, Galen, writing in the 2nd century CE, was clearly familiar with Theodosios's work. For example, while criticizing the Romans, in his *Commentary on Hippocrates' Airs, Waters, Places*, which survives only in medieval translations, Galen states that, in general they only learned Euclid's *Elements* and *Data*, while some of them knew "the science of the movement of the sphere," علم الحركة الفلك, "the science of what appears of the stars," علم ما يرى من النجوم, "the science of the inhabited earth," علم الأرض المعمورة, and "the science of night and day," علم الليل والنهار, but almost none of them studied all of geometry, such as conic theory, and they completely neglected the advanced astronomical works of Hipparchos and his successor Dioskorides, ديسقوريدس (Toomer 1985, 196, 199).[8] Although these expressions are somewhat different than the standard Arabic translations for the titles in question,[9] it is clear that the texts that Galen suggests that the Romans should have studied after the *Data* are Autolykos's *Moving Sphere*, Euclid's *Phenomena*, and Theodosios's *Habitations* and *Days and Nights* (Toomer 1985, 202). Furthermore, in his *Diagnosis and Cure of the Soul's Errors*, when discussing the type of certainty that is produced in the mathematical sciences, Galen mentions that the knowledge of the geometrician regarding the things that are proved in Euclid's *Elements* is the same as most people's knowledge that twice two is four. He then states that the same is true for the spherical theorems, σφαιρικὰ θεωρήματα, as well as conic theory and sundial theory (*De pecc. dig.* 1.4). It is tempting

---

[8] The astronomical writings of this Dioskorides are not known to us. It is also possible that the name has been misconstrued in the transmission.

[9] This difference could be explained by pointing out that Galen's treatise was translated into Syriac by Ḥunayn ibn Isḥāq and then into Arabic by Ḥubaysh ibn al-Ḥasan (Lamoreaux 2016, 105), neither of whom was, himself, a scholar of the mathematical sciences.

to read this as a reference to Theodosios's *Spherics*, since the four texts mentioned above do not fit the context very well and Menelaos's *Spherics* seems unlikely. In either case, it is certain that Galen knew of Theodosios's astronomical work, and that by the first half of the 2nd century, there was a fairly well accepted group of canonical works in spherics, which included some of Theodosios's treatises, and which was studied as a sort of middle course in the mathematical sciences.

The situation is less certain with Galen's older contemporary Ptolemy. Although Ptolemy never mentions Theodosios by name, in a number of places in his treatment of spherical astronomy, especially in *Almagest* II, he uses geometrical facts about the sphere that are demonstrated in Theodosios's *Spherics*. It is not known whether Ptolemy took these claims directly from the *Spherics*, or rather from some other work on spherical astronomy that already incorporated this material. Nevertheless, given what Galen says about the role of Theodosios's work in mathematical scholarship at that time, it is likely that Ptolemy was also familiar with the *Spherics*, and probably knew it well.

The next author that we can be certain studied the *Spherics* was Pappos, who discusses the text in his *Collection*, written in the early 4th century. In *Collection* VI, which Pappos tells us addresses the "field of astronomy," ἀστρονομούμενος τόπος, he sets out a number summaries of, as well as lemmas, extensions, and corrections to, Theodosios's *Spherics*, Autolykos's *Moving Sphere*, Theodosios's *Days and Nights*, Aristarchos's *Sizes and Distances of the Sun and Moon*, and Euclid's *Optics* and *Phenomena*. Of this material, only the first run of propositions treats the *Spherics*. Here, Pappos uses Menelaos's concept of a spherical triangle to develop a number of preliminary results and then uses these to provide new proofs for *Sph.* III.5 and III.6 (Malpangotto 2003). The ordering principle for Pappos's treatment of the works in the "field of astronomy" seems to be from more abstract to more material, but it is also the case that the *Spherics* develops material that is mathematically required in the other treatises, so that they can be understood to build on each other. Indeed, this ordering of the treatises probably began to affect the way that they were edited, and the final form in which they were transmitted to the medieval period and copied in our earliest manuscripts. At any rate, it is clear that by the Roman Imperial period, there was a fairly well established group of texts that was used to teach astronomy, and that Theodosios's *Spherics* played a key role in these.

In the manuscript that transmits Pappos's *Collection*, this group of classical astronomical treatises is twice called the *Little Astronomy*, μικρὸς ἀστρονομούμενος. The expression occurs first in a marginal scholium at the beginning of *Collection* VI, and then again in a sort of colophon following the same book (Vat.gr. 218, 87v, 118r). Moreover, this phrase is also found in the anonymous *Introduction to the Almagest*, which asserts that a certain lemma dealing with the sides and angles of a right triangle was shown by Theon in his *Commentary on the Little Astronomy* (Acerbi et al. 2010, 121),[10] and in two scholia to the so-called *Elements* XIV by Hypsikles, which state that certain steps in an argument in the *Spherics* are shown "in the little astronomy" (Heiberg 1903, 332–333). Hence, the group of treatises, including the *Spherics*, that, in the Imperial period, was studied between Euclid's *Elements* and *Data* and either gnomonics, conic theory, or texts of more technical astronomy by Hipparchos or Dioskorides, was later studied before Ptolemy's *Almagest*, or the *Great Mathematical Treatise*, μεγάλη [μαθηματικὴ] σύνταξις, as it was sometimes called by late-ancient authors (Pingree 1994, 81, 84). It was in such a context that this group of shorter astronomical works became known as the *Little Astronomy* (Acerbi 2014, 141 n. 87). Whether or not Pappos himself used this terminology, it is clear that the mathematical scholars of late antiquity encountered this grouping of texts as a sort of codicological unit, studied and taught it as such, and wrote works about it as a whole. We can include Pappos's *Collection* VI and the lost commentary by Theon among such studies, and we find evidence for this style of scholarship in the scholia that are transmitted with the treatises of the *Little Astronomy*. Furthermore, the fact that the *Spherics* was grouped with other texts, and was then understood to justify steps that were carried out in these treatises almost certainly affected the way that the *Spherics* itself was transmitted and edited in late antiquity and the middle ages (Acerbi 2014, 141–151). Hence, the *Spherics*, like all the texts of the *Little Astronomy*, was heavily interpolated (Czinczenheim 2000, 29).

In the first half of the 6th century, John Philoponos, in his *Commentary on Aristotle's Physics*,[11] discussed the *Spherics* in part of his overall discussion of Aristotle's distinction between the mathematician, ὁ μαθηματικός, and the philosopher of nature, ὁ φυσικός.[12] In the course of this discussion, Philoponos claims that the mathematician teaches about figures and their characteristics by abstracting them from matter, while the philosopher considers them as material. He provides the following example (Vitteli 1887–1888, XVI.220):[13]

> Indeed, Theodosios in the *Spherics*, when teaching the properties obtaining in the sphere, does not also take matter into

---

[10] This lemma is discussed below in the Commentary to *Sph.* III.11 (p. 341).

[11] Philoponos also provides a related discussion of the *Spherics* in his *Commentary on Aristotle's Posterior Analytics* (Wallies 1909, 300–301).

[12] The debate about the relative epistemological positions of philosophy and mathematics was carried out by a number of scholars throughout antiquity, such as Aristotle, Poseidonios, Geminos, Heron, Ptolemy, Simplikios, and Philoponos (Bowen 2007; Bowen 2013, 38–40; Feke 2013; Feke 2018, 40–44). It will not come as a surprise to many readers that philosophers argued for the epistemological superiority of philosophy, while mathematicians, such as Heron and Ptolemy, argued for that of the mathematical sciences.

[13] This translation can be compared with that of Lacey (1993, 33).

account, but separating the spherical figure from all substance thus considers the characteristics in it – that if a sphere is cut by a plane it makes a circle, and so on. But Autolykos, writing *On a Moving Sphere* and about what occurs in the moving sphere, is more concerned with the particular (μερικώτερός) than Theodosios, and more approaching to the philosopher of nature (τῷ φυσικῷ) – for movement is somehow near to substance – for if he does not consider some substance in the moving sphere, nevertheless he apprehends a composition of the figure and movement, and in this is somehow near to substance. Still more concerned with the particular than this is the *Phenomena* of Euclid, and the whole of astronomy simply, for here the substance itself is also taken into consideration. For he also takes into account the movement of heaven, the sun, and the remaining stars, for he does not consider movement of a sphere simply, but that of the fixed ⟨stars⟩, or Saturn, or some other, and the relation of these to one another. And the summit of mathematics is easy to distinguish and separate from the theory of nature (τῆς φυσιολογίας) – such as the *Spherics* of Theodosios, the 13 books of Euclid, the *Arithmetics*,[14] for in these there is absolutely no mention of matter – but its foothills are somehow near to the study of nature, for the astronomer theorizes (θεωρεῖ) as to a figure of the sun and movement of the sun, and their magnitude; and likewise the theorist of nature (ὁ φυσιολόγος).

In this passage, we see that Philoponos considers the *Spherics* to be a well-known text that he can use to make a general point about which sorts of methodological approaches are appropriate to which fields of inquiry. Although Philoponos is not describing a course of study, it is worthwhile noting that the ordering of the three books of the *Little Astronomy* that he names agrees with that in Galen,[15] Pappos, and in the 9th-century codex Vat.gr. 204 (**A**, 3r, 4r). This makes it clear, once again, that by late antiquity the *Spherics* was a well-known, canonical text in the mathematical sciences that was usually taken together as a unit, both codicologically and conceptually, with *Moving Sphere* and *Phenomena*, and which was probably transmitted in the same codices.

In the 9th century, the *Spherics* was included in a collection of treatises that begins with those of the *Little Astronomy*, but includes other mathematical works as well (Acerbi 2014, 141–151). Vat.gr. 204 (**A**), which is our earliest source for the text of the *Spherics*, includes numerous scholia

---

[14] This must be a reference to the *Arithmetics* of Diophantos.

[15] Although Galen does not mention the *Spherics*, the ordering of the other two works is the same.

to almost all of the treatises in the collection, both in the margins and in two cases compiled together as collections of scholarly notes. These scholia, which show much internal cross-referencing both within individual treatises and to other treatises in the corpus, show that the texts of the *Little Astronomy* were studied together as a unit, and were the subject of commentaries and related scholarly projects. The treatises contained in Vat.gr. 204 are as follows: Theodosios's *Spherics*, Autolykos's *Moving Sphere*, Euclid's *Optics* and *Phenomena*, Theodosios's *Habitations* and *Days and Nights*, Aristarchos's *Sizes and Distances of the Sun and the Moon*, Autolykos's *Risings and Settings*, Hypsikles's *Ascensions*, Euclid's *Catoptrics*, Eutokios's *Commentary on Apollonios's Conics*, Euclid's *Data*, with a collection of scholia, Marinos's *Introduction to the Data*, and a collection of scholia to the *Elements*. In this collection, the first three treatises may have been arranged following Philoponos's principle of decreasing levels of abstraction (Acerbi 2014, 142), after which the ordering principle becomes less clear. It is also possible that the treatises were arranged in this way for structural reasons, since it could be argued that the later treatises rely conceptually and technically on the earlier treatises. Whatever the case, this collection can be taken as a sort of model to remind us that throughout the medieval and early modern periods, most people who read the *Spherics* in Greek, would have encountered it, not as an individual work, but in a collection of astronomical treatises, for which it served as a geometrical prerequisite.

In the same century in which Vat.gr. 204 was compiled, the *Spherics* was studied, translated and adopted by Arabic speaking scholars – most likely from one or more Greek codices that grouped it with other treatises of the *Little Astronomy* – and it was apparently understood to form part of an intermediate course in the mathematical sciences.[16] The *Spherics* is explicitly included in a list of treatises to be read after the *Elements* that is provided in both a 16th-century Arabic manuscript that contains copies of mathematical treatises and a late 12th-century Latin codex containing many of Gerard of Cremona's translations from Arabic of works in the mathematical sciences. The relevant passage in the Arabic manuscript reads as follows (Beirut, St. Joseph University 223, image 64):[17]

---

[16] One of the earliest references to a text that may be the *Spherics* comes from the work of the polymath Ya'qūb ibn Ishāq al-Ṣabāḥ al-Kindī. In the course of listing works that should be studied before the *Almagest*, he gives the titles of a number of works that are related the *Little Astronomy*, although the titles that he states are often different from the later standard Arabic titles for these texts. The name that he gives for the text dealing with the material of the *Spherics*, however, is so vague that it is not certain that he is referring to the actual work of Theodosios (Rosenthal 1956, 440–441).

[17] See also the translation by Brentjes (2018, 39–40), which we follow with modifications.

Order of what is read after Euclid found in a manuscript from
the handwriting of Isḥāq ibn Ḥunayn:

Treatise on Optics by Euclid, one book; *The Spheres* by Theodosios, three books; *The Moving Sphere* by Autolykos, one book;
The Phenomena by Euclid, one book; *The Inhabited ⟨Places⟩*
by Theodosios, one book; *The Risings and the Settings* by Autolykos, one book; *The Nights and the Days* by Theodosios, one
book; *The Ascensions* by Hypsikles (انقلاوس), one book; *The Distances of the Planets and their Sizes* by Aristarchos, one book.

In the Latin manuscript, the only difference is that the name to which
this list is attributed is "Iohannicius," who is usually understood to be
Isḥāq's father, Ḥunayn ibn Isḥāq, the famous physician and medical translator (Par.lat. 9335, 28v; Czinczenheim 2000, 18). Since the younger Isḥāq
is known to have worked extensively on the treatises in this collection, and
since there are other instances in which his name appears to have been
mistaken for that of his more illustrious father,[18] it is most likely that the
original name was "Isḥāq ibn Ḥunayn" and that this was then misconstrued
in Latin as "Iohannicius."

Based on this passage, it is unknown whether or not Isḥāq provided any
discussion of the contents of these texts, or just gave a list. Nevertheless, it
is clear that he believed that the books of what had been called the *Little
Astronomy* should be read following the geometrical works of Euclid, just
as Galen had held in the 2nd century. Moreover, although the ordering of
the texts is somewhat different from some of those that we saw above, the
three texts of the *Spherics*, *Moving Sphere*, and *Phenomena*, are presented
in what we can call the standard order.

Another scholar from the same generation provides our first citation
of the name that these texts would acquire in the Islamic world. According to Samaw'al ibn Yaḥyā al-Maḡribī, writing in the 12th century, Isḥāq's
contemporary Qusṭā ibn Lūqā wrote an epistle on the collection of "what
must be read of the *Intermediaries* (المتوسّطات) before the *Almagest*," now
lost (Leiden or. 98, f.43a; Sezgin 1978, 66). The work that this title refers
to, the *Intermediaries*, is commonly called the *Middle Books* – meaning the
collection of treatises to be read between Euclid's *Elements* and Ptolemy's
Almagest. Aside from the list by Ibn Ḥunayn, just quoted, we do not know
much about the early versions of this collection.

Qusṭā ibn Lūqā is also associated with translating the *Spherics* into Arabic. Although the translator is not named in the bibliographic literature,

---

[18] For example, of eight manuscripts containing Jacob ben Makhir's Hebrew translation from Arabic of Menelaos's *Spherics*, six state the Arabic translator as Isḥāq ibn
Ḥunayn and two as Ḥunayn ibn Isḥāq (Rashed and Papadopoulos 2017, 19). See also
n. 19, below.

in the introduction to his *Revision of the Spherics*, composed in the 13th century, al-Ṭūsī tells us that the translation of the *Spherics* was ordered by Abū al-ʿAbbās Aḥmad ibn al-Muʿtaṣim, the prince or future caliph, and carried out by Qusṭā ibn Lūqā up to "the fifth proposition of the second book," and by someone else after this (Hyderabad 1939/40, 2). The whole thing was then corrected by Thābit ibn Qurra. This broadly agrees with what is found in the manuscripts of the early Arabic translation, which variously credit Qusṭā, Thābit, and Ḥunayn ibn Isḥāq (Kunitzsch and Lorch 2010, 2).[19] Certainly, the early Arabic version that is preserved in our sources includes a number of corrections and glosses that are consistent with the style of scholarship found in the known restorations by Thābit. Hence, it seems that in the 9th century two or three of the more competent mathematical scholars and translators in Baghdad worked on the *Spherics*, and a polished Arabic version of the text was put out by Thābit ibn Qurra (Kunitzsch and Lorch 2010). Due to Thābit's abilities in the mathematical sciences, comparing his restoration of the treatise with the Greek text can often shed light on the intended meaning of the Greek source, but because he had a tendency to correct the ancient texts along conceptual and mathematical lines, we should also be wary of understanding the various divergences of his version, when compared to the currently extant Greek manuscripts, as witnessing a more pristine Greek tradition. Each difference must be considered on a case-by-case basis.

There is some evidence for the study of the *Spherics* in Arabic over the next few centuries. Marginal glosses in the medieval manuscripts make it clear that some of the treatises of the *Middle Books* were studied by scholars such as ʿAlī ibn Aḥmad al-Nasawī, in the 10th century, and Ibn al-Sarī ibn al-Ṣalāḥ, in the 12th (Sidoli and Kusuba 2014, 157–158; Sidoli and Isahaya 2018, 26; Rashed and Papadopoulos 2017, 615). A 12th-century copy of Thābit's version of the *Spherics* records evidence of the scholarship of a certain al-Ḥasan ibn Saʿīd, particularly relating to the diagrams (Kunitzsch and Lorch 2010, 3–4, 313–314). The mathematical results of the treatise were also used by some authors in their own work, such as by Yūsuf al-Muʾtaman ibn Hūd in the 11th century, who produced an intrinsic generalization of the important theorem *Sph.* III.11 (Hogendijk 1986; Hogendijk 1991, 217, 254–259; Lorch 1996, 172–173; Rashed and al-Houjairi 2010), as well as by Jābir ibn Aflaḥ in the 12th century (Lorch 1996, 173).

In the middle of the 13th century, in 1253, Naṣīr al-Dīn al-Ṭūsī revised the *Spherics* and included it among his revisions of the *Middle Books*, interjecting his own commentary following some of the propositions, and incorporating into his comments a number of scholia that had been transmitted in earlier Arabic, and sometimes originally Greek, sources (Hyder-

---

[19] This last may be another slip for Isḥāq ibn Ḥunayn.

abad 1939/40, 52; Sidoli and Kusuba 2008). The great majority of the Arabic manuscripts of the *Spherics* that are now known contain al-Ṭūsī's version, and they are generally found in the many collections of his revisions of the *Middle Books*. From the 13th century on, the *Middle Books* included the Greek classics mentioned above, along with Ṭūsī's revisions of Euclid's *Data* and Menelaos's *Spherics*, as well as original works by authors writing in Arabic, such as the Banū Mūsā and Thābit ibn Qurra. Furthermore, some of al-Ṭūsī's own works were also incorporated into this collection, or at least circulated in the same manuscripts that contained the *Middle Books* (Brentjes 2018, 39–40).

In the final three decades of the 13th century, in Provence, the *Spherics* was translated twice, or rather three times, into Hebrew. Moses ibn Tibbon translated the text in 1271, and Jacob ben Makhir translated it again in 1290, which a colophon tells us was to replace a translation that he himself had made 20 years earlier, but which had been stolen (Knorr 1986b, 232–235). Moses ibn Tibbon's translation was probably made on the basis of an Arabic version that is still extant in Hebrew characters, and although somewhat different from the version used earlier by Gerard of Cremona (Lorch 2014), probably also derived from the restoration of the *Spherics* that had been produced by Thābit, and was certainly not al-Ṭūsī's more recent revision. Although these translations must have been done from Arabic collections that contained the *Middle Books* – or at least Jacob ben Makhir's earlier translation work was done at the same time as a number of other books from this collection – they do not appear to have circulated in the Hebrew tradition in a comparable collection of works (Lévy 1997, 443–444). In most of the manuscripts that we have seen, Theodosios's *Spherics* was transmitted with that of Menelaos, and sometimes also with works of Euclid (New York, JTSA, Ms. 8182; Bodleian, Hunt. 96; St. Petersburg, IOM, C(S) 12 and NLR, Ms. EVR II A 13). Although these manuscripts contain a number of scholarly notes, the extent to which the *Spherics* was studied or used by scholars working in Hebrew has yet to be studied.

There is evidence for Byzantine mathematical scholars studying the *Spherics*, along with other works of the ancient mathematical sciences throughout the centuries. Indeed, following the scholarly project of compiling Vat.gr. 204, all of our sources for the Greek text are due, directly or indirectly, to the scholarship of the Byzantines. The manuscript sources indicate that certain groups of scholars studied the mathematical treatises closely – copying and writing scholia, revising and improving the texts, regularizing the language and letter-names, inserting deductive steps and logical qualifiers, adding cases perceived to be missing and merging or splitting propositions. They also made collections of their own recensions of the ancient treatises, which, for example, in the 13th and 14th centuries included the *Spherics* with the other primary treatises of the *Little Astron-*

omy and a few other works in the mathematical sciences, usually including
the *Elements* (Acerbi 2016, 150, 154–158, 165–168). Maximos Planoudes,
working in the second half of the 13th century, mentions, in a letter to a
colleague, a certain manuscript of the *Spherics* before discussing a copy of
Diophantos's *Arithmetics* that he restored (Wendel 1940, 417; Leone 1991,
102). John Pediasimos mentions the work of Theodosios in his commentary to Cleomedes' *Heavens* (Pérez Martín and Manolova 2020, 93), and
Theodore Metochites claims to have studied Theodosios's *Spherics*, along
with a group of texts that included treatises of the *Little Astronomy*, under
Manuel Bryennios (Acerbi 2020b, 108 n. 10). The 14th-century Par.gr. 2448
(**K**), includes a revision of the *Spherics* in a grouping of treatises that only
includes a couple of the other works of the *Little Astronomy* (Czinczenheim
2000, 239–258; Acerbi 2016, 162–163). This version contains a number of
mathematical improvements over the text in Vat.gr. 204 (**A**), which misled
Heiberg into preferring it in some places when he produced his critical edition of Theodosios's treatise. In general, when the *Spherics* was studied in
Byzantine circles, it appears to have been situated within a context that was
less firmly related to the other treatises of the *Little Astronomy*, but also included other works in the mathematical sciences. This transition may have
taken place due to attempts to construct mathematical curricula based on
the then traditional concept of a quadrivium composed of arithmetic, music, geometry and astronomy (Pérez Martín and Manolova 2020, 83–89).
The scholarly approach taken by the Byzantines towards the *Spherics* was
largely, although not exclusively, philological.

In early 12th-century Toledo, Thābit's correction of *Spherics* had served
as the basis for a close translation into Latin made by Gerard of Cremona,
as part of his broad program of translating scientific works from Arabic
sources (Kunitzsch and Lorch 2010, 5). When Gerard made this translation, however, he no longer construed the *Spherics* in the context of the
Arabic *Middle Books*, but rather conceived of it as an element in a division
of the sciences that had been put forward by al-Fārābī in his *Classification
of the Sciences*, and exemplified by Aḥmad ibn Yūsuf in a parable on the
importance of approaching the sciences in the proper order (Burnett 1999;
2001) – both of which texts Gerard also translated. The underlying concept
of Gerard's translation program can be drawn from a bibliographic notice
that was written by his students, or colleagues, to which they appended a
select and organized list of the master's translations. The works were set
out in the following categories: logic (*dyaletica*), geometry, astronomy (*as-
trologia*),[20] philosophy, medicine (*phyisca, fisica*),[21] alchemy, and geomancy
(Burnett 2001, 275–281). Within this division, Theodosios's *Spherics* falls
not in "astronomy," but in "geometry," immediately following Euclid's *El-*

---

[20] The list of works makes it clear that this is what we would call astronomy.
[21] The list of works makes it clear that this is medicine.

ements, including books XIV and XV, which were added to the treatise
in the late-ancient period. Although some of the treatises of the *Middle
Books*, such as Menelaos's *Spherics* and Thābit's *Sector Theorem* are included within "geometry," others, such as Theodosios's other works and
Autolykos's *Moving Sphere* are placed in "astronomy." Moreover, since "geometry" also contains al-Khwārizmī's *Algebra*, Gerard's separation of the
treatises of the *Middle Books* into the two sciences of "geometry" and "astronomy" seems to be a different division from that which we find in Greek
sources.

The most significant early Latin translation, or revision, however, was
probably that made by Campanus of Novara, in the 13th century. Although
the attribution is not certain, this version is a heavy reworking of the text
that has similarities in style to Campanus's work on the *Elements*, and which
is credited to Campanus in a number of 15th-century manuscripts (Lorch
1996, 169–171).[22] This version contains considerable material not found in
the text or scholia of any currently known Arabic or Greek version – such
as converses, lemmas, and cases not considered in the original – but it ends
after an extra theorem that follows *Sph.* III.9, although in some manuscripts
the treatise is completed with the Gerard version.[23] It also contains internal
references to previous theorems of the treatise, with the books numbered
XVI, XVII, and XVIII – that is, as three books of the *Elements* following
the fifteen medieval books. Both of the medieval Latin versions are usually
found in manuscripts that include it with the *Elements*, as well as some
other texts in the mathematical sciences.

Although the extent to which the *Spherics* was read by medieval scholars working in Latin has yet to be fully studied, there is not much evidence
that it was studied in the universities. A 14th-century manuscript describing two cycles of works in "mathematics" includes Theodosios's *Spherics*
in a first list that contains arithmetical works, the *Elements*, Menelaos's
Spherics and the *Almagest*, and is followed by a second list that is de-
voted to works in astrology by Latin and Arabic authors (Bjørnbo 1903),
all probably as preparatory to medical studies (Lemay 1976, 210–212). It
is not clear, however, how much of this extensive list was actually taught at
any particular university. Indeed, most of the university statues that have
been studied give a much more limited scope for the mathematical sciences.
Spherical astronomy was generally taught through Johannes of Sacrobosco's
Sphere and its commentaries, while Theodosios's *Spherics* is not mentioned

---

[22] This version was attributed to Plato of Trivoli, working in the 12th century, by Pena
(1558, a iij(v)—a iiij(r)), on the authority of an anonymous author of a text on burning
mirrors. This source, however, has not been located, so it has not been possible to verify
the claim.

[23] The broad differences between this text and that preserved in most of the Greek
manuscripts can be found listed in an appendix to Nizze's (1852, 163–192) edition of the
Greek.

(Thorndike 1949, 42–43; Høyrup 2014, 115–119). Moreover, the *Spherics* was not included in the collections of works, which modern scholars have called the *Corpus astronomicum*, that are thought to have been used in the universities to study astronomy (Pedersen 1975, 73–82). This collection was centered around original Latin works by Sacrobosco, such as his *Sphere* and *Algorithm*, and various other Latin authors, such as Robert Grosseteste, along with a few translations from authors who wrote in Arabic, such as Thābit ibn Qurra and Māshā'allāh ibn Atharī. Probably most students did not read more than a few of these texts, and it is even unlikely that many students made it to the end of Euclid's *Elements*. Hence, it is improbable that large numbers of students read Theodosios's *Spherics* in the medieval universities.

The first two printed versions of the *Spherics* both appeared in 1518, as almost identical sections of two rather similar compilations of texts made by Lucantonio Giunta and the heirs of Ottavianus Scoto, respectively (Various 1518a,b, 91v–104r, 133v–139r). These were just two of a large number of compilations, from various printers in a number of European cities, that centered around Sacrobosco's *Sphere* and contained other texts that were commentaries to this, or considered to deal with related subject matter, mostly directed towards the market of university students (Valleriani 2020). The basis for these editions was the Latin version of Campanus, but their publications were not carefully executed. That there was no particular attention paid to Theodosios's *Spherics* in these editions is shown from the following facts: the treatise is found in two discontinuous parts; the text itself is corrupt, with many lacunae and repetitions; and the diagrams are sometimes referenced incorrectly, or are missing altogether (Malpangotto 2010, 77).

In 1529, Johannes Voegelin produced a corrected edition of the same version of the treatise, in the preface of which he points out that the many errors of the previous edition were so grave as to confuse the reader (Voegelin 1529, A iijr,v). Voegelin's book was a stand-alone publication of the Campanus version, based on two "old" manuscripts, that was meant to restore the treatise to its original state. He also added his own scholia, introduced by his name, following many of the propositions, in which he remarks on some mathematical considerations, and sets out the astronomical implications of the theorems. These notes are still useful for understanding the spherical astronomy implicit in the treatise.

The Arabo-Latin version of the text served as the basis of a new restoration of the treatise produced by Francesco Maurolico (1558). This publication, which was one part of Maurolico's projected program of effecting a complete "instauration" of the mathematical sciences, was again a compilation of mathematical treatises (Rose 1975, 173–174). It begins with Theodosios's *Spherics* in the "tradition of Maurolico," followed by Menelaos's

Spherics, in the same, then Maurolico's own *Spherics*, a number of other titles of the *Little Astronomy*, and finally some original Latin works on trigonometry and spherical triangles. In this context, the *Spherics* was presented as part of Maurolico's project of revising the ancient mathematical sources by presenting them freed from any mathematical errors or gaps, and including additional material that he regarded as necessary to the mathematical argument. The many changes that he introduced to the text were made on the basis of mathematical, as opposed to philological, considerations. A substantial change made to this version was the introduction of a new style of diagram, based on the techniques of perspective drawing (Malpangotto 2010). In these diagrams, the circles in the sphere are depicted with smooth curves, largely following rules that had been developed over the previous century and written about by authors such as Leon Battista Alberti and Piero della Francesca (Andersen 2007, 17–80, 115–122). In Maurolico's organization, Theodosios's *Spherics* appears as the foundation of a treatment of the mathematics of the sphere that had its roots in ancient sources, but also incorporated newer developments.

In Paris, Johannes Pena (1558) published the *editio princeps* of the Greek text in the same year as Maurolico's Latin version, and accompanied it with his own Latin translation. This project was explicitly articulated as a philological restoration to the original language and format of the text. The differences between the text found in Greek manuscripts and that in the Arabo-Latin tradition of the Campanus version were attributed to scholars working in Arabic, and all of the additions were considered to be superfluous corruptions (Pena 1558, a iiij(r)).

The Greek text served as the basis of a revised Latin version by the Jesuit Christoph Clavius (1586),[24] professor of mathematics at the Collegium Romanum. Since Clavius had worked on mathematical texts with Maurolico in Messina for some months in 1574 (Lattis 1994, 19), it is not surprising that his treatment of the text was influenced by that of Maurolico. In this presentation, Theodosios's *Spherics* is followed by a number of Clavius's own works on trigonometrical topics accompanied by tables. This edition was part of Clavius's overall program of publishing Greek mathematical texts in a modern format, including many comments, lemmas and additional material, with references to Maurolico's insights where appropriate, and with the justifications of the steps of the argument included in the margins. In general, Clavius followed the more modern perspective approach to the diagrams, but he returned to the use of schematic lens-shaped figures to depict foreshortened circles (see Figure 3, p. 41 for an example of a lens-shaped

---

[24] The various proposals for Clavius's original German name found in the secondary literature are all guesses based on the Latin word *clavius* itself. We are not aware of any certain evidence about his birth name.

figure).[25] Clavius's version of the text was translated into English in the early 18th century (Stone 1721).

In the 17th century, the *Spherics* was reworked and reprinted a number of times, usually as part of a *cursus mathematicus*, and accompanied by editorial claims that the text needed to be restored for educational purposes, along with various innovations in notation introduced for these purposes (Ver Eecke 1927, xliv–li; Lorch 1996, 163–164). The extent to which these versions of the treatise were actually read, or used in teaching, has not yet been studied.

Ancient spherics: geometry or astronomy

Modern scholars, following the organization and conception of Maurolico's publication of Theodosios's work, use the term *spherics* for ancient approaches that utilize what we call spherical geometry to address problems and topics in what we call spherical astronomy, specifically as set out in the following treatises: Autolykos's *Moving Sphere* and *Risings and Settings*, Euclid's *Phenomena*, and Theodosios's *Spherics*, *Habitations*, and *Days and Nights* (Mogenet 1947, 235; Berggren 1991, 229; Aujac 1993, 442), to which we may perhaps add Menelaos's *Spherics*, and Ptolemy's *Almagest* II, VIII.5 and VIII.6. That is, spherics is generally understood in modern scholarship to be the use of the geometry of the sphere to model the position of the sun relative to the local horizon, for the purpose of addressing questions such as the relationship between terrestrial latitude and the longest period of daytime, the length of the daytime at different times of the year, the instantaneous positions of the ecliptic, and so on. It is unclear, however, that ancient thinkers held that there was a well-defined mathematical science that groups together these texts and these problems.

The modern term *spherics* can be derived from either τὰ σφαιρικά, the title of the above mentioned books by Theodosios and Menelaos, or from ἡ σφαιρική, the name of a mathematical science used by later authors with Pythagorean or Platonist tendencies, such as Nikomachos, in his *Introduction to Arithmetic* or Proklos in his *Commentary on Elements I*. When Nikomachos and Proklos use this term, however, it seems to mean astronomy generally, perhaps as a form of synecdoche. For example, after discussing number theory (arithmetic) and music, Nikomachos says (*Int. arith.* I.3.2; Hoche 1866, 6):[26]

> Again, since of the *how much*[27] there is that in rest and in place,
> and that in motion and in revolution, two different sciences cor-

---

> responding to them will accurately investigate the *how much* –
> that which stays and is at rest, geometry; and that which is
> carried and traverses, spherics.

In this place, Nikomachos repeats a well-known Pythagorean trope about the division of the mathematical sciences into number theory, geometry, astronomy and music, except that he uses "spherics" in place of astronomy. A few passages later, while discussing the treatment of the mathematical sciences in Plato's *Republic*, he more explicitly equates the two by relaying a claim that "spherics and astronomy" are useful to farming and navigation (*Int. arith.* I.3.7), and later still he tells us that spherics uses arithmetic (*Int. arith.* I.5.2). Proklos, in his *Commentary on Elements I*, uses "spherics" in the same way when discussing the same fourfold division of the mathematical sciences (Friedlein 1873, 37, 59).

Neither the usage of *spherics* in the title of the specific works of Theodosios and Menelaos nor as a metonym for astronomy at large, however, covers precisely the same range of meanings as is usually understood by the modern usage of the term. Furthermore, the texts that modern scholars group together to form spherics, as listed above, are not so grouped by ancient and medieval scholars, since, as discussed above, we find texts that are not based on spherical geometry in both the *Little Astronomy* and the *Middle Books*.[28] Finally, spherics is not synonymous with ancient approaches to spherical geometry, since there were other ways to study the geometry of the sphere that were used in antiquity and are not addressed in the texts that we associate with spherics – as exemplified in Ptolemy's *Analemma*, dealing with sundial theory, or his *Planisphere*, which provides methods for modeling the sphere in the plane.

What is more, it is not even clear that there was any consistent agreement about where a supposed field of spherics might fit into ancient divisions of the mathematical sciences. As discussed above, there is no mention of spherics in ancient sources before the late Hellenistic period. Indeed, when we look at the division of the mathematical sciences set out by Geminos, in the 1st century BCE, it is not at all clear that there was a well-defined place for spherics (Friedlein 1873, 38–42; Evans and Berggren 2006, 43–48, 246–249). Geminos divides the mathematical sciences into those that treat intelligible objects – such as arithmetic and geometry, which are prior and more authentic – and those that treat perceptible objects, such as mechanics, astronomy (ἀστρολογία), optics, geodesy, music theory (κανονική), and computation (λογιστική). He then goes on to further subdivide the sciences, among which he divides geometry into plane theory and solid geometry (στερεομετρία), and he divides astronomy, according to the use of certain

---

[28] A possible exception to this is the division given by Galen, which may indicate a category of texts in spherics that corresponds to the modern conception.

instruments, into sundial theory (γνωμονική), which measures the hours through gnomons, meteoroscopy (μετεωροσκοπική),[29] which determines the altitudes and distances of celestial bodies, and teaches other matters, and dioptrics (διοπτρική), which investigates the positions of the celestial bodies using sighting instruments. The closest of these to our conception of spherics may be meteoroscopy, which is the science of an observational instrument that is constructed in the form of an armillary sphere, but on the whole spherics does not fall neatly, or exclusively, into any of Geminos's categories. Another possibility is that one of his divisions of mechanics, namely sphere-making (σφαιροποιΐα), was meant to include spherics. As a practical activity, however, sphere-making involved imitating some aspect of the celestial motions with a mechanism; while, as a theoretical discipline, it involved modeling the motions of each of the various celestial bodies using spheres or some other representation of the cosmos (Aujac 1970; Evans and Berggren 2006, 51–53). Such an endeavor, however, would both fall beyond the scope of spherics as represented in the texts on spherical astronomy mentioned above, and would also be more related to observation than the material that we find in the *Spherics* of Theodosios and Menelaos.

Starting from at least the 1st century CE and continuing into late antiquity, the use of "spherics" as a synonym for astronomy in general by Nikomachos and Proklos suggests that they considered the former to be a significant subset of the latter. Moreover, in the passage from his *Commentary on Aristotle's Physics* that we read in the previous section (see p. 13), Philoponos suggests both that the *Spherics* belongs together with *Moving Sphere* and *Phenomena* and also that it is is somehow purer and more mathematical than these other two, which are successively closer to astronomy as studied by the philosopher of nature. This suggests that the Platonists regarded astronomy itself as a proper branch of mathematics and held that Theodosios's *Spherics* was an example of the purer form of this discipline.

Another late-ancient reference to spherics as a discipline comes from an anonymous author who was probably a student of Olympiodoros, the younger contemporary of Philoponos who became head of the Athenian school of philosophy after Eutokios. This author, known as Pseudo-Elias/David,[30] presents an eight-fold division of the mathematical sciences in a series of lectures on Porphyry's *Introduction*. The prior and more general sciences are the traditional Pythagorean quadrivium of number theory, geometry, music and astronomy, and then each of these is associated

---

[29] A meteoroscope was a sort of armillary sphere for making observations, as discussed by Ptolemy in his *Geography* (I.3).

[30] The meaning of this name is that this is a person who we have reason to believe is neither the author of the texts attributed to Elias nor of those attributed to David, the two scholars who probably succeeded Olympiodoros as teachers of philosophy (Westerlink 1967, xi–xvi; Watts 2006, 257).

with an embodied (ἔνυλος) science: number theory with computation, music
with embodied music, geometry with geodesy, and astronomy with spherics. Following this, Pseudo-Elias/David lists the recognized masters of these
sciences, as follows (Westerlink 1967, 38–39):[31]

> Let us mention successively the inventors of these ⟨sciences⟩,
> who are the best in them. So, Nikomachos is at the head of
> number theory (τῆς ... ἀριθμητικῆς), Diophantos of computation
> (τῆς ... λογιστικῆς), again, the Pythagoreans are at the head of
> music, ⟨while the Aristoxeneans of embodied music, and Euclid
> of geometry,⟩ while Heron of geodesy, while Paul of astronomy,
> and Theodosios of spherics (τῆς σφαιρικῆς).

There are a number of aspects of this passage that warrant comment. In
the first place, the prior and more general astronomy is represented by a
Paul, who is probably Paul of Alexandria, the 4th-century author of the
Introduction to Astrological Effects, which was the subject of a surviving
course of lectures, probably delivered by Olympiodoros in 564 (Westerlink
1971). From our perspective, however, Paul's only surviving text, which is
a basic introduction to astrology, seems to be neither less material nor more
general than Theodosios's *Spherics* – rather the opposite. Hence, it will
likely come as a surprise to most readers that the anonymous author finds
Paul to be the authority on astronomy in general. Furthermore, his claim
that Theodosios is the master of an embodied science known as spherics
is hard to square with either the text of Theodosios's *Spherics* itself or
Philoponos's account of the *Spherics* that we read in the previous section.
Indeed, Philoponos explicitly asserts that the *Spherics* has nothing to do
with matter.[32] It may be that Pseudo-Elias/David is thinking of spherics
as the mathematical study of certain material objects, namely solid and
armillary spheres. Whatever the case, this text presents a rare case of
listing spherics as a distinct science. This text, however, is not by a well-
known authority and it is not clear that this was a position that was ever
widely accepted.

Without making any pretense to exhaust the divisions of the mathematical sciences put forward by ancient authors, it may be useful to survey the
various divisions that we have seen and to address where the texts in what we
call spherics might fit into them. In the first place, the division put forward
by the Pythagoreans, as exemplified in the passage above from Nikomachos
(p. 23), divides the mathematical sciences into those that study discrete

---

[31] See also the translation by Christianidis and Megremi (2019, 23), which we follow
with modifications.

[32] The disagreement between Pseudo-Elias/David and Philoponos here may be related
to a broad set of disagreements that Philoponos had with Olympiodoros and his students,
for which see Watts (2006, 232–259).

objects and those that study continuous objects, and again into those that study stationary objects and those that study moving objects. If we apply this division to a proposed field of spherics it would seem to place the *Spherics* into geometry but the related texts of *Moving Sphere* and *Phenomena* into astronomy, since, as Philoponos points out, the *Spherics* does not concern motion, while the other two texts do. Alternatively, Geminos proposes a rather different division of the mathematical sciences, into those that treat intelligible objects and those that treat perceptible objects, each of which has a number of subdivisions. As discussed above, however, it is difficult to fit a proposed field of spherics into any of the subcategories that Geminos recognized. Furthermore, this division again seems to cleave the *Spherics* from the other texts in the proposed field, because it largely treats intelligible objects,[33] while the other spherics texts treat perceptible objects. The discussion by Philoponos introduces a similar polarity between intelligible objects and embodied objects, but he seems to see this as more of a graded scale than a sharp distinction, because he sets the *Spherics*, *Moving Sphere*, and *Phenomena* as belonging together in one discipline but occupying different places on the scale from intelligible to embodied. Hence, Philoponos's distinction between works that treat intelligible objects and those that treat embodied objects is, in fact, not the same as that of Geminos, because it does not distinguish the mathematical sciences from each other, but rather operates within each of the sciences, and orientates their methods with respect to the methods of philosophy. Again, Pseudo-Elias/David presents the distinction between intelligible objects and embodied objects as delineating the sciences that study them, which, as mentioned, contradicts the position of Philoponos, but also, once again, seems to separate the text of the *Spherics* from the proposed eponymous science. Finally, there is the perspective of the educational approach, which places various texts between studies of elementary geometry and those of more advanced gnomonics and astronomy, as exemplified by Galen and Pappos; but these approaches do not separate out texts in spherics from texts related to optics and other approaches to astronomy. In fact, all of these divisions of the mathematical sciences, although related, are incompatible with each other in various key aspects and it is clear that there was no universally accepted approach to this matter. Most importantly, the modern understanding of a proposed field of spherics that encompasses a certain set of texts and applies the geometry of the sphere to solving problems of spherical astronomy does not

---

[33] The one possible exception to this is the use of the expressions "visible pole" and "non-visible pole" to denote the poles on either side of a great circle that may be taken to represent the horizon (see pp. 29 and 282). As discussed below, however, there are also mathematical reasons to introduce some technical term to denote these two poles, so that these expressions are probably not essentially observational, but just a shorthand to denote the two poles of a base great circle.

correspond to a unique and well-defined mathematical science in any of the ancient schemes that we have examined.

One of the principal reasons for this lack of agreement among ancient thinkers about the place of spherics within the mathematical sciences probably originates in the lack of any institutional place for instruction in mathematics. Since teaching the mathematical sciences was essentially private, teachers in the major cultural centers appear to have selected the texts to be studied according to increasing levels of difficulty, usefulness in setting the groundwork for more advanced topics, and significance for conveying concepts and techniques that they regarded as important. In this selection, however, considerable respect was given to tradition, so that the texts that were taught were those that were regarded as classics, and there was relatively little proliferation of new teaching materials. Since the teachers themselves were not specialized in any particular field of the mathematical sciences, they neither perceived nor advocated rigid demarcations between proposed fields. On the other hand, the philosophers, who used the mathematical sciences as preparatory to philosophical studies that they regarded as more advanced or as examples for epistemological schemes that they sought to promote, divided up the mathematical sciences in various ways according to the positions that they wanted to establish.

In fact, the external evidence for discussions of astronomy and spherics in ancient sources casts doubt on the idea that "spherics" was even recognized as a distinct area of mathematical research before Theodosios produced his text. There is no clear articulation of such a field in early sources, and even Geminos, perhaps roughly Theodosios's contemporary, does not seem to have any concept of spherics as a special area of mathematical activity. Indeed, the first authors to articulate such a notion are Nikomachos and Galen, both in the Imperial period, by which time Theodosios's works were in circulation, and apparently well known. We must acknowledge the possibility that Theodosios's works themselves played some role in these developments.

Furthermore, it should be noted that there is no evidence in the sources that any ancient thinker regarded spherical geometry as a fully autonomous branch of geometry that was then applied to studying particular spheres, whether the supposed celestial sphere, or actual solid and armillary spheres. In fact, the closest that the ancient sources come to such a conception is found in the structural organization of the two *Spherics* of Theodosios and Menelaos. Both of these texts begin with more or less general geometrical theories. Theodosios presents a solid geometry of the sphere, which only adumbrates an intrinsic spherical geometry, whereas Menelaos begins with, and extensively develops, such an intrinsic approach. In both cases, however, this material is presented as prefatory to the more advanced topics that take up a significant portion of each treatise and were given an astronomical interpretation by ancient and medieval readers. That is, they give no indication of having regarded the geometrical foundations of their work as fully independent from its astronomical applications. Nor is there any clear evidence that specialized texts in spherical geometry were ever produced by other authors in the ancient period. Hence, it appears that Theodosios and Menelaos sought to make their work more general and directed more towards intelligible objects by avoiding any mention of astronomical objects and by not explicitly introducing the concept of motion.

Despite the fact that there was no clearly articulated concept of a discipline of spherics among ancient authors, the use of such terminology, as well as the terminological distinction between spherical geometry and spherical astronomy is still useful when discussing the contents of the two *Spherics* of Theodosios and Menelaos. With the caveat that by these terms we do not designate an ancient conception of autonomous sciences, we will use *spherics* to denote the ancient sources discussed in this section and use *spherical geometry* and *spherical astronomy* to designate different aspects of these sources. We turn now to explaining the astronomical concepts that motivate the latter parts of Theodosios's *Spherics* – that is, from *Sph.* II.10, or perhaps *Sph.* II.17, to the end of the treatise.

Concepts and orientation of spherical astronomy

The astronomical subject matter of the *Spherics*, which we call spherical astronomy, was the study of the geometrical configurations produced by the motion of an assumed celestial sphere relative to the local horizon. Although no astronomical objects – except, perhaps, the visible and invisible poles, P_v and P_i [34] – are named in the *Spherics* itself, many of the propositions have clear astronomical interpretations that ancient readers would have understood. Here, we do not give a full treatment of spherical astronomy but merely provide a basic description of the configurations that are relevant to understanding the subject matter, along with terminology that will be used later.[35] As mentioned in the quote from Strabon above (p. 10), ancient readers would have learned the facts of spherical astronomy directly from an armillary or solid sphere in a "first course of the mathematical sciences," well before attempting to read a text like Theodosios's *Spherics* (*Geog.* I.1.21).

In spherical astronomy, the great circle of the local *horizon* is generally assumed to be the base plane – whether in the actual cosmos, or on a solid

[34] Although the terminology of the visible and invisible poles obviously has an astronomical origin, its usage in the *Spherics* can be explained by the need to have technical terms to designate each of the two hemispheres determined by a great circle in the sphere (see p. 282, below). Here *invisible* simply means not currently visible, not incapable of being seen.

[35] For general discussions of Greco-Roman spherical astronomy, see Neugebauer (1975, 21–52) and Evans (1998, 75–161).

or armillary sphere – and the orientation of the other circles is described in terms of the horizon (see Figure 1). The great circle of the *meridian* passes through the celestial poles, P_v and P_i, and is perpendicular to the horizon, meeting it at the north and south points. The prime *vertical*, not shown in Figure 1, is the great circle perpendicular to both the meridian and the horizon, meeting the meridian at the zenith and nadir and the horizon at the east and west points. In Figure 1, the west point, which is the intersection of the equator and the horizon, faces us, although it is not labeled as such in the diagram. The great circles of the horizon, meridian, and vertical determine the local coordinates, although the vertical plays no role in the *Spherics*.

Over the course of a single day,[36] the celestial sphere rotates once around the axis through the celestial poles, P_vP_i, such that any given point on the sphere that crosses the horizon will rise in the eastern hemisphere and set in the western hemisphere. This diurnal rotation determines the celestial *equator*, and any given point of the celestial sphere traces out an arc of right ascension, α, over some duration, which in the course of a day completes a circle parallel to the equator at some declination, δ, measured along great circles through the celestial poles, and perpendicular to the equator, known as *colures*. We can call such parallel circles of constant declination δ-circles. The system of great circles made up of the celestial equator and the colures determine the equatorial coordinates.

For horizons at the terrestrial equator, $\varphi = 0°$, P_v and P_i will lie on the horizon, the celestial equator will be perpendicular to the horizon, and an arc of a δ-circle that is bounded by two colures will have a rising time, ϱ, that is the same as that arc of the equator that is bounded by the same two colures, α. This configuration is known as the *upright sphere (sphaera recta)*, and gives rise to our expression *right ascension* for α (Figure 2 (left)).

For horizons between the terrestrial equator and poles, $0° < \varphi < 90°$, the visible pole, P_v, will be elevated above the horizon by some arc, φ, along the meridian (see Figure 1). This configuration is known as the *inclined sphere (sphaera obliqua)*. The elevation of the pole is, in fact, the same angle as the terrestrial latitude of the horizon in question, so that the angle φ is a characteristic of the latitude. In the inclined sphere, the horizon divides the δ-circles into three sets: (1) those that intersect the horizon, (2) those that are always above the horizon and (3) those that are always below the horizon. These three sets are separated by two equal and parallel circles,

---

[36] The *day* of spherical astronomy is the the period of time in which each star will return to the same position relative to the local coordinates, usually called the *stellar day*. Since the sun moves slowly relative to the stars, this differs from the *solar day*, the period of time in which the sun returns to the same position relative to the local coordinates. In fact, since spherical astronomy does not include a theory of solar motion, we can regard the sun as a point on the ecliptic carried by the motion of the sphere over the course of a single day, so that the stellar day and solar day are the same.

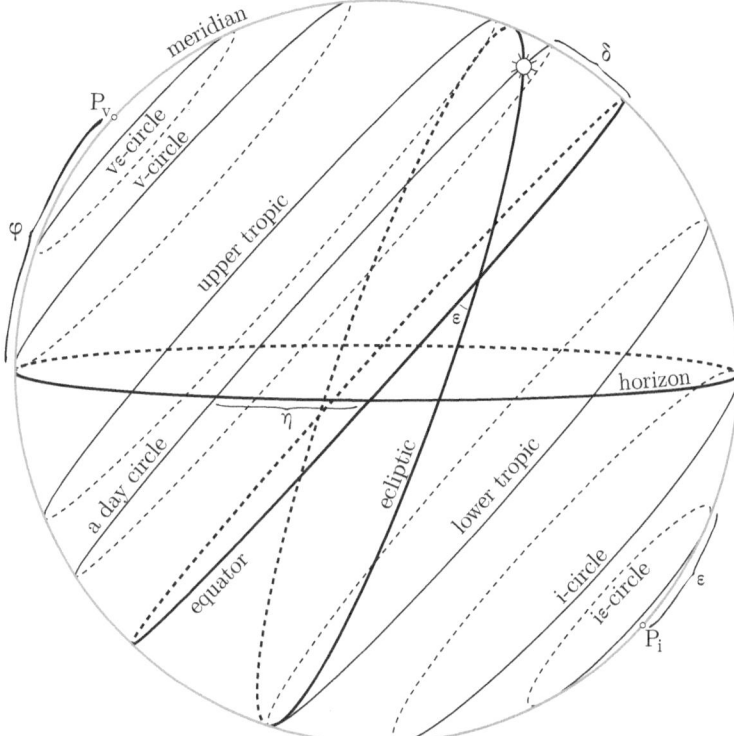

Figure 1: The celestial sphere used in spherical astronomy.

both tangent to the horizon at opposite points, the greatest of the always visible circles and the greatest of the always invisible circles. We can call these two δ-circles determined by the local horizon together the *visibility circles*, and individually the *v-circle* and the *i-circle*.[37]

Over the course of a year,[38] the sun will move along the great circle of the *ecliptic*, about 1° per day, in the opposite direction from that in which the celestial sphere rotates. The ecliptic intersects the equator in the points of the vernal and autumnal equinoxes at an angle known as the obliquity of the ecliptic, ε. In Figure 1, the vernal equinox faces us above the horizon. In this way, the ecliptic, like the horizon, will be tangent to

---

[37] This terminology is adopted from Schmidt (1943, 23). Note that these circles are called the arctic (northern) and antarctic (southern) circles by Euclid (*Phen.* 14), Geminos (*Int.* V.2), and probably by Strabon in the passage quoted above (*Geog.* I.1.21).

[38] The *year* handled by spherical astronomy is the period of time in which the sun returns to the same position relative to the intersections of the ecliptic with the celestial equator, known as the *tropical year*. Since there is no solar theory included in spherical astronomy, we simply take the sun to be at some position of the ecliptic for the course of a single revolution of the sphere. Spherical astronomy does not itself include a treatment of either the *stellar year* or the precession of the equinoxes.

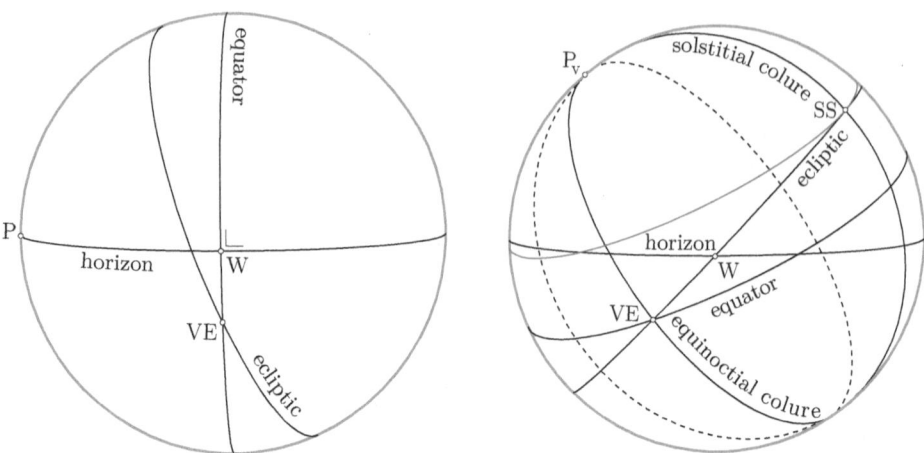

Figure 2: The celestial sphere: (left) the upright sphere, (right) the principal colures.

two equal and parallel δ-circles, the *upper tropic* and the *lower tropic*.[39] The tropic circles determine the points of greatest declination of the sun at their points of tangency with the ecliptic, the solstices. The equinoxes and solstices are the four cardinal points of the ecliptic, known as the vernal or spring and autumnal equinoxes, *VE* and *AE*, and the summer and winter solstices, *SS* and *WS*. The great circles through these four points and the celestial poles are the *equinoctial* and *solstitial colures* (Figure 2 (right)). Because the colures pass through the celestial poles they are perpendicular to the equator,[40] so that the three mutually perpendicular great circles of the ecliptic and the equinoctial and solstitial colures determine the zodiacal coordinates.

Because the ecliptic is fixed in its inclination with the equator, the system of the zodiacal coordinate circles also rotates once a day, and the ecliptic itself wobbles relative to the local horizon, returning to the same inclination and orientation once per day. One feature of this system, however, remains fixed relative to the local horizon. As the ecliptic rotates in its inclined position relative to the equator, its poles trace out a pair of small circles that are separated from the celestial poles by the same angle at which the ecliptic is inclined to the equator, ε. Since these circles are also handled by the geometrical configuration discussed in the *Spherics*, we will call that in

[39] We have chosen not to call these the tropic of Cancer and Capricorn, or the northern and southern tropics, because there is no such terminology in the *Spherics*. Indeed, there is no mention of the *tropics* in the text. Nevertheless, the use of *upper* and *lower*, in reference to an assumed base plane does agree with the terminology of this treatise, so we have used it here. In this sense, *upper* just means above the horizon and in the direction of the visible pole, while *lower* means the opposite.

[40] The solstitial colure is also perpendicular to the ecliptic.

the direction of the visible pole the *vε-circle*, and that in the direction of the invisible pole the *iε-circle*.[41]

For the purposes of spherical astronomy, we neglect the daily proper motion of the sun of about 1° and regard the sun as simply found at some daily fixed, but annually variable, point on the ecliptic. Hence, in being carried around with the motion of the celestial sphere, the sun will also determine a δ-circle, the *day-circle*. As the sphere rotates, the equator and a given day-circle will cut off between them a characteristic arc on the meridian equal to the declination, δ, of the day-circle, and when the declination of the day-circle is less than that of the visibility circle the day-circle will determine a characteristic arc on the horizon known as the ortive amplitude, η, between the intersections of the horizon with the day-circle and with the equator (Neugebauer 1975, 37–39).

In this section, we have simply given a descriptive account of the configuration of the celestial sphere with no attempt at mathematical argumentation. In the Commentary, however, we will point out how geometrical abstractions of each of the claims made above, among others, are actually shown in the *Spherics*. The basic elements of this configuration make up what modern scholars have called the *two-sphere model* (Goldstein and Bowen 1983). The two-sphere model consists of a sphere of the earth mirrored by a sphere of the heavens, which has bands at the latitudes of the tropics and at the visibility circles. As will be discussed below, it is generally held that this model was produced in the mid-4th century, probably by Eudoxos, among others.

We should emphasize that there are a number of abstractions and simplifications in this model that must have been known to Theodosios. In the first place, the sun was considered to move along the ecliptic continuously, not in a step-like fashion as modeled in spherical astronomy. More importantly, however, by the time Theodosios was writing the *Spherics*, Apollonios and Hipparchos had already proposed that the sun does not maintain a constant distance from the earth, so that in their models it could not be held to move in a sphere concentric with the earth. Hence, the model of the spherics texts must have been understood to be an abstraction and simplification of whatever actually takes place in the heavens. The geometry of Theodosios's *Spherics* could then be understood to be an abstraction of a simplified model, and hence only indirectly a mathematical model of the actual cosmos. On the other hand, the mathematics of the *Spherics* could

---

[41] These circles do not appear to be discussed in Geminos's *Introduction to the Phenomena*, but they are found fixed to the moving sphere in many of the early modern illustrations of armillary spheres – for example in the illustration accompanying Sacrobosco's treatise in *Sphaerae mundi compendium foeliciter inchoat* (Various 1488, Aiv v), Regiomontanus's *Epytoma Ioannis de Monte Regio in Almagestum Ptolomei* (1496, a3 r), the title page of Maurolico's *Theodosii sphaericorum elementorum libri III ...* (1558), and in the 1584 illustration reproduced by Evans (1998, 80).

be understood as a direct model of the actual spheres that were used to study elementary astronomy, and which are depicted in art and literature from the whole of the Greco-Roman period. Hence, we can also understand the *Spherics* to be a study of the geometry of the solid and armillary spheres that were known to every educated person.

Ancient spherics before Theodosios

Although it is difficult to say much with certainty about the history of work in spherics prior to the beginning of the Hellenistic period, when the treatises of Autolykos and Euclid were written, we can, nevertheless, attempt to make some general claims by looking at remarks made by certain philosophers on astronomical topics in the 4th century BCE. We begin by looking at the work of Plato.

It is fairly clear that Plato meant us to understand that at the time of the dramatic setting of the *Republic*, probably near the end of the 5th century BCE, there was nothing like the spherics texts that we know from later authors. In this dialog, when discussing a potential role for mathematical sciences in the education of elites, Socrates makes it clear that there was no one at that time practicing astronomy in a way that was comparable to what experts in plane geometry were doing by using "problems" to pursue their discipline (*Rep.* 527d–531d; Mourelatos 1980, 36–42; Berggren 1991, 233). Moreover there was, as yet, no systematically established discipline of solid geometry, although certain investigators had produced certain results (*Rep.* 528c; Huffman 2005, 398–399); and such a field would need to form the basis of Plato's proposed science of astronomy.[42] Since the methods of the plane geometers, including the use of problems, and a foundation of solid geometry are essential to the spherics of Autolykos, Euclid, and especially Theodosios, we can be fairly confident that nothing like these works was known at the end of the 5th century.

Some decades later, in the *Timaios*, Plato has the eponymous Pythagorean describe the construction of the cosmos by a divine Craftsman in a way that clearly invokes the imagery of the construction of an actual sphere. In the process of creating the living being of the cosmos in the shape of a sphere, the Craftsman turned it around in a circle (*Tim.* 33b), using a verb

---

[42] Tarán (1975, 93), Mueller (1980), Robins (1995, 376), and others take Plato to be using the dialog to discuss a recent historical period prior to the proper development of these sciences, which they assume took place in the early decades of the 4th century according to Socrates' hopes. This position, however, requires reading Plato as a sort of historian of mathematics and fleshing out this history with ample reconstruction, since we no longer have any of the mathematical texts in question. Plato himself, however, is rather vague about both what is wrong with these sciences at the time of the dialog, and what he thinks they should become.

that is commonly used for turning an object on a lathe. The use of this literary device means that the image of a solid sphere being turned on a lathe was so well known, or so easy to imagine, that it could be invoked as a metaphor for understanding the unknown production of the cosmos. A few passages later, the Craftsman is described as taking two pieces of the material of the cosmos, pounding it out into long flat bands (*Tim.* 35a,b), attaching them together in the shape of a χ, wrapping them around in a circle, and joining them again on the other side (36b,c). This is often taken to be an indication that Plato is thinking of an armillary sphere (Cornford 1935, 72–74; Evans 1998, 81), but the text is invoking a kind of poetic metaphor and immediately goes on to describe the two bands as moving in opposite directions – something that does not happen on an armillary sphere. It is also possible that Plato is creating a vivid literary image and does not have a specific instrument in mind. Finally, Timaios later tells us that all of the various motions involved in astronomy cannot be understood without the use of an imitation, or representation, μίμημα, which may mean an actual instrument of some kind (*Tim.* 40c,d). All of this indicates that at this time there were various models of the cosmos, of which the simplest was a turned sphere – and it is not difficult to imagine that mathematicians would have been investigating the properties of such objects. There is, however, no use of any specialized terminology in the Platonic corpus to describe the features of the spherical model of the cosmos, which leaves room for doubt that the basic model of spherical astronomy was either fully developed, or at least widely accepted, when the Platonic works were being composed (Tarán 1975, 107; Goldstein and Bowen 1983, 335).

When the *Epinomis* was written, probably shortly after Plato's death, there was no expression of doubt about the status of solid geometry. Indeed, we are even told that this discipline is called "stereometry" by its practitioners (*Epin.* 990d; Tarán 1975, 93). This indicates that in the decades between the composition of the *Republic* and the *Epinomis*, solid geometry was set on a more secure foundation and some systematic treatises had been written developing its propositions. We may suppose that this was done by Archytas, Theaitetos and Eudoxos, among others.

Archimedes, in the introductions to his *Sphere and Cylinder* and *Method*, mentions that Eudoxos did important work in solid geometry and, more specifically, that he produced proofs for stereometric facts that Demokritos had discovered, but not shown (Heiberg and Stamatis 1972, I.4; Netz et al. 2011, 71). This indicates that Eudoxos played an important role in the development of solid geometry as a deductive science that took place sometime around the middle of 4th century.

Indeed, Eudoxos is that 4th-century mathematician whose attested work is most closely related to the later treatises of spherics, and which appears to have established the model for ancient spherical astronomy. In two lost works, *Phenomena* and *Mirror*, Eudoxos apparently set out a detailed description of the celestial sphere, establishing the basic two-sphere model of spherical astronomy (Lasserre 1966, 39–67; p. 33, above). Although Eudoxos's treatises on spherical astronomy are lost, they are discussed and quoted in Hipparchos's *Commentary on the Phenomena of Aratos and Eudoxos*, mentioned above, and Aratos's famous poem was closely based on these two works by Eudoxos, as often attested by ancient authors (Gee 2000, 109–114). As seen from Hipparchos's argument, which is based on many direct quotations of the texts of *Phenomena* and *Mirror* and detailed comparisons with Aratos's poem (Manitius 1894, 8–24), the latter's *Phenomena* should be taken as indicating, even if inexactly, the overall scope and content of Eudoxos's *Phenomena* and *Mirror*. Hence, it becomes clear that in these particular treatises Eudoxos's approach to spherical astronomy was essentially descriptive (Lasserre 1964, 128). Furthermore, in one of the ancient sources that refers to this material it is unclear whether Eudoxos simply described what he saw in the heavens themselves or engraved a model of the celestial sphere (Cic. *Rep.* I.22). Although there are various indications that Eudoxos was discussing what he actually saw in the heavens, it is possible that he also constructed a celestial globe, used this to some extent in his work (Aujac 1979, 38), and assumed that his readers would follow what his text described on a globe of their own. Whatever the case, there is no indication in the evidence for these two works of any use of solid geometry, nor of any of the mathematical methods found in the early Hellenistic texts on spherics.

In discussing Eudoxos, we must also mention his homocentric theory of the cosmos, or of individual celestial bodies, which is largely regarded as his most significant contribution to mathematical astronomy.[43] The original sources for this theory have been lost and our best evidence for this work now comes from a rather short remark made by Aristotle, who was Eudoxos's somewhat younger contemporary, and a longer explanation in the *Commentary on Aristotle's Heavens* written some nine centuries later by Simplikios, working through the intermediaries of Sosigenes, in the 2nd century CE, Eudemos, in the 4th century BCE, and perhaps Sosigenes's student Alexander (*Met.* 1073a17–32; *In cael.* 493–497; Lasserre 1966, 67–74). At the most basic level, according Aristotle in the *Metaphysics*, the theory apparently consisted of the following components: for the sun and the

---

[43] Although the homocentric theory of the cosmos and spherical astronomy both use spheres as their core objects, we should distinguish them from each other. Spherical astronomy is simpler, using a single rotating sphere and assuming all celestial bodies have some position on that sphere. The homocentric theory uses multiple spheres and tries to account for the proper motion of each of the luminaries and planets. Furthermore, spherical astronomy was studied and developed by mathematicians well into the modern period, whereas homocentric theory was only occasionally worked on as an alternative theory by a small number of individuals.

moon, there were three concentric spheres, one of which moved with the
fixed stars, one with the ecliptic, and the third with a circle inclined on the
ecliptic; and for the "wandering stars," the planets, there was a sphere that
moved with the fixed stars, another that moved with the ecliptic, a third
that had its poles on the ecliptic, and a fourth that moved at an angle to the
third. According to Simplikios, the third and fourth spheres of the planets
move in such a way as to describe a curve that he tells us Eudoxos called
the *hippopede*. A great deal of scholarship has gone into explaining these
texts,[44] but the sources are sufficiently vague that it has not been possible
to reach full consensus on all aspects of the Eudoxean theory. There is debate about the specifics of the motion of the third and fourth spheres, there
is uncertainty about the role of numerical parameters in the models, there
is doubt about whether or not Eudoxos held that the sun has a latitudinal
motion (Bowen 2013, 262–264), the precise form of the so-called hippopede
is not certain (Yavetz 1998, 225–237), and so on. Indeed, our knowledge
of the evidential basis for 4th-century astronomy among Greek authors is
so spotty that it is not even known precisely what celestial phenomena Eudoxos's model was supposed to explain (Bowen 2002), although these may
have included elongations from the sun, invisibility periods, retrograde or
turning motions of some kind, movements in latitude, and so on.

From the perspective of the later history of texts in spherics, however,
the most significant uncertainty is that we do not know how Eudoxos approached this work – that is, we do not know to what extent this work,
especially as set out in his lost *Speeds*, was a work of deductive mathematics, if at all. In reconstructing the types of mathematics that might have
been used to justify the various models that have been proposed for the homocentric theory, one might suggest the use of mathematical facts that are
proven in Theodosios's *Spherics*, but we do not know that Eudoxos indeed
called upon these facts, nor whether or not he or anyone else demonstrated
them. Of course, Eudoxos could well have demonstrated such facts, but, if
he did, we do not know if he did so in a separate, unattested work, or if
he adduced theorems on the geometry of the sphere as lemmas in a more
descriptive astronomical work. It is also possible that Eudoxos developed

---

[44] The breakthrough work was done by Schiaparelli (1877), who established the core
elements of what we can regard as the standard interpretation of the models on the
whole, and the shape of the hippopede in particular. Accepting this model as Schiaparelli
formulated it, however, involved claiming that Simplikios was in error in certain aspects of
his explanation. Since then a number of scholars have added details and corrections, such
as Riddell (1979) and Heglmeier (1988; 1996). The use of computer modeling and close
textual analysis by Mendell (1998; 2000) has led to further details, some corrections, and
to new interpretations that allow us to understand Simplikios's remarks as substantially
correct, or at least compatible with what Aristotle says. Computer modeling and separate
interpretation of the two texts has also led to a significantly different interpretation of
core elements of the model by Yavetz (1998; 2001; 2003).

his homocentric theories primarily by using sensible and mechanical models, παράδειγμα, such as Plutarch tells us both Archytas and Eudoxos did, when approaching problems that were not tractable through logical and geometrical proofs (Plut. *Marcellus* XIV.5–6). As a result of Eudoxos's clear interest in the geometry of the sphere, and because of his strong reputation as a mathematician, it has often been asserted that he was the author of a supposed source for the *Spherics*, but this is simply modern speculation and is based on no direct ancient evidence (Neugebauer 1975, 750, 761).

From the same sources that describe the work of Eudoxos, we also learn of a continuation of this project carried out by one of his students, Kallippos, who added further spheres to the homocentric models. All of the cautions that were noted with regard to our knowledge of Eudoxos's work, also apply to that of Kallippos. Nevertheless, it is clear that in the second half of the 4th century, there was ongoing work that attempted to use the supposition of moving spheres to explain some range of celestial phenomena.

One of the first clear indications that the two-sphere model of spherical astronomy that had been articulated by Eudoxos in his *Phenomena* and *Mirror* was being taken up by some philosophers comes from the writings of Aristotle. In his *Meteorology*, Aristotle discusses a comet that was seen in 341 or 340 BCE, and relates that it appeared near the "equinoctial circle," or equator, and notes that comets often appear outside of the circles of the tropics (*Mete.* 345a; Goldstein and Bowen 1983, 334–335). Later in the text he describes in detail the divisions of the earth that mirror the divisions of the celestial sphere (*Mete.* 362a,b). In other places he also specifically mentions the "zodiacal circle," or "inclined circle" (*Mete.* 343a24, 345a20, 346a12; *Gen. et corr.* 336a32, *Met.* 1071a15). These references make it clear that by the second half of the 4th century, the components of the two-sphere model were well known and passing references to them could be used in explaining various facts, and developing other ideas.

A tantalizing glimpse of how another philosopher of nature reacted to this style of mathematical work comes from Epikouros, who never accepted the two-sphere model of the cosmos. In his *Nature* XI, which survives only partially in papyri, Epikouros gives an epistemological critique of the use of an instrument, ὄργανον, as an analogy for understanding some aspect of the phenomena of the sun and moon (P.Herc. 154, 1042; Sedley 1976). Although the text is fragmentary, and the description of the instrument obscure, Epikouros is attacking someone who held that "the patterns on the instrument," τὰ ἐπὶ τοῦ ὀργάνου δείγματα, furnish an analogy to the "phenomena according to the things in the sky," τὰ κατὰ τὰ μετέωρα φαινόμενα (Sedley 1976, 32). According to Epikouros, one cannot draw inferences from the object, ὑποκείμενον, which is the instrument or the conceptual model that underlies it, to the natural phenomena, because when one considers the object, or instrument, one argues about the properties of the object itself, and one has no way of knowing whether or not this object corresponds with the appearances of the things in the sky. We do not know the details of this instrument, nor of the phenomena that it was used to discuss, but these apparently included the appearances or images, φάσματα, of the sun, and the risings and settings of the sun and moon. There is also an obscure reference to the "obliquity," πλαγιασμός, related to the sun, which cannot be blamed for the fact that the sun appears to rise and set at different places if one changes location (Sedley 1976, 31).

Based on other material in the same ancient library, it has plausibly been argued that these criticisms are directed against a certain "astronomico-geometer (ἀστρολογογεωμέτρης) of Kyzikos," perhaps a member of the school that Eudoxos founded there (P.Herc. 1289; Sedley 1976, 26–30; Mueller 2004, 58–59). This connection, along with the reference to obliquity, which may correspond to a displacement oblique to the ecliptic that Simplikios tells us was involved in Eudoxos's third solar sphere, makes it tempting to read this instrument as involving spheres to model the phenomena of solar and lunar risings. It has sometimes been assumed that Epikouros's discussion of such a model should be more complex, involving the extra spheres added to the Eudoxan system by Kallippos (Sedley 1976, 36–37; Mendell 2000, 130), but there is no reason why he should address himself to this more complicated system. As Proklos points out, the Epicureans seek to discredit the starting points and principles of geometry (Friedlein 1873, 199), and the simplest possible model will serve for such a purpose. All that Epikouros needs to argue is that we have no way of knowing whether or not the detailed inferences that are made on the basis of even a very simple model are able explain the phenomena of the things in the sky. Whatever the case, it seems clear that Epikouros objected to the entire project of the ancient mathematical sciences – that is, to producing a model, whether actual or conceptual, and then studying the mathematical details of this model under the belief that its features will help explain the phenomena.

Around this time, or perhaps a decade or two later, the first surviving texts of spherics were written – namely, Autolykos's *Moving Sphere* and *Risings and Settings*, and Euclid's *Phenomena*. These two authors were almost certainly contemporaries, although Autolykos was likely the elder. Autolykos can be securely dated to the range 360–290 BCE (Aujac et al. 2002, 8–10), while, under the assumption that Euclid worked before Archimedes, or was an older contemporary, it is usually held that he was active around the turn of the 3rd century BCE. Although it is possible that the two texts were written completely independently, there are a number of places where the *Phenomena* may rely on claims made in *Moving Sphere*,[45] and it

---

[45] For example, it seems that *Phen.* 2 uses *Mov. Sph.* 10, *Phen.* 3 uses *Mov. Sph.* 1, 2, and 7, *Phen.* 7 uses *Mov. Sph.* 11, and a number of propositions in *Phenomena* appear to rely on *Mov. Sph.* 2, and so on. Neugebauer (1975, 750) argues against a strong position

is possible that Euclid expected his readers to be familiar with Autolykos's work, which likely predates the *Phenomena* by a decade or two.

Autolykos's *Moving Sphere* deals with various geometrical configurations of circles on a solid sphere that rotates about an axis. Although it is presented in a mathematical style, and makes some generalizations, there are fairly few geometrical deductions in the text, except that the arguments often rely on geometrical claims, many of which are shown in Theodosios's *Spherics*. Autolykos's *Risings and Settings* treats the phases, or appearances, of the fixed stars, morning and evening risings and settings, and after distinguishing between apparent and true phases, establishes certain conditions for the phases of stars variously placed on the sphere. Once again, the object of investigation is a solid celestial sphere. Euclid's *Phenomena* is somewhat more deliberate in attempting to relate the model that it develops to the cosmos itself – insofar as it explicitly mentions a dioptra, discusses the cosmos, uses astronomical terminology, and sometimes orientates the discussion to objects around "us." It develops propositions treating the position of the earth in the cosmos, some of the instantaneous positions of the ecliptic on the horizon, risings and settings of fixed stars, rising and setting times of arcs of the ecliptic, and comparisons of the times in which arcs of the ecliptic pass above and below the horizon. Although these treatises – in particular *Risings and Settings* and *Phenomena* – cover a number of topics of spherical astronomy, they are, in fact, treatments of the configurations of a celestial sphere, whether actual or conceptual, and, like the *Spherics*, have almost nothing to do with observational astronomy (Evans 1998, 89).

These texts are all highly structured and, hence, marked as belonging to the mathematical sciences, as these fields were understood by authors writing in Greek. Nevertheless, from our perspective, they often combine geometric reasoning that uses constructions and deductions from established geometric facts with various types of kinematic reasonings that appear to be based directly on an intuition of the objects under investigation. In fact, it often appears that Autolykos and Euclid are implicitly asking us to consider the properties of an actual sphere, whether solid or armillary.

In order to understand this point, it may be helpful to look at *Mov. Sph.* 12, which reads as follows (Aujac et al. 2002, 65–67; Figure 3):[46]

---

that the *Phenomena* makes use of *Moving Sphere* by pointing out that the explicit reference in *Phen.* 2 is likely a later addition. This point is well taken, but it does not address the many other places where Euclid appears to assume that his readers are familiar with material that is established in *Moving Sphere*.

[46] In the manuscripts, circle *EZH* is generally drawn touching circle *BDG*, but this is not necessary and may be misleading (see, for example, **A**, 43v).

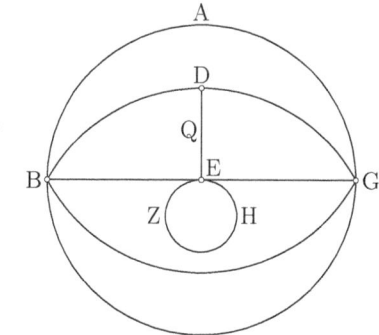

Figure 3: Diagram for *Mov. Sph.* 12.

⟨Enunciation:⟩

If a stationary circle in a sphere always bisects some carried[47] circle of those in the sphere, and neither of them are at an upright to the axis nor through the poles of the sphere, then each of them will be great.

⟨Exposition:⟩

Let there be a stationary circle, *ABG*, and let it always bisect a carried ⟨circle⟩ of the circles in the sphere, *GDB*, and let neither of them be at an upright to the axis nor through the poles of the sphere.

⟨Specification:⟩

I say that each of circles *ABG* and *GDB* is great.

⟨Construction:⟩

For, let there be their common section, *BG*. Therefore, *BG* is a diameter of circle *GDB*. Let *BG* be bisected at point *E*. Therefore, point *E* is the center of circle *GDB*. And it is obvious that point *E* is always in the plane of circle *ABG*, also throughout the whole revolution of the sphere.

⟨Local specification:⟩

Then, I say that point *E* is on the axis.

⟨Demonstration:⟩

For, if it is not on the axis, therefore point *E* will draw a circle at an upright to the axis with a rotation of the sphere. Let it draw *EZH*. Since point *E* is always in the plane of circle *ABG*, it is also carried along a circle, *EZH*. Therefore, circle *EZH* is always in the plane of circle *ABG*. And circle *EZH* is at an upright to the axis. Therefore, circle *ABG* is upright on the axis, which is

---

[47] That is carried about by the movement of the sphere.

not what was hypothesized. Therefore, point *E* is not not on the axis.

Therefore, it is on the axis.

⟨*Local specification:*⟩
Then, I say that point *E* is the center of the sphere.

⟨*Demonstration:*⟩
Indeed not, but if possible, let there be the center of the sphere, point *Q*, and let *QE* be joined. Therefore, line *EQ* is the axis of the sphere, for each of points *Q*, *E* is on the axis. And, since a circle, *GDB*, is in a sphere, while a line, *QE*, is joined from the center of the sphere, *Q*, to the center of circle *GDB*, therefore *QE* is upright on circle *GDB*, so that circle *GDB* is also upright on *QE*. And, *QE* is the axis. Therefore, circle *GDB* is upright on the axis, which is not what was hypothesized.

Therefore, point *Q* is not the center of the sphere.

Then, similarly we will show that it is not any other except *E*. Therefore, point *E* is the center of the sphere. And, it is in each of circles *ABG*, *GDB*. Therefore, each of circles *ABG*, *GDB* is great.

When we read through these passages, we find a combination of geometric constructions, inferences based on geometric properties, and spatial or kinematic inferences based on some intuitive understanding of what happens when a sphere rotates. For example, in the *construction*, *BG* is bisected at *E*, which is then asserted to be the center of a circle – which may follow from the definition of a circle. Or later, in the second *demonstration*, it is noted that since *QE* is drawn from the center of the sphere to the center of **C**(*GDB*), it must be perpendicular to the circle. This is explicitly articulated as depending on a known geometric fact, which, indeed, is demonstrated in *Sph.* I.7. On the other hand, in the first *demonstration*, the argument also depends on claims about the motion of an internal point, *E*, throughout the rotation of the sphere – a situation not previously dealt with in this treatise – namely, that it "draws" a circle that is perpendicular to the axis of the sphere. Although one might suppose that there was some lost text of mechanics or kinematics that proved this fact, a more obvious explanation is that Autolykos intended his readers to consider what happens when a sphere rotates, or, simply rotate an actual sphere and see what occurs. Indeed, when this treatise, with its peculiar diagrams, is read in conjunction with manipulating a real sphere, many of its claims become almost trivially self-evident – a circumstance that is true of many propositions in the treatises of ancient spherics. In fact, the assumption of an immobile circle, introduced in *Mov. Sph.* 4 without comment and used again here in

Mov. Sph. 12, presupposes that there is some reference circle against which the rotating sphere appears to move. This immobile circle is most simply explained as the horizontal base, depicted as holding the sphere in many of the images of solid and armillary spheres from antiquity. A number of steps in the arguments of the spherics texts of Autolykos and Euclid can be explained by an implicit appeal to the circumstances of an actual sphere. Indeed, all things considered, it seems likely that much of the mathematical work in spherics throughout the 4th century began with real objects of human manufacture and proceeded towards more general and abstract concepts only slowly and tentatively – apparently much to Plato's annoyance.

Regarding the relationships between the various spherics texts, it is well known that *Moving Sphere* and *Phenomena* include inferences from geometrical facts, many of which are shown in the *Spherics*, and they sometimes treat subject matter that is related to material covered in the *Spherics*, but in a somewhat different way. In order to discuss these relationships, we provide a list of the propositions of *Moving Sphere* and *Phenomena* followed by those in the *Spherics* to which they are related, or on which they seem to depend. The notation (∗) indicates a dependency or relationship that is uncertain, or which must be fleshed out by some further argument, and (p) indicates that the required proposition is a *problem*. The list is as follows:[48]

> *Mov. Sph.* 1: *Sph.* I.1, I.8, II.2
> *Mov. Sph.* 2: *Sph.* I.1, II.10
> *Mov. Sph.* 5: *Sph.* I.15
> *Mov. Sph.* 6: *Sph.* I.def.5, I.15, I.20(p), II.2, II.6(∗), III.1[49]
> *Mov. Sph.* 7: *Sph.* I.def.5, I.15, I.20(p), I.21(p, ∗), III.1
> *Mov. Sph.* 8: *Sph.* II.2(∗), II.13
> *Mov. Sph.* 9: *Sph.* II.20
> *Mov. Sph.* 10: *Sph.* I.15, I.20(p), II.5[50]
> *Mov. Sph.* 12: *Sph.* I.6, I.7

---

[48] This list may be compared to the notes in the various editions of the texts, and to that provided by Schmidt (1943, 7–8). This list can also be compared to those compiled by Heath (1921, 2.252), Bulmer-Thomas (1970, 320 n6), and Nikolantonakis (2016, 166), but because they only list the propositions of the *Spherics* that they think are used, and not which theorems they believe them to be used in, it can be difficult to check these dependencies. We omit the introduction to the *Phenomena*, as a late addition to the text.

[49] Schmidt (1943, 7) indicates that *Mov. Sph.* 6 also relies on *Sph.* I.21(p), but the pole is mentioned in the enunciation, and hence does not need to be taken by construction.

[50] Schmidt (1943, 8) sets *Mov. Sph.* 10 as depending on *Sph.* III.3, but in fact the same inference is made in *Sph.* II.22, prior to that proposition, and an argument for this can be made from *Sph.* I.15 and some propositions in the *Elements* (see p. 298, below).

Phen. 2: *Sph.* I.13, I.15, I.20(p), II.5, II.6, II.9, II.13, II.15(p), II.19
Phen. 4: *Sph.* II.13, II.15(p),
Phen. 5: *Sph.* II.13, II.15(p)
Phen. 6: *Sph.* II.17, II.19
Phen. 7: *Sph.* II.13, II.22(*)[51]
Phen. 8: *Sph.* II.17, II.18, III.5(*), III.7
Phen. 9: *Sph.* II.13(*), II.20
Phen. 12: *Sph.* II.13, II.15(p), II.17, II.18, III.3, III.6(*), III.8(*)
Phen. 14: *Sph.* II.7, II.13, II.14(p), II.20[52]

Because of the fact that the propositions of the *Spherics* appear to provide justifications for many of the steps in the early treatises of spherics, a number of scholars have traced back the mathematical requirements of the requisite geometric facts and reconstructed a lost elements of spherics, which is presumed to have been written in the 4th century BCE (Hultsch 1886; Heiberg 1882, 41–52; Bjørnbo 1902, 55–64). This led to an acceptance of the view that Theodosios simply adopted this earlier spherics, essentially wholesale, and was no more than a compiler and a pedantic writer of textbooks (Heath 1921, 2.252; Mogenet 1947). Although some scholars have cautioned against such an assessment (Schmidt 1943, 11–12; Neugebauer 1975, 750; Berggren 1991, 241–245), there still seems to be a tendency to read the *Spherics* as straightforwardly reflecting work of the 4th century.

While there is considerable close verbal agreement between these three texts (Aujac 1984), there are a number of reasons that may have contributed to this. In the first place, there were almost certainly earlier works of spherics containing specific technical terminology with which Autolykos, Euclid, and Theodosios were all familiar. Then, Theodosios doubtless knew the *Moving Sphere* and *Phenomena* well, and was influenced by their diction. Finally, in the editorial scholarship that grouped these texts together and treated them as a codicological unit – probably starting in the Imperial period and certainly progressing through the late-ancient period – the language of these texts was probably made even more homogenous and consistent. For example, it was once argued that the unnecessary definitions in the *Spherics* must have been carried over from a supposed earlier spherics (Mogenet 1947, 238), but, given recent scholarship on the editorial work of late antiquity, it now seems more likely that these were added to the text

---

[51] Schmidt (1943, 8) notes that *Phen.* 7 uses *Sph.* II.8, but we cannot find this dependency.

[52] Schmidt (1943, 8) includes *Sph.* III.1 for *Phen.* 14, but we cannot find where this is used. He also notes *Sph.* II.6 as used for this proposition, but we find *Sph.* II.7 to be more appropriate (see p. 253, n. 51). Menge (1916, 97) notes that *Sph.* III.3 is necessary for this proposition, but this must be a typographic error for *Sph.* II.13. *Sph.* II.22 is used in an alternate proof for *Phen.* 14, but this whole argument is a later addition.

at that time so as to make the *Spherics* itself into an introduction for all three works (Acerbi 2014, 146).

Before discussing the possible relationships between what appear to be references in the early treatises on spherics, and the propositions of the *Spherics*, we should set out some of the various ways in which Greek mathematicians mobilize previous claims. Greek mathematical texts employ a number of different strategies for bringing in a previously established assertion. Although we will elucidate these by using examples that seem to call from the *Moving Sphere* to the *Spherics*, all of them are also found internally within any single treatise.

The most obvious sort of reference is an explicit repetition of the *enunciation* of a proposition, often verbatim. For example in *Mov. Sph.* 1, the final step in the argument is justified by asserting that (Aujac et al. 2002, 44),

> οἱ ... περὶ τοὺς αὐτοὺς πόλους ὄντες ἐν σφαίρᾳ κύκλοι παράλληλοί εἰσι

> circles in a sphere being about the same pole are parallel,

which is exactly the same as the *enunciation* of *Sph.* II.2 (Aujac 1984, 103; Czinczenheim 2000, 83). In fact, such obvious references are fairly rare in mathematical works.

A more common type of reference states a mathematical fact and its conditions in general terms, intermixed with the letter-names of the objects in the current configuration, to which the fact pertains (Aujac 1984, 104–105). For example in the second *demonstration* of *Mov. Sph.* 12, as seen above, we read (Aujac et al. 2002, 67),

> ἐπεὶ ἐν σφαίρᾳ κύκλος ἐστὶν ὁ ΓΔΒ, ἀπὸ δὲ τοῦ κέντρου τῆς σφαίρας τοῦ Θ ἐπὶ τὸ κέντρον τοῦ ΓΔΒ κύκλου ἐπέζευκται εὐθεῖα ἡ ΘΕ, ἡ ΘΕ ἄρα ὀρθή ἐστι πρὸς τὸν ΓΔΒ κύκλον

> since a circle, *GDB*, is in a sphere, while a line, *QE*, is joined from the center of the sphere, *Q*, to the center of circle *GDB*, therefore *QE* is upright on circle *GDB*.

Because the *enunciation* of *Sph.* I.7 reads (Czinczenheim 2000, 61),

> ἐὰν ᾖ ἐν σφαίρᾳ κύκλος, ἀπὸ δὲ τοῦ κέντρου τῆς σφαίρας ἐπὶ τὸ κέντρον αὐτοῦ ἐπιζευχθῇ τις εὐθεῖα, ἡ ἐπιζευχθεῖσα ὀρθή ἐστι πρὸς τὸν κύκλον

> if a circle is in a sphere and some straight line joins from the center of the sphere to its center, then the joining ⟨straight line⟩ is upright to the circle,

the parallel between the terminology of these two, while not exact, is sufficiently close to mark the statement in *Moving Sphere* as a reference to some proposition that establishes the fact that is shown in *Sph.* I.7.

Another type of dependency may be asserted tacitly, with no verbal
reference to a previous proposition. For example, the final passage of
Mov. Sph. 12 reads (Aujac et al. 2002, 67),

τὸ Ε ἄρα σημεῖον κέντρον ἐστὶ τῆς σφαίρας· καὶ ἔστιν ἐν ἑκατέρῳ
τῶν ΑΒΓ ΓΔΒ κύκλων· μέγιστος ἄρα ἐστὶν ἑκάτερος τῶν ΑΒΓ ΓΔΒ
κύκλων.

therefore, point *E* is the center of the sphere, and it is in each
of circles *ABG*, *GDB*. Therefore, each of circles *ABG*, *GDB* is
great.

In this case, there is simply a statement of a geometric fact, but the inference itself is not asserted. Nevertheless, we can understand this as a silent
reference to *Sph.* I.6, or at least to the geometrical fact established in that
theorem – namely, that circles passing through the center of the sphere are
great circles. When such a statement is made in a mathematical treatise
in which a previous proposition establishes the fact in question, it seems
virtually certain that it should be taken as a tacit reference to the earlier
proposition. The same is likely the case when the statement might be taken
to refer to a proposition in a previous and well-known treatise – such as
when statements are made in the *Spherics* regarding facts that are shown in
the *Elements*. In a situation like this, however, in which the historical relationship between *Moving Sphere* and *Spherics* is itself in question, we might
be more cautious in assigning a direct reference. For example, an inference
such as this could refer, not to a proposition, but to a definition of a great
circle as passing through the center of the sphere. Nevertheless, it seems
likely that this statement in *Moving Sphere* appeals to some well-known
mathematical fact, which was probably asserted in a previous treatise.

Whether all statements of this sort, when made in early treatises, refer
to previous work or not is unknown, because we do not now posses the
treatises to which they might refer. Indeed, we do not know when this
type of referencing began, or rather, came to be used systematically. It is so
used by Euclid in the *Elements*, but even other 3rd-century mathematicians,
such as Aristarchos and Archimedes, seem to have felt less need to make
sure that everything had been explicitly established, and some mathematical
facts upon which they rely may have been taken as obvious without proof or
explicit statement. It is possible that the *Elements* itself was the model of a
strict referencing approach, and the *Spherics* was almost certainly influenced
along these lines.

Furthermore, because of the well-attested editorial work of later, and
particularly late-ancient, scholars, it is not possible to simply take all the
direct and explicit references as a verified dependance on a proposition established in a previous treatise. Hence, one should consider all of the apparent references from *Moving Sphere* and *Phenomena* to *Spherics* on a

case-by-case basis. Although we will not undertake such a project here, we do emphasize that there are various difficulties that follow upon the assumption that Autolykos and Euclid based their work on a treatise that was essentially the same as the current *Spherics*. As well as certain issues that have already been raised (Schmidt 1943, 11–12; Berggren 1991, 241–245), we give as examples the relationship between a number of Theodosios's theorems and two of the most important astronomical topics that they might be taken to address.

***Phen.*7 and *Sph.*II.22.** Although there is clearly some relationship between the third part of *Phen.*7 and *Sph.*II.22, the proposition in the *Spherics* is considerably more sophisticated (Berggren 1991, 244–245), and it is not clear that Euclid is making a direct reference to any previously established geometric theorem. The entire *demonstration* of *Phen.*7 Part 3, which concerns the orientation of the ecliptic relative to the horizon, reads as follows (Menge 1916, 40; Berggren and Thomas 1996, 70, with modifications):

καὶ φανερόν, ὅτι ἄλλοτε ἄλλως ὑπὲρ ἡμᾶς ἵσταται.

ὅταν μὲν γὰρ ἡ συναφὴ τοῦ ζῳδιακοῦ κύκλου καὶ τοῦ θερινοῦ τροπικοῦ ἐπὶ τῆς διχοτομίας ᾖ τοῦ ὑπὲρ γῆν τμήματος τοῦ θερινοῦ τροπικοῦ κύκλου, ὀρθότατος ἔσται πρὸς ἡμᾶς· ὅταν δὲ ἐπὶ τῆς διχοτομίας τοῦ ὑπὸ γῆν τμήματος τοῦ θερινοῦ τροπικοῦ, ταπεινότατος ἔσται πρὸς ἡμᾶς· καὶ ἀεὶ μὲν πορρώτερον γιγνόμενος τῆς διχοτομίας τοῦ ὑπὲρ γῆν τμήματος τοῦ θερινοῦ τροπικοῦ, μᾶλλον ἔσται κεκλιμένος· ὁμοίως δὲ ἔσται κεκλιμένος, ἴσον ἀπέχων ὁποτερασοῦν τῶν διχοτομιῶν.

And, it is obvious that it stands over us differently at different times.

For, when the ⟨point of⟩ contact of the zodiacal circle and the summer tropic is at the bisector of the circle of the summer tropic above the earth, it will be most upright to us; while when it is at the bisector of the segment of the summer tropic below the earth, it will be lowest to us; and, getting ever farther from the bisector of the segment of summer tropic above the earth, it will be more inclined; while it will be similarly inclined when it is equally distant from whichever of the bisectors.

At first glance, this does not appear to be a reference to a mathematical result of any of the three types listed above. Nevertheless, modern scholars have asserted that this passage depends on *Sph.*II.22 (Menge 1916, 41; Schmidt 1943, 8). Indeed, among other things, *Sph.*II.22 makes the claim that – with respect to great circles that can be understood to be positions of the ecliptic above the horizon at certain latitudes (p. 147, below) –

... ὀρθότατος μὲν ἔσται ὁ τὴν συναφὴν ἔχων κατὰ τὴν διχοτομίαν
τοῦ μείζονος τμήματος, ταπεινότατος δὲ ὁ τὴν συναφὴν ἔχων κατὰ
τὴν διχοτομίαν τοῦ ἐλάσσονος τμήματος, τῶν δὲ ἄλλων οἱ μὲν ἴσον
ἀπέχοντες ὁποτερασοῦν τῶν διχοτομιῶν ὁμοίως εἰσὶ κεκλιμένοι, αἰεὶ
δὲ ὁ πορρώτερον τὴν συναφὴν ἔχων τῆς διχοτομίας τοῦ μείζονος
τμήματος τοῦ ἔγγιον μᾶλλον ἔσται κεκλιμένος ...

... the most upright will be that having the ⟨point of⟩ contact at
the bisector of the greater segment, while the lowest is that having the ⟨point of⟩ contact at the bisector of the lesser segment,
while of the others, those equally distant from whichever of the
bisectors are similarly inclined, but that having the ⟨point of⟩
contact farther from the bisector of the greater segment will be
ever more inclined than the nearer ...

In fact, however, there are both stylistic and technical difficulties with reading the text in *Phen.* 7 as a reference to that in *Sph.* II.22. In the first place,
the quoted passage is the entirety of the *demonstration* of *Phen.* 7 Part 3, so
that Euclid put forward this statement itself as the argument that his claim
holds. It does not make reference to any geometric facts, nor relate such
facts to the letter-names of the objects in question. Rather, Euclid seems to
be making use of astronomical facts, expressed in terms of direct observations, or perhaps as seen on a solid or armillary sphere. Moreover, for the
claims in *Phen.* 7 to be true, some bound must be placed on the terrestrial
horizon – namely, at least $\varphi < 90° - \varepsilon$. Theodosios, indeed, introduces
a bound of $\varphi < 45°$ in his treatment (see the Commentary to *Sph.* II.22,
below), but Euclid simply refers to the situation around "us" – presumably
meaning at locations where Greek speakers live, such as at Alexandria.

In comparison to Theodosios's treatment of this configuration, Euclid's
is partial and almost purely descriptive. It is hard to believe that the proof
sketch in *Phen.* 7 Part 3 could be a reference to the fully developed, and
carefully shown, results of *Sph.* II.22. Rather, it is more likely that Euclid
simply described the facts as they appear, whether in the actual sky or on
a solid or armillary model of the celestial sphere, and that Theodosios later
developed the geometrical propositions necessary to show why and under
what conditions this is so.

Phen. 8, 12 and *Sph.* III.5–III.8. The propositions concerning the
ortive amplitudes and rising times of arcs of the ecliptic in the *Phenomena* are mathematically related to a group of theorems in *Spherics* III, but
once again Theodosios's treatment of this material is more developed and

precise than Euclid's, and there is reason to doubt that Euclid based his approach on theorems such as we now read in the *Spherics*.[53]

In *Phen.* 8, Euclid shows that the signs of the zodiac rise and set over unequal arcs of the horizon, or ortive amplitudes, and that these arcs successively decrease as the signs are taken from the equator to the tropics. In showing this, in the first few steps of the *demonstration*, Euclid makes a number of geometrical assertions of the following form (Menge 1916, 42),

... ἐπεὶ περιφέρειαι αἱ HK, KN, NΓ ἴσαι ἀλλήλαις εἰσίν, αἱ ΖΛ, ΛΞ, ΞΓ ἄρα μείζους εἰσὶν ἀλλήλων ἀρχόμεναι ἀπὸ μεγίστης τῆς ΖΛ ...

... since, arcs, *HK, KN, NG*, are equal to one another, therefore *ZL, LX*, and *XG* are greater than one another, starting from maximum *ZL* ...[54]

Although this does appear to be a reference to a previously established result, it is not certain where this fact was established or how it was shown. In fact, a result that is mathematically equivalent to this is shown in both *Sph.* III.5 and III.7, but more completely, and using somewhat different language. Euclid, however, does not tell us whether *Phen.* 8 should pertain to the upright sphere, the inclined sphere, or both. The diagram in the manuscripts of the *Phenomena* for this proposition are closer to that of *Sph.* III.5, which might imply that it is intended for the upright sphere, but it has been assumed that this proposition should pertain to the inclined sphere, and require *Sph.* III.7 (Menge 1916, 43; Schmidt 1943, 8). Furthermore, if Euclid did intend this theorem to hold on the inclined sphere, he should have mentioned a bound on the terrestrial latitude of $\varphi < 90° − \varepsilon$. With such a bound assumed, *Phen.* 8 would hold for both the upright and inclined spheres. In fact, Theodosios treats the underlying geometric configuration completely – the upright sphere in *Sph.* III.5, and the inclined sphere in III.7, with separate diagrams and the bound appropriately addressed. If Euclid had Theodosios's theorems available to him, he implemented them in a rather careless way.

A similar situation pertains to *Phen.* 12, which clearly requires a geometric result that is closely related to that shown in *Sph.* III.8. Once again, however, there are difficulties in believing that Euclid used a treatment that was substantially the same as that in the *Spherics*. *Phen.* 12 attempts to show, in part, that equal arcs of the ecliptic in the semicircle from Cancer to Sagittarius set in decreasing times from the tropics to the equator. In the course of the *demonstration*, Euclid sets out four detailed conditions, and then draws a geometrical inference from these (Berggren and Thomas 1996,

---

[53] An introduction to the rising times of arcs of the ecliptic is provided at the beginning of the next section.

[54] That is, $ZL > LX > XG$.

84–85), which corresponds to the conditions and conclusion of *Sph.* III.8, although expressed in somewhat different technical language. Although it is clear that Euclid intends this step in his argument to be based on some previous work, there is reason to doubt that this previous material was as complete as Theodosios's.

In the first place, although Euclid's proposition is enunciated in terms of any equal arcs of the ecliptic in the stated semicircle, his argument only concerns signs of the zodiac. Theodosios both states and proves his theorem in terms of any equal arcs in the pertinent region. In *Phen.* 12, the upright sphere is only addressed as a sort of afterthought, at the end of the proposition, in what is clearly an editorial addition to the text. In the *Spherics*, the upright sphere is treated fully and independently in *Sph.* III.6. Euclid makes no mention of any upper bound on the terrestrial latitude. This bound, however, must be set to $\varphi < 90° - \varepsilon$, because at $\varphi = 90° - \varepsilon$ six signs rise and set instantaneously, while at $\varphi > 90° - \varepsilon$ certain arcs of the ecliptic rise and set in the opposite directions, or, from a computational perspective, have negative rising times.[55] Theodosios correctly effects this bound in his treatment of the related geometric objects in *Sph.* III.8.

On the whole, Euclid's treatment of the rising phenomena of the arcs of the ecliptic is partial, astronomically imprecise, and, frankly, somewhat haphazard; whereas that of Theodosios is clear, systematic and geometrically more complete. Hence, it is most likely that Euclid developed his work on the basis of some earlier treatment that was much less complete than what now find in the *Spherics*, and that Theodosios later worked all of this up into a more polished and mathematically satisfying form.

Although there is no doubt that there was some work in spherics before *Moving Sphere* and *Phenomena* were composed, we do not know what form this took, nor how completely it was developed. What does seem certain is that the more advanced theorems of the *Spherics* were not available to Euclid around the turn of the 3rd century, and it is not certain that there was a systematic, purely geometric, treatise on the geometry of the sphere in circulation.

The *Spherics* and numerical methods

As we just saw, in *Spherics* III, Theodosios develops a number of theorems that bear directly on the topic of the rising and setting times of arcs of the ecliptic – a topic that had already been addressed in a more limited way by Euclid in *Phen.* 12 and 13. The geometrical understanding of this topic is best addressed by reference to a solid or armillary sphere that can rotate around its axis, which we may simulate with a perspective diagram (Figure

---

[55] Sacrobosco says they rise and set "preposterously" (Thorndike 1949, 109, 138).

4). If we imagine the sphere to rotate uniformly once a day, then arcs of
the equator will rise in times that are directly proportional to their size, 1°
every 4 minutes, and we can take those arcs of the equator that rise over
the horizon in the same time as given arcs of the ecliptic as a measure of
the rising times of the latter.

For example, on the upright sphere, Figure 4 (left), when the celestial
poles are on the horizon, with the sphere rotating from east to west and the
east point facing us, at E, an arc of the ecliptic, λ, rising before the point of
the vernal equinox, VE, such that it is determined by an earlier and later
position of the same horizon, $horizon_1$ and $horizon_2$, will rise in the same
time as the arc of the equator that is cut off between the same two horizons.
We can consider this arc of the ecliptic to be the rising time, ϱ, of λ. On
the upright sphere, the rising time of an arc of the ecliptic is equal to the
arc of its right ascension, $\varrho(\lambda, 0°) = \alpha(\lambda)$. Again, on the inclined sphere,
Figure 4 (right), where the visible pole, P_v, is above the horizon at a height
of φ, measured along the meridian, and an arc of the ecliptic, λ, rises over
the horizon, its rising time, $\varrho(\lambda, \varphi)$, is again determined by the arc of the
equator cut off between the earlier and later positions of the horizon, which
are both tangent to the v-circle.

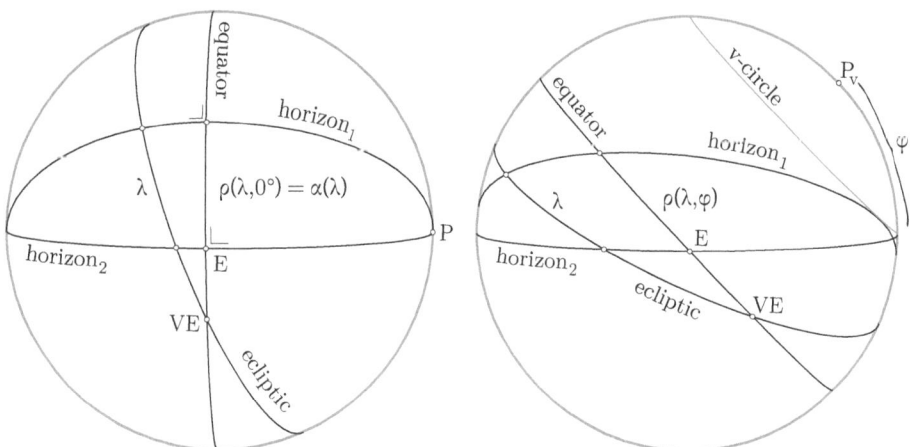

Figure 4: Diagrams for the rising times: (left) upright sphere, (right) inclined
sphere.

It is clear from these considerations, that, on both the upright and the
inclined spheres, the arc of the ecliptic diametrically opposite to λ, namely
$\overline{\lambda}$, will set in the same time as λ rises. Indeed, the symmetry of the sphere
assures us that the arc of the equator that sets with $\overline{\lambda}$ will be diametrically
opposite to that which rises with λ. That is, $\varrho(\lambda, \varphi) = \overline{\varrho}(\overline{\lambda}, \varphi)$ and $\varrho(\overline{\lambda}, \varphi) =$
$\overline{\varrho}(\lambda, \varphi)$, using $\overline{\varrho}(\lambda, \varphi)$ to denote the setting time of λ at latitude φ.

In *Phen.* 12, Euclid establishes that equal arcs of the ecliptic in the semi-circle from Cancer to Sagittarius, **semiC**(Can → Sag), have setting times such that arcs that are adjacent to the points of the tropics, *SS* and *WS*, set in the greatest time; those which are further from *SS* and *WS* set in less time; and those that are adjacent to the point of the equinox, *VE*, set in the least time (Figure 5). Furthermore, equal arcs that are equidistant from *VE* set and rise in equal times. The proof for this claim involves an argument concerning the signs of the zodiac. The implication is that, in **semiC**(Can → Sag), the change in the setting times of the signs, $\Delta\bar{\varrho}(\text{Sign})$, in going between the tropics and the equinox is monotonic. In *Phen.* 13, related claims are established for the opposite semicircle, primarily using kinematic arguments. Namely, in **semiC**(Cap → Gem) equal arcs that are adjacent to the tropics, *SS* and *WS*, rise in the greatest time; arcs which are farther from the tropics rise in less time; arcs that are adjacent to the autumnal equinox, *AE*, rise in the least time; and equal arcs that are equidistant from *AE* set in equal times. Again, the implication of the argument is that in going between the tropics and the equinox, in the other hemisphere, the change in the rising times of the signs, $\Delta\varrho(\text{Sign})$, is monotonic. It is clear that Euclid understood that the situation of the setting times of the arcs of **semiC**(Can → Sag) is the same as the rising times of the diametrically opposite arcs in **semiC**(Cap → Gem).

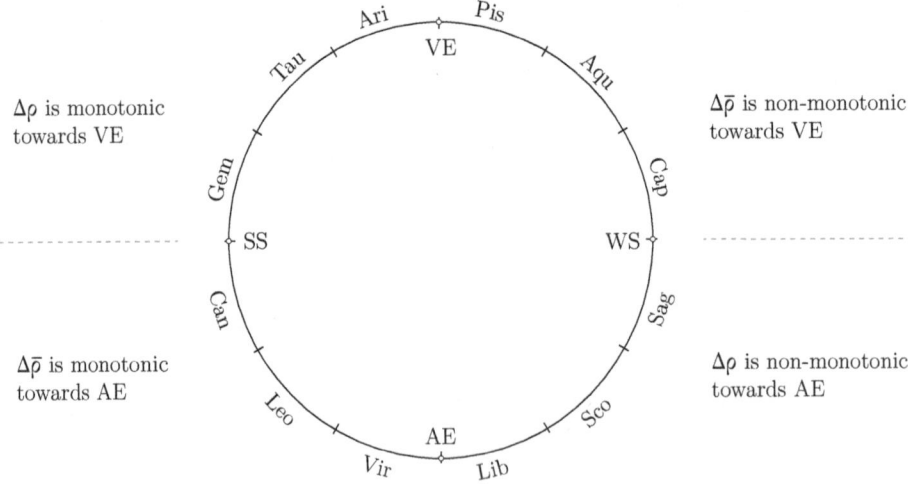

Figure 5: Schematic representation of rising and setting times adopted from that in Hypsikles's *Ascensions* (see **A**, 135r). (We do not show his numbers for the rising times.)

In the *Phenomena* nothing is said about changes in the rising times in the semicircle from Cancer to Sagittarius, $\Delta\varrho(\text{Can} \rightarrow \text{Sag})$, or about the setting times of the semicircle from Capricorn to Gemini, $\Delta\bar{\varrho}(\text{Cap} \rightarrow \text{Gem})$.

Hence we cannot be certain that Euclid knew that these are non-monotonic. Nevertheless, he must have either known that they are non-monotonic, or been unable to prove that they are monotonic. Indeed, visual inspection of a solid or armillary sphere readily confirms the situation, and the use of a caliper with a well-made sphere would allow a fairly straightforward nomographic calculation of the magnitudes of these differences.[56] Whether or not Euclid ever made such an inquiry, however, there is no indication of any computational procedure in the *Phenomena*.

In fact, precise numerical methods in Greek astronomy appear to have been stimulated by an adaptation of certain Babylonian methods around the turn of the 2nd century BCE (Jones 1991). The first indication that we have of a computational approach to addressing rising times in Greek sources comes from the *Ascensions* of Hypsikles, who worked in the early part of the 2nd century BCE (Montelle 2016, 290). The *Ascensions* is a short tract that uses three propositions on sums of numbers to justify mathematically a rising-time scheme that was structurally the same as an earlier Babylonian scheme (De Falco, Krause, and Neugebauer 1966; Neugebauer 1975, 715–718; Montelle 2016). In the course of this work, Hypsikles introduces "spatial degrees" of which there are 360° to a circle, "time degrees" of which there are 360^t in a day, so that $1^t = 4$ minutes, and sexagesimal fractions, all of which conventions derive from Mesopotamian sources. Although the methods that Hypsikles sets out could be adopted to any latitude for which the ratio of the longest to the shortest period of daytime is known, the text works out the situation for Alexandria, in which the ratio is 7 : 5. On this basis, Hypsikles determines that at Alexandria

$$\varrho(\text{Ari}) = 21;40^t, \qquad \varrho(\text{Tau}) = 25^t, \qquad \varrho(\text{Gem}) = 28;20^t,$$
$$\varrho(\text{Can}) = 31;40^t, \qquad \varrho(\text{Leo}) = 35^t, \qquad \varrho(\text{Vir}) = 38;20^t.$$

He then uses the fact that those signs which are equally distant from the equinox rise in the same time, as shown in *Phen.* 13, to assert that

$$\varrho(\text{Lib}) = 38;20^t, \qquad \varrho(\text{Sco}) = 35^t, \qquad \varrho(\text{Sag}) = 31;40^t,$$
$$\varrho(\text{Cap}) = 28;20^t, \qquad \varrho(\text{Aqu}) = 25^t, \qquad \varrho(\text{Pis}) = 21;40^t.$$

Finally, he notes that since the setting times are equal to the rising times of those signs that are "along a diameter," κατὰ διάμετρον, they will be obvious (De Falco, Krause, and Neugebauer 1966, 38). He then exhibits this material in a circular diagram with the signs arranged counterclockwise

---

[56] In regards to the possibility of using a well-made celestial sphere to calculate the rising times, Geminos in his *Introduction to the Phenomena* mentions that it is not possible to discover accurately the lengths of the days and the nights unless the sphere has been properly inscribed with the requisite circles (*Int.* V.14).

around the circumference and the values of the rising times written inside the arc for each sign (similar to Figure 5, but we do not show his numbers).[57] We may also convey this information in a graph, which clearly displays the zig-zag function of this system of rising times (Figure 6).

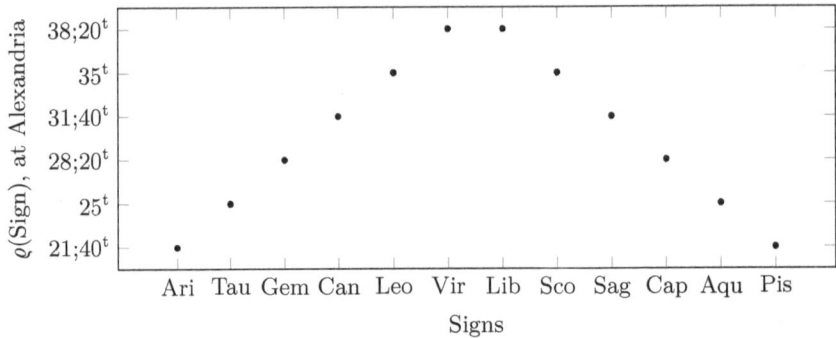

Figure 6: Graphical representation of the rising times of the signs of the zodiac at the latitude of Alexandria as set out by Hypsikles.

Hypsikles's treatment of rising and setting times reveals an interesting mix of numerical and geometrical approaches. While the mathematical propositions and derivations of the values of the rising times of the first set of signs are entirely numerical, Hypsikles then extends these values to the other signs, and also to the setting times, using geometric considerations. For example, he appears to justify his values for $\Delta\varrho(\text{Lib} \to \text{Pis})$ using a claim that the rising times of equal arcs of ecliptic that are equally distant from either side of *VE* and *AE* are equal, which was shown geometrically in *Phen.* 13. Furthermore, his extension to the rising times relies on the notion of pairs of signs being diametrically opposite to one another, using an expression common in geometrical texts and used by Autolykos and Euclid in their works on spherics. Finally, he exhibits his findings in a circle, which can be understood as that of the ecliptic. On the other hand, the values that he gives for the rising times are incompatible with a fully geometric model of the phenomena, and are unjustified by the claims of *Phen.* 12 and 13. In particular, while these propositions show that the rising times change monotonically in **Quad**(Ari $\to$ Gem) they say nothing about continuing this pattern to **Quad**(Can $\to$ Vir), which is what Hypsikles requires. Indeed, the geometry of the sphere does not justify such a continuation. Hence, it seems that Hypsikles's work provides an arithmetical justification for a way of stating rising times that was already in circulation – probably derived from Babylonian sources – but that his reference to geometric considerations were merely an expedient. There is no attempt to investigate the numerical implications of a spherical model of rising times in this work.

[57] See Montelle (2016, 289), or **A** 135r, for the actual diagram.

The first person that we may be reasonably certain calculated the rising times and produced a pattern of values that agree with what we find much later in Ptolemy's *Almagest*, is Hipparchos. Observation reports in the *Almagest* indicate that Hipparchos was active from the late 160s to the early 120s BCE (Pedersen 2010, 413–415), which would make him two or three generations younger than Hypsikles. Although Hipparchos's own work on this material does not survive, a description of the findings in his lost *Ascensions of the Twelve Zodiacal Signs* that Pappos gives in his own study of the *Little Astronomy* makes it clear that Hipparchos knew the pattern of the rising times that can be determined from a spherical model.

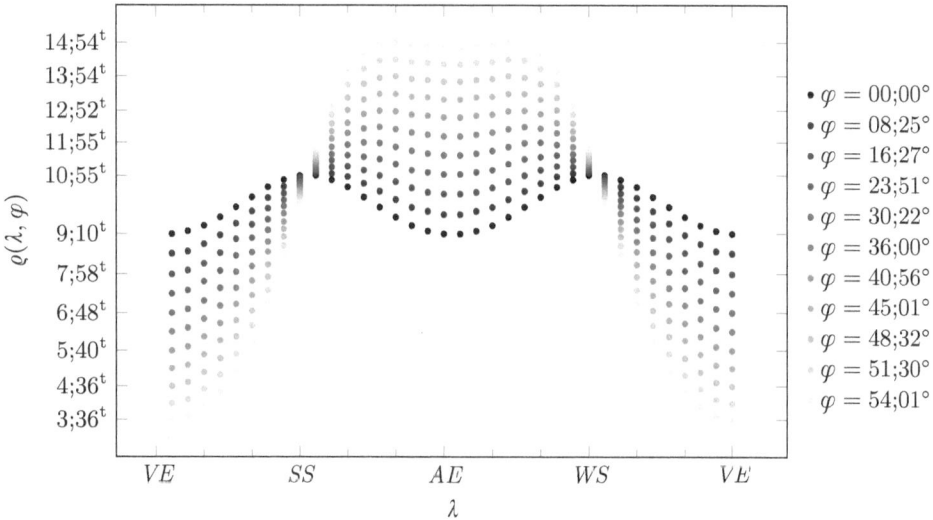

Figure 7: Graphical representation of the columns of the rising times of 10° intervals in the table of rising signs, *Alm.* II.8.

It may be useful first to introduce the actual pattern, as it was later recorded. In his table of rising times, *Alm.* II.8, Ptolemy sets out, among other things, a column of the rising times of 10° arcs of the ecliptic for each of 11 latitudes starting from the upright sphere at $\varphi = 0°$ and going north to $\varphi = 54;01°$ (Toomer 1984, 100–103). The design of this column, and its position in the overall table, allows one to see readily the pattern of the rising times that we more conveniently display with a graph (Figure 7; Sidoli 2014, 26–28). That is, as the latitude, φ, is taken farther to the north, we notice that (1) the overall minimum at the vernal equinox, *VE*, decreases while the local minimum at the autumnal equinox, *AE*, increases, and that (2) the maxima increase together and move away from the solstices, *SS* and *WS*, towards *AE*. This helps us see that (3) there is a symmetry about the equinoxes, as established in *Phen.* 12 and 13, but it also shows that (4) for more northerly latitudes the rising times at *AE* are greater than those at

SS and *WS*, which is not true for latitudes between 0° and around 23;51°.[58]
We will see that Hipparchos apparently demonstrated this final relationship
through calculation.

In his commentary to *Phen.* 12 in *Collection* VI, Pappos points out that
Euclid only demonstrates facts about the setting times for **semiC**(Can →
Sag). First he quotes Euclid's text to point out that in this interval equal
arcs "at the ⟨points of⟩ contact of the tropics" always set in more time
than those "at the equator." Then, discussing why Euclid does not treat
the rising times for **semiC**(Can → Sag), he goes on to say that (Hultsch
1876–1878, 600),

> Hipparchos, in *On the Ascensions of the 12 Zodiacal ⟨Signs⟩*,
> demonstrates, through numbers (δι' ἀριθμῶν), that it is not ⟨the
> case⟩ that just as equal circumferences of the semicircle follow-
> ing Cancer set having some relation of time to one another so
> too do they rise; for there are some locales (οἰκήσεις) in which,
> of equal circumferences of the semicircle following Capricorn,
> those nearer the equator ever rise in more time than those at
> the ⟨points of⟩ contact of the tropics.

The method that Hipparchos used to arrive at these results is described
by Pappos as "through numbers," but this expression generally refers to calculation in Greek mathematical texts (Sidoli 2006, 66). We do not know
how Hipparchos calculated the rising times of **semiC**(Can → Sag), but
there are a number of methods that may have been available to him. He
might have used the stereographic methods of the *Planisphere* (Neugebauer
1975, 868–869; Sidoli and Berggren 2007, 120–126), or the constructions
of the *Analemma* ('Id 1969; Sidoli 2020b), either of which would allow full
calculation of the rising times using chord-table trigonometry in the plane.
If the spherical chord-table trigonometric methods of the Sector Theorem
go back as far as the 2nd century BCE (Bjørnbo 1902, 83–85; Rome 1933,
42, Sidoli 2006, 61–70), Hipparchos may have used these. Some scholars believe that Hipparchos carried out most of his spherical computations
nomographically on a solid sphere (Nadal and Brunet 1984; Marx 2020),
and a well-made solid globe would certainly allow for computation of the
rising times. It should be pointed out that the methods of the planisphere
and the analemma also allow for nomographic computations on a plane.
We do not know the details of Hipparchos's computational procedure or
how he presented the numerical results, but for our purposes here it hardly
matters. Any reasonably accurate computational method based on the geometry of the sphere would have produced a set of numbers which agrees

---

[58] This upper value should be precisely equal to the obliquity of the ecliptic, $\varphi = \varepsilon$.
Since Ptolemy's value for ε is 23;51,20°, we can understand his latitude of 23;51° to be
that at which $\varphi = \varepsilon$.

with what Pappos tells us Hipparchos claimed, and which is evidenced in neither *Phen.* 12 and 13 nor in the *Ascensions*.

In particular, in the first place, Hipparchos asserted that the relationship of the rising times of **semiC**(Can → Sag) is not like that of their setting times, a fact that was probably known to both Euclid and Hypsikles, although it would have been overlooked by the latter because of the arithmetical nature of his project. Then Hipparchos supports this claim by pointing out that there are "some locales" in which for equal arcs of **semiC**(Cap → Gem) those which are nearer to AE always rise in more time than those that have an endpoint at SS or WS. The fact that this claim is modified by "some" implies that there are other latitudes for which this is not the case.

Indeed, if we consider Figure 7, we observe that, in the latitudes below $\varphi = 23;51°$, the rising times of arcs that are in the vicinity of AE are less than those of equal arcs at SS or WS, because those points are at or near the maxima, while AE is the local minimum. At the latitude equal to the obliquity of the ecliptic, $\varphi = \varepsilon \approx 23;51°$, the rising times at AE are equal to those at SS and WS, but because the maxima are generally closer to the solstice points, rising times of arcs with an endpoint at AE will still be greater than those at SS or WS. Since, however, the local minimum at AE increases, while the maxima move away from SS and WS towards AE, at some northerly locale the rising times of all arcs nearer to AE will be greater than equal arcs that have an endpoint at SS or WS. Pappos's remark indicates that Hipparchos demonstrated this "through numbers," but the overall pattern is the same as what we see in the graph, and could readily be seen in a numerical table of rising times, whether this was laid out in 10° intervals, like Ptolemy's, or in the 30° intervals of the signs of the zodiac (Sidoli 2006, 65–67).

We introduce this material in order to point out that by the time Theodosios composed his *Spherics*, probably in the early 1st century BCE, all of this was already known – whether or not it had been known to the earlier authors working on rising times, such as Euclid and any authors who may have preceded him. Hence, one might well ask why Theodosios would work on such a purely geometrical approach to this material at a time when other, in some ways more precise, methods were already in use.

It might be assumed that he was simply repeating an older precomputational approach that had been superseded by Hipparchos's work for antiquarian, or perhaps educational, purposes, but this would not explain why Theodosios expended the effort to set out the rather difficult material in *Spherics* III, which is not likely to have helped many students. Another explanation is that he sought to show that the deductive methods of the *Elements* could be used to explain how the various features of the rising times are a consequence of the geometry of the sphere itself. As discussed, the geometrical presentation of rising times in *Spherics* III is much more systematic than that in the *Phenomena*. It also offers a number of valuable insights. For example, if one attends to the details of the proof of *Sph.* III.8 one can find a geometrical explanation for why the rising times of equal arcs on **semiC**(Cap → Gem) are monotonic, while those on **semiC**(Can → Sag) are not (see Commentary, *Sph.* III.8). Moreover, *Sph.* III.9 demonstrates that, on the upright sphere, the changes in rising times on **semiC**(Cap → Gem) are strictly monotonic, something which is not actually shown in the *Phenomena*, and which is not necessarily implied by the observations that Pappos attributes to Hipparchos. Finally, in the sequence of theorems *Sph.* III.10–III.12, Theodosios demonstrates a geometric relationship between the magnitude of an arc of the ecliptic and its rising time as determined by the obliquity of the ecliptic. Hence, it is probable that Theodosios believed that a study of the geometry of the sphere still has something to tell us about *why* the rising times behave as they do, even after their numerical values are known and tabulated.

Theodosios's *Spherics*

As discussed above, when compared with the various Hellenistic texts of geometry that have come down to us, the *Spherics* most closely resembles the *Elements* in its overall structure and mathematical methods.[59] Like many of these texts, it is highly structured on a number of different levels.[60] In the first place, the text is divided into three books:

Spherics I begins with six definitions, some of which may be later additions, and then introduces the properties of small and great circles in the sphere. It then treats the configuration of a diameter of the sphere passing through the center of a circle in the sphere, and sets out theorems concerning the intersections of circles in the sphere. Finally, it develops some *problems* that are used in the rest of the treatise. (There are also two final theorems that are spurious.)

Spherics II begins with one definition and then develops theories of parallelism and tangency as relating to small and great circles on the sphere. These concepts are then used to study a configuration involving a bundle of parallel circles and a pencil of great circles that either pass through the poles of the parallel circles or are tangent to a pair of equal parallels. These two arrangements of the pencil of great circles

---

[59] In this section we will simply describe the *Spherics*. For detailed arguments that the text has the various characteristics asserted here, see the Commentary.

[60] See Decorps-Foulquier (2018) and Sidoli (2020a, 194–197) for discussions of various types of structure in Greek mathematical texts.

may be understood as a geometrical representation of instantaneous configurations of the horizon, whether upright or inclined. In the final section, another pencil of great circles with a different inclination is introduced and the various angles of the two sets of oblique great circles against each other are investigated. This final configuration can be understood as a geometrical representation of the instantaneous positions of the ecliptic relative to the horizon.

Spherics III starts with some geometric lemmas and then uses the same two-part configuration treated in the previous book to develop a number of theorems that treat various relations concerning arcs of another inclined great circle and arcs cut off by the bundle of parallel circle or the great circles either passing through their poles or tangent to a pair of equal parallels. These propositions can be interpreted as making claims about the rising times of arcs of the ecliptic over the horizon. There are then some theorems that lead to a bound on the ratio between arcs of the ecliptic and their rising times, which is itself related to the obliquity of the ecliptic.

Each book treats a fairly well-contained topic: (1) basic geometry of the sphere as it pertains to a model of the celestial sphere, (2) instantaneous positions of the horizon and the ecliptic, and (3) rising times of arcs of the ecliptic.

Structure of theories

Within the three books, propositions are grouped together into thematic units that treat certain objects or properties of objects. At the simplest level there are pairs of converses, such as *Sph.* I.16, I.17 or *Sph.* II.11, II.12, or groups of partial converses, such as *Sph.* II.3–II.5 or *Sph.* II.6–II.8. There are groups of propositions that treat key properties of certain configurations, such as a plane tangent to a sphere, *Sph.* I.3–I.5, a line as the axis of a circle in the sphere, *Sph.* I.7–I.9, or a pencil of great circles that pass through the poles of, or are tangent to a pair of circles belonging to, a bundle of parallel circles, *Sph.* II.10, II.13, and II.16. There are runs of propositions that treat certain properties such as perpendicularity, *Sph.* I.7–I.10, I.13–I.15, or tangency in the surface of the sphere, *Sph.* II.3–II.8. Finally, there are groups of propositions that have direct relevance to certain unmentioned astronomical topics, such as the instantaneous inclinations of the ecliptic on the horizon, *Sph.* II.21–II.23, or the rising times of the arcs of the ecliptic, in general, *Sph.* III.5–III.9, and as related to the obliquity of the ecliptic, *Sph.* III.10–III.12.

Although there is no terminology in the ancient mathematical texts, or their commentators, that is used to discuss these groups of texts as such,

there are a number of indications that ancient mathematicians did organize
their work into such subgroups, or theories. In the case of the *Spherics*,
most of the propositions that treat a particular topic are grouped together
in immediate succession. One exception to this is the group of theorems
Sph. II.10, II.13, and II.16, which treat a related pair of configurations, such
that *Sph.* II.16 is a converse to each of II.10 and II.13. Moreover, as well as
having a similar mathematical context, these theorems are marked as belonging together by the fact that the same or related mathematical objects
are given the same letter-names in all three propositions – an occurrence
that is otherwise uncommon in Greek geometrical texts. A group of propositions that treats objects that are related in a more complicated way are
Sph. III.10–III.12. In these propositions, related objects are assigned related
letter-names, or introduced in the *specification* in the same way, such that
the normal alphabetical order of naming is disrupted. Another indication
that propositions were intentionally grouped together into coherent theories,
is the fact that there are a number of propositions that are not used later in
the text but which are required for a mathematically complete treatment of
the topics addressed – namely, *Sph.* I.5, I.12, II.16, and perhaps III.9. The
other unused theorems all have astronomical significance, and they were
presumably included in the treatise because of their relevance to spherical
astronomy.

Structure of propositions

It is well known that the individual propositions of a Greek geometrical text
are highly structured, and that the units of this structure were discussed
already by Proklos in the 5th century (Netz 1999b; Acerbi 2011, 1–117).
In his *Commentary on Elements I*, he tells us that a proposition has the
following parts (Friedlein 1873, 203):

> *enunciation* (πρότασις),
>
> *exposition* (ἔκθεσις),
>
> *specification* (διορισμός),
>
> *construction* (κατασκευή),
>
> *demonstration* (ἀπόδειξις), and
>
> *conclusion* (συμπέρασμα).

In fact, however, this division is only fairly exact for the first book of the
Elements, and many of the propositions of later books, and of other geometrical treatises, have various modifications of this structure: dropping certain

parts, multiplying or dividing others, or intermingling steps that are usually characteristic of different parts. In fact, the structure of *Mov. Sph.* 12, which we read above (p. 40), is a modification of these, introducing multiple *specifications*, or "I say" statements, into the course of a *demonstration* that is divided into a number of different stages.

Sph. I.2, the first *problem* in the *Spherics*, exhibits a pattern that we can regard as the complete structure of a *problem*, and which was used already in *Elem.* III.1 (Sidoli 2018b, 410–418), on which *Sph.* I.2 was doubtless modeled. Looking ahead to the translation of *Sph.* I.2 (p. 92), we see that the structure is as follows:

⟨*enunciation*:⟩ "Find the center ...

⟨*exposition*:⟩ Let there be ...

⟨*problem-specification*:⟩ Then, it is required ...

⟨*problem-construction*:⟩ Let it be cut ...

⟨*proof-specification*:⟩ I say that ...

⟨*proof-construction*:⟩ In fact not ...

⟨*demonstration*:⟩ Therefore, point K ...

⟨QED *statement*:⟩ Which it was required to show."

This structure for *Elem.* III.1 and *Sph.* I.2 is necessitated by the fact that the proof in each relies on an indirect argument, but such a structure could be employed for any *problem* in which new objects must be introduced for the sake of the *demonstration* alone. We will see that certain aspects of this structure are also employed in some of the later *problems* of the *Spherics*, which are demonstrated directly.

We notice that there is no *conclusion* in *Sph.* I.2. In fact, when Proklos goes through the details of the structure that he proposes, using *Elem.* I.1 as an example, he sets out two conclusions, a "first conclusion" according to the *exposition*, and another, "general" *conclusion* (Friedlein 1873, 209–210). Although some readers take the final step of the *demonstration*, which is stated in terms of letter-names, as the conclusion of the proposition, we will reserve the designation *conclusion* as a part of a proposition, for an overall, summary statement. According to this reading, which we believe is standard (Mueller 1981, 13; Netz 1999b, 286), very few Greek geometrical propositions have a *conclusions* – despite Proklos's claim that it must always be present (Friedlein 1873, 203). There are only eight in the *Spherics* (*Sph.* I.3, I.6, I.8, I.12, II.10(2), and II.13(2)).[61]

---

[61] The *conclusion* for *Sph.* I.8 is abbreviated.

On the other hand, in the *problems* of the *Elements*, there is usually a more general concluding sentence that follows the final inference of the *demonstration* and summarizes the overall goal of the *problem*, stating that it has been accomplished, but using letter-names. Indeed, Proklos's "first conclusion" for *Elem.* I.1 is of this type, and the "general" conclusion that he gives is not in our manuscripts of the *Elements* (Friedlein 1873, 209; Heiberg and Stamatis 1969–1977, 8). In fact, from the perspective of the structure of the argument, in the texts themselves there are two different types of overall *conclusions*: (1) an overview statement of the task that has been completed, using letter-names, for *problems*, and (2) a general statement of what has been shown for *theorems*. We can call the former a *problem-conclusion*. Even under this refinement, there is no *conclusion* in *Sph.* I.2. There are, however, *problem-conclusions* in three of the seven *problems* in the *Spherics* (*Sph.* I.21, II.14, and II.15).

Although the structure of the propositions of the *Spherics* is neither fixed nor absolute, the use of these structural elements still plays an important role in producing the deductive force of the individual arguments. Hence, it is often useful to analyze the propositions in terms of their structure. For these purposes we will use the terminology introduced here, such that the italicized terms refer to the parts of a proposition, but not necessarily to their mathematical content. For example, the *construction* is that part of the proposition in which, in principle, objects not mentioned in the *exposition* are introduced, whether or not those elements are introduced through construction or by some other means. Moreover, constructions may also be employed to introduce further objects in the *demonstration*, and so on. The *demonstration* is that part of the proposition in which inferences are made based on objects introduced in either the *exposition* or the *construction*. We will use the word "proof" for the whole argument, and in particular for both the *construction* and the *demonstration*.

Orientation and directionality

Starting from the final propositions of *Spherics* II, we encounter a number of expressions – some of which are found already in *Moving Sphere* and *Phenomena* – that communicate a sense of directionality, or orientation. For example, in *Sph.* II.19 and II.20, one pole is called the "visible pole," φανερὸς πόλος, and the other the "non-visible pole," ἀφανὴς πόλος (see p. 27). In *Sph.* II.21 and II.22, considering two great circles inclined on one another, the pole of one is said to be "higher," μετεωρότερος, in the sense of more raised above the ground, than the other. Also, in *Sph.* II.22, one circle is said to be "most upright," ὀρθότατος, and its opposite is "lowest," ταπεινότατος; while another is said to be "more upright," ὀρθότερός, to which is opposed "more inclined," μᾶλλον κέκλιται, or "lower," ταπεινότερός. In all of these cases, it is clear that the frame of reference – the base great circle that orients these descriptive expressions – is the horizon. That is, despite the fact that, with the possible exception of the visible and invisible poles, these propositions are articulated in entirely geometric language, the conception is still oriented to the local horizon – whether the real horizon around us, or that on a solid or armillary sphere. This brings us to the question of what sphere forms the object of investigation in the *Spherics*.

The sphere of the *Spherics*

On one level, the sphere that is studied in the *Spherics* is a purely mathematical sphere defined as a surface that is equidistant from a point. Indeed, *Spherics* I presents a study of the solid properties of such a sphere, and shows how to perform constructions on its surface. Even here, however, the interest is not primarily with the geometrical properties of objects drawn in the surface of the sphere, as would become the case in Menelaos's later work on spherical geometry, but rather with the solid properties of great and small circles, and their various relationships with one another. In fact, these small and great circles are also the core components of the celestial sphere.

On another level, all of the objects studied in the *Spherics* arise quite naturally on a model of the ancient celestial sphere – that is, a bundle of parallel circles determined by rotation of the sphere about fixed poles, and the instantaneous positions of two great circles that are oblique to this motion, one carried and one stationary. Indeed, the underlying configuration of *Spherics* II and III, which consists of a bundle of parallel circles that are upright, or equally inclined, on a pencil of great circles, may not be particularly fundamental when we consider a sphere generally, but becomes immediately obvious when we take a model of the celestial sphere and spin it on its axis. Hence, we can think of the *Spherics* as attempting to produce an abstraction, or generalization, of the geometry of the *celestial sphere*, by stripping out the astronomical names of the various objects involved and describing their geometry in more general terms.

Furthermore, when one goes out on a clear night and looks up into the heavens, one does not have a direct apprehension of the celestial circles involved in spherical astronomy – although one can, of course, learn how to lay out one's understanding of the theoretical circles of spherical astronomy onto the night sky that one actually sees. When we look at an engraved star globe or an armillary sphere, such as we see depicted in numerous Greek and Roman artworks (Arnaud 1984; Gundel 1992, 285–293; Figures 8, 9, 10), however, we see immediately the various circles that are at stake, although never named, in the propositions of the *Spherics*. Indeed, many theorems of spherics that are difficult to the point of obscurity when expressed in the

Figure 8: Tomb at Pella, Central Macedonia, Greece, ca. 300 BCE. A fresco belonging to the Ephorate of Antiquities of Pella. It depicts a young philosopher standing on a green knoll, holding a pointer directed towards a blue sphere that rests on a stand. © Ministry of Culture and Sports, Archaeological Resources Fund, Greece.

generalized language of Greek mathematical texts, and accompanied by the nearly abstracted diagrams that we find in the manuscripts, become obvious when considered as pertaining to objects drawn on a solid sphere (Evans and Carman 2014, 153). Even the more advanced propositions of *Spherics* III become fairly straightforward if we use a well-made sphere and caliper to transfer the arcs whose magnitudes are being compared to some standard circle on the sphere.

All of this is to say that we should not think of the *Spherics* as having been composed in order to teach anyone basic facts about the celestial sphere, but rather to provide rigorous demonstrations of facts that the reader is already expected to know. In this regard, we may remember Aristotle's position that demonstrated knowledge is acquired by beginning with premises that are more known by nature, but less known to us, because they are further from sensation and more universal (for example, *An. post.* 71b29–

Figure 9: Roman mosaic from the Villa of Titus Siminius Stephanus, Pompeii, 1st c. BCE–1st c. CE, now in the Museo Archeologico Nazionale, Naples, inv. 124545. Seven sages, or philosophers, in discussion. The central figure uses a pointer to indicate a solid or armillary sphere. Photograph © Jebulon / Wikimedia Commons.

72a6). In this way, we can understand the *Spherics* to be a scientific treatise that demonstrates that the facts that were known to be true of the celestial circles and their configurations arise, necessarily, from the geometry of the sphere itself.

Reading the *Spherics* in antiquity

It is not feasible, of course, to claim that there was one unified way of reading the *Spherics* in the ancient period, nor that all readers encountered the text in the same context. In this section, which is highly speculative, we will simply use some aspects of the material culture of reading and education in the Greco-Roman world to consider various ways in which students and other readers may have approached the *Spherics*.

Some ancient readers will, perhaps, have heard and written out short passages of the text, such as the definitions and some of the first propositions, under the tutelage of a teacher at the secondary level, such as a

Figure 10: Mosaic depicting an armillary sphere, late 2nd to early 1st century BCE, now in the Area Archeologica e Antiquarium di Solunto, Sicily. Photograph © Elizabeth F. Evans.

Figure 11: Outline of the Solunto sphere.

grammarian or a geometer. The very few people who were practitioners of the mathematical sciences, such as Menelaos, doubtless owned the full treatise in bookrolls, and studied it carefully at their leisure, in conjunction with a solid globe and a writing tablet. Galen tells us that he learned the mathematical sciences from his father, Nikon, an architect who must have owned a private library of bookrolls, which may have included the *Spherics* (Sidoli 2015, 395–396). Galen probably read the *Spherics*, and almost certainly read other works by Theodosios under his father's guidance, working through the propositions on a solid globe, or armillary sphere. As discussed above, Galen mentions a sort of informal curriculum in the mathematical sciences that was probably in place for some time before the 2nd century, and through which a number of students were probably exposed to the *Spherics* by their teachers (p. 11).

A reader of the *Spherics* was expected to know already the basics of spherical astronomy, to have mastered Euclid's *Elements* I–VI and XI, and probably to have completed some more advanced texts in mathematical astronomy. As noted in the quote from Strabon's *Geography* that we read above (p. 10), the primary circles of the celestial sphere were studied in a first course of the mathematical sciences. Introductory astronomy was probably taught by grammarians or geometers through reading Aratos's poem on the celestial sphere (Mastorakou 2019), about which a number of commentaries were composed (Marrou 1956, 253–255; Cribiore 2001, 142–142). This instruction would have been carried out in conjunction with an illustrated globe. There are a number of representations in Greek and Roman artworks of philosophers and goddesses depicted with a globe, holding a pointing stick, a papyrus roll, or both. Some of these figures are labeled "Aratos," the poet, or "Ourania," the muse of astronomy (for example, Evans 2016, 145). Consider the young philosopher, one of six, who props a foot on a green knoll and directs his pointer towards a blue sphere in a tomb fresco that was completed, at Pella, around the time that Autolykos and Euclid were writing their treatises (Figure 8); or the group of seven bearded philosophers, three of whom hold papyrus rolls, and one of whom points towards a marked sphere in a Pompeii mosaic that was probably laid down not long after Theodosios composed the *Spherics* (Figure 9). They both depict a figure indicating some features on a globe with a rod. This is, naturally, an artistic trope, but it must have evoked a familiar image for many viewers, remembering their own teachers pointing out the constellations and significant circles on an engraved or decorated sphere.

Probably fairly few students went on from this to study the *Elements*, and fewer still persisted until the solid books, which would be necessary to follow the details of the arguments in the *Spherics*. Furthermore, when we say that readers of the *Spherics* were expected to know the *Elements*, we do not simply mean that they must have developed a general knowledge

of elementary geometry, but rather that they should have had command
of the text itself, memorizing much of it, or at least the enunciations of
the propositions. It is fairly clear that the formulaic language of Greek
geometry arose in an oral culture, which relied to a considerable extent
on memorization (Aujac 1984, 107–108), but there are many indications
of a persistence of oral culture in the surviving text of the *Elements* itself
(Saito 2018). We discussed above the highly verbal way in which previously
established results are brought in to justify steps in an argument – which
seems designed to recall to the reader's, or listener's, mind the intended
proposition.

Three documentary papyri containing Euclidean material, may give
evidence for how the *Elements* was studied in antiquity (P.Oxy. 29,
P.Berol. 17469, and P.Oxy. 5299). In these papyri, we find lists of enunciations, accompanied by unlabeled diagrams, which probably serve as a
visual reminder of the contents of the propositions they illustrate. While
these texts may have been copied from a book of such enunciations that
circulated privately, their diagrams are rather poorly executed, so that it is
more likely that they are the direct result of a typical process of studying
the text. For example, while the teacher presented the text – drawing out
diagrams on a tablet and reading aloud from a bookroll – the students may
have produced a private copy of the enunciations and unlettered diagrams,
for further reference. Whatever the case, such a list of enunciations and
diagrams would be useful for understanding the references in the arguments
of the *Elements* itself, and, once completed, could also be used in following
more advanced texts, such as the *Spherics*.

It is unclear what further texts Theodosios may have anticipated his
readers would have known, but, since certain steps in *Spherics* II and III
require knowledge of geometrical facts that are not shown in the *Elements*
but are used by Aristarchos and Archimedes in their works on astronomy
and statics, it is clear that he expected his readers to have advanced beyond the *Elements* in their studies of the mathematical sciences. Indeed,
if Theodosios directed the *Spherics* towards mathematicians themselves, he
would have been addressing readers already familiar with various works of
Autolykos, Euclid, Aristarchos, Eratosthenes, Archimedes, Hypsikles, Hipparchos, and others, now lost to us. Over the following decades, or centuries,
the *Spherics* was then taken up as a canonical text for introducing students
to spherical geometry and astronomy, and these more difficult steps were
then justified by arguments in scholia, which now accompany the treatise.

Teachers who lectured from the *Spherics*, and mathematicians who studied it carefully, would have owned the treatise in the form of a bookroll.
There is an early codex papyrus of the *Elements* from the 3rd century
(P.Oxy. inv. no. 105/24, unpublished), so that it is possible that codices of
the *Spherics* were also made starting from the end of the Imperial period.

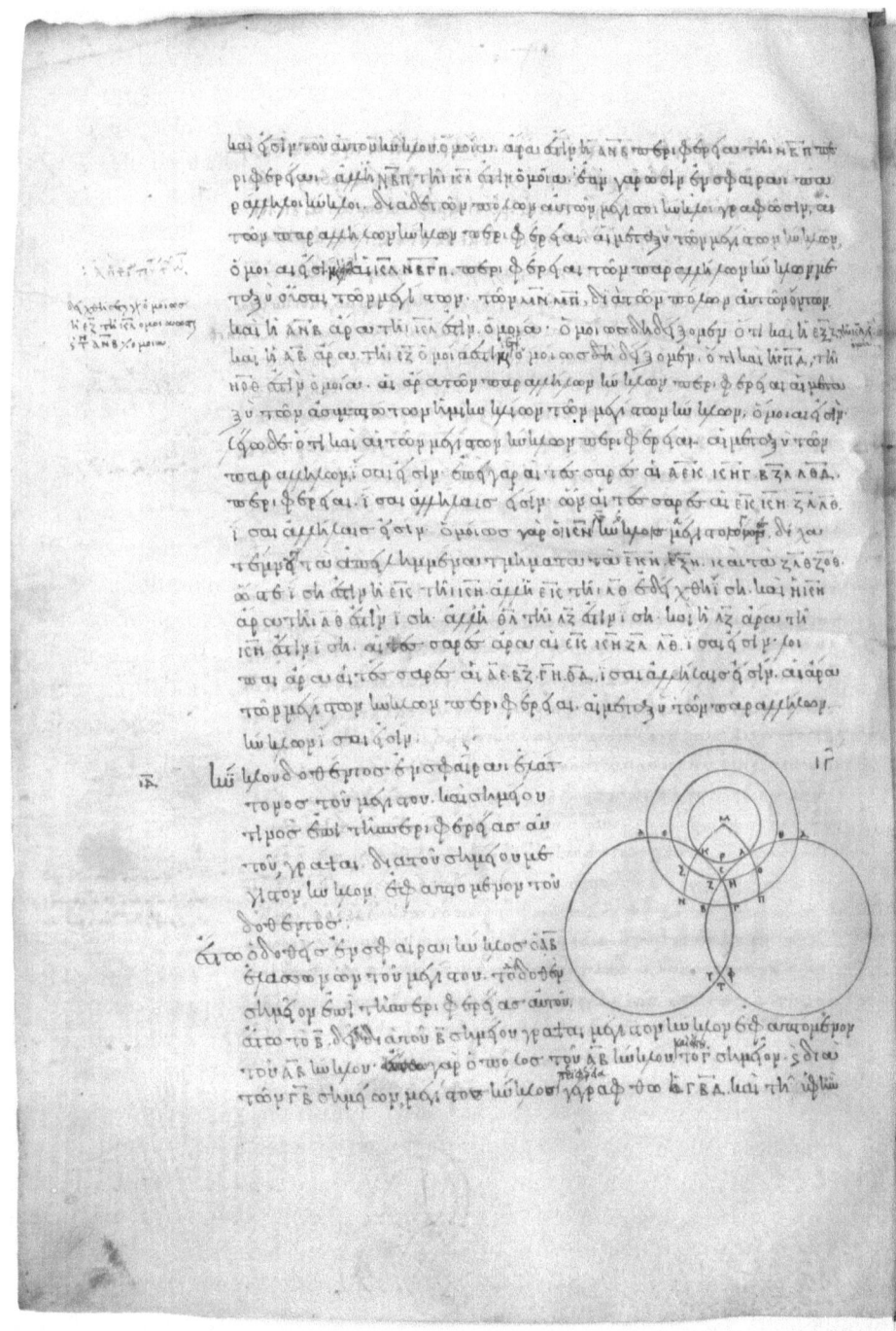

Figure 12: Vat. gr. 204 (**A**), 15v; the beginning of *Sph.* II.14 (IΔ in the margin, in red ink), with the diagram for *Sph.* II.13 (labeled IΓ in black ink) following the text of that proposition, as usual. © Biblioteca Apostolica Vaticana.

When the text was written, however, and throughout the late Hellenistic and early Imperial periods, all copies of the text itself would have been in bookrolls. We can imagine Menelaos working with a roll, and Galen's father, Nikon, owning a copy of the *Spherics* among the various bookrolls in his library. These papyrus rolls, mostly copied by professional scribes, would have carried the text in narrow columns, possibly with a slight slant to the right, as seen in the remains of elegant bookrolls that now survive from Roman Egypt (Johnson 2004, 91–90; Figures 13 and 14). The material experience of reading from such an object may be contrasted with that of the medieval manuscripts in which scholars now encounter the text (Netz 2020, 530–531). Indeed, the written lines of the medieval manuscripts show wide columns, parallel to the edges of the leaves of the codices, often wrapping around the figures, which could have been larger if the copyist had wished, in contrast to the narrow, slanting columns of the ancient works (see, for example, Figures 12, 13, and 14). Although these narrow columns help the eye to parse the unbroken, unpunctuated majuscule text, they do not leave much room for the diagrams, which should have been fitted into these narrow columns and must have been designed to do so. The diagrams should have been simple, and as robust as possible in order to survive the vicissitudes of transmission. There are no known papyri containing the *Spherics*, but the only two, currently known, fragments of the *Elements* itself, a bookroll and a codex (P.Fay. 9; P.Oxy. inv. no. 105/24), show relatively narrow columns with the diagrams set inside. We should expect the same format for ancient copies of the *Spherics* (Figure 14).

In his fantastical account of the liberal arts, the *Marriage of Philology and Mercury*, written in the 5th century, Martianus Capella describes the lady Geometry, the first of the four mathematical sciences, as accompanied by a group of handmaidens who carry a colored board, covered in glass powder, or sand, which we are told is called an *abacus*, and upon which can be drawn straight lines, circular arcs, triangles, and even the entire scope and form of the world (Mart. Cap. 575, 579). Although this is a literary allegory and rather late, the description of the abacus must have been meant to remind readers of their own education in geometry, which is not likely to have changed much in material format from the Hellenistic to the late-ancient periods. We can imagine a teacher reading from a bookroll and drawing figures on a dust board, while students followed along, perhaps drafting figures on their wax tablets and writing the enunciations and schematic diagrams on documentary papyri for memorization.

In order to draw diagrams on the board, the teacher would have used a straightedge and compass. We do not know if there were special compasses for use in geometry, but there is extensive evidence for the use of compasses among artisans and craftspeople. Numerous Greek vases show the clean line of a compass-drawn circle around the vivid indication of the circle's

Figure 13: The Arden Hyperides papyrus, P.Lond.Lit. 132, ca. 100 CE (Johnson 2004, pl. 17). © The Trustees of the British Museum.

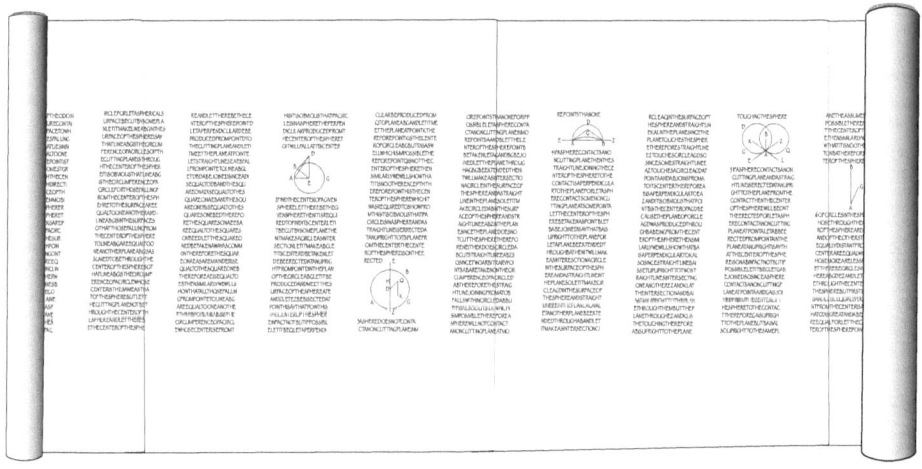

Figure 14: An imaginary bookroll of *Spherics* I from Roman Egypt. Adopted from a model image by Johnson (2004, x).

center.[62] There are many depictions of compasses and calipers in Greek and Roman funerary reliefs and other artworks (for example, Burford 1972, 182, fig. 8; Ulrich 2007, 14, 33, 40, 54–55). As we will argue in the Commentary to *Sph.* I.18–I.21, there are indications in the text of the *Spherics* that Theodosios intended his constructions to be carried out by a compass and drawn on both a plane surface, such as a dust board or wax tablet, and

---

[62] Early usage of a compass can be seen on roof tiles from Ephesos, made prior to 600 BCE (Schädler 2001).

the surface of a solid sphere.[63] For working on a solid sphere, it might have
been better to use a bowed caliper to transfer distances and draw circular
arcs, and we see such an instrument depicted in a 1st-century CE relief of
a Roman carpenter's workshop (Ulrich 2007, 11; Figure 15). When mathematicians studied the *Spherics* on their own, they probably often still found
it useful to draw out some of the more involved diagrams as they worked
through the propositions.

Figure 15: Detail of a caliper, or dividers, in a relief of the 1st century CE depicting
a carpenter's workshop, now in the Musei Capitolini, Rome, MC2743. Photograph
© Marie-Lan Nguyen / Wikimedia Commons.

Returning to the *Marriage of Philology and Mercury*, a few passages
later Capella describes Geometry herself holding a rod (*radius*) in one hand
and a solid globe in the other (Mart. Cap. 580). The following verses about
the sphere itself make it clear that this is a piece of speculative fiction,
because the globe is described as a miniature cosmos, containing an atmo-
sphere, oceans and the various moving stars (Mart. Cap. 583–585). We are
perhaps invited to imagine a divine version of the *sphaera* that Cicero, in

---

[63] The first explicit description of the use of a compass to construct figures on the
surface of a sphere, of which we are aware, was made by Abū al-Wafāʾ al-Būzjānī, writing
in the 10th century (Woepecke 1855, 352–358).

his *Republic*, has Philus tell us was made by Archimedes (Cic. *Rep.* I.21).[64] Once again, this is a literary device, but it was doubtless meant to evoke a well-known image. We may recognize under the poetic embellishments the basic image of a figure with a rod and a globe that, some eight centuries prior, had already appeared in the tomb at Pella mentioned above (Figure 8). In the Hellenistic and Imperial periods, a "geometer" was a teacher of the mathematical sciences. A stele of the 3rd–2nd century BCE shows a geometer, γεωμέτρης, by the name of Ptolemy drilling a small child on a multiplication table (Sesiano 2020, 86–87); in the Roman Imperial period, there was a profession of *geometer*, whose salary could be set by provincial governors (Cuomo 2000, 11–16, 35). The trope of the figure with rod and globe likely came to represent such a teacher. On the whole, the various iconography indicates that it was not uncommon for people to learn the mathematical sciences from a teacher who possessed a sphere of some sort.

In imagining what sort of spheres these teachers might use, we need not believe that they usually possessed a mechanical device of the sort that was attributed to Archimedes, and which formed the subject of the mathematical discipline of sphere-making, upon which Archimedes is said to have written a treatise (Gee 2000, 96–100; Evans and Berggren 2006, 27–34; Pappus, *Col.* VIII.3). For teaching spherical astronomy a simple solid or armillary sphere would suffice. It is clear that solid spheres were made on lathes (Plat. *Leg.* X.898b; Cic. *Rep.* I.22), which is a fairly simple procedure, and, moreover, the process of their manufacture was well enough known that it could be used as metaphor for explaining either the production or the movement of the cosmos (Plat. *Tim.* 33b; Pseudo-Aristotle *Mund.* 391b21–22). We are told that Thales and Anaximander each made a sphere (Cic. *Rep.* I.21; Diog. Laert. II.2). If so, these were probably solid. We do not know when the armillary sphere was developed, but whether or not Plato intended this in his *Timaios*, it is likely that articulated spheres of some kind were constructed already in the mid- to late-4th century BCE, as discussed above. Geminos refers to both solid and armillary spheres in his *Introduction to the Phenomena* (*Int.* XVI.10,12). Although there are only three surviving globes, and a fragment and modern photographs of two others, since lost (Künzl 1997/98; Künzl et al. 2000; Cuvigny 2004; Dekker

---

[64] While it has been disputed that Archimedes actually built such a spherical orrery (Keyser 1998, 244–248), that such a device *could* have been built using technologies available to Archimedes has been demonstrated by Wright (2019). We should, however, also recognize that the term σφαῖρα, or *sphaera*, may have at some point become a technical expression for any mechanical device that represented the cosmos, whether or not it had the shape of a sphere (Jones 2017, 239–241). If this was the case, then we could take the famous Antikythera Mechanism as an example of such a *sphere*, and understand the ancient discipline of sphere-making to concern the theory and practice of all such devices. That said, although Cicero may not have meant to claim that Archimedes made a sphere-shaped mechanism, Capella probably read his text in this way.

2013, 52–102), there are numerous images and representations of globes in surviving Greco-Roman artifacts (Arnaud 1984; Gundel 1992, 285–293). There are no surviving armillary spheres, but there are a number of artistic representations, such as the detailed mosaic from Solunto, Sicily, that might have been made while Theodosios was writing the *Spherics* (Figures 10 and 11).

For the purposes of reading the *Spherics*, either a solid or armillary sphere would do. If one wanted to draw out all of the diagrams on an actual sphere, one could use a lathed wooden, or perhaps ceramic, sphere, coated in wax, which could be marked with a compass or caliper. Such detailed diagrams, however, would probably only be needed by a practicing mathematician who wanted to understand the text in depth and, perhaps, produce new results. For students an inscribed celestial globe, or armillary sphere, would already depict the objects necessary to follow most of the propositions in the text. Geminos discusses the various circles engraved on ancient globes (*Int.* V.45, 62). In such a situation, the teacher could draw out the diagrams from the text, on a board, and then describe which objects on the actual sphere correspond to which objects in the diagram. In this way, it would be possible to follow the details of the argument.

The text of an ancient work in geometry can be thought of as a sort of script or screenplay for an activity of some kind (Sidoli 2009; Netz 2020, 531–535). In the case of the *Spherics*, the bookroll with its stylized diagrams showed one how to undertake mathematical practices involving a sphere and a writing surface. Most often, the context of this activity would have been teaching the mathematical contents of the text. The teacher would have read aloud from the bookroll, drawing the diagrams on the board, and pointing out with a rod the various circles as they should appear on the sphere. In some cases, this may have involved presentation by a mathematician to colleagues, perhaps in commenting on some obscure aspect of the text, or providing a justification for some step in the argument. Occasionally, it would have involved a mathematician working through details of the text to see if its ideas and methods could be pushed in new directions – such as was done by Menelaos, and to a lesser extent by Pappos. For understanding the details of the more involved theorems, it might have been useful to draw out the diagrams on an actual sphere, but this would not always have been necessary. Although the diagrams in the ancient and medieval sources strike most present-day readers as obscure, they have their own internal logic and convey visual information just as well as, if differently from, a modern diagram in linear perspective. The best way to see that this is the case is to undertake a reading of the text ourselves using the medieval diagrams, but we can also be certain that ancient and medieval readers of the text must have been able to understand these diagrams because they demonstrate their mastery of the material by adducing useful lemmas, fleshing out various cases, and applying the geometric ideas in the text in their own mathematical research.

Editions and manuscript sources for the Greek text

There are two critical editions of the text based on most or all of the known medieval and early modern manuscripts containing the *Spherics* – namely, those of Heiberg (1927) and Czinczenheim (2000).[65] Although there are many minor differences between these two editions, the most important divergence is that Czinczenheim usually choses to follow the 9th-century Vat.gr. 204 (**A**, Figure 12), whereas Heiberg sometimes prefers the reading of the 14th-century Par.gr. 2448 (**K**). As mentioned above, however, **K** is a Byzantine edition of the treatise, not an accurate copy of an ancient source. Hence, we have followed Czinczenheim in taking **A** as generally the more accurate witness – a position that is also supported by the readings of Thābit's Arabic version, which must have been originally based on another Greek manuscript extant in the 9th century. Hence, our translation is made on the basis of Czinczenheim's text.

In a few cases, we have preferred Heiberg's text and note this where it occurs. There are also even fewer places where we have used a variant that is found in the manuscripts but is not preferred by either of the editors – these are also noted. In those places where Heiberg's text is substantially different from that of Czinczenheim, we have translated it in footnotes.

Czinczenheim's edition indicates with square brackets many passages that she believes should be removed from the text as later interpolations. While we agree with many of these, there are some others that are, in our opinion, undecidable, and others with which we disagree. Rather than trying to decide whether or not each such passage belongs to an assumed original composition, we have decided not to include these brackets in our translation, instead setting off likely additions with dashes and indicating material that is more conspicuously extraneous with footnotes.

Translation practices

The overall goal of the translation is to reproduce faithfully the concepts involved in the Greek text, without trying to recreate the particularities of the Greek language. We try to render individual words and phrases with something that we believe has the same meaning in English. For example, we have translated ἀσύμπτωτος with *non-intersecting*, as opposed

---

[65] Czinczenheim (2000, 631–650) provides a discussion of all of the Greek editions of the text.

to "asymptotic," despite the fact that in context it denotes something rather more specific than simply non-intersecting (see *Sph.* II.13, p. 127; Mugler 1959, 86). We believe that a reader of the Greek will also have to determine the technical meaning of the expression from the context.

We have tried to translate technical terminology consistently, even when this may result in awkward English. That is, where possible, we use the same expression in English to render a certain expression in Greek. We have translated cognates and closely related words in Greek with cognates in English. This means that we have had to choose an English expression that could work for all occurrences of any given Greek terminology. For example, we have rendered the substantive adjective κάθετος with *perpendicular* and the adjective ὀρθός with *upright*, because an English phrase involving "upright" could be found for all of the various Greek phrases involving ὀρθός (Mugler 1959, 235–236, 312–314). Here we may mention that in the interest of the consistency of the translation we have always translated κεκλιμένος with *inclined* (Mugler 1959, 252–252), despite the fact that in *Sph.* II.21 (see p. 144, and following) this results in English expressions that some readers will find to be inconsistent with modern mathematical usage. In using *inclined*, however, we are following a long tradition of modern translators.

In general, we do not attempt to translate the Greek expressions into modern idiomatic expressions used in technical English, although we do this when the meaning of the Greek and English is close. For example, περιφέρεια is consistently translated with *circumference*, whether or not it denotes the whole circumference or an arc. On the other hand, δίχα τέμνειν is translated with *to bisect* (Mugler 1959, 413–416), because we did not think that there would be any benefit in translating more literally with something like "to cut in two."

We have not tried to translate the same preposition in Greek everywhere with the same preposition in English. Nevertheless, we have applied a principle of consistency when a certain preposition conveys the same meaning in certain set constructions, and we have consistently translated the same prepositional phrases in the source with the same phrase in the translation. When the Greek preposition has a more specific meaning than is generally conveyed in English with a single preposition, we have translated with other grammatical forms. For example, πρός [+ an object in the accusative] is translated with *abutting* when it conveys the sense of being immediately adjacent.

One of the more important aspects of our translation is the way that we have chosen to handle the Greek uses of indefinite expressions, which are essential to conveying the generality intended by the text. Although it has often been maintained that proofs in Greek geometrical texts concern specific objects with given letter-names, recent scholarship has convincingly shown that this interpretation is based on a misunderstanding of the role of indefinite expressions in conveying generality (Federspiel 1995; Acerbi 2011, 39-57; Acerbi 2020a). In particular, when an object is introduced in the *exposition* this is done with an expression like "let there be an object, *AB*," using the verb of being, in the 3rd person imperative, and an indefinite noun followed by a definite letter-name. The meaning of such an expression is that the object in question is any one of the sort of object under consideration, which is then assigned a certain letter-name. The letter-name is a sign that points, not to a specific object, but, rather, to any member of the class of objects that the proposition concerns. Hence, the *exposition* begins with full generality. Furthermore, this generality is maintained throughout the argument by the use of indefinite expressions whenever the text refers back to something that has been introduced in the *exposition* and *construction* or established in a previous result (Acerbi 2011, 26–32). We have sought to convey this use of the indefinite in our translation, although because of the different ways that Greek and English use articles, this has not always been possible. In general, however, an expression like μέγιστος κύκλος ὁ ΑΒΓ is translated with *a great circle, ABG,* or *great circle, ABG,* while ὁ ΑΒΓ μέγιστος κύκλος is translated with *great circle ABG,* since the latter may be understood as definite in English. In more extended noun phrases, we sometimes include "the" for clarity – such as *the square on AB* for τὸ ἀπὸ τῆς ΑΒ.

Related to this, we try to distinguish clearly between the existential and copulative uses of εἶναι, the verb of being (Federspiel 2010). In general, when objects are being introduced, we translate with the existential sense, although this sometimes produces awkward English. For example, in *Sph.* II.9 we translate ἔστω ... τοῦ μὲν ΑΓΒΔ κύκλου καὶ τοῦ ΖΑΕΒ κοινὴ τομὴ ἡ ΑΒ with *let there be the common section of circle AGBD and ZAEB, AB* – whereas one might be tempted to render the Greek as "let *AB* be the common section of circle *AGBD* and *ZAEB*," which sounds more natural in English (Czinczenheim 2000, 89). When we look back at the first time that such an expression is used, however, in *Sph.* II.3, the text reads ἔστωσαν ... αὐτῶν κοιναὶ τομαὶ τοῦ μὲν ΑΓΕ καὶ τοῦ ΑΒΓ ἡ ΑΓ, τοῦ δὲ ΑΓΕ καὶ τοῦ ΓΔΕ ἡ ΗΓΖ, which must be read as an existential usage of εἶναι (Czinczenheim 2000, 83).[66] Hence, we understand the later introductions of common sections to be existential as well.

We have silently supplied many elided words, when we believe that someone reading the Greek would have no doubt of the intended meaning. For example, as mentioned, we translate τὸ ἀπὸ τῆς ΑΒ with *the square on AB,* and ἡ ὑπὸ ΑΒΓ with *angle ABG.* We render both εὐθεῖα and εὐθεῖα γραμμή with *straight line,* because we believe that there was no intended differ-

---

[66] When common sections are introduced in *Spherics* I, a different idiom is used.

ence in meaning (Mugler 1959, 201–203). On the other hand, where the
elided material is subject to some interpretation but is required by English
grammar or sense, we have included it in ⟨angled brackets⟩.

We have introduced punctuation and paragraph formatting according
to what we think makes the mathematical argument most clear. Paragraph
formatting is used to highlight the structure of the propositions themselves,
and to exhibit the shifts of indirect arguments. Greek particles that control
the flow of the discourse are sometimes translated and sometimes incorporated into the English punctuation and paragraphing. We have tried to be
consistent in the translation of particles, unless this would make the English
too strained. For example, γάρ is translated as *indeed* when it introduces
an indirect argument but *for* elsewhere, and whether this "for" is understood as introducing a reason, or simply introducing the next step in the
discourse, is indicated by the punctuation.

There are two layers of footnotes, both keyed to the page. Numbered
footnotes provide the justifications for steps in the argument and make other
comments of a mathematical nature. Lettered footnotes provide comments
on textual and linguistic issues. Although the text itself does not make
references by proposition number, the practice of providing such explicit
justifications for steps in the argument is found in the scholia accompanying the text in the 9th-century Vat.gr. 204, and very likely goes at least as
far back as the scholarship of the Imperial and late-ancient periods. Nevertheless, since the practice of referencing by reciting all or part of the enunciation of the intended proposition is not always used, as discussed above,
it is not always possible to know what justifications Theodosios intended
for his claims. In particular, sometimes we cannot find a single proposition
in the *Spherics* or *Elements* that justifies the claim in question. In these
cases, we reference a number of propositions from which an argument can
be made, and note such propositions as "by implication." Hence, the dependencies that we have included are sometimes a matter of interpretation.
In the more interesting cases, we have discussed some of these issues in
the Commentary. Even in those cases where we cannot be certain what
mathematical facts Theodosios thought should justify his claims, however,
we think that references to sufficient propositions in ancient works will help
modern readers better understand the context of mathematical knowledge
in which the *Spherics* was written.

Diagrams

The diagrams in the texts of spherical astronomy, and particularly those
in the *Spherics*, provide us with an important source for studying the di-
agrammatic practices of ancient mathematicians (Neugebauer 1975, 751–
755). Research on the geometrical diagrams in ancient and medieval sources

has shown that they have two primary characteristics that distinguish them from the diagrams produced from the early modern period and on, namely overspecification and indifference to visual, particularly metrical, accuracy (Saito and Sidoli 2012). By overspecification one denotes the tendency of the diagrams to depict objects with more regularity, or symmetry, than is required by the text. For example, rectangles represent parallelograms, isosceles or equilateral triangles represent arbitrary triangles, squares represent rectangles, right angles represent arbitrary angles, regular figures represent arbitrary figures, the overall figure exhibits more symmetry then is required by the text, the lines may be aligned with the margins, and so on. By calling such diagrams indifferent to visual, and particularly metrical, accuracy one indicates that the diagrams make little attempt to be an accurate visual portrayal of the objects at issue. For example, equal lines may be depicted as unequal, unequal angles may appear to be equal, the bisection of a line may not fall at its center, a special curve may be depicted with the arc of a circle, straight lines may be represented with arcs, and so on.

Although these characteristics are nearly ubiquitous in the ancient and medieval sources, we must recognize that certain aspects of this visual culture have probably been produced by the accidents of transmission. In particular, if an error were introduced in copying a diagram, it was likely be in the direction of introducing more symmetry than had been found in the source diagram (Acerbi 2017, 145–146; Carman 2018; 2020). Indeed, after a diagram was copied a relatively small number of times, it was likely to become more symmetrical than it may have been originally. Hence, when we encounter a mathematically incorrect, or impossible, figure in our manuscript sources, we should first ask if we can attribute such an error to the process of transmission rather than the ancient author.[67]

It remains the case, however, that there are key aspects of the ancient and medieval diagrams that are different from modern practice, but which are not mathematically incorrect. For example, it is not necessarily an error to depict a parallelogram as a rectangle, or a triangle as an isosceles triangle, since a rectangle *is* a parallelogram, an isosceles triangle *is* a triangle. Indeed, such conventions could involve a different understanding of the role of a visual image in conveying the concept of generality. For example,

---

[67] A conspicuous example is a diagram from Ptolemy's lunar theory, *Almagest* V (Vat. gr. 180, f.129r; Carman 2018, 221). In the Greek manuscripts, this diagram is drawn with a chord parallel to the lines of the text, which results in a mathematically absurd figure. In all the Arabic manuscripts of the various versions of the *Almagest* that we have checked, however, this diagram is correctly drawn. An important exemplar is a 1272 CE copy of al-Ṭūsī's *Revision of the Almagest* (Tehran Sipahsalar 4727, p. 486), which was copied by one of Ṭūsī's students from a copy that was checked by the master. In the same manuscript and by the same hand, however, overspecification that does not involve mathematical error, of the sort mentioned above, is widely used.

when we see a triangle with unequal sides we believe that it represents an arbitrary triangle, not a triangle of the specific shape that is depicted, not because it necessarily does so, but because we have become accustomed to such a convention. That is, we believe that we should be able to take information about the apparent equality of the sides and angles directly from the figure, but not information about the precise ratios between the sides and the angles. But this belief is clearly just a result of our visual culture – it is not an objective fact of observing the image itself. The depiction of generality in a visual image is a matter of visual style, and there is no absolute criterion for such decisions. Furthermore, there is little indication that ancient and medieval images were used to convey visual information in the ways to which we have become accustomed. Hence, it is unlikely that the ancient sources were all metrically accurate and presented generality in the same style as modern depictions, and then that these habits were simply lost in the medieval copies.

Finally, it must be pointed out that the earliest manuscripts that we possess are not many times removed from the scholars of late antiquity who edited the Hellenistic texts (Acerbi 2017, 144–147; Netz 2020, 516–523), and, moreover, the few diagrams that are preserved in ancient papyri also exhibit overspecification and indifference to visual accuracy. For these reasons, we must acknowledge that the visual culture of the ancient bookroll and codex involved the use of schematic images, not visually accurate pictorial illustrations (Saito and Sidoli 2012, 157–160; Netz 2020). In such a visual culture, even if an individual mathematical author chose to adorn a mathematical text with visually accurate drawings, these would quickly be lost in the study and transmission of the text, so that the vast majority of all readers who encountered such a treatise throughout the ancient and medieval periods would have read it accompanied with the overspecified, schematic diagrams that are now found in nearly all our evidence from these times. In fact, because the vast majority of the mathematical diagrams that ancient authors saw were of this schematic variety, and because they knew that their own works would be transmitted in such a visual culture, they probably designed their diagrams to be resilient to the types of changes that might be brought about by the process of transmission – especially in the case of spherics, in which elaborate perspective diagrams would quickly be rendered unintelligible by copying errors.

In fact, Theodosios's *Spherics* provides us with an interesting case study for the choices made by a Greek mathematician in deciding what kinds of information to transmit visually, because the diagrams of all of the treatises in ancient spherics show signs of deliberate choices in this regard (Neugebauer 1975, 751–755; Malpangotto 2010; Le Meur 2012). Here we set out the primary characteristics of the diagrams of the *Spherics*, which are conveyed in the diagrams that accompany our translation.

The first indication that the diagrams in the *Spherics* are not meant to be illustrations of the objects under discussion is the fact that they never portray the sphere itself. In fact, the diagrams accompanying the later propositions do convey the visual impression of a sphere (Malpangotto 2010, 80–89), but even here the outer circle is not the sphere itself, but rather one of its great circles (Neugebauer 1975, 752; Le Meur 2012, 185–188). In the earlier diagrams, however, the objects that are under discussion are simply drawn in the plane of the figure, and the sphere itself is not visually indicated in any way. A striking example of this practice is the diagram to *Sph.* I.5 (p. 95), which simply uses two lines meeting at a point to depict a diameter of a sphere and another line inside the sphere that meets the diameter at one of its endpoints on the surface of the sphere. The plane and the sphere, which play a role in both the *enunciation* and the proof, have no visual representation at all.

In order to understand the type of information that the diagrams convey, it may be useful to provide a taxonomy of their visual characteristics. The diagrams of the works in spherics can be divided broadly into two classes, those which appear to be fairly flat, and those which appear to be more three-dimensional (Neugebauer 1975, 752–753; Malpangotto 2010, 80–89). Because this distinction is based on our psychological impression of what sorts of information the image coveys, for the purpose of studying the diagrams in the *Spherics*, we will also introduce two other dichotomies that can be found in the figures.

Before looking at the special features of the diagrams of the *Spherics*, however, we should recognize that there are a number of diagrams in the *Spherics* that depict objects that are not, or not necessarily, in the sphere, such as those diagrams accompanying theorems that act as geometric lemmas to later propositions. These diagrams exhibit the same characteristics that we find in the diagrams of the *Elements*, and we will refer to these figures as being *standard planar* and *standard solid* diagrams. They are as follows:

Planar (standard)	*Sph.* I.18(2nd)°, I.19(2nd)°
Solid (standard)	*Sph.* I.5, II.11, II.12, II.21, III.1, III.2

° *Sph.* I.18 and I.19 both have two diagrams.

Setting these aside, in order to classify the diagrams of the *Spherics* that represent objects *in the sphere*, we will use three independent dichotomies. The first of these is between diagrams that contain (1a) circles and arcs that are intrinsic to the surface of the sphere *as well as* straight lines that are extrinsic to the spherical surface, and those that contain (1b) *only* intrinsic objects. While this is essentially an objective criteria, since we can simply check the diagram to see whether or not it contains lines, it is not actually based on a significant mathematical difference, because there are some propositions, such as *Sph.* II.13, which use straight lines in the argument, but neither construct nor name them, so that they do not appear in the figure.

The second dichotomy is between diagrams that depict (2a) a fairly local region of the spherical surface and those that show objects on (2b) a more global surface. Specifically, this distinction is between those that contain small circles and, perhaps, arcs of great circles, and those that must contain at least one full great circle. Again, this is a fairly objective criteria, because we can check the text to see if any of the full circles in the diagram must represent a great circle. Once again, this is a visual as opposed to mathematical distinction, because some diagrams represent a great circle with an arc of a circle, such as *Sph.* II.9, while others represent it with a full circle although this is not needed for the argument, such as *Sph.* II.13. Indeed, the full medieval transmission of the treatise, including sources in Arabic, Hebrew and Latin, shows some variations along these lines, which were likely introduced by the scholars who studied and transmitted the text.

The third dichotomy is that which we raised above between diagrams that have the appearance of (3a) a flat, planar surface, and those that appear to depict (3b) a solid spherical surface. This characteristic is the most subjective, since it only relies on the viewer's impression of how flat a diagram appears. For this reason, some diagrams could be categorized as either (3a) or (3b), namely those for *Sph.* II.10, II.19, and III.13. It should be clear that all three of these dichotomies are based on visual, as opposed to mathematical, characteristics.

With these as our categories, we can set out an array of the diagrams in the *Spherics* as in Table 1. The first thing to notice about this table is that some diagrams appear in two places. This is largely because of the subjectivity involved in judging the dichotomy between planar and solid, as mentioned above. For one group, *Sph.* I.13–I.15, we include the Arabic diagrams in our table, which are different in such a way as to change the diagrams from solid to planar. In fact, in the whole medieval tradition, there are many local variations of individual diagrams that might change their placement in this array, but we find the situation of the diagrams for *Sph.* I.13–I.15 to be interesting and historically significant, and will return to it later.

Of the three dichotomies, the easiest distinction to make is that between diagrams that contain both intrinsic and extrinsic objects, and those that contain only intrinsic objects, (1a) or (1b), because we simply make a visual check for the presence or absence of straight lines. About 58% of the diagrams in the text contain both types of objects. In these diagrams, there are often false crossings between the straight lines themselves and between both the straight lines and the circles or circular arcs that lie in the surface of the sphere. In all diagrams of the various spherics texts, however, intrinsic

	Intrinsic and extrinsic	Only intrinsic
Local, planar	*Sph.* I.1, I.2, I.7, I.8, I.9, I.10, I.18(1st), II.1, II.2, II.3, II.5, II.9, II.10°, II.15, III.3°,	*Sph.* II.4, II.14, II.16, III.13°
Local, solid	*Sph.* II.10°, III.3°, III.4*	*Sph.* III.13°
Global, planar	*Sph.* I.3, I.4, I.6, I.11, I.12, *Sph.*$_A$ I.13§, I.14§, I.15§, *Sph.* I.16, I.17, I.19(1st), I.20	*Sph.* I.21, II.6, II.7, II.8, II.13, II.22, II.23
Global, solid	*Sph.* I.13, I.14, I.15, II.17*†, III.11*	*Sph.* II.18, II.19†, II.20, II.5, III.6, III.7†, III.8†, III.9, III.10, III.12†, III.14†

° These diagrams are ambiguous between planar (3a) and solid (3b).

* These diagrams give the impression of a figure made using the principles of linear perspective.

§ These are the three diagrams from Thābit's Arabic version.

† These diagrams include an exceptional feature.

Table 1: Array of diagrams of spherical objects in the *Spherics*.

objects make no false crossings, so that the topology of their intersections and tangencies is correctly depicted.

In order to make a decision regarding the next dichotomy between local and global, (2a) or (2b), it is necessary to read the text to determine whether or not any circle depicted in the diagram must be a great circle. In a number of cases, such as *Sph.* II.1–II.5, at least one of the circles *could* be a great circle, but if none of the circles must be great, we consider the diagram to be local. Of the diagrams accompanying our translation, again about 58% of them are global.

The most interesting dichotomy, however, is that between diagrams that appear to be planar, and those that appear to be solid, (3a) or (3b). This is not only because the visual style is so different between these two classes of figures, but also because different sets of conventions, or encodings, are used to visually convey information about the objects depicted. In planar diagrams, (3a), it is often not possible to know whether the plane of the figure is meant to be a flattened-out depiction of the spherical surface itself, or simply a drawing in the plane of the various objects that occur on the surface of the sphere. Whatever the case, except for depicting circles on the sphere with circles or arcs in the diagram, there are no mathematical rules, or procedures, for producing spherical or solid objects in the plane, such as we find in the diagrams of Ptolemy's *Planisphere* or *Analemma*. Nevertheless, the planar diagrams of the *Spherics* make it possible to encode information about the various geometrical objects into the diagram itself, so that certain relationships can be conveyed visually (Neugebauer 1975, 753; Malpangotto 2010, 81–86). For example, in these diagrams, a circle in the spherical surface is represented with a circle or an arc in the figure, and any intersection between such objects in the diagram represents an actual intersection – there are no false crossings between these objects (*Sph.* I.11–I.15, I.17, I.19–I.21, II.3–II.6, II.9, II.13–II.16, II.22, II.23, III.1–III.4, III.13). Two great circles can be depicted as bisecting one another with a pair of equal circles that cut each other into two pairs of pairwise equal arcs that are themselves unequal, so that proportionality of arc length is not preserved (*Sph.* I.11, I.12, I.17, I.20, I.21, II.13). Parallel circles are encoded with concentric circles (*Sph.* II.1, II.2, II.4, II.8, II.10, II.13, II.15, II.16, II.22, II.23). Tangent circles are encoded by circles that touch one another, internally or externally (*Sph.* II.3–II.8, II.13–II.16, II.22, II.23). A circle can be understood to divide the plane into two regions, inside and outside, just as a circle in the sphere divides it into two parts (*Sph.* I.11–I.15, I.17, I.19–I.21, II.1–II.3, II.10, II.13–II.16, II.22, II.23). In terms of spherical astronomy, in planar diagrams the horizon is depicted as the primary circle of the diagram, the base circle (Le Meur 2012, 185–192), inside of which the visible portion of the horizon is depicted (*Sph.* II.22, II.23).

Some of the planar diagrams with both intrinsic and extrinsic objects, (3a) together with (1a), appear to convey a sense of depth, which comes close to being a perspective drawing (*Sph.* I.1, I.7–I.10, I.16).[68] Because the primary circle is depicted with a circle and not an ellipse or a lens-shaped figure, however, we can also understand these as simply depicting all of the objects laid out flat in the plane of the diagram, and categorize them as planar, as opposed to solid. In these diagrams, circles on the sphere are depicted with circles and circular arcs, while solid objects are depicted with straight lines, which allows for the depiction of solid figures such as triangles in the plane of the diagram, although angles are not preserved (*Sph.* I.1, I.6–I.10, I.17, I.18(1st), I.19(1st), II.3, II.5, II.9, II.15, III.3, III.4). The relationship between a circle in the sphere and its axis is depicted with a circle in the plane of the figure and an extended radius or diameter of the circle that has one or more false crossings with the circle (*Sph.* I.1, I.2, I.6–I.10, I.16, I.19, II.1, II.2).[69] The common diameter of a pair of intersecting

---

[68] Indeed, one might argue that some of these diagrams are standard solid diagrams, of the sorts that we find in the *Elements*.

[69] We do not count the figure to *Sph.* I.23 among these because it breaks the pattern, and accompanies a proposition that is spurious.

great circles is represented with the chord joining the intersection of the two equal circles in the plane (*Sph.* I.11, I.12, I.17).

The other major category of diagrams, which appear to be solid, is introduced for the three theorems *Sph.* I.13–I.15, but then does not reappear until again until *Sph.* II.17. Following this, such diagrams are then used for the rest of the treatise with the exception of *Sph.* II.22, II.23 – and possibly III.13, because this diagram might be interpreted as either solid or planar. The most conspicuous feature of solid diagrams is that they convey a visual impression of the sphere, although, as mentioned, the outer circle of the diagram is not the sphere itself, but a great circle in it, which we can call the *base circle* (Le Meur 2012, 185–192). When interpreted from the perspective of spherical astronomy, the base circle can usually be understood to represent the horizon, inside of which the visible portion of the celestial sphere is depicted (*Sph.* II.17–II.20, III.5–III.11, III.14). In another configuration, the base circle can be interpreted as the solstitial colure (*Sph.* III.12).

In solid diagrams, circles that are inclined on, or perpendicular to, the base circle are represented with circular arcs that meet the base circle at two points. These arcs can be understood as depicting the lens-shaped figure that is often used to represent a foreshortened circle in ancient and medieval mathematical diagrams (for example, Figure 3, p. 41).[70] The use of a lens-shaped figure making a cusp, drawn either in whole or in half, is a characteristic feature of ancient and medieval diagrams of spherics that must not be overlooked. Moreover, we should not think that the use of this cusp was a general characteristic of ancient Greek and Roman art. Except for a few exceptions, the vast majority of the many chariot wheels, circular shields, drinking vessels, and serving plates that are depicted as seen from the side in ancient pottery, painting and mosaics, show foreshortened circles as smooth curves.[71] Moreover, the ancient mathematicians from Euclid, through Ptolemy, to Pappos certainly understood that circles actually appear as curves when viewed from the side (*Optics* **B** 34–37; *Geog.* VII.6; *Coll.* VI.41–52; Knorr 1992; Jones 2000). Hence, the use of the lens-shaped figure must have been a deliberate choice on the part of ancient mathematicians. Indeed, this visual device allows the intersection of two circles to be encoded with a clearly distinct point at the cusp where the lens meets the

---

[70] There are no diagrams of the sort that Le Meur (2012, 192–195) calls "à base orientée" in the *Spherics*.

[71] For some exceptions using cusps from the early to mid-5th century BCE, see the image of the shield carried by Menelaos on an Attic red-figure amphora, London, British Museum E263, or that carried by an owl on an Attic mug, Paris, Louvre CA2192 (Oakley 2014, pl.30B; Mitchell 2012, 135, fig.60). For an example from the late 5th century BCE of a foreshortened shield drawn with a smooth curve, see that carried by Menelaos on an Attic red-figure nestoris, Malibu, California, Getty Museum, Getty Villa 81.AE.183.2 (Oakley 2014, 250).

circle. In fact, ancient mosaics that depict the solid and armillary spheres that were used in spherics clearly portray the cusps of these lens-shaped figures – see, for example, the Roman mosaics depicting the armillary sphere from Solunto, or that of the sphere surrounded by philosophers at Pompeii (Figures 10, 11, and 9). Since these may be directly contrasted with contemporary mosaics of the same or nearby regions that depict, for example, drinking and serving vessels seen from the side with smooth curves (Dunbabin 1999, 44, 46), it seems clear that the lens-shaped figure, in full or in half, was so well known in antiquity that it served as a visual motif that signified the study of spherics itself.

In solid diagrams information about a bundle of parallel circles is encoded with a set of half-lenses facing the same direction – that is, arcs of concentric, or nearly concentric, circles (*Sph.* II.17–II.20, III.5–III.8, III.11, III.12, III.14). In a few cases, as well as the primary bundle of parallels that is depicted with concentric arcs or half-lenses, another parallel, about the pole, is depicted with a full circle about a different center, either entirely inside the base circle, or cutting it (*Sph.* III.7, III.8, III.12, III.14). With the exceptions of *Sph.* I.13–I.15, which do not support this encoding, the part of the sphere on the other side of the primary great circle is always depicted outside of the base circle, following the convention developed in the diagrams using surface representation (*Sph.* II.17, II.19, III.12).

In solid diagrams a pencil of great circles that passes through the pole of another great circle can be conveyed with a set of large circular arcs each one of which passes through a single point and stand upright on a large half-lens that is orientated horizontally and represents the great circle through whose pole the pencil passes (*Sph.* III.6, III.9, III.10). A pencil of great circles that are tangent to the same parallel circle and, hence, are inclined at the same angle on the greatest of the parallels can be conveyed by a set of circular arcs of large circles that meet the parallel circle that is represented with a full circle and stand on a large half-lens that is orientated horizontally (*Sph.* III.7, III.8, III.12, III.14).

An important subcategory of solid diagrams are those that are also global and show both intrinsic and extrinsic objects, (3b) with (1a) and (2b). These diagrams show both surface and solid objects and have the appearance of an articulated or armillary sphere (*Sph.* I.13, I.14, I.15, II.17, III.11). These diagrams depict the diameters of small or parallel circles, or some other internal objects. Although there are few diagrams of this type in the *Spherics*, there are a few more in Theodosios's *Habitations*. Since, however, there are no texts of ancient spherics that have this sort of diagram by any other ancient authors, it is likely that Theodosios deliberately chose them. Furthermore, they make it clear that he was perfectly capable of drawing diagrams that convey the three-dimensional configuration of the sphere, but generally did not chose to do so (Malpangotto 2010, 89).

Because the ancient diagrams exhibit a clear pattern of deliberation on the part of the mathematician, and because they convey information that the mathematician probably regarded as crucial to understanding the propositions themselves, we accompany the translation with diagrams in the pre-modern style. Since there are sometimes errors in the manuscript diagrams, however, we have not simply reproduced the diagrams from one of the manuscripts. Instead, we have used the above stated principles, in consultation with the most important Greek manuscripts, as well as a few manuscripts from the Arabic and Latin traditions, to produce a diagram that we believe could have accompanied the text in antiquity. Although, techniques have been developed for producing a *stemma codicum* for geometrical diagrams (Raynaud 2014), these rely on tracing mathematical errors as the main points of divergence, whereas much of the variation in the diagrams of texts in spherics is of a purely visual nature. For this reason, we have not attempted to trace historical relationships among the variants, nor given a critical apparatus for the diagrams (for which, see Czinczenheim 2000, 608–720). For the most part, we have been guided by the diagrams in Vat.gr. 204, except where a certain diagram contained a mathematical error, or some other issue that calls its authenticity into question. We do not attempt to reproduce exactly what is found in the manuscript – in any case, images of Vat.gr. 204 and a number of other important manuscripts are now available online and can be consulted directly. Any place where we deviate significantly from the manuscripts has been noted in the Commentary for that proposition, in the section called "Textual Comments."

A final note should be made about the way that we portray points. In the diagrams accompanying our translation some letter-names appear near a point, denoted with a small circle, while others simply lie near a geometrical object or an unmarked intersection. Because the use of such marked points is not a consistent feature of the Greek manuscripts, these diagrams may convey more information than is found in the Greek sources, at least visually. Nevertheless, they follow a principle that is compatible with the information conveyed by the text itself, and which is depicted in certain manuscripts of the Arabic tradition that were copied by competent mathematicians. In particular, when a letter-name is introduced, it may simply denote a mathematical object, such as circle, but not necessarily any specific point that lies on that object. In such cases, we simply place the letter-name near the object, but do not draw a point, as *A*, *B*, and *G* in the diagram for *Sph.* I.2 (p. 93). We only use a small circle to denote a point when the point itself is either introduced and named in the text, or when the letter appears in the letter-name of two or more named objects, so that it must be their intersection. Finally, we often silently move free letters away from false crossings where they tend to cluster in the manuscript diagrams,[72] because such letters do not denote intersections. For example, in the diagram for *Sph.* I.2 in Vat.gr. 204, the letter B appears near the false crossing of line EZ and circle ABΓ, as though denoting an intersection (**A**, 2r). In our diagram, we have moved *B* away from this false crossing so as to remove any confusion about what it denotes.

---

[72] For example, see Υ and Φ in the diagram for *Sph.* II.13 in Vat.gr. 204 (Figure 12, above).

Part II

Translation

⟨Def. 1⟩ A solid figure contained by a single surface, to which all straight lines falling from a single point that lies inside the figure are equal to one another, is a *sphere*.

⟨Def. 2⟩ The point is the *center of the sphere.*

⟨Def. 3⟩ Some straight line produced through the center and bounded in each direction by the surface of the sphere, around which as an immobile straight line the sphere rotates, is an *axis of the sphere.*[a]

⟨Def. 4⟩ The extremities of the axis are *poles of the sphere.*

⟨Def. 5⟩ A *pole of a circle in a sphere* is a point on the surface of the sphere, from which all the straight lines falling on the circumference of the circle are equal to one another.[b]

⟨Def. 6⟩ A plane is said to *incline similarly* to a plane, one to another, when, in each of the planes, straight lines being produced at an upright to the common section of the planes contain equal angles at the same points.

1 ⟨Theorem⟩

⟨C53⟩

If a spherical surface is cut by some plane, then the ⟨curved⟩ line in the surface of the sphere is the circumference of a circle.

For, let a spherical surface be cut by some plane. Let it make ⟨curved⟩ line *ABG* in the surface of the sphere.

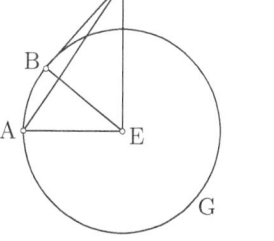

I say that ⟨curved⟩ line *ABG* is the circumference of a circle.

So, if the cutting plane is through the center of the sphere, it is obvious that ⟨curved⟩ line

⟨Case 1⟩

ABG is the circumference of a circle, for the ⟨straight lines⟩ falling from

⟨H4⟩

---

[a] In some of the manuscripts, including the oldest, **A** (1r), this is a definition of a *diameter of the sphere.*

[b] In some of the manuscripts, but not the oldest **A** (1r) which is followed by Heiberg (1927, 2), this definition is stated as asserting what is *called* the pole of the circle.

the center of the sphere to the surface are equal to one another,[1] and ⟨curved⟩ line ABG is in the surface – so that the ⟨straight lines⟩ falling from the center of the sphere to ⟨curved⟩ line ABG are equal to one another. Plane ABG is assumed to be through the center of the sphere, so that ⟨curved⟩ line ABG is the circumference of a circle, whose center is also the same as that of the sphere.[2]

⟨Case 2⟩   But, let the cutting plane not be through the center of the sphere, and let there be the center of the sphere, point D.[3] Let a perpendicular, DE, be produced from point D to the cutting plane,[4] and let it meet the plane at point E. Let straight lines, EA and EB, fall from point E to ⟨curved⟩ line ABG.[5] Let DB and DA be joined.[6]

⟨C54⟩   Since AD is equal to DB,[7] and the squares on AE and ED are equal to the square on AD, and the squares on BE and ED are equal to the square on DB,[8] therefore, the squares on AE and ED are equal to the squares on BE and ED.[9]   Let the square on ED be taken away, as common. Therefore the square on AE, as remainder, is equal to the square on EB.[10] Therefore, AE is equal to EB.[11] Then, similarly, we will show that all ⟨straight lines⟩ falling from point E to ⟨curved⟩ line ABG are equal to one another. Therefore, ⟨curved⟩ line ABG is the circumference of a circle whose center is E.[12]

⟨Corollary⟩

From this it is obvious that, if a circle is in a sphere, the perpendicular produced from the center of the sphere to it will fall at its center.

2 ⟨Problem⟩

Find the center of a given sphere.

Let there be a given sphere.

Then, it is required to find its center.

---

[1] *Sph.* I.def.1, I.def.2.   [2] *Elem.* I.def.15.   [3] A method for finding the center of a sphere is shown in the next proposition, *Sph.* I.2. Here we simply assume that some point is the center. See Commentary.   [4] *Elem.* XI.11.   [5] These are any straight lines, which now specify the two points A and B, as endpoints – which had previously simply been parts of the name of $\mathbf{C}(ABG)$.   [6] Both by *Elem.* I.post.1.   [7] *Sph.* I.def.1, I.def.2.   [8] Both by *Elem.* I.47.   [9] *Elem.* I.c.n.1.   [10] *Elem.* I.c.n.3.   [11] *Elem.* VI.20.cor., but also see Commentary.   [12] *Elem.* I.def.15.

Let it be cut by some plane.[1] Then, it makes a circle as intersection[2]   ⟨H6⟩
– let it make *ABG*. Let its center, *D*, be taken.[3] Let *DE* be erected
at an upright from point *D* in the plane of the circle *ABG*.[4] Let it be
produced and meet the surface of the sphere at *E* and *Z*.[5] Let *EZ* be
bisected at point *H*.[6]

I say that point *H* is the center of the sphere.

In fact not, but, if possible, let it be *Q*.[7] Let   ⟨C55⟩
a perpendicular be produced from *Q* to plane
ABG,[8] and let it meet the plane at point *K*.

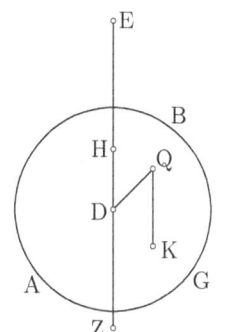

Therefore, point *K* is the center of circle
ABG.[9] But, *D* is as well, which is impossible.[10]

Therefore, point *Q* is not the center of the
sphere. Then, similarly, we will show that it is no
other, except *H*. Therefore, point *H* is the center of the sphere. Which
it was required to show.

⟨Corollary⟩

From this it is obvious that, if a circle is in a sphere and a straight
line is erected at an upright to its plane from its center, the center of
the sphere is on the erected ⟨straight line⟩.

3 ⟨Theorem⟩

A sphere does not contact a non-cutting plane in more points than one.

For, if possible, let a sphere contact a non-cutting plane in more
points, *A* and *B*. Let the center of the sphere, point *G*, be taken.[11]
Let *AG* and *BG* be joined.[12] Let the plane through *AG* and *BG* be
extended.[13] Then, it will make as intersection a circle in the surface of
the sphere and a straight line in the plane.[14] So, let it make circle *DAB*
in the surface of the sphere and straight line *EABZ* in the plane.

---

[1] This is the assumption of the existence of such a plane. See Commentary.   [2] *Sph.* I.1.
[3] *Elem.* III.1.   [4] *Elem.* XI.12.   [5] *Elem.* I.post.2.   [6] *Elem.* I.10.   [7] This is counterfactual. Here, as in *Sph.* I.1, it is simply the assumption that some other point is
the center of the sphere.   [8] *Elem.* XI.11.   [9] *Sph.* I.1.cor.   [10] *Elem.* I.def.15, I.def.16.
[11] *Sph.* I.2.   [12] *Elem.* I.post.1.   [13] *Elem.* XI.2, an unstated construction postulate.
See Commentary.   [14] *Sph.* I.1, *Elem.* XI.3.

⟨C56⟩

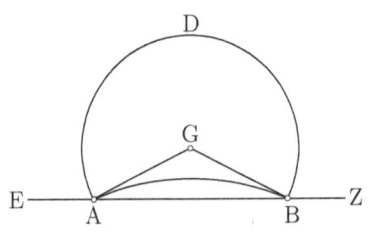

Since the ⟨assumed⟩[a] plane does not cut the sphere, therefore neither does circle *DAB* cut straight line *EABZ*. So, since two arbitrary points, *A* and *B*, are taken on the circumference of a circle, *DAB*, therefore the straight line joining from *A* to *B* falls within circle *DAB*.[1] But, it is also outside,[2] which is impossible.

⟨H8⟩   Therefore, a sphere will not contact a non-cutting plane at more points than one.

4 ⟨Theorem⟩

If a sphere contacts a non-cutting plane, then the straight line joining from the center of the sphere to the ⟨point of⟩ contact is a perpendicular to the plane.

For, let a sphere contact some non-cutting plane at point *A*. Let the center of the sphere be taken as point *B*. Let *BA* be joined.

I say that *BA* is upright to the plane.

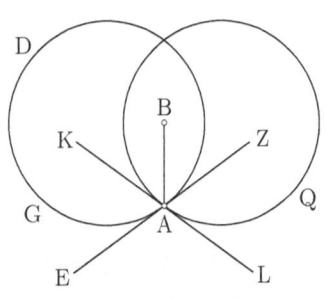

For, let a plane be extended through *BA*.[3] Then, it will make, as intersection, a circle in the surface of the sphere and a straight line in the plane.[4] So, let it make circle *AGD* in the surface of the sphere and straight line *ZE* in the plane. Again, let another plane be extended through *AB*,[5] and let it make, as intersection, circle *AQ* in the surface of the sphere and straight line *KAL* in the ⟨assumed⟩ plane.[6]

⟨C57⟩   Since the plane touches the sphere, therefore straight line *EZ* touches circle *AGD*.[7] So, since some straight line, *EAZ*, touches a circle, *AGD*,

---

[1] *Elem.* III.2.   [2] *Elem.* XI.def.2.   [3] An unstated construction postulate. See Commentary. [4] *Sph.* I.1, *Elem.* XI.3. [5] An unstated construction postulate. See Commentary. [6] *Sph.* I.1, *Elem.* XI.3. [7] *Sph.* I.3, *Elem.* III.def.2.

[a] This adjective is used in *Sph.* I.5.

at point *A*, and *AB* joins from *A* to its center, therefore *AB* is a perpendicular to *EAZ*,[1] and it is obvious that point *B* is the center of circle
AGD, because the plane of circle *AGD* was produced through *BA*, being a radius[a] of the sphere.[2] Then, similarly, we will also show that
BA is a perpendicular to *KAL*. So, since a straight line, *BA*, is set up
upright to two straight lines intersecting one another, *EZ* and *KL*, at
the intersection, and *BA* is at an upright to the plane through them,[3]
but the plane through *EZ* and *KL* is the touching ⟨plane⟩, therefore *AB*
is upright to the plane touching the sphere.

5 ⟨Theorem⟩

If a sphere contacts a non-cutting plane and a straight line is erected at
an upright to the plane from the ⟨point of⟩ contact, then the center of
the sphere will be on the erected ⟨straight line⟩.

For, let a sphere contact a non-cutting plane ⟨H10⟩
at point *A*. Let *AB* be erected from point *A* in
the plane, at an upright.

I say that the center of the sphere is on *AB*.

In fact not, but, if possible, let it be *G*.[4] Let
GA be joined.[5]

So, since a sphere contacts a non-cutting plane ⟨C58⟩
at point *A*, and *GA* is joined from the center of the sphere to the ⟨point
of⟩ contact, therefore *GA* is upright to the plane.[6] But, *BA* is also
upright to the assumed plane. Therefore, from the same point, *A*, there
are two ⟨straight lines⟩ erected, straight lines *AB* and *AG*, at an upright
to the same plane, the assumed, which is impossible.[7]

Therefore, *G* is not the center of the sphere. Then, similarly, we will
show that it is no other, except on *BA*. Therefore, the center of the
sphere is on *BA*.

---

[1] *Elem.* III.18.   [2] *Sph.* I.def.1, I.def.2, *Elem.* I.def.15.   [3] *Elem.* XI.4.   [4] As in *Sph.* I.1
and I.2, this is the counterfactual assumption that some point not on the line is the
center. See Commentary, *Sph.* I.2.   [5] *Elem.* I.post.1.   [6] *Sph.* I.4.   [7] *Elem.* XI.13.

[a] Here and below, literally, "the ⟨distance⟩ from the center." See Commentary.

6 ⟨Theorem⟩

Of the circles in the sphere, those through the center of the sphere are
great, and of the others, those equally distant from the center are equal,
while those more ⟨distant⟩ are lesser.

For, let there be circles in a sphere, AB, GD, and EZ. Let GD be
through the center of the sphere. But, first, let AB and EZ be equally
distant from the center.[1]

I say that GD is great, and AB and EZ are equal.

For, let the center of the sphere, point H, be taken.[2] Therefore, it
is also the center of circle GD.[3] Let perpendiculars, HQ and HK, be
produced from point H to the planes of the circles AB and EZ,[4] and
let them meet the planes of the circles at points K and Q. Therefore,
points Q and K are the centers of AB and EZ.[5] Let straight lines, QL,
HM and KN, be extended out from Q, H and K to circles AB, GD and
EZ.[6] Let HL and HN be joined.[7]

⟨H12,
C59⟩
⟨Part 1⟩

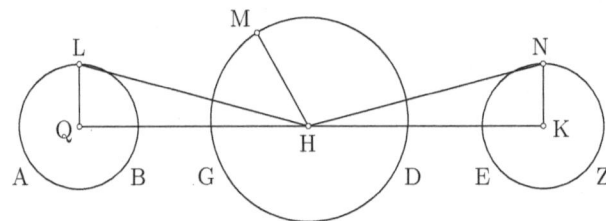

Since HQ is up-
right to the plane of
circle AB, therefore
it will also make an-
gles upright to all
⟨straight lines⟩ con-
tacting it and being in the plane of the circle AB.[8] But QL, being in
the plane AB, contacts it. Therefore, angle LQH is upright. Then, similarly, we will show that angle HKN is upright. Again, since angle LQH
is upright, therefore angle LHQ is less than upright,[9] therefore, angle
LQH is greater than angle LHQ. Therefore, side LH is greater than side
LQ,[10] and LH is equal to HM – through the ⟨fact that⟩ point H is the
center of the sphere, and LH and HM fall from it to the surface of the
sphere.[11] Therefore, HM is greater than LQ. And HM is a radius of

---

[1] This is the *exposition* of Parts 1 and 2, but Part 3 also shares the first part of
this exposition.   [2] *Sph.* I.2.   [3] *Sph.* I.def.1, I.def.2, *Elem.* I.def.15.   [4] *Elem.* XI.11
[5] *Sph.* I.1.cor.   [6] Arbitrary straight lines are joined from the centers to the circumference. See Commentary.   [7] *Elem.* I.post.1.   [8] *Elem.* XI.def.3.   [9] *Elem.* I.17.
[10] *Elem.* I.18, in **T**(LQH).   [11] *Sph.* I.def.1, I.def.2.

circle *GD*, and *QL* a radius of circle *AB*, therefore circle *GD* is greater than circle *AB*.[1] Then, similarly, we will show that also of all circles in the sphere that are not also through the center of the sphere.

I say also that circles *AB* and *EZ* are equal.   ⟨Part 2⟩

For, since circles *AB* and *EZ* are equally distant from the center of the sphere, therefore *HQ* is equal to *HK*.[2] Since point *H* is the center of the sphere, *HL* is equal to *HN*.[3] Therefore, the square on *LH* is equal to the square on *HN*.[4] But, the squares on *LQ* and *QH* are equal to the square on *LH*, and the squares on *NK* and *KH* are equal to the square on *HN*.[5] Therefore, the squares on *LQ* and *QH* are equal to the squares on *NK* and *KH*,[6] of which the square on *QH* is equal to the square on   ⟨C60⟩ *HK*. Therefore, the square on *QL*, as remainder, is equal to the square on *KN*, as remainder.[7] Therefore, *QL* is equal to *KN*.[8] And *QL* is a radius of circle *AB*, and *NK* is a radius of circle *EZ*. Therefore, the radius of circle *AB* is equal to the radius of circle *EZ*.[9]

But again, let circle *AB* be more distant from the center of the   ⟨Part 3⟩ sphere than *EZ*.

I say that circle *AB* is less than circle *EZ*.

For, with the same things having been constructed, since circle *AB* is more distant from the center of the sphere than *EZ*, therefore *QH* is greater than *KH*.[10] Since *HL* is equal to *HN* – for point *H* is the center of the sphere and *L* and *N* are on its surface[11] – therefore, the square on *HL*, that is to say the squares on *LQ* and *QH*,[12] is equal to the squares   ⟨H14⟩ on *NK* and *KH*,[13] of which the square on *QH* is greater than the square on *KH*.[14] Therefore, the square on *LQ*, as remainder, is less than the square on *NK*, as remainder.[15] Therefore, *LQ* is less than *NK*.[16] And

---

[1] *Elem.* XII.2, VI.20.cor.   [2] *Elem.* III.def.4 (generalized to the sphere). [3] *Sph.* I.def.1, I.def.2.   [4] *Elem.* VI.20.cor. See Commentary, *Sph.* I.1.   [5] Both by *Elem.* I.47.   [6] *Elem.* I.c.n.1.   [7] *Elem.* I.c.n.3.   [8] *Elem.* VI.20.cor. See Commentary, *Sph.* I.1.   [9] Hence, by *Elem.* III.def.1, the circles are equal.   [10] *Elem.* III.def.5 (generalized to the sphere).   [11] *Sph.* I.def.1, I.def.2.   [12] *Elem.* I.47.   [13] *Elem.* I.c.n.1, I.47.   [14] *Elem.* VI.20.cor.   [15] Since $\mathbf{S}(LQ) + \mathbf{S}(QH) = \mathbf{S}(NK) + \mathbf{S}(KH)$, while $\mathbf{S}(QH) > \mathbf{S}(KH)$.   [16] *Elem.* VI.20.cor.

QL is a radius of *AB*, and *NK* is a radius of circle *EZ*. Therefore, circle *AB* is less than circle *EZ*.[1]

Therefore, of the ⟨circles⟩ in the sphere, those through the center of the sphere are great, and of the others, those equally distant from the center are equal, while those more ⟨distant⟩ are lesser.

⟨C61⟩   7 ⟨Theorem⟩

If a circle is in a sphere and some straight line joins from the center of the sphere to its center, then the joining ⟨straight line⟩ is upright to the circle.

Let there be a circle in a sphere, *ABGD*. Let there be the center of the sphere, point *E*, and the center of the circle, *Z*. Let *EZ* be joined.

I say that *EZ* is upright to circle *ABGD*.

For, let ⟨straight lines⟩ *AZG* and *BZD* be produced through the center of the circle.[2] Let *BE* and *ED* be joined.[3]

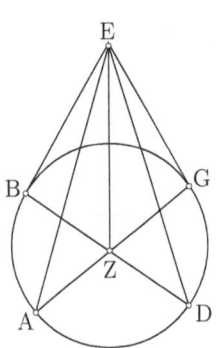

Since *BZ* is equal to *ZD*,[4] while *ZE* is common, then the two *BZ* and *ZE* are respectively equal to the two *DZ* and *ZE*, and *BE*, as base, is equal to *ED*, as base[5] – for point *E* is the center of the sphere and *B* and *D* are on its surface[6] – therefore, angle *BZE* is equal to angle *DZE*.[7] But, when a straight line on a straight line makes the adjacent angles equal to one another, each of the equal angles is upright.[8] Therefore, each of the angles *BZE* and *DZE* is upright. Therefore, *EZ* is at an upright to *BD*. Then, similarly, we will show also that ⟨*EZ* is at an upright⟩ to *AG*. So, since a straight line, *EZ*, is set up at an upright to two straight lines, *AG* and *BD*, cutting one another, at the intersection, it is upright to the plane through *AG* and *BD*.[9] But, the plane through *AG* and *BD* is circle *ABGD*. Therefore, *EZ* is upright to the plane of circle *ABGD*.

---

[1] *Elem.* III.def.1.   [2] Production of a straight line through an arbitrary point. See Commentary.   [3] *Elem.* I.post.1.   [4] *Elem.* I.def.15.   [5] That is, in **T**(*BZE*) and **T**(*DZE*). See Commentary.   [6] *Sph.* I.def.1, I.def.2.   [7] *Elem.* I.8.   [8] *Elem.* I.def.10.   [9] *Elem.* XI.4.

8 ⟨Theorem⟩ ⟨H16,

If a circle is in a sphere and from the center of the sphere a perpendicular C62⟩

is produced to it and extended in both directions, then it will fall on the

poles of the circle.

Let there be a circle in a sphere, *ABG*. Let the center of the sphere, point *D*, be taken. Let a perpendicular, *DE*, be produced from point *D* to the plane of the circle *ABG*, and let it meet the plane at point *E*. Therefore, point *E* is the center of circle *ABG*.[1] Let *DE* be produced in both directions, and let it meet the surface of the sphere at points *Z* and *H*.

I say that points *Z* and *H* are poles of circle *AB*.

For, let *AEG* and *BEQ* be produced through ⟨circle *ABG*⟩.[2] Let *AZ, ZG, AH,* and *HG* be joined.[3]

Since *ZE* is upright to circle *ABG* – and, therefore, it will also make angles upright to all straight lines contacting it and and being in the plane of the circle *ABG*[4] – therefore, each of the angles *ZEA, ZEG, ZEB,* and *ZEQ* is upright. Since *AE* is equal to *EG*,[5] while *EZ* is common and at an upright, therefore *AZ*, as base, is equal to *GZ*, as base.[6] Then,

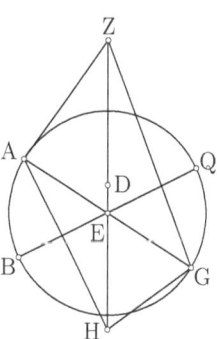

similarly, we will show also that all straight lines falling from point *Z* to ⟨curved⟩ line *ABG* are equal to one another. Therefore, point *Z* is a pole of circle *ABG*.[7] Then, similarly, we will show also that point *H* ⟨is a pole⟩. Therefore, points *Z* and *H* are poles of circle *ABG*.

Therefore, if a circle is in a sphere, and from the center of the sphere, and so on.

---

[1] *Sph.* I.1.cor.   [2] Production of a straight line through an arbitrary point. See Commentary, *Sph.* I.7.   [3] *Elem.* I.post.1.   [4] *Elem.* XI.def.3.   [5] *Elem.* I.def.15.   [6] *Elem.* I.4, in **T**(*AEZ*) and **T**(*GEZ*).   [7] *Sph.* I.def.5.

⟨C63⟩

9 ⟨Theorem⟩

If a circle is in a sphere and from either of its poles a perpendicular
to it, a straight line, is produced, then it will fall on the center of the
circle, and being extended out it will fall on the other pole of the circle.

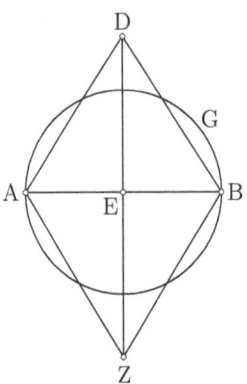

Let there be a circle in a sphere, *ABG*. Let
a perpendicular to it, *DE,* be produced from
either of its poles, *D.* Let it meet the plane
of the circle at point *E.* Let *DE* be extended
and meet the surface of the sphere at point *Z.*

I say that point *E* is the center of circle
AB, and *Z* is the other pole of the circle.

⟨H18⟩

⟨Part 1⟩

For, let *AE* and *EB* be produced through
from *E.*[1] Let *AD, DB, AZ,* and *ZB* be joined.[2]

Since *DE* is upright to the circle *ABG,* therefore it will also make upright angles to all straight lines contacting it and being in circle *ABG.*[3]
And each of *AE* and *EB* contacts it, being in the plane of *ABG.* Therefore, each of the angles *DEA* and *DEB* is upright. And, since *AD* is
equal to *DB,*[4] therefore the square on *AD* is equal to the square on *DB.*[5]
But, the squares on *DE* and *EA* are equal to the square on *AD,* and the
squares on *DE* and *EB* are equal to the square on *DB.*[6] Therefore, the
squares on *AE* and *ED* are equal to the squares on *BE* and *ED.*[7] Let the
square on *DE* be taken away, as common. Therefore, the square on *AE,*
as remainder, is equal to the square on *EB,* as remainder.[8] Therefore,
AE is equal to *EB.*[9] Similarly, we will also show that all straight lines
falling from point *E* to the ⟨curved⟩ line *ABG* are equal. Therefore,
point *E* is the center of the circle *ABG.*[10]

⟨Part 2⟩

Then, I say that point *Z* is the other pole of circle *ABG.*

⟨C64⟩

For, since *AE* is equal to *EB,*[11] while *EZ* is common and at an
upright, therefore *AZ,* as base, is equal to *BZ,* as base.[12] Then, similarly,

---

[1] Production of a straight line through an arbitrary point. See Commentary, *Sph.* I.7.
[2] *Elem.* I.post.1 (four times).   [3] *Elem.* XI.def.3.   [4] *Sph.* I.def.5.   [5] *Elem.* VI.20.cor.
See Commentary, *Sph.* I.1.   [6] Both by *Elem.* I.47.   [7] *Elem.* I.c.n.1.   [8] *Elem.* I.c.n.3.
[9] *Elem.* VI.20.cor.   See Commentary, *Sph.* I.1.   [10] *Elem.* I.def.15.   [11] Part 1.
[12] *Elem.* I.4, in **T**(*AEZ*) and **T**(*BEZ*).

we will also show that all straight lines falling from point Z to the ⟨curved⟩ line ABG are equal. Therefore, point Z is a pole of circle ABG.[1]

But, it was also shown that E is the center.

Therefore, point E is the center of ABG and Z is the other pole.

10 ⟨Theorem⟩

If a circle is in a sphere, then the straight line produced through its poles is upright to the circle and will go through its center and also the sphere's.

Let there be a circle in a sphere, $ABGD$. Let there be its poles, points E and Z. Let the ⟨straight line⟩ produced through the poles be joined, and let it be EZ.

I say that EZ is upright to circle $ABGD$, and it will go through its center and also the sphere's.

For, let EZ meet the plane of circle $ABGD$ at point H.[2] Let AHG ⟨Part 1⟩
and BHD be produced through from point H.[3] Let BE, ED, BZ, and ZD be joined.[4]

Since BE is equal to ED,[5] while EZ is common, then the two BE and EZ are respectively equal to the two ZE and ED, and BZ, as base,   ⟨H20⟩
is equal to ZD, as base,[6] therefore angle BEZ is equal to angle DEZ.[7] Again, since BE is equal to ED, while EH is common, then the two BE and EH are respectively equal to the two HE and ED, and angle BEH is equal to angle DEH, therefore BH, as base, is equal to HD, as base – and triangle EBH is equal to triangle EDH, and the remaining angles   ⟨C65⟩
will be equal to the remaining angles, which the equal sides subtend.[8] Therefore, angle BHE is equal to angle DHE. Therefore, each of angles BHE and DHE is upright.[9] Therefore, EH is at an upright to BD. Then, similarly, we will show that EH is at an upright to AG, and to

---

[1] *Sph.* I.def.5.   [2] This simply names the meeting point.   [3] Production of a straight line through an arbitrary point, twice. See Commentary, *Sph.* I.7.   [4] *Elem.* I.post.1.
[5] *Sph.* I.def.5.   [6] *Sph.* I.def.5.   [7] *Elem.* I.8, in $\mathbf{T}(BEZ)$ and $\mathbf{T}(DEZ)$.   [8] *Elem.* I.4, in $\mathbf{T}(EBH)$ and $\mathbf{T}(EDH)$.   [9] *Elem.* I.def.10.

the plane through BD and AG^1 – that is to say, EHZ is at an upright to circle $ABGD$. Therefore, EZ is upright to circle $ABGD$.

⟨Part 2⟩

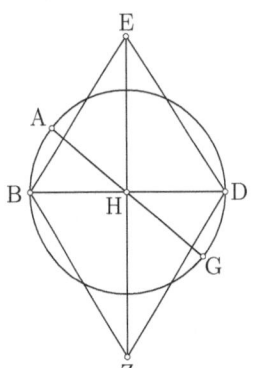

Then, I say that it will also pass through its center and the sphere's.

For, since a circle, $ABGD$, is in a sphere, while a perpendicular, EH, is produced from either of its poles to it,[2] and is meeting the plane at point H, therefore H is the center of circle $ABGD$.[3]

⟨Part 3⟩

Then, I say that it is also through the center of the sphere.

For, since there is a circle in a sphere, $ABGD$, and EHZ is erected upright to the plane of the circle from its center H,[4] therefore the center of the sphere is on EHZ.[5] Therefore, EZ is through the center of the sphere.

Therefore, EZ is upright to circle $ABGD$ and through its center and also the sphere's.

⟨C66⟩

11 ⟨Theorem⟩

Great circles in a sphere bisect one another.

For, let two great circles in a sphere, AB and GD, cut one another at points E and Z.

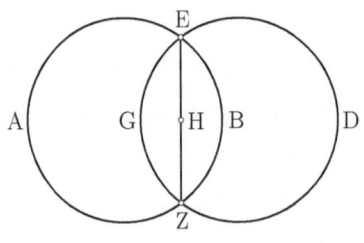

I say that circles AB and GD bisect one another.

For, let their center be taken, and let it be point H.[6]

It is also the ⟨center⟩ of the sphere.[7,a] Let EH and HZ be joined.[8]

---

[1] *Elem.* XI.4.   [2] Part 1.   [3] *Sph.* I.9.   [4] Part 1.   [5] *Sph.* I.2.cor.   [6] *Elem.* III.1.
[7] *Sph.* I.6 (converse).   [8] *Elem.* I.post.1.

[a] This phrase is thought to be a later interpolation, since a converse to *Sph.* I.6 is included in the later manuscripts, but is found in neither the earliest Greek manuscript, **A**, nor Thābit's Arabic version (Kunitzsch and Lorch 2010, 28–29). See Ver Eecke (1927, 10–11, 17 n. 5). Notice that this claim plays no role in the proof.

Since points *E*, *H*, and *Z* are in plane *AB*, while they are also in   ⟨H22⟩
⟨plane⟩ *GD*, therefore points *E*, *H*, and *Z* are in both the planes of circles
AB and *GD*.[1] Therefore, they are on their common section.[2] But,
the common section of every two planes is a straight line.[3] Therefore,
EHZ is a straight line. And since point *H* is the center of circle *AB*,
therefore *EZ* is a diameter of it.[4] Therefore, each of *EAZ* and *EBZ* is a
semicircle.[5] Again, since point *H* is the center of circle *GD*, therefore *EZ*
is a diameter of it.[6] Therefore, each of *EGZ* and *EDZ* is a semicircle.[7]
Therefore, circles *AB* and *GD* bisect one another.

12 ⟨Theorem⟩   ⟨C67⟩

Circles in a sphere that bisect one another are great.

For, let two circles in a sphere, *AB* and *GD*, bisect one another at
points *E* and *Z*.

I say that *AB* and *GD* are great circles.

For, let *EZ* be joined.[8] Therefore, *EZ* is a diameter of circles *AB*
and *GD*.[9] Then, I say, it is also ⟨a diameter⟩ of the sphere. Let *EZ* be
bisected at *H*.[10] Therefore, point *H* is the center of the circles.[11] Let
HQ be erected, from point *H*, at an upright to the plane of circle *GD*,
and *HK* ⟨erected⟩ at an upright to ⟨circle⟩ *AB*.[12]

Since *GD* is a circle in a sphere,
and *HQ* is erected from its center at
an upright to the plane of the circle,
therefore the center of the sphere is on
HQ.[13] Then, similarly, we will show
that it is on *HK*. Therefore, the com-
mon section of *HQ* and *HK* is the center of the sphere.[14] But the common section is point *H*, therefore point *H* is the center of the sphere.

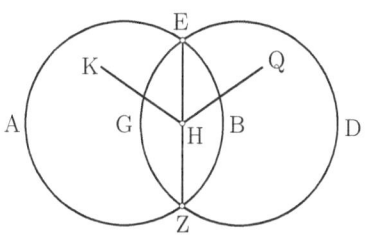

---

[1] They are two points on the circumference, and the center, of the two
circles. [2] Since each of the three points belongs to each of the two planes.
[3] *Elem* XI.3. [4] *Elem* I.def.17. [5] *Elem.* I.def.18. [6] *Elem.* I.def.17. [7] *Elem.* I.def.18.
[8] *Elem.* I.post.1. [9] *Elem.* I.def.17. [10] *Elem.* I.10. [11] *Elem.* I.def.15, I.def.16, I.def.17.
[12] Both by *Elem.* XI.12. [13] *Sph.* I.2.cor. [14] Since the center must be on each line.

But point *H* is also the center of *AB* and *GD*, and those circles, being about the same center as the sphere, are great.[1]

Therefore, circles in a sphere that bisect one another are great.

⟨H24,   13 ⟨Theorem⟩

C68⟩   *If a great circle in a sphere cuts some circle of those in the sphere at an*

upright, then it will bisect it and be through the poles.

For, let a great circle in a sphere, *ABGD*, cut some circle of those in the sphere, *EBZD*, at an upright.

I say that it will bisect it and be through the poles.

⟨Part 1⟩

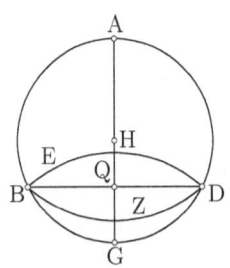

For, let their common section, *BD*, be joined.[2] Let the center of circle *ABGD*, point *H*, be taken.[3] It is also the center of the sphere.[4] Let a perpendicular, *HQ*, be produced from point *H* to *BD*,[5] and let it be extended in both directions,[6] meeting the surface of the sphere at points *A* and *G*.

Since two planes – that of circle *ABGD* and that of *EBZD* – are upright to one another, and *QA* is produced at an upright to their common section, *BD*, being in one of the planes, *ABGD*, therefore *AG* is at an upright to the plane *EBZD*.[7] So, since *EBZD* is a circle in a sphere, and a perpendicular to it, *HQ*, is produced from the center of the sphere and intersects the plane of circle *EBZD* at point *Q*, therefore point *Q* is the center of the circle *EBZD*.[8] Therefore, each of *BED* and *BZD* is a semicircle.[9] Therefore, circle *ABGD* bisects circle *EBZD*.

⟨C69⟩   Then, I say that it is also through its poles.

⟨Part 2⟩   For, since *EBZD* is a circle in a sphere, and a perpendicular to it, *HQ*, is produced from the center of the sphere, and it is extended in both directions, and it intersects the surface of the sphere at *A* and *G* – while, if a circle is in a sphere and from the center of the sphere a perpendicular is produced to it and extended in both directions, it will

---

[1] *Sph.* I.6.   [2] *Elem.* XI.3. See Commentary.   [3] *Elem.* III.1.   [4] *Sph.* I.6.   [5] *Elem.* I.12.
[6] *Elem.* I.post.2.   [7] *Elem.* XI.4.   [8] *Sph.* I.1.cor.   [9] *Elem.* I.def.18.

fall on the poles of the circle[1] – therefore points *A* and *G* are the poles of circle *EBZD*. Therefore, circle *ABGD* cuts circle *EBZD* through the poles.

But, it also bisects it.[2]

Therefore, circle *ABGD* both bisects circle *EBZD* and is through the poles.

14 ⟨Theorem⟩   ⟨H26⟩

If a great circle in a sphere bisects some circle of those the sphere, not being great, then it will both cut it at an upright and be through the poles.

For, let a great circle in a sphere, *ABGD*, bisect some circle of those in the sphere, not being great, *EBZD*,

I say that it both cuts it at an upright and is through the poles.

For, let their common section, *BD*, be joined.[3]   ⟨Part 1⟩

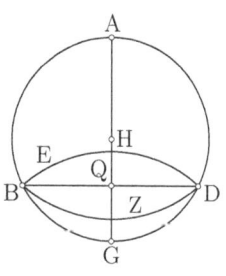

Since circle *ABGD* bisects circle *EBZD*, therefore each of *BED* and *BZD* is a semi-circle.[4] Therefore, *BD* is a diameter of circle *BZDE*.[5] So, let *BD* be bisected at point *Q*.[6] Therefore, point *Q* is ⟨the⟩ center of circle *EBZD*.[7] Let the center of circle *ABGD* be taken as point *H*.[8] But,   ⟨C70⟩
it is also ⟨the center⟩ of the sphere.[9] And *HQ* being joined, let it be extended in each direction and intersect the surface of the sphere at points *A* and *G*.[10]

Since *EBZD* is a circle in a sphere, and a straight line, *HQ*, is joined through the center of the sphere to its center, therefore *HQ* is upright to *EBZD*.[11] Therefore, all planes through *HQ* are upright to *EBZD*.[12] But, circle *ABGD* is one of the planes through *HQ*. Therefore, *ABGD* is upright to *EBZD*. Therefore, circle *ABGD* cuts circle *EBZD* at an upright.

---

[1] *Sph.* I.8.   [2] Part 1.   [3] *Elem.* XI.3. See Commentary to *Sph.* I.13.   [4] *Elem.* I.def.18.
[5] *Elem.* I.def.17.   [6] *Elem.* I.10.   [7] *Elem.* I.def.16.   [8] *Elem.* III.1.   [9] *Sph.* I.6.
[10] *Elem.* I.post.1, I.post.2.   [11] *Sph.* I.7.   [12] *Elem.* XI.18.

⟨Part 2⟩   Then, I say that it is also through the poles.

For, since *EBZD* is a circle in a sphere, and a perpendicular, *HQ*, is produced from the center of the sphere to it, and is extended in both directions and intersects the surface of the sphere at points *A* and *G*, therefore points *A* and *G* are poles of *EBZD*.[1] Therefore, circle *ABGD* cuts circle *EBZD* through the poles.

But it also cuts it at an upright.[2]

Therefore, circle *ABGD* cuts circle *EBZD* at an upright, and through the poles.

⟨H28,   15 ⟨Theorem⟩

C71⟩   *If a great circle in a sphere cuts some circle of those in the sphere through the poles, then it will bisect it at an upright.*[a]

For, let a great circle in a sphere, *ABGD*, cut some circle of those in the sphere, *EBZD*, through the poles.

I say that it bisects it at an upright.

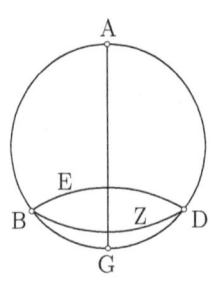

For, let there be the poles of circle *EBZD*, points *A* and *G*[3] – it is obvious that they are on circle *ABGD*, for *ABGD* cuts circle *EBZD* through the poles. Let *AG* be joined.[4]

Since *EBZD* is a circle in a sphere, and a straight line, *AG*, is produced between the poles – while, if a circle is in a sphere, the straight line produced through its poles is upright to the circle, and will go through its center also the sphere's[5] – therefore *AG* is upright to circle *EBZD*. Therefore, all planes through *AG* are upright to *EBZD*.[6] But, circle *ABGD* is one of the planes through *AG*. Therefore, circle *ABGD* is upright to circle *EBZD*. Therefore, circle *ABGD* cuts *EBZD* at upright angles. And it bisects it.[7] Therefore, circle *ABGD* bisects circle *EBZD* at an upright.

---

[1] *Sph.* I.8.   [2] Part 1.   [3] See Commentary.   [4] *Elem.* I.post.1.   [5] *Sph.* I.10.
[6] *Elem.* XI.18.   [7] *Sph.* I.13.

[a] Here and elsewhere, literally, "cuts it in two and at an upright."

16 ⟨Theorem⟩ ⟨C72⟩

If a great circle is in a sphere, then the polar distance is equal to the side of the square inscribed in the great circle.[a]

Let there be a great circle in a sphere, *ABGD*.

I say that its polar distance is equal to the side of the square inscribed in the great circle.

For, let two diameters of circle *ABGD*, *AG* and *BD*, have been produced at an upright to one another.[1] From point *E*, let ⟨straight line⟩ *EZ* be erected at an upright to the plane of circle *ABGD*,[2] and let it meet the surface of the sphere at point *Z*. Therefore, point *Z* is a pole of circle *ABGD*.[3] Let *BA* and *ZA* be joined.[4] Therefore, *BA* is a side of a square inscribed in circle *ABGD*,[5] and *ZA* is the polar distance.[6]

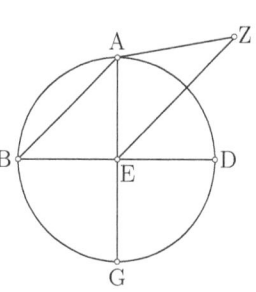

⟨H30⟩

I say that *ZA* is equal to *AB*.

For, since *ZE* is upright to *ABGD* – and, therefore, will make angles upright to all ⟨straight lines⟩ contacting it and being in the plane of circle *ABGD*[7] – therefore, *ZE* is upright to each of *AE*, *BE*, *GE*, and *DE*. And since point *E* is the center of the sphere,[8] *BE* is equal to *EZ*,[9] while *EA* is common.[10] Then, the two *BE* and *EA* are respectively equal to the two *AE* and *EZ*, and angle *BEA*, as upright, is equal to angle *AEZ*, as upright, therefore, *BA*, as base, is equal to *AZ*, as base.[11] And *AZ* is the polar distance of circle *ABGD*, while *BA* is a side of the square inscribed in the great circle. Therefore, the polar distance of circle *ABGD* is equal to the side of the square inscribed in the great circle.

---

[1] *Elem.* III.1, producing a line through a given point, *Elem.* I.11. See Commentary.
[2] *Elem.* XI.12. [3] *Sph.* I.2.cor., I.8. [4] *Elem.* I.post.1. [5] *Elem.* I.def.22, IV.6.
[6] *Sph.* I.def.5. [7] *Elem.* XI.def.3. [8] *Sph.* I.6. [9] *Sph.* I.def.1, I.def.2. [10] That is, in **T**(*AEB*) and **T**(*AEZ*). [11] *Elem.* I.4.

[a] For a discussion of the new terminology see Commentary.

⟨C73⟩

17 ⟨Theorem⟩

If a circle is in a sphere and its polar distance is equal to the side of the square inscribed in the great circle, then it will be great.

Let there be a circle in a sphere, *ABG*, and its pole, point *D*. Let its polar distance, *DG*, be equal to the side of the square inscribed in the great circle.

I say that *ABG* is great.

For, let a plane be extended through *DG* and the center of the sphere.[1] It will make as intersection some great circle in the surface of the sphere,[2] let it make *BDGE*. Let there be their common section, *BG*.[3] Let *DB* be joined.[4]

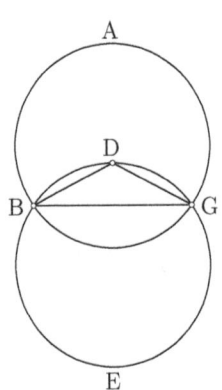

Therefore, *DG* is equal to *DB*,[5] and therefore each of *DG* and *DB* is equal to the side of the square inscribed in the great circle.[6] Therefore, *BDG* is the circumference of a semicircle.[7] Therefore, *BG* is a diameter of ⟨great⟩ circle *DE*.[8] And since point *D* is a pole of circle *ABG*, therefore circle *DBEG* cuts circle *ABG* through the poles. So, since a great circle in a sphere, *DE*, cuts some circle of those in the sphere, *ABG*, through the poles, it will bisect it at an upright.[9] And *BG* is their common section. Therefore, *BG* is a diameter of circle *ABG*,[10] and also of the sphere.[11] Therefore, circle *ABG* is great.[12]

⟨H32⟩

⟨C74⟩

18 ⟨Problem⟩

Set out the diameter of a given circle in a sphere.

Let there be a given circle in a sphere, *ABG*.

Then, it is required to set out the diameter of circle *ABG*.

Let arbitrary points, *A*, *B*, and *G*, be taken on the circumference of the circle.[13] Let a triangle, *DEZ*, be put together from three straight

---

[1] *Elem.* XI.2. See Commentary.   [2] *Sph.* I.1, I.6.   [3] *Elem.* XI.3. See Commentary to *Sph.* I.13.   [4] *Elem.* I.post.1.   [5] *Sph.* I.def.5.   [6] *Sph.* I.16.   [7] *Elem.* III.31, I.def.18.
[8] *Elem.* I.def.18.   [9] *Sph.* I.15.   [10] *Elem.* I.def.17.   [11] *Sph.* I.def.3 (alternate text).
[12] *Sph.* I.6.   [13] See Commentary.

lines, so that *DE* is equal to the ⟨straight line⟩ from *A* to *B*, *DZ* is ⟨equal⟩ to the ⟨straight line⟩ from *A* to *G*, and *EZ* is ⟨equal⟩ to the ⟨straight line⟩ from *B* to *G*.[1] Let *EH* be produced at an upright to *ED* from point *E*, and *ZH* produced at an upright to *DZ* from *Z*.[2] Let *DH* be joined.[3] Then, let diameter *AQ* of circle *ABG* be produced.[4] Let *AB*, *BG*, *GA*, and *GQ* be joined.[5]

So, since the two *AB* and *BG* are respectively equal to the two *DE* and *EZ*, and *AG*, as base, is equal to *DZ*, as base, therefore angle *ABG* is equal to angle *DEZ*.[6] But,

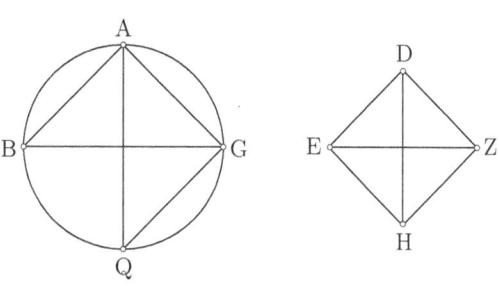

angle *ABG* is equal to angle *AQG*,[7] while angle *DEZ* is equal to angle *DHZ*,[8] therefore, angle *AQG* is equal to angle *DHZ*.[9] But, angle *AGQ*, as upright, is equal to angle *DZH*, as upright,[10] and *AG* is equal to *DZ*, therefore *AQ* is equal to *DH*.[11] And *AQ* is a diameter of the circle. Therefore, *DH* is equal to the diameter of the circle.

⟨C75⟩

19 ⟨Problem⟩

⟨H34⟩

Set out the diameter of a given sphere.

For, let the sphere whose diameter it is necessary to set out be imagined. Let two arbitrary points, *A* and *B*, be taken on the surface of the sphere,[12] and, with a pole, *A*, and a distance, *AB*, let a circle, *BGD*, be drawn.[13] Therefore, it is possible to set out the diameter of circle *DBG*.[14] So, from three straight lines – two equal to the polar distance of ⟨circle⟩ *BGD*[15] and one ⟨equal⟩ to the diameter – let a triangle, *EZH*, be put together so that each of *ZE* and *EH* is equal to

---

[1] *Elem.* I.post.3 (transfer), I.22. See Commentary.　[2] Both by *Elem.* I.12. That is, *H* is the point of intersection of the two perpendiculars, *EH* ⊥ *DE* and *ZH* ⊥ *DZ*. [3] *Elem.* I.post.1.　[4] *Elem.* III.1, I.post.1, I.post.2.　[5] *Elem.* I.post.1.　[6] *Elem.* I.8, in **T**(*ABG*) and **T**(*DEZ*). [7] *Elem.* III.21.　[8] *Elem.* III.22, III.21.　[9] *Elem.* I.c.n.1. [10] *Elem.* III.21, I.post.4.　[11] *Elem.* I.26.　[12] See Commentary.　[13] *Elem.* I.post.3 (spherical). See Commentary.　[14] *Sph.* I.18.　[15] *Elem.* I.post.3 (transfer).

⟨C76⟩ the polar distance from the pole, A,[1] and ZH to the diameter ⟨of circle DBG⟩.[2] Let straight lines ZQ and HQ be produced through Z and H at an upright angle to straight lines ZE and EH.[3] Let EQ be joined.[4]

I say that EQ is equal to the diameter of the sphere.

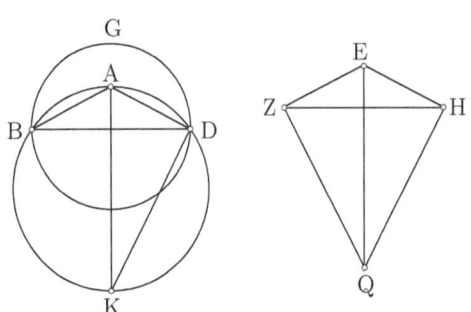

For, let the diameter of the sphere, AK, be imagined,[5] and through AK a plane.[a] It will make as intersection a great circle,[6] and let it make ADB.[7] Let AD, AB, BD, and DK be joined.[8]

Since the two AB and BD, are respectively equal to the two EZ and ZH, and AD, as base, is equal to EH, as base, therefore angle ABD is equal to angle EZH.[9] But, angle ABD is equal to angle AKD,[10] while angle EZH is equal to angle EQH,[11] while upright angle ADK is equal to upright angle EHQ.[12] Then, the two triangles AKD and EQH, having the two angles ADK and DKA respectively equal to the two angles EHQ and HQE and having one side, AD, equal to one side, EH, subtended by one of the equal angles, will therefore have the remaining sides respectively equal to the remaining sides.[13] Therefore, AK is equal to EQ. And, AK is the diameter of the sphere. Therefore, EQ is equal to the diameter of the sphere.

---

[1] Literally, "the ⟨distance⟩ from the pole, A."   [2] *Elem.* I.22.   [3] *Elem.* I.12. That is, Q is the point of intersection of the two perpendiculars, $ZQ \perp ZE$ and $HQ \perp EH$.   [4] *Elem.* I.post.1.   [5] *Sph.* I.2, *Elem.* I.post.1, I.post.2. See Commentary. [6] *Sph.* I.1, I.6.   [7] *Elem.* XI.2 (by implication). See Commentary.   [8] *Elem.* I.post.1. [9] *Elem* I.8, in **T**(ABD) and **T**(DEH).   [10] *Elem.* III.21.   [11] *Elem.* III.22, III.21. [12] *Elem.* I.post.4.   [13] *Elem.* I.26.

[a] For the terminology of imagining, see Commentary.

20 ⟨Problem⟩   ⟨H36,

Draw a great circle through two given points that are on a spherical   C77⟩
surface.

Let there be two given points that are on a spherical surface, *A*
and *B*.

Then, it is required to draw a great circle through points *A* and *B*.

So, if points *A* and *B* are along a diameter of the sphere, it is obvious   ⟨Case 1⟩
that unlimited great circles will be drawn through points *A* and *B*.[1]

Then, let points *A* and *B* not be along   ⟨Case 2⟩
a diameter of the sphere. With a pole, *A*,
and a distance as the side of the square
inscribed in the great circle, let a circle, *GDE*, be drawn.[2] Therefore, *GDE*
is great, for its polar distance is equal
to the side of the square inscribed in the
great circle.[3] Again, with a pole, *B*, and
a distance as the side of the square in-

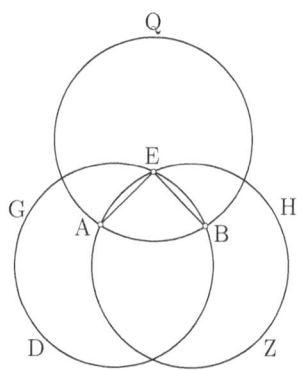

scribed in the great circle, let a circle, *ZEH*, be drawn.[4] Therefore,
circle *ZEH* is great, for its polar distance is equal to the side of the
square inscribed in the great circle.[5] Let straight lines, *EA* and *EB*,
be joined from point *E* to *A* and *B*.[6] Therefore, each of *AE* and *EB* is   ⟨C78⟩
equal to the side of the square ⟨inscribed⟩ in the great circle.[7] Therefore, *EA* is equal to *EB*.[8] Therefore, the circle drawn with a pole, *E*,
and a distance, *EB*, will also pass through *A* – through the ⟨fact that⟩
EA is equal to *EB*.[9] Let it go,[10] and let it be as *BAQ*. Therefore, circle
ABQ is great, for its polar distance is equal to the side of the square
inscribed in the great circle.[11]

Therefore, a great circle, *ABQ*, is drawn through two given points,
A and *B*, that are on a spherical surface.

---

[1] *Elem.* I.def.17, *Sph.* I.6.   [2] *Sph.* I.19, *Elem.* I.10, I.post.3, IV.6, I.post.3 (transfer, spherical). See Commentary.   [3] *Sph.* I.17.   [4] *Sph.* I.19, *Elem.* I.10, I.post.3,
IV.6, I.post.3 (transfer, spherical). See Commentary.   [5] *Sph.* I.17.   [6] *Elem.* I.post.1.
[7] *Sph.* I.16.   [8] *Elem.* I.c.n.1.   [9] *Sph.* I.def.5.   [10] *Elem.* I.post.3 (spherical).
[11] *Sph.* I.17.

21 ⟨Problem⟩

Find the pole of a given circle in a sphere.

Let there be a given circle in a sphere, *ABG*.

Then, it is required to find the pole of circle *ABG*.

⟨H38⟩   For, let some arbitrary point, *A*, be taken on its circumference.[1] Let two equal circumferences, *DA* and *AE*, be cut off.[2] Let the remaining, *DE*, be bisected at point *Z*.[3]

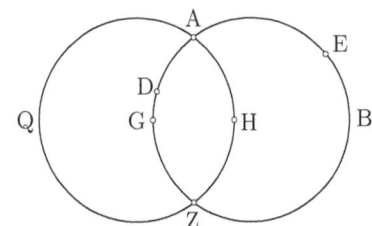

Then, either circle *ABG* is great, or not.

⟨Case 1⟩

First, let it not be great, and through two given points, *Z* and *A*, that are on the spherical surface, let a great circle, *ZAQ*, be drawn.[4]

⟨C79⟩   Since circumference *DA* is equal to circumference *AE*, while *DZ* is equal to *ZE*, therefore *ADZ*, as a whole, is equal to *AEZ*, as a whole.[5] Therefore, circle *ZAQ* bisects circle *ABG*.[6] So, since, a great circle in a sphere, *AZQ*, bisects some one of those in the sphere that are not great, it both cuts it at an upright and through the poles.[7] Therefore, ⟨circle⟩ *ZQA* both cuts ⟨circle⟩ *ABG* at an upright and through the poles.

Let ⟨circumference⟩ *ZA*[8] be bisected at point *H*.[9] Therefore, point *H* is a pole of circle *ABG*.[10]

⟨Case 2⟩   But, then again let circle *ABG* be great.

Then, similarly, we will show that circumference *ADZ* is equal to circumference *AEZ*.[11] And let circumference *A*⟨*D*⟩*Z* be bisected at point *G*.[12] Therefore, each of the circumferences *AG* and *GZ* is a quadrant.[13] Therefore, the circle drawn with a pole, *G*, and a distance, *GZ*, will also

---

[1] See Commentary.   [2] *Elem.* I.post.3 (transfer). See Commentary.   [3] *Sph.* I.18, *Elem.* I.10, I.post.3, I.post.3 (transfer), III.30, I.post.3 (transfer). See Commentary. [4] *Sph.* I.20.   [5] *Elem.* I.c.n.2.   [6] *Elem.* I.def.17, I.def.18.   [7] *Sph.* I.14.   [8] That is, of **C**(*ZQA*).   [9] *Sph.* I.19, *Elem.* I.10, I.post.3, I.post.3 (transfer), III.30, I.post.3 (transfer). See Commentary.   [10] *Elem.* III.29, *Sph.* I.def.5.   [11] As in Case 1.   [12] *Sph.* I.18, *Elem.* I.10, I.post.3, I.post.3 (transfer), III.30, I.post.3 (transfer). See Commentary. [13] *Sph.* I.16.

pass through point A, since A to Z is a diameter.[1] Let it go,[2] and let it be as as ZAQ.

Therefore, circle ZAQ is great, for its polar distance is equal to the side of the square inscribed in the great circle.[3] And since point G is a pole of circle ZAQ, therefore circle ABG cuts circle ZAQ through the poles. So, since a great circle in a sphere, ABG, cuts some one of those in the sphere, ZAQ, through the poles, it bisects it at an upright.[4] Therefore, circle ABG is upright to circle ZAQ, therefore, circle ZAQ ⟨C80⟩ is upright to circle ABG.[5] So, since a great circle in a sphere, ZAQ, cuts some one of those in the sphere, ABG, at an upright, therefore it bisects it and is through the poles.[6] Therefore, circle ZAQ bisects circle ⟨H40⟩ ABG and is through the poles.

Let circumference ZA be bisected at point H.[7]

Therefore, point H is a pole of circle ABG.[8]

22 ⟨Theorem⟩[a]

If in a sphere some straight line of those being through its center bisects some line not being through the center, then it cuts it at an upright. And if it cuts it at an upright, then it bisects it.

The proof is as in those concerning circles.[b]

For, if a plane is extended through through the diameter being through the center of the sphere and through ⟨one⟩ not being through the center, it makes as intersection a ⟨curved⟩ line, a circle.[9] And, in the plane, some straight line through the center will bisect some straight line not through the center, so that it will cut it at an upright. And, if it cuts it at an upright, then it will bisect it.[10]

---

[1] *Elem.* I.def.17, *Sph.* I.16.   [2] *Elem.* I.post.3 (spherical).   [3] *Sph.* I.17.   [4] *Sph.* I.15.
[5] *Elem.* I.post.4.   [6] *Sph.* I.13.   [7] *Sph.* I.19, *Elem.* I.10, I.post.3, I.post.3 (transfer),
III.30, I.post.3 (transfer). See Commentary.   [8] *Elem.* III.29, *Sph.* I.def.5.   [9] *Sph.* I.1.
[10] *Elem.* III.3.

[a] This proposition is generally held to be a late addition to the text for the following reasons: the content is trivial, the wording unusual, there is no diagram, it is unused in the rest of the text and it is not found in the Arabic versions.
[b] This might be a reference to *Elements* III – that is, to *Elem.* III.3; or, it might simply mean "in the circular ⟨figures⟩" (Ver Eecke 1927, 29 n. 8).

⟨C81⟩   23 ⟨Theorem⟩[a]

If a circle is in a sphere, and some straight line is joined from the center
of the sphere to the center of the circle, then the joined ⟨straight line⟩
is a perpendicular to the circle.

For, let there be a circle in the sphere, *ABG*, and let there be the
center of the sphere, point *D*, and the center of the circle, point *E*. Let
DE be joined.

I say that *DE* is upright to circle *ABG*.

For, let some straight line, *AEB*, be produced as arbitrary in it
through point *E*.[1] Let *DA* and *DB* be joined.[2]

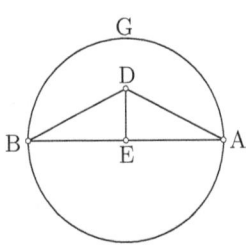

Since *AE* is equal to *EB*,[3] and *ED* is common, and the two *AE* and *ED* are respectively
equal to the two *BE* and *ED*, and *AD*, as base,
is equal to *DB*, as base,[4] therefore angle *AED*
is equal to angle *DEB*.[5] Therefore, each of angles *AED* and *BED* are upright.[6] Therefore,
DE is at an upright to *AB*. Then, similarly, we will show that straight
line *DE* is also at an upright to all straight lines produced through *E* in
the plane of circle *ABG*. Therefore, straight line *DE* is at an upright to
the plane of circle *AG*.[7]

End of ⟨Book⟩ I.

[1] See Commentary, *Sph.* I.7.   [2] *Elem.* I.post.1.   [3] *Elem.* I.def.15.   [4] *Sph.* I.def.1,
I.def.2. [5] *Elem.* I.8, in **T**(*AED*) and **T**(*BED*). [6] *Elem.* I.def.10. [7] *Elem.* XI.def.3.

[a] This proposition is also thought to be a late addition to the text: it is repetitive,
unused and is not found in the Arabic versions.

⟨Def. 1⟩ Circles in a sphere are said to *touch* one another when the common section of their planes touches both circles.

1 ⟨Theorem⟩

Parallel circles in a sphere are about the same poles.

For, let there be parallel circles in a sphere, *ABG* and *DEZ*.

I say that circles *ABG* and *DEZ* are about the same poles.

For let the poles of circle *ABG* be taken, and let them be *H* and *Q*.[1] Let *HQ* be joined.[2]

Since, if a circle is in a sphere, the straight line produced through its poles is upright to the circle and is through both its center and also the sphere's,[3] therefore *HQ* is upright to circle *ABG*, and is through both its center and also the sphere's. So, since *HQ* is upright to circle *ABG*, and *ABG* is parallel to circle *DEZ*, therefore *HQ* is also upright to circle *DEZ*.[4] So, since *DEZ* is a circle in a sphere, while a perpendicular to it from the center of the sphere, *HQ*, is produced and extended in both directions, and intersects the surface of the sphere at points *H* and *Q* – while if there is a circle in a sphere, and a perpendicular to it is produced from the center of the sphere and produced in both directions, it falls on the poles of the circle[5] – therefore points *H* and *Q* are poles of the circle *DEZ*. But, they are also ⟨poles⟩ of circle *ABG*. Therefore, *ABG* and *DEZ* are about the same poles.

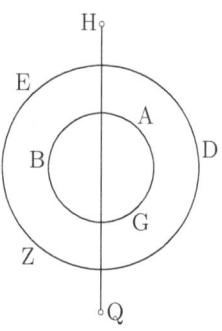

2 ⟨Theorem⟩   ⟨C83⟩

Circles in a sphere being about the same poles are parallel.

Let there be circles in a sphere, *ABG* and *DEZ*, about the same poles, *H* and *Q*.

---

[1] *Sph.* I.21.   [2] *Elem.* I.post.1.   [3] *Sph.* I.10.   [4] *Elem.* XI.16 (by implication). See Commentary.   [5] *Sph.* I.8.

I say that circles *ABG* and *DEZ* are parallel.

For, let *HQ* be joined.[1]

⟨H44⟩

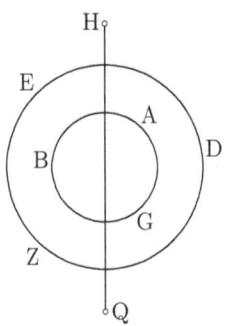

Since *ABG* is a circle in a sphere, and a straight line, *HQ*, is produced through its poles, therefore *HQ* is upright to circle *ABG*.[2] Then, similarly, we will show that *HQ* is also upright to circle *DEZ*.[3] But, the same straight line being upright to planes, the planes being extended will not intersect.[4]

Therefore, the planes through circles *ABG* and *DEZ*, being extended, will not meet. Therefore, circle *ABG* is parallel to circle *DEZ*.[5]

3 ⟨Theorem⟩

If two circles in a sphere cut the circumference of some great circle at the same point, having their poles on it, then the circles will touch one another.

For, let two circles in a sphere, *ABG* and *GDE*, cut the circumference of some great circle, *AGE*, at the same point, *G*, having their poles on it.

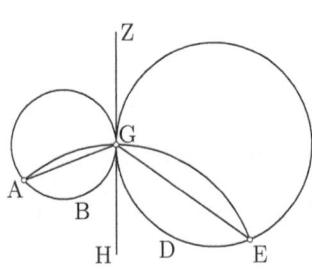

I say that circles *ABG* and *GDE* will touch one another.

For, let there be their common sections, *AG* of *AGE* and *ABG*, *GE* of *AGE* and *GDE*, and *HGZ* of *ABG* and *GDE*.[6]

Since a great circle in a sphere, *AGE*, cuts some circle in the sphere, *ABG*, through the poles, it bisects it at an upright.[7] Therefore, ⟨straight line⟩ *AG* is the diameter of circle *ABG*.[8] Then, similarly, we will show also that *GE* is the diameter of circle *GDE*. And since *AGE* is upright to each of circles *ABG* and *GDE*, therefore each of ⟨planes⟩ *ABG* and *GDE*

⟨C84⟩

[1] *Elem.* I.post.1.   [2] *Sph.* I.10.   [3] *Sph.* I.10.   [4] *Elem.* XI.14, XI.def.8.   [5] *Elem.* XI.def.8.
[6] *Elem.* XI.3. See Commentary.   [7] *Sph.* I.15.   [8] *Elem.* I.def.17.

is upright to circle *AGE*.[1] Therefore, their common section is upright
to circle *AGE*.[2] But, *ZGH* is their common section. Therefore, *ZGH* is
upright to circle *AGE*, so that it is also upright to all straight lines contacting it and being in the plane of circle *AGE*.[3] But it contacts each
of *AG* and *GE*, being in the plane of *AGE*. Therefore, *ZH* is upright
to each of *AG* and *GE*. So, since *ZH* is produced from the extremity of
the diameter *AG* of circle *ABG* at an upright, *ZH* touches circle *ABG*   ⟨H46⟩
at point *G*.[4] Then, similarly, we will show also that *ZH* touches circle
GDE at point *G*. But, circles in a sphere are said to touch one another
when the common section of the planes touches both circles,[5] and *ZH*
touches both circles at point *G*. Therefore, circles *ABG* and *GDE* will
touch one another.

4 ⟨Theorem⟩

If two circles in a sphere touch one another, then the great circle drawn
through their poles will go through their ⟨point of⟩ contact.

For, let two circles in a sphere, *ABG* and *GDE*, touch one another
at point *G*. Let there be a pole of circle *ABG*, point *Z*, and a pole of
circle *GDE*, point *H*.

I say that the great circle drawn through the poles *Z* and *H* will pass
through point *G*.

In fact not, but if possible, let it go,[6]   ⟨C85⟩
and let it be as *ZBH*.[7] And with a pole,
H, and a distance, *HB*, let a circle, *BKQ*,
be drawn.[8]

Therefore, circle *GDE* is parallel to
circle *BKQ*, for they are about the same
poles.[9]   And since two circles in a sphere, *ABG* and *BQK*, cut the
circumference of some great circle, *ZBH*, at the same point, *B*, having

---

[1] *Elem.* XI.def.4.   [2] *Elem.* XI.19.   [3] *Elem.* XI.def.3.   [4] *Elem.* III.16.cor.
[5] *Sph.* II.def.1.   [6] *Sph.* I.20 (counterfactual), using the two poles.   [7] That is, let
it go through some other point, say *B*.   [8] *Elem.* I.post.3 (spherical).   [9] *Sph.* II.2.

their poles on it, circles ABG and BKQ touch one another.[1] But they also cut ⟨one another⟩,[2] which is impossible.[3]

Therefore, the great circle drawn through poles Z and H will not not pass through point G. Therefore, the great circle drawn through the poles of circles ABG and GDE will pass through the ⟨point of⟩ contact of the circles.

5 ⟨Theorem⟩

If two circles in a sphere touch one another, then the great circle drawn through the poles of one and the ⟨point of⟩ contact will pass through the poles of the other.

For, let two circles in a sphere, ABG and GDE, touch one another at point G. Let there be a pole of circle ABG, point Z, and a pole of circle GDE, point H.

I say that the great circle drawn through points Z and G will also go through point H.

⟨H48⟩ In fact not, but if possible, let it go as ZGQ.[4] Let a great circle, ZGH, be drawn through poles Z and H.[5] Then, it will pass through G.[6] Let ZG be joined.[7]

⟨C86⟩

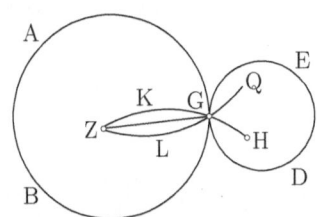

Since each of ZGH and ZGQ is great, therefore they bisect one another.[8] Therefore, each of ZKG and ZLG is a semicircle.[9] Therefore, ZG is a diameter of the sphere, since ⟨it⟩ is also ⟨a diameter⟩ of great circles ZGH and

ZGQ.[10] But, it is the polar distance of circle ABG, which is impossible.[11]

Therefore, the great circle drawn through points Z and G will also pass through point H.

---

[1] *Sph.* II.3.   [2] See Commentary.   [3] *Elem.* III.def.3, applied to a spherical surface. [4] *Sph.* I.20 (counterfactual).   [5] *Sph.* I.20.   [6] *Sph.* II.4.   [7] *Elem.* I.post.1.   [8] *Sph.* I.11. [9] *Elem.* I.def.18.   [10] *Elem.* I.def.17, *Sph.* I.6.   [11] *Sph.* I.def.5.

6 ⟨Theorem⟩

If a great circle in a sphere touches some circle of those in the sphere,
then it will also touch another ⟨circle⟩ equal and parallel to it.

For, let a great circle in a sphere, *ABG*, touch some circle of those
in the sphere, *GD*, at point *G*.

I say that circle *ABG* will also touch another circle equal and parallel
to circle *GD*.

Let the pole of circle *GD* be taken,[1] and let it be point *E*. Let a
great circle, *GEDBZH*, be drawn through points *G* and *E*.[2] Let *BZ*
equal to circumference *GE* be cut off.[3] Let a circle, *BH*, be drawn with
a pole, *Z*, and a polar distance, *ZB*.[4]

Since two circles in a sphere, *ABG* and
GD, touch one another, while a great circle,
GEDBZH, has been drawn through one of the
poles, *E*, and the ⟨point of⟩ contact, therefore
GEDBZH will pass through the poles of circle
ABG.[5] And since two circles in a sphere, *ABG*
and *BH*, cut the circumference of some great
circle, *GH*, at the same point, *B*, having their

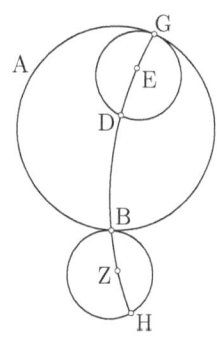

⟨C87⟩

poles on it, therefore circles *ABG* and *BH* touch one another.[6] And
since circumference *GE* is equal to circumference *BZ*, let *EB* be added,
as common. Therefore, *GB*, as a whole, is equal to circumference *EZ*.[7]
But, *GB* is a semicircle.[8] Therefore, *EZ* is a semicircle.[9] Therefore,
E is along a diameter with *Z*.[10] And point *E* is a pole of circle *GD*,   ⟨H50⟩
and therefore *Z* is the other pole of circle *GD*.[11] Again, since *EZ* is a
semicircle and *Z* is a pole of circle *BH*, therefore *E* is the other pole of
circle *BH*. Therefore, circles *GD* and *BH*, having the same poles, are
parallels.[12] Therefore, circle *GD* is parallel to circle *BH*.[a] And since

---

[1] *Sph.* I.21.   [2] *Sph.* I.20.   [3] *Elem.* I.post.3 (transfer, spherical). See Commentary.
[4] *Elem.* I.post.3 (spherical).   [5] *Sph.* II.5.   [6] *Sph.* II.3.   [7] *Elem.* I.c.n.2.   [8] *Sph.* I.11.
[9] *Elem.* I.c.n.1.   [10] *Sph.* I.6. That is, on a diameter of the sphere.   [11] *Sph.* I.9.
[12] *Sph.* II.2.

[a] Heiberg (1927, 50) marks this sentence as interpolated.

GE is equal to BZ, therefore circle GD is equal to circle BH[1] – but also parallel.[a] Therefore, circle ABG touches another circle, BH, equal and parallel circle to GD.

7 ⟨Theorem⟩

If there are two equal and parallel circles in a sphere, then the great circle touching one of them will also touch the other.

Let there be two equal parallel circles in a sphere, AB and GD.

I say that the great circle touching AB will also touch GD.

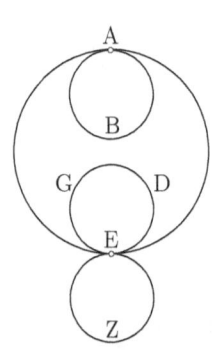

⟨C88⟩

For, if possible, let a great circle, AE, touch AB at point A, but not touch GD. And since a great circle in a sphere, AE, touches some circle of those in the sphere, AB, therefore it will also touch another ⟨circle⟩ equal and parallel to AB – so, let it touch EZ.[2] So, since AB is equal and parallel to EZ, but AB is equal and parallel to GD, therefore GD is also equal and parallel to EZ.[3] Therefore, there will be three equal and parallel circles in a sphere, which is impossible.[4]

Therefore, the great circle touching AB will not not also touch GD. Therefore, it will touch ⟨it⟩.

8 ⟨Theorem⟩

If a great circle in a sphere is oblique to some circle of those in the sphere, then it will touch two circles, equal to one another and parallel to the aforementioned.

For, let a great circle in a sphere, ABG, be oblique to some circle of those in the sphere, BD – that is,[b] let it not be through its poles.

I say that circle ABG will touch two circles equal to one another and parallel to BD.

⟨H52⟩   Since circle ABG is oblique to BD, the pole of BD is not on circle

[1] *Sph.* I.15, I.def.5, *Elem.* III.29, III.27, I.32, I.26, III.def.1. See Commentary.
[2] *Sph.* II.6. See Commentary.   [3] *Elem.* I.c.n.1; *Sph.* II.1, II.2. See Commentary, *Sph.* II.1 and II.2.   [4] *Sph.* II.1, I.8, I.6. See Commentary.

[a] Czinczenheim (2000, 87) marks this phrase as interpolated.
[b] Heiberg (1927, 51 n. 1) and Czinczenheim (2000, 88) mark this sentence as interpolated.

ABG.[1] So, let the pole of circle *BD* be taken,[2] and let it be point *E*. Let a great circle, *AEGH*, be drawn through point *E* and one of the poles of *ABG*.[3] Let a circle, *AZ*, be drawn with a pole, *E*, and a polar distance, *AE*.[4] Therefore, *AZ* is parallel to *BD*, for it has the same poles.[5]

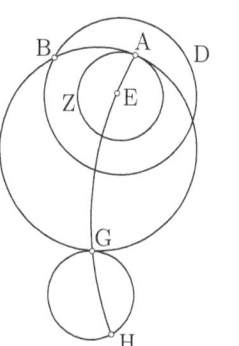

⟨C89⟩

Since two circles in a sphere, *ABG* and *AZ*, cut the circumference of some great circle at the same point, *A*, having their poles on it, therefore they touch one another.[6] Therefore, circle *ABG* touches circle *AZ*. So, since a great circle in a sphere, *ABG*, touches some circle of those in the sphere, *AZ*, therefore it also touches another ⟨circle⟩ both equal and parallel to *AZ* – so, let it touch *GH*.[7] So, since *AZ* is both equal and parallel to *GH*, but *AZ* is parallel to *BD*, therefore *GH* is also parallel to *BD*.[8] Therefore, circle *ABG* touches two circles, *AZ* and *GH*, equal to one another and parallel to *BD*.

9 ⟨Theorem⟩

If two circles in a sphere cut one another and a great circle is drawn through their poles, then it will bisect the cut-off segments of the circles.

For, let two circles in a sphere, *ZAEB* and *ZGED*,[a] cut one another at points *Z* and *E*. Let a great circle, *AGBD*, be drawn through their poles.

I say that circle *AGBD* will bisect the cut-off segments of the circles – that is, that circumference *ZA* is equal to circumference *AE*, and circumference *ZB* is equal to circumference *BE*, and *ZG* to *GE*, and *ZD* to *DE*.

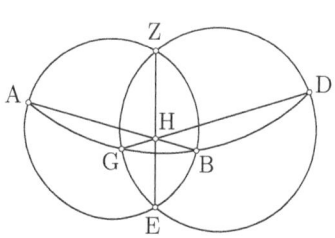

---

[1] *Sph.* I.13 (indirect).   [2] *Sph.* I.21.   [3] *Sph.* I.20.   [4] *Elem.* I.post.3 (spherical).
[5] *Sph.* II.2.   [6] *Sph.* II.3.   [7] *Sph.* II.6. See Commentary.   [8] *Sph.* II.1, II.2.

[a] These circles may have originally been called *AB* and *GD*. See Commentary.

For, let there be the common section of circle $AGBD$ and $ZAEB$, AB, and the common section of $AGBD$ and $ZGED$, GD.[1] Let ZH and HE be joined.[2]

Since points Z, H, and E are in the plane of circle $AEBZ$, but are also in the plane of circle $ZDEG$, therefore points Z, H, and E are on

⟨H54⟩   the common section of the two planes. But the common section of every two planes is a straight line.[3] Therefore, ZH is in a straight line with HE. And since a great circle in a sphere, $AGBD$, cuts some circle

⟨C90⟩   of those in the sphere, $ZAEB$, through the poles, it will bisect it at an upright.[4] Therefore, AB is a diameter of circle $ZAEB$.[5] Then, similarly, we will show that GD is a diameter of circle $ZDEG$. And since circle $AGBD$ is upright to each of circles $ZAEB$ and $ZDEG$, therefore each of $ZAEB$ and $ZDEG$ is upright to circle $AGBD$ – but, if two planes are perpendicular to some plane, then their common section is also perpendicular to the plane[6] – therefore the common section of $ZAEB$ and $ZDEG$ is also at an upright to plane $AGBD$. But, ZHE is their common section, therefore ZHE is also upright to circle $AGBD$, so that it will make an upright angle to all straight lines contacting it and being in the plane of circle $AGBD$.[7] But, each of AB and GD contacts ZHE, being in the plane of circle $AGBD$. Therefore, ZHE is upright to each of AB and GD. Therefore, each of AB and GD is also upright to ZHE. So, since in a circle, $ZAEB$, some straight line through the center, AB, cuts some straight line not being through the center, ZHE, at an upright, it bisects it.[8] Therefore, ZH is equal to HE. And HA is common and at an upright, therefore circumference ZA is equal to circumference AE.[9] Then, similarly, we will show also that ZB is equal to BE, and ZG to GE, and ZD is equal to DE. Therefore, circle $AGBD$ will bisect the cut-off segments of the circles.

---

[1] *Elem.* XI.3. See Commentary.   [2] *Elem.* I.post.1.   [3] *Elem.* XI.3.   [4] *Sph.* I.15.
[5] *Elem.* I.def.17.   [6] *Elem.* XI.19.   [7] *Elem.* XI.def.3.   [8] *Elem.* III.3.   [9] *Elem.* I.4,
III.29, considering $\mathbf{T}(HAZ)$ and $\mathbf{T}(HAE)$, which are uncompleted in the figure.

10 ⟨Theorem⟩

If there are parallel circles in a sphere and through their poles great circles are drawn, then the circumferences of the parallel circles between the great circles are similar, and[a] the circumferences of the great circles between the parallel circles are equal.   ⟨C91⟩

Let there be parallel circles in a sphere, *ABGD* and *EZHQ*, and through their poles let great circles, *AEHG* and *BZQD*, be drawn.   ⟨H56⟩

I say that the circumferences of the parallel circles between the great circles are similar – that is, that circumference *BG* is similar to circumference *ZH*, and *GD* to *HQ*, and *DA* to *QE*, and *AB* to *EZ* – and the circumferences of the great circles between the parallel circles are equal – that is, that the four *ZB*, *HG*, *QD*, and *EA* are equal to one another.

For, let there be a common section of circle *ABGD* and *BZQD*, *BD*, and a common section of circle *ABGD* and *AEHG*, *AG*, and a common section of circle *EZHQ* and *ZKQ*, *ZQ*, and a common section of *EZHQ* and *EKH*, *EH*.[1]

Since a great circle in a sphere, *AEHG*, cuts some circle of those in the sphere, *ABGD*, through the poles, it bisects it at an upright.[2] Therefore, *AG* is a diameter of circle *ABGD*.[3] Then, similarly, we will show also that *BD* is a diameter of circle *ABG*. Therefore, point *L*[b] is the center of circle *ABGD*.[4]

⟨Part 1⟩

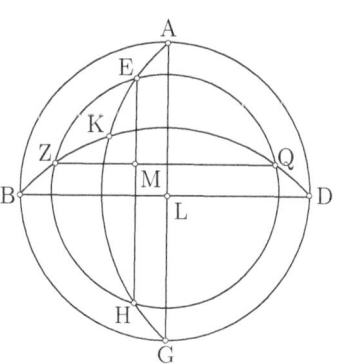

Again, since a great circle in a sphere, *AEHG*, cuts some circle of those in the sphere, *EZHQ*, through the poles, it bisects it at an upright.[5] Therefore, *EH* is a diameter of circle *EZHQ*.[6] Then, similarly, we will

---

[1] All by *Elem.* XI.3. See Commentary, *Sph.* II.9. Although *K* is not explicitly introduced it must be a pole of the parallel circles.   [2] *Sph.* I.15.   [3] *Elem.* I.def.17.
[4] *Elem.* I.def.17.   [5] *Sph.* I.15.   [6] *Elem.* I.def.17.

[a] Part 2 of this proposition may have been added to the text in late antiquity. See Commentary.
[b] This is our first introduction to the letter-name *L*.

⟨C92⟩   show also that *ZQ* is a diameter of circle *EZHQ*. Therefore, point *M*[a] is the center of circle *EZHQ*.[1] And since two parallel planes, *ABGD* and *EZHQ*, cut some plane *BZQD*, therefore their common sections are parallel.[2] Therefore, *BD* is parallel to *ZQ*. Then, similarly, we will show also that *AG* is parallel to *EH*. So, since two straight lines contacting one another, *BL* and *LG*, are parallel to two straight lines contacting one another, *ZM* and *MH*, not being in the same plane, they will contain equal angles.[3] Therefore, angle *ZMH* is equal to angle *BLG*. And since, at the centers, angle *ZMH* stands upon circumference *ZH*, and angle *BLG* upon circumference *BG*, therefore circumference *BG* is similar to circumference *ZH*.[4] Then, similarly, we will show also that circumfer-

⟨H58⟩   ence *GD* is similar to circumference *HQ*, and *AD* to *EQ*, and *AB* to *EZ*.

Therefore, the circumferences of the parallel circles between the great circles are similar.

Then, I say also that the circumferences of the great circles between the parallel circles are equal.[b]

⟨Part 2⟩   For, since point *K* is the pole of circle *ABGD*, therefore the four *KA*, *KB*, *KG*, and *KD* are equal to one another. Again, since point *K* is the pole of circle *EZHQ*, therefore the four *KE*, *KZ*, *KH*, and *KQ* are equal to one another.[5] Therefore, the four *EA*, *ZB*, *HG*, and *QD*, as remainders, are equal to one another.[6]

⟨C93⟩   Therefore, the circumferences of the great circles between the parallel circles are equal.

11 ⟨Theorem⟩

If equal and upright segments of circles have been set up on di-ameters in equal circles, and from them equal circumferences are cut off abutting the extremities of the segments, being less than half of the whole ⟨segments⟩, and from the determined points equal straight lines

---

[1] *Elem.* I.def.17.   [2] *Elem.* XI.16.   [3] *Elem.* XI.10.   [4] *Elem.* III.def.11.   [5] Both by *Sph.* I.def.5, *Elem.* III.28.   [6] *Elem.* I.c.n.3.

[a] This is our first introduction to the letter-name *M*.   [b] Again, Part 2 may have been a late addition. See Commentary.

are extended forth to the circumferences of the initial circles, then they will cut off from the initial circles equal circumferences abutting the extremities of the diameters.

For, on diameters, *AG* and *DZ*, in equal circles, *ABG* and *DEZ*, let equal segments of circles, *AHG* and *DQZ*, be set up upright. Let equal circumferences, *AH* and *DQ*, be cut off from them abutting the extremities, *A* and *D*, being less than half of the whole of *AHG* and *DQZ*. Let equal straight lines, *HB* and *QE*, be extended forth from points *H* and *Q* to the circumferences of the initial circles, *ABG* and *DEZ*.

I say that circumference *AB* is equal to circumference *DE*.

For, let perpendiculars be produced from points *H* and *Q* to the planes of circles *ABG* and *DEZ*.[1] Then, they will fall on the common sections[2] – that

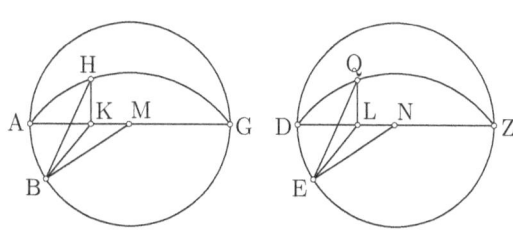

is, on *AG* and *DZ*. Let them fall, and let them be *HK* and *QL*. And let the centers of the circles be taken,[3] and let them be points *M* and *N*. Let *KB*, *BM*, *LE*, and *EN* be joined.[4]

Since *HK* is upright to the plane of circle *ABG*, therefore it will make an upright angle with all ⟨straight lines⟩ contacting it and being in the plane of circle *ABG*.[5] But, *KB* contacts it, being in the plane of circle *ABG*. Therefore, angle *HKB* is upright. Then, similarly, we will show also that angle *QLE* is upright. And since two segments, *AHG* and *DQZ*, are equal, and *AH* and *DQ* are equal cut-off circumferences, and *KH* and *QL* are produced perpendiculars, therefore *AK* is equal to *DL* and *HK* to *QL*.[6] And since *BH* is equal to *QE*, therefore the square on *BH* is equal to the square on *QE*.[7] But the squares on *HK*

⟨C94⟩

⟨H60⟩

[1] *Elem.* XI.11. [2] *Elem.* XI.def.4. Ver Eecke (1927, 45 n. 3) treats this as Peyrard's *Elem.* XI.38, but Heiberg does not include this proposition in his edition (Heiberg and Stamatis 1969–1977). See Commentary. [3] *Elem.* III.1. [4] *Elem.* I.post.1. [5] *Elem.* XI.def.3. [6] Scholium 190. See introductory Commentary to both *Sph.* II.11 and II.12. [7] *Elem.* VI.20.cor. See Commentary, *Sph.* I.1.

and KB are equal to the square on BH, and the squares on QL and LE are equal to the square on QE.[1] Therefore, the squares on HK and KB are equal to the squares on QL and LE,[2] of which the square on HK is equal to the square on QL.[3] Therefore, the square on KB, as remainder, is equal to the square on LE, as remainder.[4] Therefore, KB is equal to LE.[5] And since AM is equal to DN,[6] of which AK is equal to DL, therefore KM, as remainder, is equal to LN, as remainder.[7] But, BM is equal to NE.[8] Then, the two KM and MB are respectively equal to the two LN and NE, and KB, as base, is equal to LE, as base, therefore angle KMB is equal to angle LNE.[9] But, in equal circles equal angles stand on equal circumferences, whether they are standing at the centers or at the circumferences.[10] Therefore, circumference AB is equal to circumference DE.

⟨C95⟩   12 ⟨Theorem⟩

If equal and upright segments of circles have been set up on diameters in equal circles, and from them are cut off equal circumferences abutting the extremities ⟨of the segments⟩, being less than half of the whole segments, and in the same directions equal circumferences are cut off from the ⟨initial⟩ circles abutting the extremities of the diameters, then the straight lines joining the determined points will be equal to one other.

For, in equal circles, ABG and DEZ, let equal and upright segments, AHG and DQZ, be set up on diameters, AG and DZ. Let equal cir-
⟨H62⟩   cumferences, AH and DQ, be cut off from them in the same directions abutting the extremities of the segments, being less than half of the whole segments. Let equal circumferences, AB and DE, be taken away from circles ABG and DEZ in the same directions abutting the extremities of the diameters. Let HB and QE be joined.

I say that HB is equal to QE.

---

[1] Both by *Elem.* I.47.   [2] *Elem.* I.c.n.1.   [3] *Elem.* VI.20.cor. See Commentary, *Sph.* I.1.   [4] *Elem.* I.c.n.3.   [5] *Elem.* VI.20.cor. See Commentary, *Sph.* I.1.   [6] *Elem.* I.def.15, I.def.16.   [7] *Elem.* I.c.n.3.   [8] *Elem.* I.def.15, I.def.16.   [9] *Elem.* I.8, in **T**(KMB) and **T**(LNE).   [10] *Elem.* III.26.

For, let perpendiculars, *HK* and *QL*, be produced from points *H* and *Q* to the planes of circles *ABG* and *DEZ*.[1] Then, they will fall on the common sections[2] – that is on *AG* and *DZ*. Let the centers of the circles be taken,[3] and let them be points *M* and *N*. Let *KB, BM, LE,* and *EN* be joined.[4]

Since circumference *AB* is equal to circumference *DE*, angle *AMB* is also equal to angle *DNE*.[5] And since two segments of circles, *AHG* and *DQZ*, are

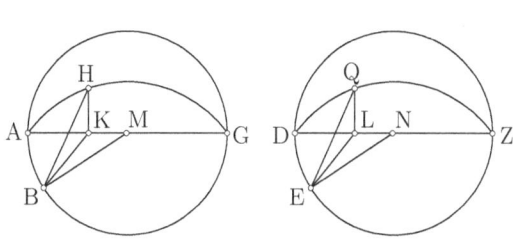

equal, and *AH* and *DQ* are equal cut-off circumferences, and *KH* and *QL* are produced perpendiculars, therefore *AK* is equal to *DL* and *HK* to *QL*.[6] So, since *AM* is equal to *DN*,[7] of which *AK* is equal to *DL*, therefore *KM*, as remainder, is equal to *LN*, as remainder.[8] But, *BM* is also equal to *NE*.[9] Then, the two *KM* and *MB* are respectively equal to the two *LN* and *NE*, and angle *KMB* is equal to angle *LNE*, therefore *KB*, as base, is equal to *LE*, as base.[10] And, since *HK* is upright to the plane of circle *ABG*, therefore it will make an upright angle on all straight lines contacting it and being in the plane of circle *ABG*.[11] But *KB* contacts it, therefore angle *HKB* is upright. Then, similarly, we will show that angle *QLE* is upright. So, since *HK* is equal to *QL*, and *KB* is equal to *LE*, then the two *HK* and *KB* are respectively equal to the two *QL* and *LE*, and they contain upright angles. Therefore *HB*, as base, is equal to *QE*, as base.[12]

13 ⟨Theorem⟩

If there are parallel circles in a sphere and great circles are drawn touching one of them and cutting the remaining ⟨parallel circles⟩, then the circumferences of the parallel circles between the non-intersecting[a]

---

[1] *Elem.* XI.11. [2] *Elem.* XI.def.4. [3] *Elem.* III.1. [4] *Elem.* I.post.1. [5] *Elem.* III.27.
[6] Scholium 190. See introductory Commentary to both *Sph.* II.11 and II.12.
[7] *Elem.* I.def.15, I.def.16. [8] *Elem.* I.c.n.3. [9] *Elem.* I.def.15, I.def.16. [10] *Elem.* I.4, in **T**(*KMB*) and **T**(*LNE*). [11] *Elem.* XI.def.3. [12] *Elem.* I.4, in **T**(*HKB*) and **T**(*QLE*).

semicircles of the great circles are similar, and the circumferences of
the great circles between the parallels are equal.

Let there be parallel circles in a sphere, *ABGD*, *EZHQ*, and *KL*.
Let great circles, *AEKHGFT*[a] and *BZLQDU*,[b] be drawn touching one
of them, *KL*, at points *K* and *L* and cutting the remaining ⟨parallel
circles⟩.

⟨Part 1⟩

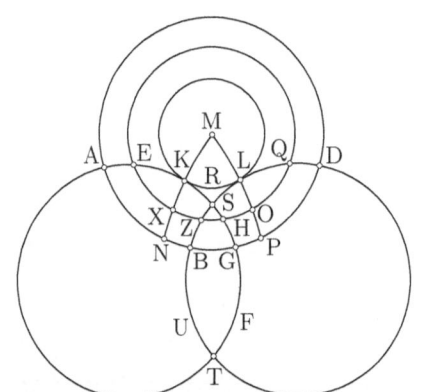

I say that the circumferences
of the parallel circles between
the non-intersecting semicircles of
the great circles are similar –
we know[c] some are between the
non-intersecting semicircles thus:
Since great circles in a sphere bisect one another, therefore *SKAT*
is a circumference of a semicircle.

Therefore, *KAT* is less than a semicircle. Again, since *TFGS* is a circumference of a semicircle, therefore *KSHGFT* is greater than a semicircle. But, truly, *KAT* is less than a semicircle. So, let *KATF* be
of a semicircle. Again, since *SBT* is a circumference of a semicircle,
therefore *LZBT* is greater than a semicircle. So, let *LSBU* be laid out
of a semicircle. Therefore, the semicircle from *K*, that is *KATF*, is
non-intersecting with the semicircle from *L*, that is *LSBU*. Then, similarly, also semicircle *KSF* is non-intersecting with semicircle *LDTU*.
Therefore, the circumferences between non-intersecting semicircles of
the great circles, being circumferences of the parallel circles, are circumferences *KRL*, *EXZ*, *ANB*, *HOQ*, and *GPD*. So, I say – that *KL*,
EZ, and *AB* are similar, and, furthermore, *KL*, *HQ*, and *GD* are similar

⟨C98⟩

[a] See the Commentary for a discussion of this term.

[a] This is the letter-name according to the manuscripts, but we would replace with
AEKHG (ΑΕΚΗΓ), because the Φ and Τ are probably later additions. See Commentary.

[b] Again following the manuscripts, but we would drop the *U* (Υ), making the letter--
name *BZLQD*.

[c] This long gloss discussing the non-intersecting semicircles is almost certainly an
interpolation, inserted before the 9th-century – as witnessed by both **A** and the
Arabic translation (Kunitzsch and Lorch 2010, 128–130).

to one another, and the circumferences of the great circles between the
parallel circles are equal – that is, that the four *EK, KH, ZL,* and *LQ*
are equal to one another, and the four *AE, BZ, HG,* and *QD* are equal
to one another.

For, let the pole of the parallel circles be taken,[1] and let it be point
M. Let great circles, *MKXN*[a] and *MLOP*, be drawn through *M* and
each of points *K* and *L*.[2]

Since two circles in a sphere, *AEKHGT*[b] and *KL*, touch one another at point *K*, while a great circle, *MKXN*, is drawn through the
poles of one of them, *KL*, and the ⟨point of⟩ contact, therefore circle
MKXN will pass through the poles of *AEKHGT*,[3] and it will be upright   ⟨H66⟩
to it.[4] Then, similarly, we will show that *MLOP* will pass through the
poles of *BZLQDT*,[c] and it will be upright to it.[5] And since equal and
upright segments of circles, *KM* and *LM* and what is continuous with
them, are set up on diameters in equal circles, *AEKHGT* and *BZLQDT*,
from points *K* and *L*, while *KM* and *LM* are equal circumferences cut   ⟨C99⟩
off from them,[6] being less than half of the whole, and the straight
line joining from *M* to *A* is equal to the straight line joining from *M*
to *D*,[7] therefore, the cut-off circumferences are equal.[8] Therefore, circumference *AK* is equal to circumference *LD*. Then, through the same
⟨reasons⟩, *EK* is also equal to *LQ*. And since two circles in a sphere,
ABGD and *AEKHGT*, cut one another, while a great circle, *MKXN*,
is drawn through their poles, therefore *MKXN* bisects the cut-off segments.[9] Therefore, circumference *AEK* is equal to circumference *KHG*,
and *AN* to *NG*. Then, similarly, we will show also that *BL* is equal
to *LD* and *BP* to *PD*. So, since circumference *AEK* is equal to circumference *LQD*, and circumference *AEKHG* is double circumference
AEK and *DQLB* is double *LQD*, therefore ⟨circumference⟩ *AEKG* is

---

[1] *Sph.* I.21.   [2] *Sph.* I.20.   [3] *Sph.* II.5.   [4] *Sph.* I.15.   [5] *Sph.* II.5, I.15.   [6] *Sph.* I.def.5,
Elem. III.28.   [7] *Sph.* I.def.5.   [8] *Sph.* II.11.   [9] *Sph.* II.9.

[a] Here we would expect *MKNX* (MKNΞ), so as to preserve the alphabetic order of
the letter-names.
[b] Again, here and below, we would drop the *T*, reading *AEKHG*.
[c] Again, here and below, we would drop the *T*, reading *BZLQD*.

equal to ⟨circumference⟩ *DQLB*.[1] And the circles are equal, for they are great.[2] Therefore, the straight line joining from *A* to *G* is equal to the straight line joining from *D* to *B*,[3] and through this ⟨reason⟩, circumference *ABG* is equal to circumference *DGB*.[4,a] And *AN* is half of circumference *ABG*, and *BP* is half of *BPD*, therefore *AN* is equal to

⟨H40⟩   *BP*.[5] Let ⟨circumference⟩ *NB* be added, as common. Therefore, *ANB*, as a whole, is equal to *NBP*, as a whole.[6] And they are of the same circle, therefore circumference *ANB* is similar to circumference *NBP*.[7]

⟨C100⟩ But, *NBP* is similar to *KL*, for if there are parallel circles in a sphere and through their poles great circles are drawn, then the circumferences of the parallel circles between the great circles are similar.[8] And circumferences *KL* and *NBGP* are of parallel circles, being between great circles, *MN* and *MP*, being through their poles. Therefore, *ANB* is similar to *KL*.[9] Then, similarly, we will show also that *EXZ* is similar to *KL*. Therefore, *AB* is similar to *EZ*.[10] Then, similarly, we will also show that *GPD* is similar to *HOQ*.

Therefore, the circumferences of the parallel circles between the non-intersecting semicircles of the great circles are similar.

⟨Part 2⟩   But, I say also that the circumferences of the great circles between the parallels are equal.

For, since the four circumferences *AEK*, *KHG*, *BZL*, and *LQD* are equal to one another, of which the four *EK*, *KH*, *ZL*, and *LQ* are equal to one another – for great circle *KN* similarly bisects the cut-off segments *EKH* and *EXH* and also *ZLQ* and *ZOQ*, so that *EK* is equal to *KH*; but *EK* has been shown to be equal to *LQ*, therefore *KH* is equal to *LQ*; but *QL* is equal to *LZ*, therefore *LZ* is equal to *KH*; therefore the four *EK*, *KH*, *ZL*, and *LQ* are equal – therefore, the four *AE*, *BZ*, *GH*, and *QD* are equal to one another, as remainders.[11]

---

[1] *Elem.* V.9.   [2] *Sph.* I.6, I.def.2, I.def.1, *Elem.* III.def.1.   [3] *Elem.* III.29.
[4] *Elem.* III.28. [5] *Elem.* V.9. [6] *Elem.* I.c.n.2. [7] *Elem.* III.27, III.def.11. [8] *Sph.* II.10.
[9] *Elem.* III.def.11, I.c.n.1.   [10] *Elem.* III.def.11, I.c.n.1.   [11] *Elem.* I.c.n.3.

[a] For the letter-names of these circumferences, we follow Heiberg (1927, 66) and the majority of manuscripts, because this makes it clearer which circumferences are under discussion.

Therefore, the circumferences of the great circles between the parallel circles are equal.

14 ⟨Problem⟩   ⟨C101⟩

Given a circle less than great in a sphere and some point on its circumference, draw through the point a great circle touching the given ⟨circle⟩.

Let there be a given circle in a sphere, *AB*, less than great, and let there be a given point on its circumference, *B*.

Then, it is required to draw a great circle through point *B* touching circle *AB*.

For, let there be the pole of circle, *AB*, point *G*.[1] Let a great circle, *GBD*, be drawn through points *G* and *B*.[2] Let there be cut off a ⟨circumference⟩ equal to that which the side of the square inscribed in a great circle subtends, *DB*.[3] Let a circle, *EBZ*, be drawn with a pole, *D*, and a distance, *DB*.[4]   ⟨H68⟩

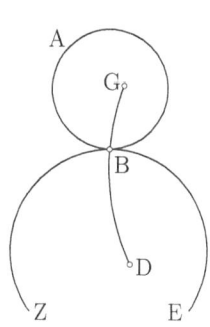

Therefore, ⟨circle⟩ *EBZ* is great, for its polar distance is equal to the side of the square inscribed in a great circle.[5] And since two circles in a sphere, *AB* and *EBZ*, cut the circumference of some great circle, *GBD*, at the same point, *B*, having their poles on it, the circles touch one another.[6] Therefore, circle *AB* will touch circle *EBZ*.

Therefore, a great circle, *EBZ*, has been drawn through a given point, *B*, touching circle *AB* at point *B*.

---

[1] *Sph.* I.21.   [2] *Sph.* I.20.   [3] *Sph.* I.19, *Elem.* I.10, I.post.3, IV.6, I.post.3 (transfer). See Commentary, *Sph.* I.20.   [4] *Elem.* I.post.3 (spherical).   [5] *Sph.* I.17.   [6] *Sph.* II.3.

⟨C102⟩

15 ⟨Problem⟩[a]

Given a circle less than great in a sphere and some point on the surface of the sphere between it and the ⟨circle⟩ equal and parallel to it, draw through the point a great circle touching the given circle.

Let there be a given circle less than great in a sphere, *AB*, and a given point on the surface of the sphere that is between *AB* itself and the ⟨circle⟩ equal and parallel to it – let it be *G*.

Then, it is required to draw through point *G* a great circle touching circle *AB*.

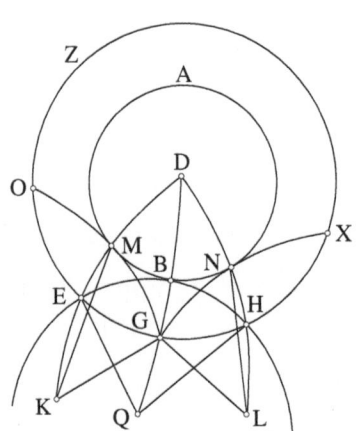

For, let the pole of circle *AB* be taken,[1] and let it be point *D*. Let a circle, *GEZH*, be drawn with a pole, *D*, and a distance, *DG*.[2] Let a great circle, *DGQ*, be drawn through points *D* and *G*.[3,b] Let there be cut off a ⟨circumference⟩ equal to that which the side of the square inscribed in a great circle subtends, *BQ*.[4] Let a circle, *EBH*, be drawn with a pole,

⟨H72⟩

⟨C103⟩

Q, and a distance, *BQ*.[5] Therefore, *EBH* is a great circle, for its polar distance is equal to the side of the square inscribed in a great circle.[6] And it touches *AB*, for two circles, *EBH* and *AB*, cut the circumference of the great circle *DBGQ* at the same point, *B*, having their poles on it.[7] Let great circles, *DMEK* and *DNHL*, be drawn through point *D*

---

[1] *Sph.* I.21. [2] *Elem.* I.post.3 (spherical). [3] *Sph.* I.20. [4] *Sph.* I.19, *Elem.* I.10, I.post.3, IV.6, I.post.3 (transfer). See Commentary, *Sph.* I.20. [5] *Elem.* I.post.3 (spherical). [6] *Sph.* I.17. [7] *Sph.* II.3.

---

[a] The texts produced by Heiberg (1927, 70–76) and Czinczenheim (2000, 102–105) for this proposition are substantially different. We have followed Czinczenheim, giving the translation of some of the more substantial differences in footnotes.
[b] Following this passage, Heiberg (1927, 70) reads, "Then, *BG* is either less than the ⟨circumference⟩ which the side of the square inscribed in a great circle subtends, or equal, or greater. First, let it be less..."

and each of points E and H.[1] Let each of EK and HL be cut off equal to circumference GQ.[2]

Since two circles in a sphere, EBH and $ZEGH$, cut one another, and a great circle, $DBGQ$, has been drawn through their poles, therefore it bisects the cut-off segments of the circles.[3] Therefore, circumference EG is equal to GH, and circumference EB to circumference BH. And since the three ⟨circumferences⟩ DE, DG, and DH are equal to one another, for they are from the pole of circle ZEH, while DM, DB, and DN are equal to one another,[4] therefore ME, BG, and NH, as remainders, are equal to one another.[5] But, EK, GQ, and HL are also equal to one another, therefore MK, BQ, and NL, as wholes, are equal to one another.[6] And BQ is equal to the ⟨circumference⟩ that the side of the square inscribed in a great circle subtends. Therefore, ⟨distances⟩ MK and NL are also equal to the side of the square.[7,a] And since a great circle in a sphere, $DBGQ$, cuts some circle in the sphere, $ZEGH$, through the poles, it bisects it at an upright.[8] Therefore, $DBGQ$ is upright to circle $ZEGH$. Then, similarly, we will show that $DNHL$ is upright to $ZEGH$, and, furthermore, $DMEK$ is upright to $ZEGH$. Let LN, LG, and QE be joined.[9] Since equal and upright segments of circles, GQ ⟨C104⟩ and HL and what is continuous with them, are set up on diameters of a circle, $ZEGH$, those from points G and H, and from them equal circumferences, GQ and HL, are cut off, being less than half of the wholes, and circumference EG is equal to circumference GH, therefore ⟨straight line⟩ QE is equal to ⟨straight line⟩ LG.[10] But, QE is ⟨the side⟩ of a square.[11] Therefore, LG is equal to the side of the square inscribed in   ⟨H74⟩ the great circle.[12] But LN is also the side of a square inscribed in the great circle, therefore GL is equal to LN.[13] Therefore, a circle drawn with a pole, L, and a distance, LG, will pass through point N.[14] Let it

---

[1] *Sph.* I.20.   [2] *Elem.* I.post.3 (transfer).   [3] *Sph.* II.9.   [4] Both by *Sph.* I.def.5, *Elem.* III.28.   [5] *Elem.* I.c.n.3.   This fact was already shown in *Sph.* II.10.   See Commentary.   [6] *Elem.* I.c.n.2.   [7] *Elem.* I.c.n.1.   [8] *Sph.* I.15.   [9] *Elem.* I.post.1.   [10] *Sph.* II.12.   [11] *Sph.* I.def.5.   [12] *Elem.* I.c.n.1.   [13] *Elem.* I.c.n.1.   [14] *Sph.* I.def.5.

[a] Here and below, "the side of the square" means "the side of the square inscribed in the great circle" through ellipsis.

go,[1] and let it be as *GNX*. Therefore, ⟨circle⟩ *GNX* is great, for its polar distance is equal to the side of the square inscribed in the great circle.[2] And since two circles in a sphere cut the circumference of some great circle, *DNHL*, at the same point, *N*, having their poles on it, therefore they will touch one another.[3] Therefore, ⟨great⟩ circle *GNX* touches circle *AB*.

Then, similarly, we will show also that the circle drawn with a pole, *K*, and a polar distance, *KG*, will pass through point *M* – for,[a] if we ⟨C105⟩ joined *GK* and *QH*, they would be equal to one another;[4] and *QH* is a side of a square, for it is the polar distance of a great circle, *EBH*;[5] therefore *GK* is a side of a square,[6] but also *KM*, therefore *KM* is equal to *KG*;[7] therefore, the circle drawn with a pole, *K*, and a distance, *KG*, will pass through point *M* – and it will be as *GMO*, and it will touch circle *AB*. And, the problem is done in two ways.

Therefore, through a given point, *G*, which is between *AB* and the ⟨circle⟩ equal and parallel to it, a great circle has been drawn, *GMX* and *GMO*.

But,[b] if someone says the cut-off ⟨circumference⟩, *BG*, is equal to the ⟨circumference⟩ that the side of the square inscribed in the great ⟨H76⟩ circle subtends, we will prove it thus.[c] Since *DG* is equal to each of *DE* and *DH*, of which *DB* is equal to each of *DM* and *DN*, therefore *BG* is equal to each of *NH* and *EM*, as remainder. But *BG* is ⟨the side⟩ of a square. Therefore, each of *NH* and *EM* is ⟨the side⟩ of a square. So, since *NH*, and *HG*, is ⟨the side⟩ of a square, for the polar distance of circle *EBH* is *HG*, therefore *NH* is equal to *GH*. Therefore, the circle drawn with a pole, *H*, and a polar radius, *GH*, will pass through point

---

[1] *Elem.* I.post.3 (spherical).   [2] *Sph.* I.17.   [3] *Sph.* II.3.   [4] *Sph.* I.def.5.   [5] *Sph.* I.16.
[6] *Elem.* I.c.n.1.   [7] *Elem.* I.c.n.1.

[a] The argument that immediately follows is probably an interpolation.

[b] This final paragraph is probably a late interpolation, although it is contained in our earliest Greek manuscript, **A**, as well as the Arabic translation (Kunitzsch and Lorch 2010, 144–146).

[c] In Heiberg (1927, 74), this sentence reads, "But if *BG* is equal to the ⟨circumference⟩ that the side of the square inscribed in the great circle subtends, we, constructing the same things as in the foregoing, will prove it thus."

N. Then, similarly, we will show also that the circle drawn with a pole, *E*, and a polar distance, *EG*, will pass through point *M*. And, the problem is ⟨done⟩ in two ways.[a]

16 ⟨Theorem⟩   ⟨C106⟩

Great circles in a sphere cutting off similar circumferences of some parallel circles are either through the poles of the parallels or they touch the same ⟨one⟩ of the parallels.

For, let great circles in a sphere, *AHG* and *BQD*, cut off similar circumferences of some parallel circles, *ABGD* and *EZHQ* – that is, *AB* is similar to *EZ*.

I say that *AHG* and *BQD* either are through the poles of the parallels or touch the same ⟨one⟩ of the parallels.

For, either circle *AHG* is through the poles of the parallels or not.

First, let it be through the poles of the parallels.   ⟨Case 1⟩

I say that circle *BQD* is also through the poles of the parallels – that is, that point *K*[b] is the pole of the parallel circles, *ABGD* and *EZHQ*.

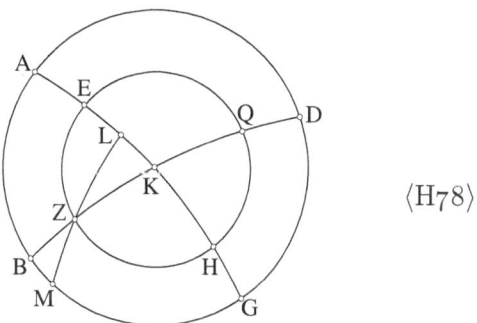

In fact not, but, if possible, let point *L* be the pole of the parallels.[1] Let a great circle, *LZM*,[c] be drawn through points *L* and *Z*.[2]   ⟨H78⟩

---

[1] This is the assumption that any other point, say *L*, is the pole.   [2] *Sph.* I.20.

[a] Following this passage, Heiberg (1927, 76) includes the following paragraph, found in **K**: "But, if *BG* is greater than a quadrant, we will fill in circle *DBGQ* until the other pole, and a ⟨circumference⟩ becoming less than a quadrant from *G*, that until the pole, *F*, of the circle parallel to *AB*, for the ⟨circumference⟩ from the pole of circle *AB*, *D*, to the other pole, *F*, is of a semicircle. But, *BG* is greater than a quadrant. We will take away a ⟨circumference⟩ equal to that which the side of the square subtends, and joining and constructing the same things as in the foregoing, we will show the circle drawn through point *G* touching the given circle, plainly *AB*."
[b] This is our only introduction to point *K*, which must be the intersection of **gC**(*AHG*) and **gC**(*BQD*).
[c] This introduces point *M*, which must be the intersection of **gC**(*LZ*) with **pC**(*ABG*).

Therefore, circumference ABM is similar to EZ.[1] But, EZ is similar to AB. Therefore, MA is similar to AB. And they are of the same circle. Therefore, circumference MA is equal to circumference AB,[2] which is impossible.[3]

Therefore, point L is not the pole of the parallel circles. Then, similarly, we will show that it is no other point, except K. Therefore, point K is the pole of the parallels.

Therefore, circles AHG and BQD are through the poles of the parallels.

⟨C107⟩ But, again, let AHG not be through the poles of the parallels.

⟨Case 2⟩ Then, either it touches circle $EZHQ$, or is oblique to it.

⟨Case 2.1⟩ First, let it touch at E – as holds in the second diagram.[a]

I say that ZB will also touch ⟨it⟩.[4]

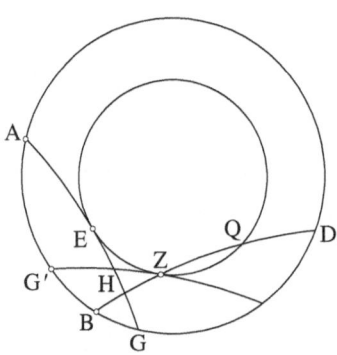

For, if possible, let it not touch. Let a great circle touching EZQ, ZG',[b] be drawn through point Z,[5] making a semicircle from ZG' non-intersecting with the semicircle from EA.

Therefore, circumference $G'A$ is similar to circumference EZ.[6] But, EZ is similar to BA. Therefore, circumference $G'A$ is similar to AB. And they are of the same circle. Therefore, circumference $G'A$ is equal to circumference AB,[7] which is impossible.[8]

Therefore, circle BZ will not not touch circle EZQ. Therefore, it will touch ⟨it⟩.

⟨Case 2.2⟩ But, then, let AHG be oblique to the parallels – as holds in the third

---

[1] *Sph.* II.10, with assumed pole L.   [2] *Elem.* III.def.11, III.26.   [3] *Elem.* I.c.n.5.   [4] That is, touch $\mathbf{pC}(EZQ)$.   [5] *Sph.* II.14 (counterfactual).   [6] *Sph.* II.13.   [7] *Elem.* III.def.11, III.26.   [8] *Elem.* I.c.n.5.

[a] The text refers to the second diagram from this point until the beginning of Case 2.2.

[b] Here and below, what we call G' is simply G (Γ) in the text, but there are problems with both the diagram and the letter-names for Case 2.1. See Commentary.

diagram. Therefore, it will touch two circles, equal to one another and parallel to *ABGD* and *EZHQ*.[1]

I say that *BZQD* is oblique to the parallels and that it will touch the same ⟨circle⟩.

For, if possible, let circle *AEHG* touch one of the parallels, *MX*, at point *L*,[2] and let *BZQD* not touch ⟨it⟩.[a]

Through point *Z*, being between circle *ML* and the ⟨circle⟩ equal and parallel

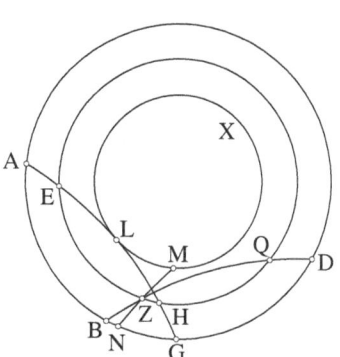

to it, let a great circle, *NZM*, be drawn touching *LM* at point *M*,[3] making the semicircle from *LA* non-intersecting with the semicircle from *MN*.

⟨H80, C108⟩

Therefore, circumference *ABN* is similar to circumference *EZ*.[4] But, circumference *EZ* is similar to *AB*. Therefore, *NA* will also be similar to *AB*. And they are of the same circle. Therefore, circumference *NA* is equal to circumference *AB*,[5] which is impossible.[6]

Therefore, circle *DQB* will not not touch ⟨it⟩.[7] Therefore, *BZQD* will touch ⟨it⟩.

Therefore, circles *AEHG* and *BZQD* touch the same ⟨one⟩ of the parallels.

17 ⟨Theorem⟩

Parallel circles in a sphere cutting off from some great circle equal circumferences abutting the greatest of the parallels are equal, and the greater ⟨the circumferences⟩ the lesser ⟨the circles⟩.

For, let parallel circles in a sphere, *AB* and *GD*, first, cut off equal   ⟨Part 1⟩

---

[1] *Sph.* II.8.   [2] *Sph.* II.8 (constructively).   See Commentary   [3] *Sph.* II.15 (counterfactual).   [4] *Sph.* II.13.   [5] *Elem.* III.dcf.11, III.26.   [6] *Elem.* I.c.n.5.   [7] That is, touch **pC**(*EZHQ*).

[a] This is a combined statement of both the straightforward assumption of Case 2.2 and the counterfactual supposition of the indirect argument. In fact, although it grammatically governs the whole sentence, the "if possible" clause only pertains mathematically to the assumption that **gC**(*BZQD*) does not touch **pC**(*LMX*).

circumferences, *DZ* and *ZB*, of some great ⟨circle⟩, *ABGD*, abutting the greatest circle of the parallels, *EZ*.

I say that circle *AB* is equal to *GD*.

For, let there be a common section of circle *AB* and *ABGD*, *AB*, and a common section of circle *EZ* and *ABGD*, *EZ*, and a common section of circle *GD* and *ABGD*, *GD*.[1]

⟨C109⟩

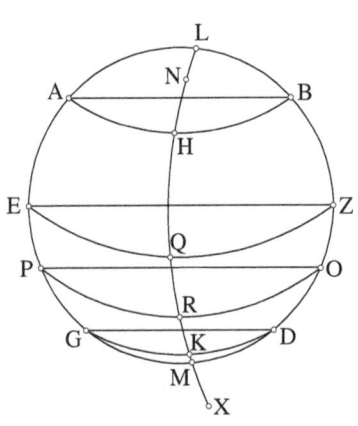

Since two parallel planes, *EQZ* and *GKD*,[a] are cut by some plane, circle *ABGD*, therefore their common sections are parallel.[2] Therefore, *EZ* is parallel to *GD*. Then, similarly, we will show also that *AB* is parallel to *EZ*. And since *EZ* and *GD* are two produced parallel ⟨straight lines⟩ in circle *ABGD*, therefore circumference *DZ* is equal to circumference *EG* – for[b] if we join *ED*, the alternate angles will be equal;[3] while in equal circles equal angles cut off equal circumferences;[4] therefore circumference *EG* will be equal to circumference *ZD*. Then, similarly, we will show also that *BZ* is equal to *AE*. But *BZ* is equal

⟨H82⟩   to *ZD*. Therefore, *AE* is equal to *EG*.[5] Therefore, *AE* and *BZ*, as sum, are equal to *EG* and *ZD*, as sum.[6] So, since circumference *EALBZ*, as a whole, is equal to circumference *EGMDZ*, as a whole – for *EQZ* and *ABDG* are great[7] – of which *AE* and *BZ*, as sum, is equal to *EG* and *ZD*, as sum, therefore *ALB*, as remainder, is equal to *GMD*, as ⟨r⟩emain-

---

[1] *Elem.* XI.3.   See Commentary, *Sph.* II.9.   [2] *Elem.* XI.16.   [3] *Elem.* I.27.
[4] *Elem.* III.26.   [5] *Elem.* I.c.n.1.   [6] *Elem.* I.c.n.2.   [7] *Sph.* I.11.

[a] This introduces points *Q* and *K*, which are now included in the names of **pC**(*EZ*) and **pC**(*GD*).

[b] The following argument is probably a later interpolation, but it may represent the justification that Theodosios intended. Another argument is to consider the diameter perpendicular to both parallels, which will bisect the arcs, by *Elem.* III.3, so that if we take away the smaller arcs from the greater, **Arc**(*DZ*) = **Arc**(*EG*), by *Elem.* I.c.n.3.

der.[1] And they are of the same circle. Therefore, straight line AB is equal to straight line GD.[2]

Then, either circle $ABGD$ cuts circles AHB[a] and GKD through the poles or not.

First, let it cut ⟨them⟩ through the poles. ⟨C110⟩

Therefore, it also bisects them.[3] Therefore, AB is a diameter of cir- ⟨Case 1⟩ cle AHB, and GD of circle GKD.[4] And AB is equal to GD.[5] Therefore, circle AHB is equal to circle GKD.[6]

But, then again, let circle $ABGD$ not cut circles AHB and GKD ⟨Case 2⟩ through the poles. Let the pole of the parallels be taken,[7] and let it be point N. Let a great circle, $LQKMX$, be drawn through point N and one of the poles of circle $ABDG$.[8] Let MX be cut off equal to circumference LN.[9]

So, since circumference LN is equal to circumference MX, let circumference NKM be added, as common. Therefore, circumference LKM, as a whole, is equal to circumference $NKMX$, as a whole.[10] But LKM is a semicircle,[11] therefore $NKMX$ is a semicircle.[12] Therefore, point N is along a diameter with point X. And N is a pole of the parallel circles, therefore, point X is the other pole of the parallel circles.[13] And since in a sphere two circles, $ABDG$ and GKD, cut one another, while a great circle, $LQKX$, is drawn through their poles, therefore $LQKX$ bisects the cut-off segments of the circles.[14] Therefore, circumference GM is equal ⟨H84⟩ to circumference MD. Therefore, circumference GMD is double circumference DM.[15] Then, similarly, we will show also that circumference BLA is double circumference AL. And circumference GMD is equal to ⟨C111⟩ circumference ALB, therefore circumference MD is equal to circumference AL.[16] Then, a segment of a circle, $LQMX$, is set up upright on a diameter,[17] that from L to M, of some circle $ABGD$, and there are equal

[1] *Elem.* I.c.n.3. [2] *Elem.* III.29. [3] *Sph.* I.15. [4] *Elem.* I.def.17. [5] Part 1.
[6] *Elem.* III.def.1. [7] *Sph.* I.21. [8] *Sph.* I.20. [9] *Elem.* III.post.3 (transfer).
[10] *Elem.* I.c.n.2. [11] *Sph.* I.11. [12] *Elem.* I.c.n.1. [13] *Sph.* I.10. [14] *Sph.* II.9.
[15] *Elem.* V.15. [16] *Elem.* V.15. [17] *Sph.* I.15.

[a] This introduces point H, which is normally introduced before Q but is now included in the letter-name of **pC**(AHB). See Commentary.

circumferences, *LN* and *MX*, cut off from it, being less than half the whole, while there are equal circumferences, *AL* and *DM*, cut off from the initial circle, therefore the straight line joining from *N* to *A* is equal to the straight line joining from *X* to *D*.[1] And the straight line joining from *N* to *A* is the polar distance of circle *AHB*, and the straight line joining from *X* to *D* is the polar distance of circle *GKD*.[2] Therefore, the polar distance of circle *AHB* is equal to the polar distance of circle *GKD*.[3] But, those circles of which the polar distances are equal, are equal.[4] Therefore, circle *AHB* is equal to circle *GKD*.

⟨Part 2⟩   But, then again, let circumference *DZ* be greater than circumference *ZB*.

I say that circle *GDK* is less than circle *AHB*.

⟨C112⟩   For, since circumference *DZ* is greater than circumference *ZB*, let an equal to *ZB* be cut off from *DZ*, *ZO*.[5] Let a parallel ⟨circle⟩ to *EQZ* be drawn through *O*, *ORP*.[6]

Therefore, circle *PRO* is equal to circle *AHB*, for circumference *OZ* is equal to circumference *ZB*.[7] But circle *PRO* is greater than circle *GKD*, for circle *PRO* is nearer to the center of the sphere than circle *GKD*.[8] Therefore, circle *AHB* is greater than circle *GKD*, so that circle *GKD* is less than circle *AHB*.

18 ⟨Theorem⟩

Equal and parallel circles in a sphere cut off of some great circle equal circumferences abutting the greatest of the parallels, and the greater ⟨the circles⟩ the lesser ⟨the circumferences⟩.

⟨H86⟩   For, let equal and parallel circles in a sphere, *AB* and *GD*, cut off
⟨Part 1⟩   circumferences, *BZ* and *ZD*, of some great circle, *ABDG*, abutting the greatest of the parallels *EZ*.

I say that circumference *BZ* is equal to circumference *ZD*.

---

[1] *Sph* II.12, where **gC**(*LQMX*) is considered as two perpendicular segments, one in each hemisphere produced by **gC**(*ABGD*). See Commentary. [2] Both by *Sph.* I.def.5. [3] *Elem.* I.c.n.1. [4] See Commentary. [5] *Elem.* I.post.3 (transfer). [6] *Elem.* I.post.3 (spherical). That is, with pole *N*, and distance *NO*. [7] Part 1. [8] *Sph.* I.6.

For, if circumference *BZ* is not equal to *ZD*, therefore neither is circle *AB* equal to circle *GD*.[1] But, it is.

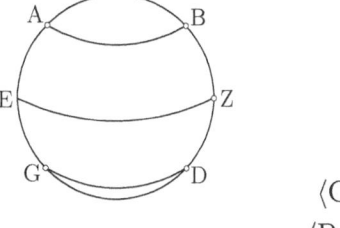

Therefore, *BZ* is equal to *ZD*.

But, again, let circle *AB* be greater than circle *GD*.

⟨C113⟩

⟨Part 2⟩

I say that *BZ* is less than circumference *ZD*.

For, if circumference *BZ* is not less than circumference *ZD*, therefore neither is circle *AB* is greater than circle *GD*.[2] But, it is.

Therefore, circumference *BZ* is smaller than circumference *ZD*.

19 ⟨Theorem⟩

If a great circle in sphere cuts some parallel circles of those in the sphere not through their poles, then it will cut them unequally except for the greatest of the parallels. And, of the cut-off segments in one of the hemispheres, as many as are between the greatest of the parallels and the visible pole[a] are greater than semicircles, and the remaining – as many as are between the greatest of the parallels and the non-visible pole – lesser. And the alternate segments of the equal and parallel circles are equal to one another.

For, let a great circle in a sphere, *ABGD*, cut some parallel circles of those in the sphere, *AD*, *EZ*, and *BG*, not through the poles. Let *EZ* be the greatest of the parallels.

I say that it will cut them unequally except for the greatest of the parallels, *EZ*. And, of the cut-off segments in one of the hemispheres, as many will be greater than semicircles as are between *EZ* and the visible pole, and the remaining, lesser. And the alternate segments of the equal and parallel circles are equal to one another.

⟨C114⟩

For, let there be the visible pole of the parallels, point *Π*.[3] Let a great circle, *HEQ*, be drawn through *E* and *H*.[4] Therefore, *HEQ* filled out will also pass through the point *Z* – for *E* is along a diameter with

⟨H88⟩

---

[1] *Sph.* II.17.   [2] *Sph.* II.17.   [3] See Commentary.   [4] *Sph.* I.20.

[a] The "visible" and "non-visible poles" are technical terms. See Commentary.

Z, through the ⟨fact that⟩ each of *EZ* and *ABGD* is great. Let it go, and let it be as *HNZK*. Let circle *BG* be filled out to points *Q* and *K*.[1]

⟨Part 1⟩

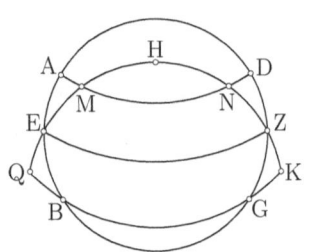

So, since a great circle in a sphere, *QEMHNZK*, cuts some circles of those in the sphere, *AMND*, *EZ*, and *QBGK*, through the poles, it will bisect them at an upright.[2] Therefore, each of *MN*, *EZ*, and *QBGK* is a semicircle.[3] So, since *MN* is a semicircle, therefore *AMND* is greater than a semicircle.[4] Then, similarly, we will show also that all segments between circle *EZ* and the pole *H* are greater than a semicircle. Again, since *QBGK* is a semicircle, therefore *BG* is less than a semicircle.[5] Then, similarly, we will show also that all segments between circle *EZ* and the non-visible pole, and being in the same hemisphere, are less than semicircles.

But again, let circle *AD* be equal and parallel to circle *BG*.

I say that the alternate segments of circles *AD* and *BG* are equal to one another.

⟨C115⟩
⟨Part 2⟩
For, since circle *AD* is both equal and parallel to circle *BG*, therefore circumference *AE* is equal to circumference *EB*, and circumference *DZ* is equal to circumference *ZG*[6] – but *AE* is equal to *DZ*, and *EB* to *ZG*[7] – therefore *AE* and *DZ*, as sum, are equal to *EB* and *ZG*, as sum.[8] And, *EADZ*, as a whole, is equal to *EBGZ*, as a whole,[9] therefore *AD*, as remainder, is equal to *BG*, as remainder[10] – through the ⟨fact that⟩ each of *EADZ* and *EBGZ* is a semicircle, for *ABGD* and *EZ* are great ⟨circles⟩. And circumferences *AD* and *BG* are of the same circle.[11] Therefore, the straight line joining from *A* to *D* is equal to the straight line joining from *B* to *G*.[12] And, the straight line joining from *A* to *D* is subtended by the circumference *AD*, and the straight line joining from *B* to *G* is subtended by the circumference *BG*.[13] But in equal circles

[1] *Elem.* I.post.3 (spherical). See Commentary.   [2] *Sph.* I.15.   [3] *Elem.* I.def.18.
[4] *Elem.* I.c.n.5. [5] *Elem.* I.c.n.5. [6] Both by *Sph.* II.18. [7] *Sph.* II.13. [8] *Elem.* I.c.n.2.
[9] They are both semicircles, *Sph.* I.11. [10] *Elem.* I.c.n.3. [11] That is, **Arc**(*AD*) and
Arc(*BG*) in **gC**(*ABCD*). [12] *Elem.* III.29. [13] That is, **Arc**(*AD*) in **pC**(*AD*) and
Arc(*BG*) in **pC**(*BG*).

equal straight lines cut off equal circumferences, the greater ⟨equal⟩ to the greater and the lesser ⟨equal⟩ to the lesser.[1] Therefore, the greater circumference of circle *AD* is equal to the greater circumference of circle *BG*, while the smaller circumference of circle *AD* is equal to the smaller   ⟨H90⟩ circumference of circle *BG*. And *AD* is greater than a semicircle, while *BG* is less. Therefore, the alternate segments of circles *AD* and *BG* are equal to one another.

20 ⟨Theorem⟩   ⟨C116⟩

If a great circle in a sphere cuts some parallel circles in the sphere not through the poles, then of the cut-off circumferences in one of the hemispheres, those nearer to the visible pole will be ever greater than similar to those more distant.

For, let a great circle in a sphere, *ABDG*, cut some parallel circles in the sphere, *AB*, *GD*, and *EZ*, not through the poles.

I say that, of the cut-off circumferences in one of the hemispheres, those nearer to the visible pole will be ever greater than similar to those farther away – that is, *AB* is greater than similar to *GD*, and *GD* to *EZ*.

For, let there be the visible pole of the parallels, point *H*.[2] Let great circles, *HQG* and *HKD*, be drawn through *H* and each of *G* and *D*.[3] Therefore, *HQG* and *HKD* cut off similar circumferences between themselves.[4] Therefore, *QK* is similar to *GD*.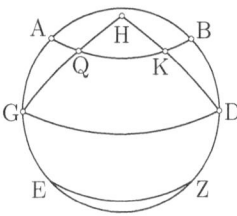

Therefore, *AQKB* is greater than similar to *GD*.[5] Then, similarly, we will show also that *GD* is greater than similar to *EZ* – with[a] us having drawn great circles through *H* and each of *E* and *Z*; and, it is possible to show ⟨this⟩ also with us not having drawn ⟨them⟩, only filling out *EZ*, as in the ⟨proposition⟩ before this.

---

[1] *Elem.* III.28.   [2] See Commentary, *Sph.* II.19.   [3] *Sph.* I.20.   [4] *Sph.* II.10.
[5] *Elem.* III.def.11, I.c.n.5. See Commentary.

[a] This final passage is probably a late interpolation.

⟨C117⟩

21 ⟨Theorem⟩

If great circles in equal spheres are inclined upon great circles, whichever one's pole is higher will be more inclined, and those circles of which the pole is equally distant from the plane are similarly inclined.[a]

⟨H92⟩ For, let there be great circles in equal spheres, *BKD* and *ZLQ*, inclined upon great circles, *ABGD* and *EZHQ*. Let there be the pole of circle *BKD*, point *M*, and the pole of circle *ELQ*, point *N*. Let *M* be higher than *N*.

I say that circle *BKD* is more inclined to circle *ABGD* than circle *ZLQ* is to circle *EZHQ*.

⟨Part 1.1⟩ For, let a great circle, *AKMG*, be drawn through *M* and one of the poles of circle *ABGD*, and a great circle, *ELNH*, through *N* and one of the poles of circle *EZHQ*.[1] And let there be a common section of circle *ABGD* and *BKD*, *BD*, and a common section of circle *ABGD* and *AKMG*, *AG*, and a common section of *BKD* and *AKMG*, *KX*, and, furthermore, a common section of circle *EZHQ* and *ZLQ*, *ZQ*, a common section of circle *EZHQ* and *ELNH*, *EH*, a common section of *ZLQ* and *ELNH*, *LO*.[2]

⟨C118⟩

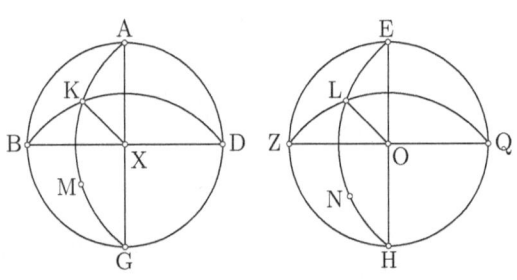

Since a great circle in a sphere, *AKMG*, cuts some circles of those in the sphere, *ABGD* and *BKD*, through the poles, it will bisect them at an upright.[3] Therefore, circle *AKMG* is upright to each of circles *ABGD* and *BKD*, so that each of *ABGD* and *BKD* is upright to circle *AKMG*. But, if two planes cutting one another are at an upright to some plane, their common section will also be at an upright to the same plane.[4] Therefore, the common sec-

---

[1] Both by *Sph.* I.20.   [2] All by *Elem.* XI.3. See Commentary, *Sph.* II.9.   [3] *Sph.* I.15.
[4] *Elem.* XI.19.

[a] See the Commentary for a discussion of the technical terminology of this proposition.

tion of them, of *ABGD* and *BKD*,[a] is upright to circle *AKMG*, so that
it will make angles upright to all straight lines contacting it and being
in the same plane, that of circle *AKMG*.[1] But each of *KX* and *XA*,
being in the plane of circle *AKMG*, contacts *BXD*. Therefore, *BX* is
upright to each of *KX* and *XA*. And since two planes, *ABGD* and *BKD*,
cut one another, and *KX* and *XA* are produced at an upright to their
common section, *BD*, and *KX* is in the plane of circle *BKD*, while *XA*
is in the plane of circle *ABGD*, therefore angle *KXA* is the inclination
with which plane *BKD* is inclined to plane *ABGD*.[2] Then, similarly,
we will show also that angle *LOE* is the inclination with which plane   ⟨H94⟩
ZLQ is inclined to plane *EZHQ*.

Then, I say that angle *KXA* is less than angle *LOE*.

For, since *M* is higher than *N*, therefore the perpendicular produced   ⟨C119⟩
from *M* to the plane of circle *ABGD* is greater than the perpendicular   ⟨Part 1.2⟩
produced from *N* to the plane of circle *EZHQ*.[3] But the perpendicular
produced from *M* to the plane of circle *ABGD* falls on the common
section of circle *AKMG* and *ABGD*, that is *AG* – through the ⟨fact
that⟩ planes *AKMG* and *ABGD* are upright to one another – while the
perpendicular produced from *N* to the plane of circle *EZHQ* falls on
EH.[4] Therefore, the perpendicular produced from *M* to *AG* is greater
than the perpendicular produced from *N* to *EH*. So, since two segments
of circles, *AKMG* and *ELNH*, are equal, and *M* and *N* are arbitrary
points, and the perpendicular produced from *M* to *AG* is greater than
the perpendicular produced from *N* to *EH*, therefore circumference *MG*
is greater than circumference *NH*.[5] But, circumference *MK* is equal to
circumference *NL*, for each of them is equal to that which the side of the
square inscribed in the great circle subtends.[6] Therefore, circumference
KMG, as a whole, is greater than circumference *LNH*, as a whole.[7] So,
since *AKMG*, as a whole, is equal to *ELNH*, as a whole,[8] of which

---

[1] *Elem.* XI.def.3.   [2] *Elem.* XI.def.6(∗).   [3] *Elem.* III.15. See Commentary.   [4] Both
by *Elem.* XI.def.4.   [5] Scholium 274. See Commentary.   [6] *Sph.* I.16, *Elem.* III.28.
[7] *Elem.* I.c.n.2 (extended to inequalities).   [8] *Sph.* I.11.

[a] Heiberg (1927, 92), following **K**, reads "the common section of *ABGD* and *BKD*,
BXD."

KMG is greater than *LNH*, therefore circumference *AK*, as remainder, is less than circumference *EL*, as remainder.[1] And an angle, *KXA*, stands on *AK*, and an angle, *LOE*, on *LE*. Therefore, angle *KXA* is less than angle *LOE*.[2] And angle *KXA* is the inclination with which plane *BKD* is inclined to the plane of circle *ABGD*, and angle *LOE* the inclination with which plane *ZLQ* is inclined to plane *EZHQ*. Therefore, the inclination of circle *BKD* is less than the inclination of circle *ZLQ* to circle *EZHQ*. Therefore, circle *BKD* is more inclined to circle *ABGD* than circle *ZLQ* to circle *EZHQ*.

⟨C120⟩

⟨H96⟩
⟨Part 2⟩

But, again, let the poles of circles *BKD* and *ZLQ* be equally distant from the planes – that is, let the perpendicular produced from *M* to the plane of circle *ABGD* be equal to the perpendicular produced from *N* to the plane of circle *EZHQ*.

I say that circles *BKD* and *ZLQ* are similarly inclined to *ABGD* and *EZHQ* – that is,[a] that angle *KXA* is equal to angle *LOE*.[3]

For, with the same constructions, similarly, we will show that angle *KXA* is the inclination with which the plane of circle *BKD* is inclined to the plane of circle *ABGD*, and angle *LOE* is the inclination with which the plane of circle *ZLQ* is inclined to the plane of circle *EHQ*.

So, I say that angle *KXA* is equal to angle *LOE*.

⟨C121⟩

For, since the perpendiculars produced from *M* and *N* to the planes of circles *ABGD* and *EZHQ* are equal, and the perpendiculars produced from *M* and *N* to the planes of circles *ABGD* and *EZHQ* fall to *AG* and *EH*,[4] therefore the perpendiculars produced from points *M* and *N* to straight lines *AG* and *EH* are equal. So, since two segments of circles, *AKMG* and *ELNH*, are equal, and *M* and *N* are arbitrary points, and the perpendicular produced from *M* to *AG* is equal to the perpendicular produced from *N* to *EH*, therefore circumference *MG* is equal to circumference *NH*.[5] But, *MK* is also equal to *LN*, for each of them is

---

[1] *Elem.* I.c.n.3 (extended to inequalities).   [2] *Elem.* III.27.   [3] *Sph.* I.def.6. See Commentary.   [4] *Elem.* XI.def.4.   [5] Scholium 274. See Commentary.

[a] This clause is probably an interpolation. Notice that it is repeated below.

equal to that which the side of the square inscribed in the great circle subtends.[1] Therefore, *KMG*, as a whole, is equal to circumference *LNH*, as a whole,[2] and *AKMG*, as a whole, is equal to *ELNH*, as whole, therefore circumference *AK*, as a remainder, is equal to circumference *EL*, as remainder.[3] And an angle, *KXA*, stands on circumference *AK*, and an angle, *LOE*, on *LE*. Therefore, angle *KXA* is equal to angle *LOE*.[4] And angle *KXA* is the inclination with which the plane of circle *BKD* is inclined to the plane of circle *ABGD*, while angle *LOE* is the inclination with which the plane of circle *ZLQ* is inclined to the ⟨plane⟩ of circle *EZHQ*. Therefore, the inclination with which circle *BKD* is inclined to circle *ABGD* is equal to the inclination with which ⟨the plane⟩ of circle *ZLQ* is inclined to circle *EZHQ*. Therefore, circles ⟨H98, *BKD* and *ZLQ* are similarly inclined to circles *ABGD* and *EZHQ* – for[a]   C122⟩ we learned that a plane is said to be similarly inclined to a plane as another to another when the straight lines produced at upright angles to the common section of the planes in each of the planes contains equal angles.[5]

22 ⟨Theorem⟩

If a great circle in a sphere touches some circle of those in the sphere, and it cuts another parallel to it, being between the center of the sphere and that ⟨circle⟩ which the great circle touches, and, furthermore, the pole of the great ⟨circle⟩ is between the parallels,[6] and great circles are drawn touching the greater of the parallels, then they will be inclined to the ⟨initial⟩ great circle, and the most upright will be that having the ⟨point of⟩ contact at the bisector of the greater segment,[7] while the lowest is that having the ⟨point of⟩ contact at the bisector of the lesser segment,[8] while of the others, those equally distant from whichever of the bisectors are similarly inclined, but that having the ⟨point of⟩ contact farther

---

[1] *Sph.* I.16, *Elem.* III.28.   [2] *Elem.* I.c.n.2.   [3] *Elem.* I.c.n.3.   [4] *Elem.* I.27.
[5] *Sph.* I.def.6. See Commentary.   [6] See introductory Commentary to both *Sph.* II.22 and II.23 for a discussion of these bounds.   [7] That is, the bisector of the greater segment of the greater parallel circle.   [8] Again, of the greater parallel circle.

[a] This final clause is probably an interpolation. See Commentary.

from the bisector of the greater segment[1] *will be ever more inclined than the nearer, and, furthermore, the poles of the great ⟨circles⟩ will be on one circle both parallel to and less than ⟨the circle⟩ that the initial great circle touches.*

⟨C123⟩ For, let a great circle in a sphere, *ABG*, touch some circle of those in the sphere, *AD*, at point *A*. Let it cut another parallel to it, *EZHQ*, being between the center of the sphere and circle *AD*. Furthermore, let the pole of circle *ABG* be between *AD* and *EZHQ*, and let it be point *K*. Let great circles, *MNX*, *BZG*, *OPR*, *ST*, and *UQ*, be drawn touching the greater of the parallels, *EZHQ*. Let circle *BZG* be at the bisector of the greater segment of *EZHQ* at point *Z*, *UQ* at the bisector of the smaller segment of *EZHQ* at point *Q*, and *MNX* and *OPR* equally distant from whichever of the bisector, and let *TS* be as arbitrary.

I say that circles *MNX*, *BZG*, *OPR*, *ST*, and *UQ* will be inclined
⟨H100⟩ to circle *ABG*, and *BZG* will be the most upright of them, and *UQ* the lowest, *MNX* and *OPR* are similarly inclined, and *ST* will be more inclined to *ABG* than *OPR*. And, furthermore, the poles of *MNX*, *BZG*, *OPR*, *ST*, and *UQ* will be on one circle both parallel to and less than *AD*.

⟨C124⟩ For, let the pole of the parallels *AD* and *EZHQ* be taken,[2] and let it be point *L*. Let a great circle, *AL*, be drawn through points *A* and *L*.[3] And, since two circles in a sphere, *ABG* and *AD*, touch one another and a great circle, *AL*, is drawn through the poles of one and the ⟨point of⟩ contact, therefore *AL* will pass through the poles of circle *ABG*,[4] and will be upright to it.[5] And point *K* is a pole of circle *ABG*. Therefore, *AL* filled out will pass through *K*. Let it go, and let it be as *ALK*. And, since two circles in a sphere, *ABG* and *EZHQ*, cut one another and a great circle, *ALK*, is drawn through their poles, therefore *ALK* bisects the cut-off segments of the circles.[6] And *Z* is the bisector of segment *EZH*, while *Q* is bisector of *EQH*. Therefore, *ALK* filled out will pass through points *Z* and *Q*. Let it go, and let it be as *QALKZ*.

---

[1] Again, of the greater parallel circle.   [2] *Sph.* I.21.   [3] *Sph.* I.20.   [4] *Sph.* II.5.
[5] *Sph.* I.15.   [6] *Sph.* II.9.

And, since point K is a pole of circle ABG, and ABG is great, therefore
AK is that which the side of the square inscribed in the great circle
subtends.[1] Therefore, AKZ is greater than that which the side of the
square inscribed in the great circle subtends.[2] And, since ⟨circle⟩ $EZHQ$
is less than great, and it is between the center of the sphere and AD,
and point L is its pole, therefore LZ is less than that which the side
of the square inscribed in the great circle subtends.[3] So, since AKZ is   ⟨C125⟩
greater than that which the side of the square inscribed in the great
circle subtends, while LZ is less, therefore if we cut off a circumference
equal to what the side of the square inscribed in the great circle subtends
from point Z, it will fall between points A and L. Let it be cut off equal   ⟨H102⟩
to it,[4] and let it be ZF. Let a circle, $FCUW$, be drawn with a pole,
L, and a distance, LF.[5] Therefore, it is parallel to AD and $EZHQ$.[6]
Let great circles, NLW, PLC, and TLY, be drawn through point L and
each of points P, N, and T.[7]

Since NL is equal to LZ –   ⟨Part 1⟩
for ⟨they are⟩ from a pole of cir-
cle $EZHQ$ – and LW is equal to
LF – for ⟨they are⟩ from a pole
of YWC[8] – therefore NLW, as
a whole, is equal to ZLF, as a
whole.[9] And ZLF is equal to that
which the side of the square inscribed in the great circle subtends, therefore NLW is equal to
that which the side of the square
inscribed in the great circle sub-

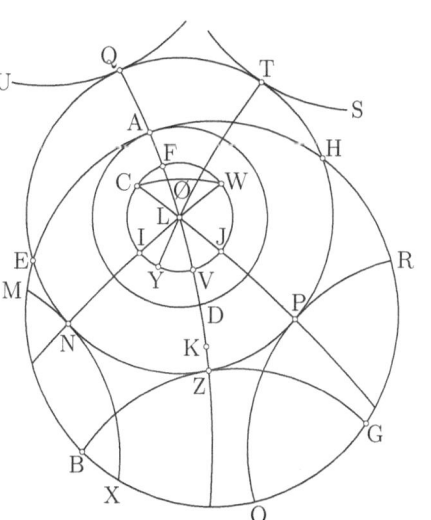

tends.[10] Then, similarly, we will show also that each of CLP, YLT, and
VLQ is equal to that which the side of the square inscribed in the great
circle subtends. And, since two circles in a sphere, MNX and $EZHQ$,

---

[1] *Sph.* I.16.   [2] *Elem.* I.c.n.5.   [3] *Sph.* I.def.5, I.16.   [4] *Elem.* I.post.3 (transfer).
[5] *Elem.* I.post.3 (spherical).   [6] *Sph.* II.2.   [7] *Sph.* I.20.   [8] Both by *Sph.* I.def.5.
[9] *Elem.* I.c.n.2.   [10] *Elem.* I.c.n.1.

touch one another and a great circle, *NLW*, is drawn through the poles
of one and the ⟨point of⟩ contact, therefore *NLW* will also pass through
the poles of *MNX*,[1] and will be upright to it.[2] And since *MNX* is a great
⟨C126⟩   circle, therefore its polar distance is equal to that which the side of the
square inscribed in the great circle subtends.[3] And *NLW* is equal to
that which the side of the square inscribed in the great circle subtends,
therefore the ⟨straight line⟩ from *N* to *W* is the polar distance of circle
MNX. Therefore, point *W* is a pole of circle *MNX*.[4] Then, similarly,
we will show also that point *F* is a pole of circle *BZG*, and *C* of *OPR*,
and *Y* of *ST*, and *V* of *UQ*. Therefore, circles *MNX*, *BZG*, *OPR*, *ST*,
and *UQ* have their poles on one circle, parallel to and less than *AD*.

Then, I say that circles *MNX*, *BZG*, *OPR*, *ST*, and *UQ* are inclined
to circle *ABG*, and that of them *BZG* is the most upright, *UQ* is lowest,
MNX and *OPR* are similarly inclined, and *ST* is more inclined to circle
ABG than *OPR* to *ABG*.

⟨Part 2.1⟩   For, since circumference *NZ* is equal to *ZP*, and they are of the same
circle, therefore circumference *NZ* is similar to circumference *ZP*.[5] But
⟨H104⟩   *NZ* is similar to *IV*, while circumference *ZP* is similar to circumference
JV.[6] And, they are of the same circle. Therefore, circumference *IV*
⟨C127⟩   is equal to circumference *JV*.[7] But, circumference *IV* is equal to circumference *FW* – for ⟨they are⟩ vertical – and *JV* to *CF*.[8] Therefore,
circumference *FW* is equal to *CF*.[9] Then, a segment of a circle, *VKZ*
and what is continuous with it, is set up upright on a diameter, that
from *F* to *V*, of some circle, *FCYW*, and *VK* is a cut-off circumference,
being shorter than half of the whole segment, and *CF* and *FW* are equal
circumferences cut off from the initial circle, therefore the straight line
joining from point *K* to *C* is equal to the straight line joining from *K* to
W.[10] Therefore, a circle drawn with a pole, point *K*, and a distance, *KC*,
will also pass through *W*.[11] Let it go,[12] and let it be as *CW*. Therefore,
CW is parallel to *ABG*, for it is about the same poles with it, for point

---

[1] *Sph.* II.5.   [2] *Sph.* I.15.   [3] *Sph.* I.16.   [4] *Sph.* I.def.5.   [5] *Elem.* III.27, III.def.11.
[6] Both by *Sph.* II.10. This sentence introduces points *I* and *V*.   [7] *Elem.* III.def.11,
I.c.n.1.   [8] *Sph.* I.15, *Elem.* I.15, III.26.   See Commentary.   [9] *Elem.* I.c.n.1.
[10] *Sph.* II.12.   [11] *Sph.* I.def.5.   [12] *Elem.* I.post.3 (spherical).

K is the pole of circle *ABG*.[1] And, since the circle *CW* is parallel to circle *ABG*, therefore the perpendicular produced from *C* to the plane of circle *ABG* is equal to the perpendicular produced from *W* to the plane of circle *ABG*[2] – then,[a] similarly, to the perpendicular produced from Ø to the plane of circle *ABG*. And, the perpendicular produced from *C* to the plane of circle *ABG* is greater than the perpendicular produced from *F* to the plane of circle *ABG*.[3] Therefore, the perpendicular produced from *W* to the plane of circle *ABG* is greater than the perpendicular produced from *F* to the plane of circle *ABG* – for[b] each ⟨C128⟩ of them is equal to that produced from Ø. Therefore, *W* is higher then *F*. And pole *W* is of circle *MNX*, and *F* of *BZG*. Therefore, the pole of *MNX* is higher than the pole of circle *BZG*. But, circles the poles of which are higher are more inclined.[4] Therefore, circle *MNX* is more inclined than circle *BZG*. Therefore, circle *BZG* is more upright than circle *MNX*. Then, similarly, we will show also that, of all the circles touching *EZHQ*, *BZG* is the most upright. Therefore, *BZG* is the most upright.

Then, I say that also *UQ* is the lowest. ⟨H106⟩

For, since the perpendicular produced from *V* to the plane of circle ⟨Part 2.2⟩ *ABG* is greater than the perpendicular produced from *Y* to the plane of circle *ABG*,[5] therefore *V* is higher than *Y*. And *V* is the pole of circle *UQ*, while *Y* is the pole of *ST*. But, circles the poles of which are higher are more inclined.[6] Therefore, circle *UQ* is more inclined to circle *ABG* than *ST*. Therefore, *UQ* is lower than *ST*. Then, similarly, we will show also that *UQ* is lower than all of the ⟨great circles⟩ touching circle *EZHQ*. Therefore, *UQ* is the lowest.

Since the perpendicular produced from *W* to the plane of the circle ⟨C129⟩ *ABG* is equal to the perpendicular produced from *C*, therefore points ⟨Part 2.3⟩

---

[1] *Sph.* II.2.   [2] *Elem.* XI.6, XI.16, I.33 (Scholium 300).   [3] See Commentary.
[4] *Sph.* II.21.   [5] See Commentary.   [6] *Sph.* II.21.

[a] This phrase is probably an interpolation. This passage and that immediately below are the only uses of the letter-name Ø (I) in the treatise. See Commentary.
[b] Again, this phrase is probably an interpolation.

W and C are equally distant from the plane. And point W is a pole of circle MNX, and point C is a pole of circle OPR. Therefore, the poles of circles MNX and OPR are equally distant from the plane. But circles the poles of which are equally distant from the plane are similarly inclined.[1] Therefore, circles MNX and OPR are similarly inclined to circle ABG.

⟨Part 2.4⟩   Again, since the perpendicular produced from Y to the plane of circle ABG is greater than the perpendicular produced from C to the plane of circle ABG,[2] therefore Y is higher than C. And Y is a pole of ST, while C is a pole of OPR. Therefore, the pole of circle ST is higher than the pole of circle OPR. But circles the poles of which are higher are more inclined.[3] Therefore, circle ST is more inclined to circle ABG than OPR.

Therefore, circles MNX, BZG, OPR, ST, and UQ are inclined to circle ABG, and BZG is the most upright of them, and UQ the lowest, and MNX and OPR are similarly inclined, and circle ST is more inclined to circle ABG than OPR, and, furthermore, their poles are on a circle both parallel to and less than AD.

⟨H108,   23 ⟨Theorem⟩
C130⟩
With the same assumptions, if the circumferences passing down from the junctions are equal, then the aforementioned great circles will be similarly inclined.

For, let the circumferences, NM and PR, passing down from the junctions, N and P, be equal.

I say that circles MNX and OPR are similarly inclined to circle ABG.

For, let the pole, point L, of the parallel circles, AD and EZHQ, be taken.[4] Let a great circle, QALKZU, be drawn through points A and L.[5] Then, it is obvious that it will go through point K, being a pole of circle ABG.[6] Let great circles, LNB and LPG, be drawn through L and each of N and P.[7]

---

[1] *Sph.* II.21. [2] See Commentary. [3] *Sph.* II.21. [4] *Sph.* I.21. [5] *Sph.* I.20. [6] *Sph.* II.5.
[7] *Sph.* I.20.

Since two circles in a sphere,
EZHQ and *MNX*, touch one another, and a great circle, *LNB*, is
drawn through the poles of one and
the ⟨point of⟩ contact, therefore
LNB will both pass through the
poles of *MNX*,[1] and will be upright
to it.[2] Then, similarly, we will show
also that *LPG* is through the poles
of *PR*, and is upright on it. And

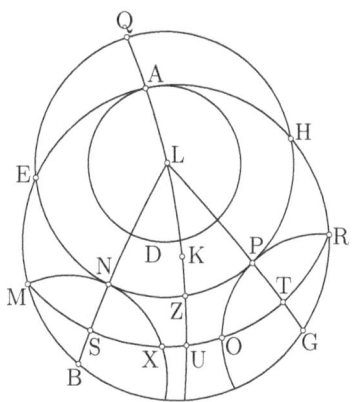

since, in equal circles, *MNX* and *OPR*, equal and upright segments of
circles, *NL* and *PL* and what is continuous with them, are set up on
diameters, those from points *N* and *P*, and there are equal circumferences taken away from them, *NL* and *PL*, being less than half of the
whole, while there are equal circumferences taken away from the initial   ⟨C131⟩
circles, *MN* and *PR*, therefore the straight line joining from *L* to *M* is
equal to the straight line joining from *L* to *R*.[3] Therefore, the circle
drawn with a pole, *L*, and a distance, *LM*, will also pass through *R*.[4]
Let it go,[5] and let it be as *MXOR*. And it is a circle parallel to *AD*
and *EZHQ*, for it is about the same poles as them.[6] And since two circles in a sphere, *ABG* and *MXOR*, cut one another, and a great circle,
QALKZU, is drawn through their poles, therefore *QALKZU* will bisect   ⟨H110⟩
the cut-off segments of the circles.[7] Therefore, circumference *MU* is
equal to circumference *UR*. Again, since two circles in a sphere, *MNX*
and *MXUR*, cut one another, and a great circle, *LNB*, is drawn through
the poles, therefore *LNB* will bisect the cut-off segments of the circles.[8]
Therefore, circumference *MN* is equal to circumference *NX*, and *MS* to
SX. Then, similarly, we will show also that *OP* is equal to *PR*, and
OT to *TR*. So, since circumference *MN* is equal to circumference *PR*,
and *MNX* is double *MN*, and *OPR* double *PR*, therefore *MNX* is also
equal to *OPR*.[9] And the circles are equal. Therefore, the straight line

---

[1] *Sph.* II.5.   [2] *Sph.* I.15.   [3] *Sph.* II.12.   [4] *Sph.* I.def.5.   [5] *Elem.* I.post.3 (spherical).
[6] *Sph.* II.2.   [7] *Sph.* II.9.   [8] *Sph.* II.9.   [9] *Elem.* V.15.

⟨C132⟩

subtending circumference *MNX* is equal to the straight line subtending circumference *OPR*.[1] But, the straight line subtending circumference *MNX* also subtends *MSX*, and the straight line subtending *OPR* also subtends *OTP*. Therefore, the straight line subtending circumference *MSX* is equal to the straight line subtending *OTP*. And they are of the same circle. Therefore, circumference *MSX* is equal to circumference *OTR*.[2] And *MS* is half of circumference *MSX* and *RT* half of *OTR*. Therefore, *MS* is equal to *PT*.[3] But, also, *MSXU*, as a whole, is equal to *UOTR*, as a whole. Therefore, *SXU*, as remainder, is equal to *UOT*, as remainder.[4] And they are of the same circle. Therefore, circumference *SXU* is similar to circumference *UOT*.[5] But, circumference *SXU* is similar to *NZ*, while circumference *UOT* is similar to circumference *PZ*,[6] therefore *NZ* is similar to *ZP*. And they are of the same circle, therefore circumference *NZ* is equal to circumference *ZP*.[7] Therefore, circles *MNX* and *OPR* are equally distant from whichever of the bisectors. But, ⟨circles⟩ equally distant from whichever of the bisectors are similarly inclined.[8] Therefore, circles *MNX* and *OPR* are similarly inclined to circle *ABG*.

End of ⟨Book⟩ II.

---

[1] *Elem.* III.29.   [2] *Elem.* III.28.   [3] *Elem.* V.15.   [4] *Elem.* I.c.n.3.   [5] *Elem.* III.27, III.def.11.   [6] Both by *Sph.* II.10.   [7] *Elem.* III.27, III.def.11.   [8] *Sph.* II.22.

1 ⟨Theorem⟩

If some straight line is produced through in a circle, cutting the circle unequally, and an upright segment of a circle not greater than a semicircle is set up on it, while the circumference of the set-up segment is divided unequally, then the straight line subtending the lesser circumference is the least of all straight lines falling from the same point to the greater circumference of the initial circle. But, if the produced-through ⟨straight line⟩ is a diameter of the circle, while the rest is supposed the same, the aforementioned straight line will be the least of all straight lines falling from the same point to the circumference of the initial circle, while the straight line subtending the greater ⟨circumference⟩ will be the greatest.

For, let some straight line, *BD*, be produced through in a circle, *ABGD*, cutting the circle unequally, and let circumference *BGD* be greater than circumference *BAD*. Let an upright segment of a circle, *BED*, not greater than a semicircle be set up on *BD*. Let circumference *BED* be divided unequally at point *E*. Let *EB* be joined.

I say that *BE* is the least of all straight lines falling from point *E* to circumference *BGD*.

For, let a perpendicular, *EZ*, be produced from point *E* onto the plane of circle *ABG*.[1] Then, it will fall on the common section of planes *ABGD* and *BED* – on the straight line *BD* – because segment *BED* is upright to

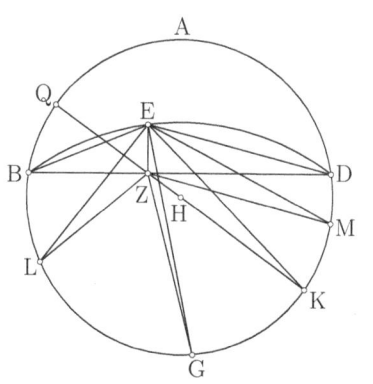

⟨C134⟩
⟨Case 1⟩
⟨Part 1⟩

circle *ABGD*.[2] Let the center of circle *ABGD* be taken,[3] and let it be *H*. Let a joined *ZH* be extended in directions *Q* and *K*.[4] From point *E*, let *EL* be extended forth to circumference *BGD*.[5] Let *ZL* be joined.[6]

---

[1] *Elem.* XI.11. [2] *Elem.* XI.def.4. [3] *Elem.* III.1. [4] *Elem.* I.post.1, I.post.2.
[5] *Elem.* I.post.1 (extended to object). See Commentary. [6] *Elem.* I.post.1.

⟨H114⟩

Since *EZ* is upright to the plane of circle *ABGD*, therefore it will also make angles upright to all straight lines contacting it and lying in the plane of circle *ABGD*.[1] But each of *ZB* and *ZL*, being in the plane of circle *ABGD*, contacts *EZ*. Therefore, each of angles *BZE* and *LZE* is upright. And since *BZ* is less than *ZL*,[2] therefore the square on *BZ* is less than the square on *ZL*.[3] Let the square on *EZ* be added, as common. Therefore, the squares on *EZ* and *ZB* are less than the squares on *EZ* and *ZL*.[4] But, the square on *BE* is equal to the squares on *EZ* and *ZB*, while the square on *LE* is equal to the squares on *EZ* and *ZL*.[5] Therefore, the square on *BE* is less than the square on *LE*. Therefore, *BE* is less than *LE*.[6] Then, similarly, we will show that *BE* is the least of all the straight lines falling from point *E* to circumference *BGD*. Therefore, *BE* is the least of all the straight lines falling from point *E* to circumference *BGD*.

⟨C135⟩

I say also that, of the straight lines produced through from point *E* between points *B* and *K*, that nearer to it is ever less than that more distant.

⟨Part 2⟩

For, let some other, *EG*, be produced through.[7] Let *ZG* be joined.[8]

Since *LZ* is less than *ZG*,[9] therefore the square on *ZL* is also less than the square on *ZG*.[10] Let the square on *ZE* be added, as common. Therefore, the squares on *EZ* and *ZL* are less than the squares on *EZ* and *ZG*.[11] But, the square on *LZ* and *ZE* are equal to the square on *LE*, while the squares on *GZ* and *ZE* are equal to the square on *EG*.[12] Therefore, the square on *LE* is less than the square on *EG*. Therefore, *LE* is less than *EG*.[13] Then, similarly, we will also show that, of the straight lines produced through from the point *E* between *B* and *K*, that nearer to *EB* is ever less than that more distant.

Then, let *EK* and *ED* be joined.[14]

---

[1] *Elem.* XI.def.3.   [2] *Elem.* III.7.   [3] *Elem.* VI.20.cor.   [4] *Elem.* I.c.n.2 (extended to inequalities).   [5] Both by *Elem.* I.47.   [6] *Elem.* VI.20.cor.   [7] *Elem.* I.post.1 (extended to object).   See Commentary.   [8] *Elem.* I.post.1.   [9] *Elem.* III.7.   [10] *Elem.* VI.20.cor.   [11] *Elem.* I.c.n.2 (extended to inequalities).   [12] Both by *Elem.* I.47.   [13] *Elem.* VI.20.cor.   [14] *Elem.* I.post.1.

Then, again, I say that *EK* is greatest of all the straight lines falling from point *E* to circumference *KD*, while *ED* is the least of all the straight lines produced through from point *E* between points *D* and *K*.

Since *KZ* is greater than *GZ*,[1] therefore the square on *KZ* is also 〈Part 3〉 greater than the square on *ZG*.[2] Let the square *EZ* be added, as common. Therefore, the squares on *KZ* and *ZE*, that is the square on *EK*, is greater than the squares of *EZ* and *ZG*, that is the square on *EG*.[3] Therefore, *KE* is greater than *EG*.[4] Then, similarly, we will show that, 〈C136〉 *EK* is greater than all the straight lines falling from point *E* to circumference *KD*. Therefore, *EK* is the greatest of all straight lines falling from point *E* to circumference *BGD*.[a]

I say also that *ED* is the least of all straight lines falling from point 〈H116〉 *E* between points *K* and *D*.

For, let another, *EM*, be produced through.[5] Let *MZ* be joined.[6] 〈Part 4, 5〉

Since *DZ* is less than *ZM*,[7] therefore the square on *DZ* is less than the square on *ZM*.[8] Let the square on *ZE* be added, as common. Therefore, the squares on *EZ* and *ZD*, that is the square on *ED*, is less than the squares on *EZ* and *ZM*, that is the square on *EM*.[9] Therefore, *DE* is less than *EM*.[10] Then, similarly, we will show also that *ED* is the least of all straight lines falling from point *E* to circumference *KD* between points *K* and *D*.

Therefore, *ED* is the least of all the straight lines falling from point *E* to circumference *KD* between points *K* and *D*, and, of the straight lines produced through between points *K* and *D*, that nearer to it is ever less than that more distant.

But, then, let the produced-through 〈straight line〉, *BD*, be a diam- 〈Case 2〉 eter of circle *ABGD*, and let the rest be assumed the same.

---

[1] *Elem.* III.7.   [2] *Elem.* VI.20.cor.   [3] *Elem.* I.c.n.2 (extended to inequalities), I.47.   [4] *Elem.* VI.20.cor.   [5] *Elem.* I.post.1 (extended to object). See Commentary.
[6] *Elem.* I.post.1.   [7] *Elem.* III.7.   [8] *Elem.* VI.20.cor.   [9] *Elem.* I.c.n.2 (extended to inequalities), I.47.   [10] *Elem.* VI.20.cor.

[a] Heiberg (1927, 114) reads this arc as *KD*, following **K** (f. 126v). In fact, the letter-name is also somewhat unclear in **A** (f. 26r), which appears to read ΒΛΓΚΜ, followed by a lower case δ in a different ink.

⟨C137⟩   I say that EB is the least all the straight lines from point E to the circumference of circle $ABGD$, and ED is the greatest.

For, with the same construction, since circumference DE is greater than circumference EB, and a perpendicular, EZ, is produced, therefore DZ is greater than ZB.[1] And BD is a diameter of circle $ABGD$, therefore the center of the circle is on ZD.[2] Therefore, DZ is greater than ZG, and ZG than ZB,[3] so that the square on DZ is greater than the square on ZG and the square on ZG than the square on ZB.[4] Let the square on ZE be added, as common. Therefore, the squares on DZ and ZE, that is the square on ED, is greater than the squares on GZ and ZE, that is the square on GE.[5] And the squares on GZ and ZE, that is the square on GE, is greater than the squares on BZ and ZE, that is the square on BE.[6] Therefore, DE is greater than EG, and EG than EB.[7] Then, similarly, we will show also that of all the straight lines falling from point E to the circumference of circle $ABGD$, DE is greatest and EB least. Therefore ED is the greatest of all the straight lines falling from point E to the circumference of circle $ABGD$, and EB the least.

⟨H118,
C138⟩   2 ⟨Theorem⟩

If some straight line is produced through in a circle cutting off a segment not less than a semicircle, and on it a segment of a circle not greater than a semicircle is set up inclined to the ⟨segment⟩ not greater than a semicircle, while the circumference of the set-up segment is divided unequally, then the straight line subtending the lesser circumference is the least of all straight lines falling from the same point to the circumference not less than a semicircle.

For, let some straight line, AG, be produced through in a circle, $ABGD$, cutting off a segment not less than a semicircle, ABG. On AG, let a segment of a circle not greater than a semicircle, AG,[8] be set up, inclined to the ⟨segment⟩ not greater than a semicircle, ADG. Let cir-

---

[1] See Commentary.   [2] *Elem.* I.def.17.   [3] *Elem.* III.7.   [4] *Elem.* VI.20.cor.
[5] *Elem.* I.c.n.2 (extended to inequalities), I.47.   [6] *Elem.* I.c.n.2 (extended to inequalities), I.47.   [7] Both by *Elem.* VI.20.cor.   [8] That is, **Seg**(AEG).

cumference *AEG* be divided unequally at point *E*, and let circumference *GE* be greater than circumference *EA*. Let *EA* be joined.

I say that *EA* is the least of all straight lines falling from point *E* to circumference *ABG*.

For, let a perpendicular be produced from point *E* onto the plane
of circle *ABGD*.[1] Then, it falls between straight line *AG* and circumference *ADG* – through the ⟨fact that⟩
segment *AEG* is inclined to segment
ADG.[2] Let it fall, and let it be *EZ*.
And, let it meet the plane of the cir-

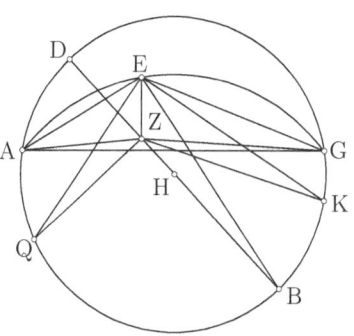

cle at point *Z*. Let the center of *ABGD* be taken.[3] Then, the center of ⟨C139⟩
circle *ABGD* is either on *AG* or between straight line *AG* and circumference *ABG* – through the ⟨fact that⟩ segment *ABG* is assumed as not
less than a semicircle.

First, let it be between straight line *AG* and circumference *ABG*, ⟨Case 1⟩
and let it be point *H*. And let a joined *ZH* be extended in directions ⟨Part 1⟩
D and *B*.[4] From point *E*, let a straight line, *EQ*, be extended forth to
circumference *ABG*.[5] Let *AZ* and *ZQ* be joined.[6]

Since *EZ* is upright to the plane of circle *ABGD*, therefore it also
makes angles upright to all straight lines contacting it and being in the
plane of circle *ABGD*.[7] But, each of *AZ* and *ZQ* contacts *EZ*, being
in the plane of circle *ABGD*. Therefore, each of angles *AZE* and *QZE*
is upright. And since *AZ* is less than *ZQ*,[8] therefore the square on *AZ* ⟨H120⟩
is less than the square on *QZ*.[9] Let the square on *ZE* be added, as
common. Therefore, the squares on *AZ* and *ZE*, that is, the square on
AE, is less than the squares on *EZ* and *ZQ*, that is, the square on *QE*.[10]
Therefore, *AE* is less than *EQ*.[11] Then, similarly, we will show also that,
of all the straight lines falling from point *E* to circumference *AB*, *AE*

---

[1] *Elem.* XI.11. [2] See Commentary. [3] *Elem.* III.1. [4] *Elem.* I.post.1, I.post.2.
[5] *Elem.* I.post.1 (extended to object). See Commentary, *Sph.* III.1. [6] *Elem.* I.post.1.
[7] *Elem.* XI.def.3. [8] *Elem.* III.7. [9] *Elem.* VI.20.cor. [10] *Elem.* I.c.n.2 (extended to
inequalities), I.47. [11] *Elem.* VI.20.cor.

is least. Therefore, AE is the least of all the straight lines falling from point E to circumference AB.

⟨C140⟩
⟨Part 2⟩   And again, similarly we will show that, of the straight lines produced through from point E to circumference AB between points A and B, that nearer is ever less than that more distant.

Then, let EB be joined.[1]

I say that EB is the greatest of all the straight lines falling from point E to circumference ABG.

For, since BZ is greater than ZQ,[2] therefore the square on BZ is greater than the square on ZQ.[3] Let the square on ZE be added, as common. Therefore, the squares on EZ and ZB, that is, the square on EB, is greater than the squares on EZ and ZQ, that is, the square on EQ.[4] Therefore, BE is greater than EQ.[5] Then, similarly, we will show also that, of all the straight lines falling from point E to circumference ABG, EB is the greatest. Therefore, EB is the greatest of the straight lines falling from point E to circumference ABG.

And, let EG also be joined.[6]

Then, I say that EG is the least of all straight lines falling from point E to circumference BG between points B and G.

⟨Part 3, 4⟩   For, let another, EK, be produced through.[7] Let ZG and ZK be joined.[8]

⟨C141⟩   Since GZ is less than ZK,[9] therefore the square on GZ is less than the square on ZK.[10] Let the square on ZE be added, as common. Therefore, the squares on GZ and ZE, that is, the square on EG, is less than the squares on KZ and ZE, that is, the square on EK.[11] Therefore, GE is less than EK.[12] Then, similarly, we will show also that, of all the straight lines falling from point E to circumference BKG between points B and G, EG is the least.

---

[1] *Elem.* I.post.1.   [2] *Elem.* III.7.   [3] *Elem.* VI.20.cor.   [4] *Elem.* I.c.n.2 (extended to inequalities), I.47.   [5] *Elem.* VI.20.cor.   [6] *Elem.* I.post.1.   [7] *Elem.* I.post.1 (extended to object).   See Commentary, *Sph.* III.1.   [8] *Elem.* I.post.1. [9] *Elem.* III.7.   [10] *Elem.* VI.20.cor.   [11] *Elem.* I.c.n.2 (extended to inequalities), I.47. [12] *Elem.* VI.20.cor.

Then again, similarly, we will show that, of the straight lines falling from point E to circumference BG between B and G, that nearer is ever less than that more distant.

Then, similarly, it will be shown if ABG is a semicircle that AE is less than all straight lines falling from E to circumference ABG.   ⟨H122⟩
⟨Case 2⟩

3 ⟨Theorem⟩

If great circles in a sphere cut one another, and equal circumferences are cut off from each of them successively both ways from the point at which they cut one another, then the straight lines joining the extremities of the circumferences in the same directions are equal to one another.

For, let two great circles in a sphere, AB and GD, cut one another at point E, and let equal circumferences be cut off successively both ways from point E, AE to EB and GE to ED. Let GA and BD be joined.   ⟨C142⟩

I say that GA is equal to BD.

For, the circle drawn with a pole, E, and a distance, EA, will pass through B.[1] Then, either it will also pass through G, or not.

First, let it go also through G. Therefore, it will also pass through D, for circumference GE is equal to circumference ED.[2] Let it go.[3] And let there be a common section   ⟨Case 1⟩

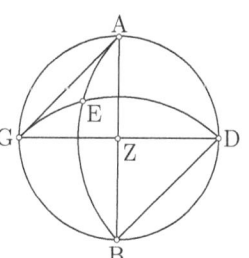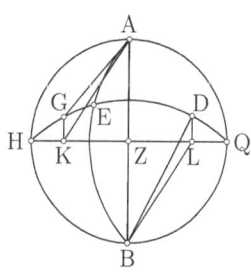

of $AGBD$ and AEB, AB, and a common section of $AGBD$ and GED, GD.[4]

Since a great circle in a sphere, AEB, cuts some circle in the sphere, $AGBD$, through the poles, it bisects it at an upright.[5] Therefore, AB is a diameter of circle $AGBD$.[6] Then, similarly, we will show that GD is a diameter of circle $AGBD$. Therefore, the four AZ, ZG, ZB, and ZD are equal.[7] So, since the two AZ and ZG are equal to DZ and ZB,

---

[1] *Sph.* I.def.5.   [2] *Sph.* I.def.5.   [3] *Elem.* I.post.3 (spherical).   [4] Both by *Elem.* XI.3.
[5] *Sph.* I.15.   [6] *Elem.* I.def.17.   [7] This introduces point Z, which must be the center of the circle, by *Elem.* I.def.15, I.def.16.

respectively, and angle AZG is equal to angle DZB, therefore AG, as base, is equal to DB, as base.[1]

⟨C143⟩

⟨Case 2⟩

⟨H124⟩

But again, then, let the circle drawn with a pole, E, and a diameter, EA, not go through G, but let it fall beyond it.[2] Therefore, it will pass through B, but also fall beyond D.[3] Let it go,[4] and let it be as $AHBQ$. Let circle GED be filled out to points H and Q.[5] Let there be a common section of circle $AHBQ$ and AEB, AB, and a common section of circle $AHBQ$ and HEQ, HQ.[6] Then, similarly we will again show that point Z is the center of circle $AHBQ$, and that each of AEB and HEQ is upright to circle $AHBQ$. Then, let perpendiculars, GK and DL, be produced from points G and D onto the plane of circle $AHBQ$.[7] Let AK and LB be joined.[8]

Since circumference EH is equal to circumference EQ – for point E is a pole – of which GE is equal to ED,[9] therefore GH, as remainder, is equal to DQ, as remainder.[10] So, since HEQ is an upright segment of a circle, and HG and DQ are equal cut-off circumferences, and GK and DL are produced perpendiculars, therefore GK is equal to DL and HK to QL.[11] But HZ, as a whole, is equal ZQ, as a whole.[12] Therefore, KZ, as remainder, is equal to ZL, as remainder.[13] And AZ is equal to BZ.[14] Therefore, AK and LB are equal.[15] So, since AK is equal to LB, and GK to DL, then the two AK and KG are equal to the two BL and LD, respectively. And, angle GKA is equal to angle DLB, for each of them is upright. Therefore, AG, as base, is equal to DB, as base.[16]

⟨C144⟩

4 ⟨Theorem⟩

If two great circles in a sphere cut one another, and from one of them equal circumferences are cut off successively both ways from the point at which they cut one another, and through the points brought about parallel planes are extended, of which one intersects the common section of the

---

[1] *Elem.* I.4, in $\mathbf{T}(AZG)$ and $\mathbf{T}(DZB)$.   [2] See Commentary.   [3] *Sph.* I.def.5.
[4] *Elem.* I.post.3 (spherical).   [5] *Sph.* I.20.   [6] *Elem.* XI.3.   [7] *Elem.* XI.11.
[8] *Elem.* I.post.1.   [9] *Sph.* I.def.5, *Elem.* III.28.   [10] *Elem.* I.c.n.3.   [11] Scholium 190.
See Commentary.   [12] *Elem.* III.28.   [13] *Elem.* I.c.n.3.   [14] *Elem.* I.def.15, I.def.16.
[15] *Elem.* I.4, in $\mathbf{T}(AZK)$ and $\mathbf{T}(BZL)$.   [16] *Elem.* I.4, in $\mathbf{T}(GKA)$ and $\mathbf{T}(DLB)$.

planes outside the surface of the sphere as towards the aforementioned point, and one of the equal circumferences is greater than each of the ⟨two circumferences⟩ cut off by the produced planes abutting the same point, then the ⟨circumference⟩ between the point and the non-intersecting plane is greater than that ⟨circumference⟩ of the same circle between the point and the intersecting plane.

For, let two great circles in a sphere, *AEB* and *GED*, cut one another at point *E*. And, from one of them, *EAB*, let equal circumferences, *AE* and *EB*, be cut off successively both ways from point *E*. Through points *A* and *B* let parallel planes, *AD* and *GB*, be produced through, of which *AD* intersects the common section of *AEB* and *GED* outside the surface of the sphere, as towards point *E*. And, let one of the equal   ⟨H126⟩ circumferences, *AE* and *EB*, be greater than each of *GE* and *ED*.

I say that circumference *GE* is greater than circumference *ED*.

For, the circle drawn with a pole,   ⟨C145⟩
E, and a distance, *EA*, will pass
through *B*,[1] but fall beyond points *G*
and *D* – through the ⟨fact that⟩ each
of *AE* and *EB* is greater than each
of *GE* and *ED*. Let it go,[2] and let
it be as *AHBZ*. Let the circles be
filled out.[3] Let circle *AD* intersect
circle *AHBZ* at point *Q*, and ⟨let⟩ *BG*
⟨intersect it⟩ at point *K*. Let there be

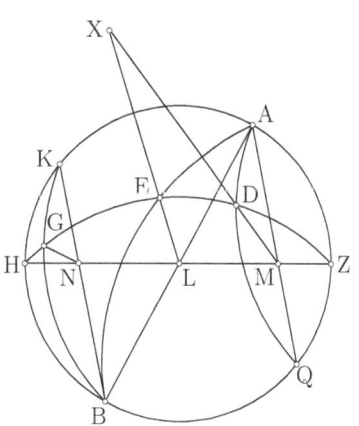

a common section of *AHBZ* and of *AEB* and *HEZ*, *AB* and *HZ*, and a common section of circle *ADQ* and *AHBZ*, *AQ*, and a common section of *KGB* and *AHBZ*, *KB*, and a common section of *HEZ* and *ADQ*, *DM*, and a common section of *KGB* and *HEZ*, *GN*.[4] And since plane *AD* intersects the common section of planes *HEZ* and *AEB*, that is *EL*, outside the surface of the sphere as towards point *E*, let it intersect it at *X*. Let *EL* be produced to *X*.[5] Therefore, point *X* is in plane *ADQ*,

---

[1] *Sph.* I.def.5.   [2] *Elem.* I.post.3 (spherical).   [3] *Elem.* I.post.3 (spherical), *Sph.* I.20.
[4] *Elem.* XI.3.   [5] *Elem.* I.post.2.

but also in *HEZ*. But also, points *D* and *M* are in both planes *ADQ*
and *HEZ*.[1] Therefore, *MD* intersects outside the surface of the sphere
as towards *E*. Then, it intersects at *X*.

Since a great circle in a sphere, *AEB*, cuts some circle of those in
the sphere, *AHBZ*, through the poles, it will bisect it at an upright.[2]
Therefore, *AB* is a diameter of circle *AHBZ*.[3] Then, similarly, we will
show that *HZ* is a diameter of circle *AHBZ*. Therefore, *L* is its center.[4]
And since two parallel planes, *KGB* and *ADQ*, are cut by some plane,

⟨C146⟩   *AHBZ*, therefore their common sections are parallel.[5] Therefore, *KB* is
parallel to *AQ*. Again, since two parallel planes, *KGB* and *ADQ*, are
cut by some plane, *HEZ*, therefore their common sections are parallel.[6]
Therefore, *GN* is parallel to *DM*. And since each of planes *AEB* and

⟨H128⟩   *HEZ* is upright to plane *AHBZ*, therefore their common section is upright to plane *AHBZ*.[7] But, *EL* is their common section, therefore *EL*
is also upright to plane *AHBZ*, so that it also makes angles upright to
all straight lines contacting it and being in the plane of circle *AHBZ*.[8]
But each of *AB* and *HZ* contacts *EL*, being in the plane of circle *AHBZ*.
Therefore, *EL* is perpendicular to each of *AB* and *HZ*. And, since an
angle, *XLN*,[a] is outside triangle *XLM*, it is greater than the interior and
opposite angle, *XML*.[9] But angle *XLN* is right, therefore angle *XML* is
acute.[10] Therefore, angle *XMZ* is obtuse.[11] And since *GN* is parallel to
DM, and *HZ* meets them, therefore angle *GNH* is equal to angle *XML*.[12]
But angle *XML* is acute, therefore angle *GNH* is also acute.[13] And *AM*
is parallel to *NB*, and *AB* and *MN* are two produced-through ⟨straight
lines⟩, and *AL* is equal to *LB*, therefore *NL* is equal to *LM*.[14] But, *HL*,
as a whole, is equal to *LZ*, as a whole,[15] therefore *HN*, as remainder,
is equal to *MZ*, as remainder.[16] So, since *HEZ* is a segment of a circle,
and *HN* and *MZ* are equal cut off ⟨straight lines⟩, and *GN* and *DM* are

---

[1] *Elem.* XI.3.   [2] *Sph.* I.15.   [3] *Elem.* I.def.17.   [4] *Elem.* I.def.15, I.def.16.   [5] *Elem.* XI.16.
[6] *Elem.* XI.16.   [7] *Elem.* XI.19.   [8] *Elem.* XI.def.3.   [9] *Elem.* I.16.   [10] *Elem.* I.def.12.
[11] *Elem.* I.13, I.def.11.   [12] *Elem.* I.28.   [13] *Elem.* I.c.n.1, I.def.12.   [14] *Elem.* VI.2
(generalization). See Commentary.   [15] *Elem.* I.def.15, I.def.16.   [16] *Elem.* I.c.n.3.

[a] Here and below, literally, "an angle, the ⟨angle⟩ under *XLN*."

produced-through parallels, and angle *GNH* is acute while angle *DMZ*   〈C147〉
is obtuse, therefore circumference *HG* is less than circumference *DZ*.[1]
So, since circumference *HE*, as a whole, is equal to circumference *ZE*,
as a whole,[2] of which *HG* is less than *DZ*, therefore circumference *GE*,
as remainder, is greater than circumference *ED*, as remainder.[3] Which
it was required to show.

5 〈Theorem〉

If the pole of the parallels is on a circumference of a great circle, and
two great circles, of which one is of the parallels and the other is oblique
to the parallels, cut this at an upright, and equal circumferences are cut
off from the oblique circle successively on the same side of the greatest
of the parallels, and parallel circles are drawn through the points brought
about, then they will cut off between them unequal circumferences of the
initial great circle, and that nearer to the greatest of the parallels will
be ever greater than that farther off.

For, let the pole of the parallels, point *A*, be on a circumference of   〈H130〉
a great circle, *ABG*. Let two great circles, *BZG* and *DZE*, cut it at an
upright, of which *BZG* is one of the parallels and *DZE* is oblique to the
parallels. And, from the oblique circle, *DZE*, let equal circumferences,
KQ and *QH*, be cut off successively on the same side of the greatest
circle of the parallels, *BZG*. And, let parallel circles, *OKP*, *NQX*, and
LHM, be drawn through points *K*, *Q*, and *H*.

I say that circles *OKP*, *NQX*, and   〈C148〉
LHM cut off unequal circumferences
of the initial great circle, *ABG*, and
that nearer *BZG* is ever greater than
that farther off. So, I say that circumference *ON* is greater than circumference *NL*.

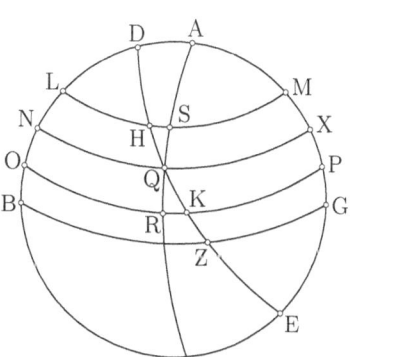

For, let a great circle, *AQR*, be
drawn through points *A* and *Q*.[4]

---

[1] See Commentary.   [2] *Sph.* I.def.5, *Elem.* III.28.   [3] *Elem.* I.c.n.3 (extended to
inequalities).   [4] *Sph.* I.20.

Since point A is a pole of circle OKP, circumference ANO is equal to circumference AQR.[1] Again, since point A is a pole of circle NQX, circumference ALN is equal to circumference ASQ.[2] Therefore, NO, as remainder, is equal to QR, as remainder.[3] Then, similarly, we will show also that circumference NL is equal to SQ. Therefore, NO is equal to QR, and LN to SQ.[4] And, since a great circle in a sphere, AQR, cuts some circle of those in the sphere, OKP, through the poles, it will bisect it at an upright.[5] Therefore, circle AQR is upright to circle OKP. Then, an upright segment of a circle, RQ and what is continuous with it, is set up on a diameter, that from R, of some circle, OKP, and a circumference, RQ, is cut off, being less than half of the set-up segment, therefore the straight line joining from Q to R is the least of all the straight lines falling from point Q to circle OKP.[6] Therefore, the straight line joining from Q to R is less than the straight line joining from Q to K. And the circles are equal, for they are great.[7] Therefore, circumference QR is less than circumference QK.[8] Then, similarly, we

⟨C149⟩   will show also that QS is less than QH. Stated[a] similarly: then, an upright segment of a circle, SQ and what is continuous with it, is set up on the diameter of some circle, LHM, that from point S, and a circumference, SQ, is cut off, being less than half of the set-up segment.

⟨H132⟩   And KQ is equal to QH. Therefore, each of KQ and QH is greater than each of RQ and QS. And since BZG is parallel to LHM, and BZG intersects the common section of circles HQK and AQR inside – that is, as at the center of – the sphere, therefore circle LHM also intersects the common section of circles HQK and AQR outside the surface of the sphere, as towards point Q.[9] So, since two great circles in a sphere, HQK and SQR, cut one another, and there are equal circumferences, KQ and QH, cut off from one of them, HQK, successively both ways

---

[1] *Sph.* I.def.5, *Elem.* III.28.   [2] *Sph.* I.def.5, *Elem.* III.28.   [3] *Elem.* I.c.n.3.   [4] Once again, this was already demonstrated in *Sph.* II.10. See Commentary.   [5] *Sph.* I.15.   [6] *Sph.* III.1.   [7] *Sph.* I.6, I.def.2, I.def.1.   [8] *Elem.* III.28.   [9] See Commentary.

[a] This sentence is bracketed as a later interpolation by both Heiberg (1927, 130) and Czinczenheim (2000, 149).

from point *Q*, and through points *H* and *K* parallel planes, *LHM* and *OKP*, have been extended, of which *LHM* intersects the common section of planes *HQK* and *SQR* outside the surface of the sphere as towards point *Q*, and one of the equal circumferences *KQ* and *QH* is greater than each of *RQ* and *QS*, therefore circumference *RQ* is greater than *QS*.[1] But *RQ* is equal to *ON*, and *QS* to *NL*. Therefore, circumference *ON* is greater than circumference *NL*.

6 ⟨Theorem⟩   ⟨C150⟩

If the pole of the parallels is on a circumference of a great circle, and two great circles, of which one is of the parallels and the other is oblique to the parallels, cut this at an upright, and equal circumferences are cut off from the oblique circle successively on the same side of the greatest of the parallels, and great circles are drawn through the points brought about and the pole, then they will cut off between them unequal circumferences of the greatest of the parallels, and that nearer to the initial great circle will be ever greater than that farther off.

For, let the pole of the parallels, point *A*, be on a circumference of a great circle, *ABG*. And, let two great circles, *BZG* and *DZE*, cut circle *ABG* at an upright, of which *BZG* is of the parallels and *DZE* is oblique to the parallels. Let equal circumferences, *KQ* and *QH*, be cut off from the oblique circle, *DZE*, successively on the same side of the greatest of the parallels, *BZG*. Let great circles, *AHL*, *AQM*, and *AKN*, be drawn through *A* and each of the points *H*, *Q*, and *K*.

I say that circumference *LM* is greater than circumference *MN*.

For, let parallel circles *XHO*, *PQR*, and *SKT* be drawn through   ⟨H134,
points *H*, *Q*, and *K*.[2] Therefore, circumference *SP* is greater than cir-   C151⟩
cumference *PX*[3] – through the previously proven theorem. But, *SP* is
equal to *UQ*, and *PX* to *QF*.[4] Therefore, *UQ* is also greater than *QF*.
So, let *QC* be laid out equal to *QF*.[5] But, *HQ* is equal to *QK*. Therefore, the straight line joining from *H* to *F* is equal to the straight line

---

[1] *Sph.* III.4.   [2] *Elem.* I.post.3 (spherical).   [3] *Sph.* III.5.   [4] *Sph.* II.10 (but see Commentary).   [5] *Elem.* I.post.3 (transfer).

joining from C to K.[1] Then, through C let a circle, $ICYW$, be drawn parallel to the initial ⟨circle⟩.[2]

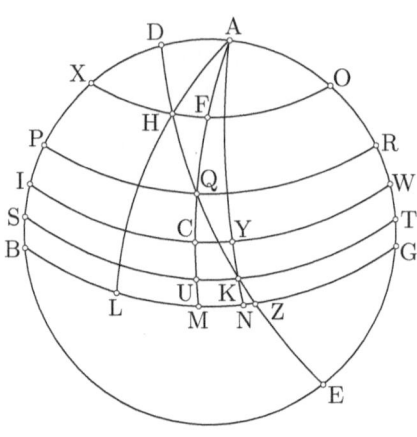

Since a great circle, $AYKN$, cuts some circle of those in the sphere, CYW, through the poles, it bisects it at an upright.[3] Therefore, circle $AYNK$ is upright to circle CYW, so that circle CYW is upright to circle $AYKN$. And since two parallel planes, BZG and CYW, cut some plane, $AYKN$, therefore their common sections are parallel.[4] Therefore, the common section of $AYKN$ and BZG – which is the diameter from point N, being of circle $AYKN$[5] – and the common section of $AYKN$ and CYW are parallel, so that the common section of $AYKN$ and CYW is parallel to the diameter of $AYKN$. Then, some straight line, the common section of circles $AYKN$ and CYW, has been produced through in a circle, $AYKN$, cutting the

⟨C152⟩   circle unequally – for it is parallel to the diameter of circle $AYNK$[6] – and on it an upright segment of a circle, CY and what is continuous with it, is set up, and the circumference of the set-up segment is divided unequally at C, and circumference UC is less than half of the set-up segment. Therefore, the straight line joining from C to Y is the least of all straight lines falling from point C to circumference YKN.[7] Therefore, the ⟨straight line⟩ from C to Y is less than that from C to K, so that that from C to K is greater than that from C to Y. But,

⟨H136⟩   the ⟨straight line⟩ from C to K is equal to that from H to F. Therefore, that from H to F is also greater than that from C to Y. And since circle CYW is nearer to the center than XHO, therefore circle CYW is greater than circle XHO.[8] So, since CYW and XHO are two unequal circles, and XHO is the lesser, and there are produced-through straight lines in

---

[1] *Sph.* III.3.   [2] *Elem.* I.post.3 (spherical).   [3] *Sph.* I.15.   [4] *Elem.* XI.16.   [5] *Sph.* I.11, *Elem.* I.def.17.   [6] This follows immediately from *Elem.* I.def.17.   [7] *Sph.* III.1. [8] *Sph.* I.6.

them, that from *H* to *F* in *XHO* and that from *C* to *Y* in *CYW*, and that from *H* to *F* is greater than that from *C* to *Y*, therefore circumference *HF* is greater than similar to circumference *CY*.[1] But *HF* is similar to *LM*, and *CY* is similar to *MN*.[2] Therefore, circumference *ML* is greater than similar to circumference *MN*. And, they are of the same circle. Therefore, circumference *ML* is greater than circumference *MN*.[3]

7 ⟨Theorem⟩

⟨C153⟩

If a great circle in a sphere touches some circle of those in the sphere, and some other great circle, being oblique to the parallels, touches greater ⟨circles⟩ than those the initial ⟨circle⟩ touches, and, furthermore, the ⟨points of⟩ contact are on the initial great circle, and equal circumferences are cut off from the oblique circle successively on the same side of the greatest of the parallels, and through the points brought about parallel circles are drawn, then they will cut off between them unequal circumferences of the initial great circle, and that nearer the greatest of the parallels will be ever greater than that farther off.

For, let a great circle in a sphere, *ABG*, touch some circle of those in the sphere, *AD*, at point *A*. And, let some other great circle, *EZH*, being oblique to the parallels, touch greater ⟨circles⟩ than those *ABG* touches. And, furthermore, let the ⟨points of⟩ contact be on circle *ABG* at points *E* and *H*. And, let *BZG* be the greatest of the parallels. And, let equal circumferences, *LK* and

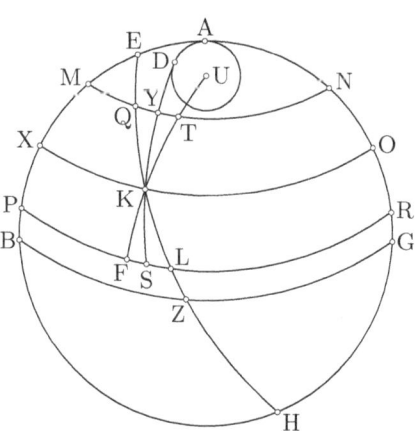

KQ, be cut off from the oblique circle, *EZH*, successively on the same side of the greatest of the parallels, *BZG*. And, through points *Q*, *K*, and *L* let parallel circles, *MQN*, *XKO*, and *PLR*, be drawn.

I say that circumference *PX* is greater than circumference *XM*.

---

[1] See Commentary.   [2] *Sph.* II.10.   [3] *Elem.* III.def.11, III.26.

⟨C154⟩   For, let a great ⟨circle⟩, *DKS*, touching circle *AD* be drawn through
point *K*,[1] so that the semicircle from *A*, as in direction *B*, is non-
⟨H138⟩   intersecting with the semicircle from *D*, as in direction *S*. Let the pole
of the parallels be taken,[2] and let it be point *U*. Let a great circle,
UKF,[a] be drawn through points *U* and *K*.[3]

Since a great circle in a sphere, *UKF*, cuts some circle of those in the
sphere, *PLR*, through the poles, it bisects it at an upright.[4] Therefore,
circle *UF* is upright to circle *PLR*. Then, an upright segment of a circle,
FKU and what is continuous with it, is set up on a diameter, that from
F, of some circle, *PLR*, and the circumference of the set-up segment is
divided unequally at point *K*, and *FK* is less than half. Therefore, the
straight line joining from *K* to *F* is the least of all straight lines falling
from point *K* to circumference *PLR*, and that nearer to it is ever less
than that more distant.[5] Therefore, the straight line joining from *K* to
S is less than the straight line joining from *K* to *L*, so that the straight
line joining *K* to *L* is greater than the straight line joining from *K* to
S. And, circles *DS* and *ELH* are equal, for they are great.[6] Therefore,
circumference *KL* is greater than circumference *KS*.[7] Then, similarly,
⟨C155⟩   we will show also that circumference *QK* is greater than circumference
KY. And *QK* is equal to *KL*. Therefore, one of *QK* and *KL* is greater
than each of *YK* and *KS*. And since circle *BZG* is parallel to circle *MQN*,
and circle *BZG* intersects the common section of *QKL* and *YKS* inside
the surface of the sphere – as at the center of the sphere – therefore
circle *MQN* produced[8] also intersects the common section of *QKL* and
YKS outside the surface of the sphere, as towards point *K*.[9] So, since
great circles in a sphere, *QKL* and *YKS*, cut one another at point *K*,
and *QK* and *KL* are equal circumferences cut off from one of them,
QKL, successively both ways from the point at which they cut one
another, and through points *Q* and *L* parallel planes *PLR* and *MQN*

---

[1] *Sph.* II.15.   [2] *Sph.* I.21.   [3] *Sph.* I.20.   [4] *Sph.* I.15.   [5] *Sph.* III.1.   [6] *Sph.* I.6, I.def.2,
I.def.1, *Elem.* III.def.1.   [7] See Commentary.   [8] That is, the plane of the circle.   [9] See
Commentary, *Sph.* III.5.

[a] Here and below, this circle was probably originally called *UTKF*. See Commentary.

are extended, of which *MQN* intersects the common section of circles   ⟨H140⟩
QKL and *YKS* outside the surface of the sphere, as towards point *K*,
and one of circumferences *QK* and *KL* is greater than each of *SK* and
KY, therefore *SK* is greater than *KY*.[1] But, *SK* is equal to *PX*, and
KY to *XM*.[2] Therefore, *PX* is greater than *XM*.

8 ⟨Theorem⟩   ⟨C156⟩

If a great circle in a sphere touches some circle of those in the sphere,
and some other great circle, being oblique to the parallels, touches greater
⟨circles⟩ than those the initial ⟨circle⟩ touches, and, furthermore, the
⟨points of⟩ contact are on the initial great circle, and equal circumfer-
ences are cut off from the oblique circle successively on the same ⟨side⟩
of the greatest of the parallels, and through the points brought about
great circles also touching that ⟨circle⟩ which the initial circle touches
are drawn, cutting off between them similar circumferences of the paral-
lel circles,[a] then they will cut off between them unequal circumferences
of the greatest of the parallels, and that nearer the initial great circle
will be ever greater than that farther off.

For, let a great circle in a sphere, *ABG*, touch some circle of those
in the sphere, *AD*, at point *A*. And, let some other great circle, *EZG*,
being oblique to the parallels, touch greater ⟨circles⟩ than those *ABG*
touches. And, furthermore, let the ⟨points of⟩ contact be on circle *ABG*,
points *E* and *G*. And let *BZ* be the greatest of the parallels. And, let   ⟨C157⟩
equal circumferences, *HQ* and *QK*, be cut off from the oblique circle,
EZG, successively on the same side of the greatest of the parallels, *BZ*.
Through points *H*, *Q*, and *K*, let great circles *DHL*, *MQN*, and *XKO*
be drawn touching circle *AD* at points *D*, *M*, and *X*, cutting off similar
circumferences of the parallel circles between them.

---

[1] *Sph.* III.4.   [2] *Sph.* II.13.

[a] Here Heiberg (1927, 138) includes the following passage, which is a scholium found
in the text of a number of manuscripts: "making the semicircles from the ⟨points
of⟩ contact as in the direction of the points through which they were drawn, non-intersecting with the semicircle of the initial great circle, on which is the ⟨point of⟩
contact of the oblique circle between the visible pole and the greatest of the parallels."

I say that circumference *LN* is greater than circumference *NO*.

For, let parallel circles *PHR*, *SQ*, and *TUK* be drawn through points *H*, *Q*, and *K*.[1] Therefore, circumference *ST* is greater than circumference *SP*.[2] But *ST* is equal to *QU*, and *SP* is equal to *QR*.[3] Therefore, *UQ* is also greater than *QR*. So, let *QF* be laid out equal *QR*.[4] But, *HQ* is equal to *QK*. Therefore, the straight line joining from *H* to *R* is equal to the straight line joining from *F* to *K*.[5] Then, through *F* let there be drawn a circle, *CFY*, parallel to whichever of *PHR*, *SQ*, *TUK*, or *BZ*.[6] Let the pole of the parallel circles, point *W*, be taken.[7] Let a great circle, *WO*, be drawn through point *W* and *O*.[8]

⟨H142⟩

⟨C158⟩

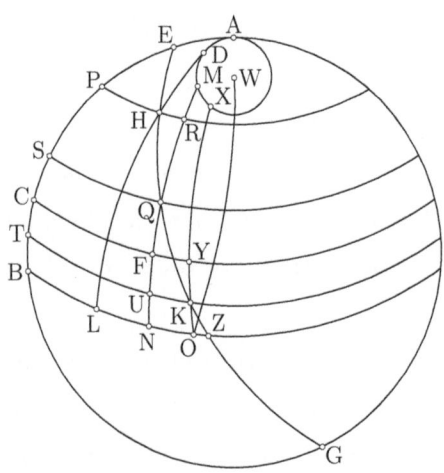

Since a great circle in a sphere, *WO*, cuts some circle in the sphere, *BZ*, through the poles, it bisects it at an upright.[9] Therefore, circle *WO* is upright to circle *BZ*. Therefore, *XKO* is also inclined to *ZB* as in directions *A*, *E*, and *B*.[10] Therefore, *ZB* is inclined to *XKO* as in directions *A*, *E*, and *B*. And *BZ* is parallel to *CFY*. Therefore, *CFY* is also inclined to *XKO* as in direction *X*. And, since two parallel planes, *BZ* and *CFY*, cut some plane, the oblique one, *XKO*, therefore their common sections are parallel.[11] Therefore, the common section of *XKO* and *CFY* is parallel to the common section of *XKO* and *BZ*. But, the common section of *XKO* and *BZ* is the diameter of circle *XKO* from point *O*. Therefore, the common section of *XKO* and *CFY* – that which is produced through from point *Y* – is

---

[1] *Sph.* I.21, *Elem.* I.post.3 (spherical). See Commentary.   [2] *Sph.* III.7.   [3] *Sph.* II.13.
[4] *Elem.* I.post.3 (transfer).   [5] *Sph.* III.3.   [6] *Sph.* I.21, *Elem.* I.post.3 (spherical).
[7] *Sph.* I.21. Note that this point must already have been found for any of the parallel circles to have been drawn. See Commentary.   [8] *Sph.* I.20.   [9] *Sph.* I.15.   [10] Since **gC**(*XKO*) is between **gC**(*WO*) and **gC**(*BZ*).   [11] *Elem.* XI.16.

parallel to the diameter of circle *XKO* from *O*.[1] Then, some straight line, the common section of *XKO* and *CFY*, is produced through in a circle, *XKO*, cutting the circle unequally – for it is parallel to the diameter of circle *XKO*[2] – and on it a segment of a circle, *CY* and what is continuous with it, is set up inclined to the ⟨segment⟩ not greater than a semicircle, and the circumference of the set-up segment is divided unequally at point *F*, and *FY* is less than half of the set-up segment. Therefore, the straight line joining from *F* to *Y* is the least of all of the straight lines falling from point *F* to the circumference not less than a semicircle.[3] Therefore, the ⟨straight line⟩ from *F* to *Y* is less than that from *F* to *K*, and the ⟨straight line⟩ from *F* to *K* is equal to that from *H* to *R*, so that that from *H* to *R* is greater than that from *F* to *Y*. And since circle *YFC* is nearer to the center of the sphere than *PHR*, circle *CFY* is greater than circle *PHR*.[4] So, since there are two unequal circles, *CFY* and *PHR*, and circle *PHR* is the lesser, and there are produced-through straight lines in them, that from *H* to *R* in *PHR* and that from *F* to *Y* in *CFY*, and the produced-through ⟨straight line⟩ in the lesser circle is greater than the produced-through straight line in the greater circle – that is, that from *H* to *R* is greater than that from *F* to *Y* – therefore circumference *HR* is greater than similar to circumference *YF*.[5] But, circumference *HR* is similar to circumference *LN*, and *FY* is similar to *NO*. Therefore, *LN* is greater than similar to *NO*. And they are of the same circle. Therefore, circumference *LN* is greater than circumference *NO*.[6]

⟨H144⟩
⟨C159⟩

9 ⟨Theorem⟩ ⟨C160⟩

If the pole of the parallels is on a circumference of a great circle, and two great circles, of which one is of the parallels and the other oblique to the parallels, cut this at an upright, and equal circumferences are cut off, not successively, from the oblique circle on the same side of the greatest circle of the parallels, and great circles are drawn through the

---

[1] *Elem.* XI.9.   [2] This follows immediately from *Elem.* I.def.17.   [3] *Sph.* III.2.
[4] *Sph.* I.6.   [5] See Commentary, *Sph.* III.6.   [6] *Elem.* III.def.11, III.26. See Commentary.

points brought about and the pole, then they will cut off between them unequal circumferences of the greatest of the parallels, and that nearer to the initial great circle will be ever greater than that farther off.

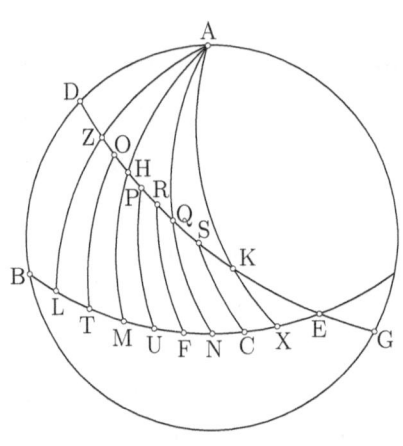

For, let the pole of the parallels, point A, be on a circumference of a great circle, ABG. Let two great circles, DEG and BE, cut circle ABG at an upright, of which BE is one of the parallels and DEG is oblique to the parallels. And, from the oblique circle, DEG, let equal circumferences, ZH and QK, be cut off, not successively, on the same side of the greatest of the parallels, BE. Let great circles, AZL, AHM, AQN, and AKX, be drawn through points Z, H, Q, and K and pole A.

I say that circumference LM is greater than circumference NX.

For, HQ is either commensurable with ZH and QK, or not.

⟨H146,
C161⟩
⟨Case 1⟩
First, let HQ be commensurable with ZH and QK. Let ZH, HQ, and QK be divided into the measures[1] at points O, P, R, and S.[2] Let great circles, OT, PU, RF, and SC, be drawn through points O, P, R, and S and pole A.[3]

So, since circumferences ZO, OH, HP, PR, RQ, QS and SK are successively equal to one another, therefore the circumferences LT, TM, MU, UF, FN, NC, and CX are successively greater than one another, starting from the greatest, LT.[4] So, since LT is greater than NC, and TM than CX, therefore LM, as a whole, is greater than NX, as a whole.

⟨Case 2⟩
Then, let HQ not be commensurable with ZH and QK.

I say that, similarly, circumference LM is greater than circumference NX.

---

[1] That is, the wholes are divided into parts that are themselves the measures, or units, that fit into the wholes.   [2] *Elem.* X.4. See Commentary.   [3] *Sph.* I.20.   [4] *Sph.* III.6.

For, if *LM* is not greater than circumference *NX*, it is either less than it or equal.

First, if possible, let *LM* be less than *NX* – as holds in the second diagram. Let *LM* be laid out equal to *NO*.[1] Let a great circle, *PO*, be drawn through pole *A* and *O*.[2] And, with the three circumferences of the same kind,[a] *KQ*, *QP*, and *HQ*, being unequal, let some circumference, *QR*, be taken, being greater than *QP*, less than *QK*, and commensurable with *HQ*.[3] And, let *SH* be laid out equal to *QR*.[4] Let great circles, *ST* and *RU*, be drawn through points *S* and *R* and through pole *A*.[5]

⟨Case 2.1⟩

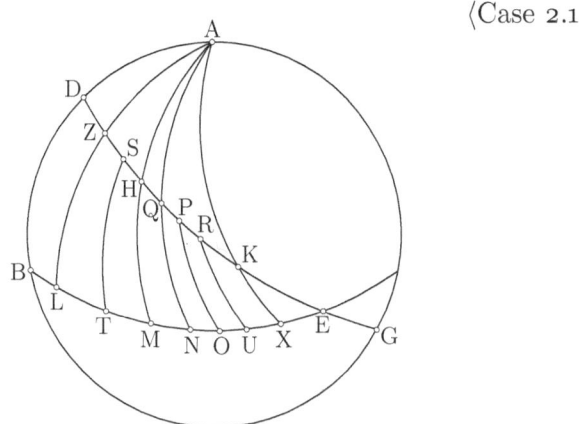

So, since *SH* is equal to *QR*, and *HQ* is commensurable with each of *SH* and *QR*, therefore *TM* is greater than *NU*.[6] But, *LM* is greater than *TM*.[7] Therefore, *ML* is much greater than *NO*.[8] But, they are also equal, which is impossible.

⟨C162⟩

Therefore, *LM* is not less than *NX*.

But, I say that *LM* is also not equal to *NX*.

⟨Case 2.2⟩

For, if possible let it be[9] – as holds in the 3rd diagram. Let *ZH* and

⟨H148⟩

---

[1] *Elem.* I.post.3 (transfer).   [2] *Sph.* I.20.   [3] Scholium 416. See Commentary.
[4] *Elem.* I.post.3 (transfer).   [5] *Sph.* I.20.   [6] Case 1.   [7] *Elem.* I.c.n.5.   [8] Since *NU* > *NO*, by *Elem.* I.c.n.5.   [9] That is, let **Arc**(*LM*) = **Arc**(*NX*).

[a] This is the only time that this qualification is used in the *Spherics*. See Commentary.

QK be bisected at points *O* and *P*.[1] Let great circles, *OR* and *PS*, be drawn through points *O* and *P* and pole *A*.[2]

So, since *ZO* and *OH* are successively equal to one another, therefore *LR* and *RM* are successively greater than one another, starting from the greatest, *LR*. Therefore, *LR* is greater than *RM*.[3] Therefore, *LM* is greater than double *MR*. Again, since *QP* and *PK* are successively equal to one another, therefore *NS* and *SX* are successively greater than one another, starting from the greatest, *NS*.[4] Therefore, *NS* is greater than *SX*, so that *XN* is less than double *NS*. So, since *LM* is equal to *NX*,[5] of which *LM* is greater than double *MR*, while *XN* is less than double *NS*, therefore *RM* is less than *NS*, with *OH* and *QP* being assumed equal, which is impossible.[6]

Therefore, *LM* is not equal to *NX*.

But, it has been shown that it is not lesser.

Therefore, circumference *LM* is greater than circumference *NX*.

⟨C163⟩

10 ⟨Theorem⟩

If the pole of the parallels is on a circumference of a great circle, and two great circles, of which one is of the parallels and the other oblique to the parallels, cut this at an upright, and two arbitrary points are taken on the oblique circle on the same side of the greatest of the parallels, and great circles are drawn through the points and the pole, then it will be as the circumference of the greatest of the parallels between the initial great circle and the next ⟨great circle⟩ through the poles to the circumference of the oblique circle between the same circles so is the next circumference of the greatest of the parallel circles – that between the great circles through the pole and the taken points – to some circumference less than the circumference of the oblique circle between the two taken points.

For, let the pole of the parallels, point *A*, be on a circumference of a

⟨H150⟩

great circle, *ABG*. Let two great circles, *DEG* and *BE*, cut circle *ABG* at an upright, of which *BE* is of the parallels and *DEG* is oblique to

---

[1] *Sph.* I.19, *Elem.* I.post.3 (transfer), III.3, I.post.3 (transfer); or *Elem.* I.post.3 (spherical), *Sph.* I.20. See Commentary. [2] *Sph.* I.20. [3] *Sph.* III.6. [4] *Sph.* III.6.
[5] By hypothesis. [6] Since, by Case 1, **Arc**(*OH*) = **Arc**(*QP*) ⇒ **Arc**(*RM*) > **Arc**(*NS*).

the parallels. And, furthermore, let two arbitrary points, *Z* and *H*, be taken on the oblique circle, *DEG*, on the same side of the greatest of the parallels, *BE*. Let great circles, *AZQ* and *AKH*, be drawn through points *Z* and *H* and pole *A*.

I say that it is as circumference *BQ* to circumference *DZ* so circumference *QK* to some circumference less than circumference *ZH*.

For, either *ZH* is commensurable with *DZ*, or not.   ⟨C164⟩

First, let it be commensurable.   ⟨Case 1⟩

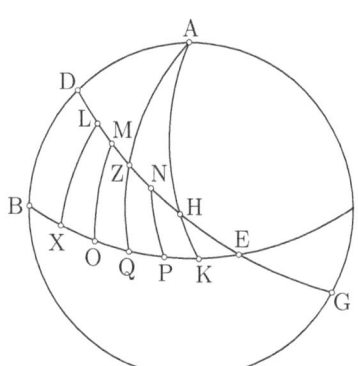

Let *DZ* and *HZ* be divided into the measures at points *L*, *M*, and *N*.[1] Let great circles, *LX*, *MO*, and *NP*, be drawn through points *L*, *M*, and *N* and pole *A*.[2]

So, since *DL*, *LM*, *MZ*, *ZN*, and *NH* are successively equal to one another, therefore *BX*, *XO*, *OQ*, *QP*, and *PK* are successively greater than one another, starting from the greatest, *BX*.[3] So, since *BX*, *XO*, *OQ*, *QP*, and *PK* are successively greater than one another, and *DL*, *LM*, *MZ*, *ZN*, and *NH* are successively equal to one another, and the multitude of *BX*, *XO* and *OQ* is equal to the multitude of *DL*, *LM* and *MZ*, while the multitude of *QP* and *PK* is equal to the multitude of *ZN* and *NH*, therefore *BQ* has to *DZ* a ratio greater than *QK* to *ZH*.[4] Therefore, if we make as *BQ* to *DZ* so *QK* to some other,[5] it will be to a lesser than *ZH*. Therefore, it is as circumference *BQ* to circumference *DZ*, so *QK* to some circumference less than *ZH*.

Then, let *ZH* not be commensurable with *DZ*.   ⟨Case 2⟩

I say that it is also so – as circumference *BQ* to circumference *DZ*, so circumference *QK* to some ⟨circumference⟩ less than circumference *ZH*.

For, if not, it will be either to a greater than *ZH*, or to it.

---

[1] See Commentary, *Sph.* III.9.   [2] *Sph.* I.20.   [3] *Sph.* III.6.   [4] *Elem.* V.8, V.12 (generalized to ratio inequalities). See Commentary.   [5] See Commentary.

⟨H152,
C165⟩

⟨Case 2.1⟩

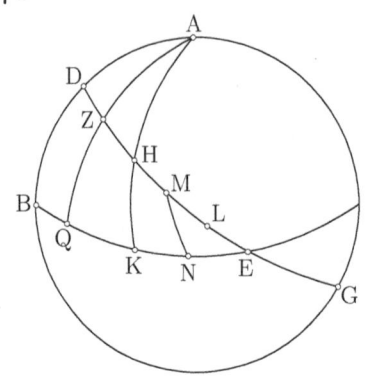

First, if possible, let it be to a greater than ZH, LZ[1] – as holds in the 2nd diagram. And, with there being three unequal circumferences, LZ, ZH, and ZD, let some circumference, ZM, be taken being less than ZL, greater than ZH, and commensurable with ZD.[2] And let a great circle, MN, be drawn through M and pole A.[3]

So, since ZM is commensurable with ZD, therefore it is as BQ to DZ so QN to some ⟨other⟩ less than ZM.[4] But, as BQ to DZ so is QK to ZL.[5] Therefore, as QK to ZL so is QN to a lesser than ZM.[6] And alternately; therefore as QK to QN so ZL to a lesser than ZM.[7] But, QK is less than QN.[8] Therefore, LZ is also less than the lesser than ZM.[9] But it is also greater, which is impossible.

Therefore, it is not as BQ to DZ so QK to some circumference greater than circumference ZH.

And, I say that it is not to it.

⟨Case 2.2⟩

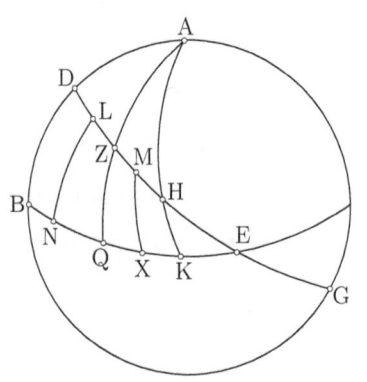

⟨C166⟩

For, if possible let it be as BQ to DZ so QK to ZH – as holds in the third diagram. Let each of DZ and ZH be bisected at points L and M.[10] Let great circles, LN and MX, be drawn through each of the points L and M and pole A.[11]

So, since DL and LZ are successively equal to one another, therefore BN and NQ are successively greater than one another, starting

---

[1] That is, **Arc**(BQ) : **Arc**(DZ) = **Arc**(QK) : **Arc**(ZL). [2] Again, see Commentary, *Sph.* III.9. [3] *Sph.* I.20. [4] Case 1. [5] The hypothesis of Case 2.1. [6] *Elem.* V.11.
[7] *Elem.* V.16. [8] *Elem.* I.c.n.5. [9] *Elem.* V.def.5. [10] *Sph.* I.19, *Elem.* I.post.3 (transfer), III.3, I.post.3 (transfer); or *Elem.* I.post.3 (spherical), *Sph.* I.20. See Commentary, III.9. [11] *Sph.* I.20.

from the greatest, *BN*.[1] Therefore, *BQ* is greater than double *QN*.
Then, similarly, we will also show that *KQ* is less than double *QX*. So,
since *BQ* is greater than double *QN* and *KQ* is less than double *QX*,
therefore *BQ* has to *QN* a ratio greater than *KQ* to *QX*.[2] And alternately; therefore *BQ* has to *QK* a ratio greater than *NQ* to *QX*.[3] But,
as *BQ* to *QK* so is *DZ* to *ZH*.[4] Therefore, *NQ* has to *QX* a ratio less
than *DZ* to *ZH*. And, as *DZ* to *ZH* so is *LZ* to *ZM*.[5] Therefore, *NQ*
has to *XQ* a ratio less than *LZ* to *ZM*.[6] And alternately; therefore *NQ*
has to *LZ* a ratio less than *QX* to *ZM*.[7] Therefore, if we make as *NQ*
to *LZ* so *QX* to some other, it will be to a greater than circumference ⟨H154⟩
ZM, which is impossible.[8]

Therefore, it is not as *BQ* to *DZ* so *QK* to *ZH*.

But it has been shown that it is not to a greater.[9] Therefore, it is
to a lesser.

Therefore, it is as *BQ* to *DZ* so *QK* to a lesser than circumference
ZH.

11 ⟨Theorem⟩   ⟨C167⟩

If the pole of the parallels is on a circumference of a great circle, and
two great circles of which one is of the parallels and the other oblique to
the parallels, cut this at an upright, and some other great circle through
the poles of the parallels cuts the oblique circle between the greatest of
the parallels and that ⟨parallel⟩ which the oblique ⟨great circle⟩ touches,
then the diameter of the sphere has to the diameter of the circle that
the oblique circle touches a greater ratio than the circumference of the
greatest of the parallels between the initial great circle and the next ⟨great
circle⟩ through the pole to the circumference of the oblique circle between
the same circles.

For, let the pole of the parallels, point *A*, be on a circumference of
a great circle, *ABG*. Let two great circles, *BEG* and *DEZ*, cut circle
ABG at an upright, of which *BEG* is of the parallels and *DEG* is oblique

---

[1] *Sph.* III.6.   [2] *Elem.* V.def.7.   [3] *Elem.* V.16 (generalized to ratio inequalities).
[4] The hypothesis of Case 2.2.   [5] *Elem.* V.12.   [6] *Elem.* V.11 (generalized to ratio
inequalities).   [7] *Elem.* V.16 (generalized to ratio inequalities).   [8] Case 1.   [9] Case
2.1.

to the parallels. Let some other great circle through the poles of the parallels, AHK,[a] cut DEZ between BEG and that ⟨parallel⟩ which DEZ touches. Let DLM be that ⟨parallel⟩ which DEZ touches.

I say that the diameter of the sphere has to the diameter of circle DLM a greater ratio than circumference BQ to circumference DH.

⟨C168⟩ For, let a parallel circle, NHX, be drawn through H.[1] Let there be common sections of the planes, AK, DZ, BG, NX, DM, QO, HP, OH, and HR.

So, since a great circle in a sphere, ABG, cuts some circles in the sphere, DLM, NHX, and BEG, through the poles, it bisects them at an upright.[2] Therefore, DM, NX, and BG are diameters of circles DLM, NHX, and BEG,[3] and circle ABG is upright to each of circles DLM, NHX, and BEG. So, since there are are parallel circles in a sphere, DLM,

⟨H156⟩ NHX, and BEG, and some straight line, AK, is produced through their poles, therefore AK is upright to each of circles DLM, NHX, and BEG, and is through their centers and also the sphere's.[4] Therefore, points S, P, and O are the centers of circles DLM, NHX, and BEG. And, since parallel planes, DLM, NHX, and BEG, cut some plane, ABG, therefore their common sections are parallel.[5] Therefore, DM, NX, and BG are parallel to one another. Again, since two parallel planes, NHX and BEG, cut some plane, AHK, therefore their common sections are parallel.[6] Therefore, HP is parallel to QO. So, since two straight lines contacting each other, NP and PH, are parallel to two straight lines contacting one another, BO and OQ, not being in the same plane, they contain equal angles.[7] Therefore, angle NPH is equal to angle BOQ.

⟨C169⟩ And, since NHX and DEZ are upright to circle ABG, therefore the common section of NHX and DEZ is upright to circle ABG.[8] But their common section is HR. Therefore, HR is also upright to circle ABG, and makes angles upright to all straight lines contacting it and being in

---

[1] *Elem.* I.post.3 (spherical).   [2] *Sph.* I.15.   [3] *Elem.* I.def.17.   [4] *Sph.* I.10.
[5] *Elem.* XI.16.   [6] *Elem.* XI.16.   [7] *Elem.* XI.10.   [8] *Elem.* XI.19.

[a] This great circle may have originally been called $AHQK$. See Commentary.

the plane of circle *ABG*.[1] But, *HR* contacts each of *PR* and *RO*, being in the plane of circle *ABG*. Therefore, each of angles *HRP* and *HRO* is upright. And, since *AK* is upright to *NX*, angle *RPO* is upright. So, since angle *RPO* is upright, therefore angle *POR* is acute.[2] Therefore, *OR* is greater than *RP*.[3]

So, let *RT* be laid out equal to *PR*.[4] Let *HT* be joined.[5]

So, since *PR* is equal to *RT*, while *HR* is common, then the two *PR* and *RH* are respectively equal to the two *RT* and *RH*, and angle *PRH*, an upright, is equal to angle *TRH*, an upright,[6] therefore *HP*, as base, is equal to *HT*, as base, and triangle *PRH* is equal

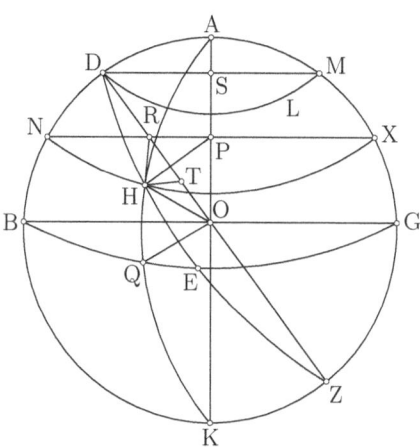

to triangle *TRH*, and the remaining angles are equal to the remaining angles that the equal sides subtend.[7] Therefore, angle *HPR* is equal to angle *HTR*. But, angle *HPR* is equal to angle *QOB*, and therefore angle *HTR* is equal to angle *QOB*.[8] And, since there is a triangle, *HOR*, having the angle at *R* upright, and some ⟨straight line⟩, *HT*, is produced through, therefore *OR* has to *RT* a greater ratio than angle *RTH* to angle *ROH*.[9] But *RT* is equal to *RP*, and angle *RTH* to angle *QOB*. Therefore, *OR* also has to *RP* a greater ratio than angle *BOQ* to angle *ROH*. But, as *RO* to *RP* so is *OD* to *DS*,[10] that is, *DZ* to *DM*.[11] But, as angle *BOQ* is to angle *HOR* so is circumference *BQ* to circumference *DH*.[12] Therefore, *ZD* also has to *DM* a greater ratio than circumference *BQ* to circumference *DH*. And *DZ* is a diameter of the sphere, and *DM* is a diameter of circle *DLM*. Therefore, the diameter of the sphere has to the diameter of circle *DLM* a greater ratio than circumference *BQ* to circumference *DH*.

⟨H158⟩

⟨C170⟩

[1] *Elem.* XI.def.3. [2] *Elem.* I.32, in **T**(*RPO*), not shown in the diagram. [3] *Elem.* I.19, in **T**(*RPO*). [4] *Elem.* I.3. [5] *Elem.* I.post.1. [6] *Elem.* I.post.4. [7] *Elem.* I.4. [8] *Elem.* I.c.n.1. [9] See Commentary. [10] *Elem.* VI.2. [11] *Elem.* V.15. [12] *Elem.* VI.33.

12 ⟨Theorem⟩

If great circles in a sphere touch the same one of the parallels, cutting off similar circumferences of parallel circles between them, while some other great circle, being oblique to the parallels, touches greater ⟨circles⟩ than those the initial ⟨circle⟩ touches and cuts the ⟨great circles⟩ touching the same ⟨circle⟩ between the greatest of the parallels and that ⟨parallel⟩ which the initial ⟨great circles⟩ touch, then the double of the diameter of the sphere has to the diameter of the circle that the oblique circle touches a greater ratio than the circumference of the greatest of the parallels between ⟨great circles⟩ touching the same circle to the circumference of the oblique circle between the same circles.

⟨C171⟩ For, let great circles in a sphere, *AB* and *GD*, touch the same one of the parallels, *AG*, at points *A* and *G*, cutting off between them similar circumferences of the parallel circles. And, let some other oblique great circle, *EZ*, being oblique to the parallels, touch ⟨circles⟩ greater than those which *AB* and *GD* touch, and let it cut *AB* and *GD* between the greatest of the parallels and that which *AB* and *GD* touch, circle *AG*. And, let *MBZ*[a] be the greatest of the parallel circles. And, let *EH* be that which circle *EZ* touches.

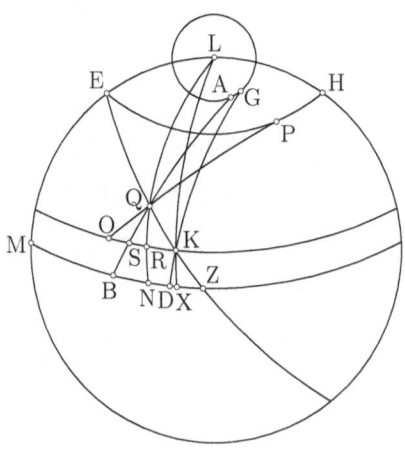

I say that the double of the diameter of the sphere has to the diameter of circle *EH* a greater ratio than circumference *BD* to circumference *QK*.

⟨H160⟩ For, let the pole of the parallels be point *L*.[1] Let great circles *LEM*, *LQN*, and *LKX* be drawn through *L* and each of points *E*, *Q*, and *K*.[2] And, let parallel cir-

---

[1] *Sph.* I.21.   [2] *Sph.* I.20.

[a] This letter-name was probably originally *BZ*. See Commentary.

cle *OK* be drawn through *K*.[1] And, through *Q*, let great circle *QP* be drawn touching circle *EH* at *P*.[2]

So, since there are are two parallel circles in a sphere, *OK* and *EPH*, and there are two great circles, *EQKZ* and *OQP*, drawn touching *EPH* at *E* and *P*, and through the point *Q* and pole *L* a great circle, *LQR*, is drawn, therefore *OR* is equal to *RK*.[3] Therefore, *RS* is less than *RK*.[4] Therefore, *SK* is less than double *KR*. But, *SK* is similar to *BD*,[5] and ⟨C172⟩ *KR* to *NX*.[6] Therefore, *BD* is also less than double *NX*. And since the diameter of the sphere has to the diameter of circle *EH* a greater ratio than circumference *MN* to circumference *EQ*,[7] and also circumference *MN* has to circumference *EQ* a greater ratio than circumference *NX* to *QK*,[8] therefore the diameter of the sphere has to the diameter of circle *EPH* a greater ratio than circumference *NX* to circumference *QK*, and doubles of the antecedents.[9] Therefore, double the diameter of the sphere has to the diameter of circle *EPH* a greater ratio than double circumference *NX* to circumference *QK*. But double circumference *NX* has to circumference *QK* a greater ratio than circumference *BD* to *QK*, for double circumference *NX* is greater than *BD*.[10]

Therefore, the double of the diameter of the sphere has to the diameter of circle *EPH* a ratio much greater than circumference *BD* to circumference *QK*.

13 ⟨Theorem⟩ ⟨C173⟩

If parallel circles in a sphere cut off of some great circle equal circumferences abutting the greatest of the parallels, and through the points brought about great circles are drawn either through the poles of the parallels or touching the same one of the parallels, then they will cut off ⟨H162⟩ *between them equal circumferences from the greatest of the parallels.*

For, let parallel circles in a sphere, *AB* and *GD*, cut off equal circumferences, *AE* and *ED*, of some great circle, *AD*, abutting the greatest of the parallels, *ZEH*. Through points *A*, *E*, and *D*, let great circles, *AZG*,

---

[1] *Elem.* I.post.3 (spherical).   [2] *Sph.* II.15.   [3] Scholium 456. See Commentary.
[4] *Elem.* I.c.n.5.   [5] *Sph.* II.13.   [6] *Sph.* II.10.   [7] *Sph.* III.11.   [8] *Sph.* III.10.   [9] *Elem.* V.4
(generalized to ratio inequalities). See Commentary.   [10] *Elem.* V.8.

QEK, and *BHD*, be drawn either through the poles of the parallels or touching the same one of the parallels.

I say that circumference *ZE* is equal to circumference *EH*.

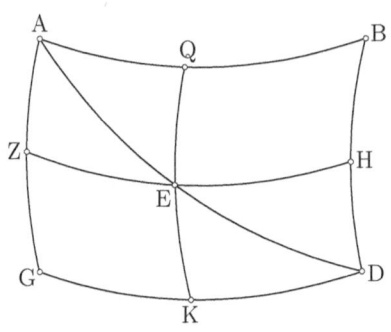

For, since parallel circles in a sphere, *AB* and *GD*, cut off equal circumferences, *AE* and *ED*, of some great circle, *AD*, abutting the greatest of the parallels, *ZH*, therefore circle *AB* is equal to circle *GD*.[1] So, since equal and parallel circles in a sphere, *AB* and *GD*, cut off circumferences, *QE* and *EK*, of some great circle, *KQ*, abutting the greatest of the parallel circles, *ZH*, therefore *QE* is equal to *EK*.[2] But, *AE* is equal to *ED*. Therefore, the straight line joining *A* to *Q* is also equal to the straight line joining *K* to *D*.[3] Therefore, circumference *AQ* is also equal to circumference *KD*.[4] And the circles are equal.[5] Therefore, circumference *AQ* is similar to circumference *KD*.[6] But, *AQ* is similar to *ZE*, and *KD* is similar to *EH*.[7] Therefore, *ZE* is also similar to circumference *EH*. And, they are of the same circle. Therefore, circumference *ZE* is equal to circumference *EH*.[8]

⟨C174⟩

14 ⟨Theorem⟩

If a great circle in a sphere touches some circle of those in the sphere, and some other great circle, being oblique to the parallels, touches greater ⟨circles⟩ than those the initial ⟨great circle⟩ touches, then they will cut off between them dissimilar circumferences of the parallel circles, and those nearer to whichever of the poles will be ever greater than similar to those farther off.

For, let a great circle in a sphere, *ABG*, touch some circle of those in the sphere, *ADX*, at point *A*. And, let some other great circle, *BEG*,

---

[1] *Sph.* II.17.   [2] *Sph.* II.18.   [3] *Sph.* III.3.   [4] *Elem.* III.28.   [5] *Sph.* I.6, I.def.2, I.def.1, *Elem.* III.def.1.   [6] *Elem.* III.27, III.def.11.   [7] *Sph.* II.10, II.13.   [8] *Elem.* III.def.11, III.26.

being oblique to the parallels, touch greater ⟨circles⟩ than those *ABG* touches.

I say that they will cut off between them dissimilar circumferences ⟨H164⟩ of the parallels, and those nearer to whichever of the poles will be ever greater than similar to those farther off.

For, let two arbitrary points, *E* and *K*, be taken on the oblique circle ⟨C175⟩ *BG*. Let parallel circles, *ZEH* and *QKL*, be drawn through points *E* and *K*.[1]

I say that circumference *EH* is greater than similar to circumference *KL*, and circumference *QK* is greater than similar to *ZE*.

For, through points *E* and *K* let great circles, *DEM* and *XNK*, be drawn touching *ADX*,[2] so that the semicircle from *D*, as in direction *M*, is non-intersecting with the semicircle from *A*, as in direction *Q*, and the semicircle from *X*, as in direction *K*, ⟨is non-intersecting⟩ with the semicircle from *A*, as in direction *L*.

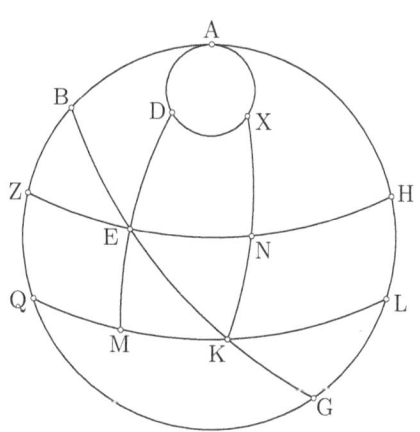

So, since semicircles *AL* and *XK* are non-intersecting, and *NH* and *KL* are circumferences of the same parallel circles between them, therefore, circumference *NH* is similar to circumference *KL*.[3] Then, through the same ⟨reasons⟩, *ZE* is similar to *QM*.[4] And since circumference *NH* is similar to circumference *KL*, therefore circumference *EH* is greater than similar to circumference *KL*.[5] Then, through the same ⟨reasons⟩, *QK* is also greater than similar to circumference *ZE*.

[1] Sph. I.21, *Elem.* I.post.3 (spherical). See Commentary.   [2] *Sph.* II.15.   [3] *Sph.* II.13.
[4] *Sph.* II.13.   [5] *Elem.* I.c.n.5.

Part III

Commentary

Commentary

This Commentary is meant to accompany a reading of our translation or paraphrase of the *Spherics*. Following an introduction to the subject matter of each book, groups of textual units – definitions or propositions – that treat related mathematical or astronomical topics are introduced and described. Each proposition is then treated separately. The commentaries to the individual propositions have, or can have, the following topics:

(1) a brief description of the key ideas in the proof, often simply by reference to the propositions that are used to establish it, and the proposition's role in the rest of the treatise,

(2) a discussion of technical terminology (optional),

(3) notes on the structure of the proposition (optional),

(4) notes on the specificity of the letter-names (optional),

(5) issues with the *construction* or *demonstration* (optional),

(6) lemmas and corollaries (optional),

(7) astronomical implications, including practical implications for working with globes and armillary spheres (optional),[1]

(8) a list of the propositions that are used for the proof,[2]

(9) a list of the later propositions in the *Spherics* that rely on this proposition,

(10) A description of the diagram accompanying the translation, using the taxonomy developed in the Introduction (pp. 78–88),

---

[1] Voegelin (1529) gives explanations for the astronomical meaning of the individual propositions, with which ours can be compared.

[2] Such justifications are provided in the scholarly notes of many of the medieval and early modern versions of the text. Heiberg (1927) gives internal references in his edition of the text. Internal dependencies are displayed graphically by Czinczenheim (2000, 910–914) and Nikolantonakis (2016, 136, 143, 157), and in tabular format by Berggren (1991, 244). Ver Eecke (1927) provides full justifications for the individual steps of the argument, with which ours often agree. Note, however, that he refers to Peyrard's (1814–1818) edition of the *Elements*, which sometimes has different proposition numbers from that of Heath (1908), who follows Heiberg (1883–1888).

(11) textual comments (optional),

(12) comparison with Thābit's Arabic version (optional), and

(13) comparison with Gerard's Latin translation from the Arabic (optional).

As noted, many of these topics are optional and the only ones that are stated for each proposition are (1), (8), (9), and (10). Astronomical implications are also sometimes explained in the introduction to a group of theorems, and then omitted for the individual theorems. A thorough comparison between the Greek text and Thābit's Arabic version is provided by Kunitzsch and Lorch (2010, 343–419) in the notes to their mathematical summary. Here we simply note aspects of the Arabic version that shed light on our choices for the translation or diagram, or which relate directly to a topic raised in our Commentary. In trying to determine what propositions are used to complete (8) it sometimes happens that the argument makes an inference that is not directly implied by a proposition in the *Elements* or the *Spherics*, but can be readily inferred from such propositions. In these cases, we note that such propositions are used "by inference." Of course, we do not claim certainty in these situations, since it is often possible to find other justifications for such steps.

Spherics I

It is well known that *Spherics* I is developed through analogies with *Elements* III, Euclid's treatment of circles in the plane (Heath 1921, 247–248). These analogies function, however, not exclusively by considering objects on the surface of the sphere constructed with the arcs of great circles that are analogous to rectilinear objects in the plane – such as in the approach developed by Menelaos in the Roman Imperial period – but also by considerations that were probably originally based on the ancient analemma, the core configuration of Greek sundial theory.[3] In the analemma configuration, circles on a sphere are treated by considering their orthogonal projections in the analemma plane, which is perpendicular to the planes of the solid circles – that is, solid circles are represented by their diameters in the plane of the analemma.[4]

For example, in the analemma, a great circle, gC, is treated as one of its diameters, while a small circle, sC, that is in a plane perpendicular to

---

[3] For general discussions of the analemma, and analemma constructions, see Luckey (1927), Neugebauer (1975, 839–856), Evans (1998, 133–136), and Sidoli (2020b).

[4] The second major configuration of the analemma – the orthogonal rotation, or folding, of solid circles into the analemma plane – does not feature in *Spherics* I, but can be used to justify a number of steps in *Sph.* II.22, and to explain the motivation for *Sph.* III.11. See the Commentary to those propositions, below.

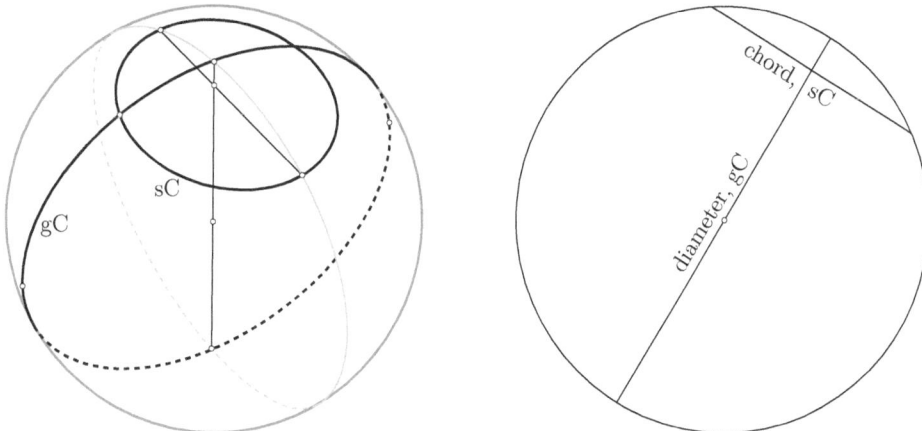

Figure 1: Analogy between a small circle and a great circle on the sphere and in the plane: (left) perspective figure for the primary analemma configuration, (right) objects constructed directly in the plane of the analemma.

that of gC, is treated as a chord perpendicular to the diameter, which is itself a diameter of the small circle (Figure 1 (left)). Both of these lines occur in the receiving plane of the analemma, which is orthogonal to both of the solid circles, and depicted here in gray. In this way, in the analemma diagram, a line in the circle can play a number of roles – at the same time, it represents a diameter of the sphere (if it is a diameter of the analemma circle), a diameter of a circle, and the circle itself, as will be explained (Figure 2 (right)).

Likewise, two great circles that intersect on the sphere, gC_1 and gC_2, can be represented by the diameters in the plane of the analemma, which is the plane passing through the great circle whose poles are the intersections of gC_1 and gC_2 (Figure 2 (left)). Once again, in the analemma diagram, the plane diameters represent diameters of the sphere, diameters of the great circles, and the great circles themselves (Figure 2 (right)).

This multiplicity of representations is also a feature of the development of *Spherics* I, and plays on the fact that in the *Spherics* circles in a sphere are ambiguously understood to be both solid disks cutting the sphere, and curved lines drawn on the surface of the sphere. Despite the fact that there is no clear linguistic distinction in the text, in discussing these differences in the Commentary we will call a circle considered as a solid disk a *solid* circle, and a circle considered purely as a curved line a *surface* circle. Likewise, constructions and arguments involving objects inside the sphere, or on various planes, will be called *solid*, while those that involve only objects that occur on the surface of the sphere will be called *surface*.

In fact, the analogies between *Elements* III and *Spherics* I go through three stages (see the table below). In the first part of *Spherics* I, *Sph.* I.1–

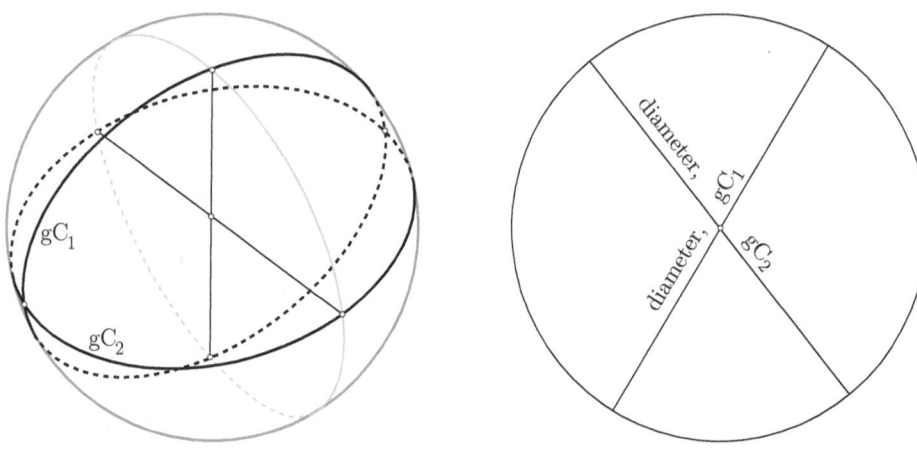

Figure 2: Analogy between two great circles on the sphere and in the plane: (left) perspective figure for the primary analemma configuration, (right) objects in the plane of the analemma.

I.10, the analogies relate solid circles and planes to their correspondents in the analemma diagram, and then in the second part, *Sph.* I.11–I.15, this shifts to setting small and great surface circles analogous to chords and diameters in the analemma, respectively. Finally – in fact only *Sph.* I.20 and I.21,[5] but continuing into the problems of *Spherics* II – surface objects correspond directly to plane objects – great circles to lines and small circles to circles – and the representation through the analemma model no longer obtains.

There are also three groups of propositions in *Spherics* I that have no analogs in the *Elements*, because they treat aspects of the geometry of the sphere that are fundamentally solid and have no interesting analogy with the plane situation – *Sph.* I.8–I.10 deal with poles, I.16 and I.17 treat the polar radius of a great circle, and I.18 and I.19 establish construction lemmas directly in the plane that allow for working with surface spherical circles. Furthermore, since *Spherics* I deals only with objects on a single sphere, there are no analogs to any of those propositions from *Elements* III that deal with two circles, such as *Elem.* III.5, III.6, and III.10–III.13. The analogies between *Spherics* I and *Elements* III are summarized in the following table:

---

[5] As discussed below in the Commentary to *Sph.* I.22 and I.23, these two final propositions are spurious, so that *Sph.* I.21 should be regarded as the final proposition of the book.

Spherics I	*Elements* III	Analogies
I.1	—	
I.2	III.1	
I.3	III.def.2	sphere ↔ base circle
I.4	III.18	diameter of sphere ↔ diameter
I.5	III.19	plane (solid circle) ↔ line
I.6	III.14, III.15	
I.7	III.3	
I.8	—	
I.9	—	
I.10	—	
I.11	III.def.17	
I.12	III.def.17	
I.13	III.3	sphere ↔ base circle
I.14	III.3	surface great circle ↔ diameter
I.15	III.3	surface small circle ↔ chord
I.16	—	
I.17	—	
I.18	—	
I.19	—	
I.20	I.post.1, I.post.2	surface great circle ↔ any line
I.21	III.1	surface small circle ↔ any circle

The shift between solid and surface analogies can be illustrated by considering the analogs to *Elem.* III.1, to find the center of a given circle, and *Elem.* III.3, a diameter and chord are perpendicular if and only if the diameter bisects the chord. The first time these propositions are modeled in the *Spherics* the analogy is between solid objects and plane objects. Namely, *Sph.* I.2 corresponds to *Elem.* III.1 and shows how to find the center of a given sphere, while *Sph.* I.7 corresponds to *Elem.* III.3 and shows that a small circle in the sphere is perpendicular to the diameter of the sphere passing through the center of the circle. In *Sph.* I.2, the analogy is between the sphere itself and a circle. In the analemma diagram, we may take the base circle to be a great circle representing the sphere. In *Sph.* I.7, the analogies are a solid small circle with a chord and a diameter of the sphere with a diameter of the plane circle. In the analemma diagram, we can take the base circle to represent the sphere, a chord to represent the solid small circle, and the perpendicular diameter of the analemma circle to represent the perpendicular diameter of the sphere. This configuration is extended beyond the plane analogy in *Sph.* I.8–I.10, in which the poles of the small circle play an important role, for which the endpoints of a diameter have no significant counterpart. The next time these propositions are modeled, however, the analogy has shifted to one between surface objects on the sphere and plane objects. Specifically, *Sph.* I.13–I.15 are analogous to *Elem.* III.3 such that, now, a surface great circle corresponds to a diameter, and a surface small circle corresponds to a chord. Finally, *Sph.* I.21 corresponds to *Elem.* III.1, but now a surface small circle corresponds to a plane circle and the pole of the small circle corresponds to the center of the plane circle.

Another major goal of *Spherics* I is the production of the *problems*, *Sph.* I.2, I.18–I.21, which are applied later in the treatise, both in *theorems* and in other *problems*. As with the theorems, there is a shift in the problems between solid and surface considerations. In *Sph.* I.2, the *problem-construction* involves passing a cutting plane and producing internal lines. In *Sph.* I.18–I.21, however, the operations that complete the problem itself can all be carried out on the surface of the sphere and in an external plane, while internal, solid objects are sometimes produced for use in the *demonstrations*. The problems have both theoretical and practical interpretations, the details of which will be individually discussed below. Here it will suffice to point out that *Sph.* I.2 provides an operation that is used a few times later in the treatise, in theorems, and is hence purely theoretical; *Sph.* I.18 and I.19 provide lemmas that are used to produce the elements of spherical circles in the plane; and *Sph.* I.20 and I.21 are used to produce surface elements that have both theoretical and practical applications. Furthermore, *Sph.* I.20 and I.21 establish the analogies of a great surface circle with any line in the plane and of a small surface circle with any circle in the plane, which will be continued in the problems of *Spherics* II, but which are, in fact, a minority theme in the text as a whole.

In terms of spherical astronomy, the results of *Spherics* I are rather modest. The theorems of this book have some implications for comparing the sizes of δ-circles, for the relationships between δ-circles and the celestial axis and certain great circles, for the relationship between all of the principal great circles in the sphere, and for the orientation of the celestial sphere in problems of gnomonics. That said, before we have well-defined conditions of parallelism and tangency we are not fully able to treat the δ-circles as a bundle of parallel circles about the celestial poles and tangent to principal great circles. The theories of parallelism and tangency will be developed in *Spherics* II. Hence, the spherical astronomical implications of *Spherics* I remain merely suggestive and the primary concern of the book seems to be the geometrical transition between solid geometry and geometry on the surface of the sphere. Nevertheless, the astronomical topics pertinent to the theorems of this book will be briefly discussed below, in connection with the relevant passages in the treatise.

Definitions

The key definitions of this book, *Sph.* I.def.1, I.def.2, and I.def.5, also follow the general pattern of the rest of the book of developing analogies with the *Elements*. Specifically, they correspond to *Elem.* I.def.15 and I.def.16, of the circle and its center, such that, in the first case, the sphere corresponds to a circle and its center to the circle's center, and, in the second case, a surface circle corresponds to a circle and its pole to the circle's center.

Of the remaining definitions, *Sph.* I.def.3, I.def.4, are unused in the text, or are, at least, mathematically unnecessary, and *Sph.* I.def.6 has some textual issues that are discussed below. Hence, it is it is likely that these three definitions did not form part of the original composition.

Comparison to Arabic: After *Sph.* I.def.5, Thābit's version has six definitions, *Sph.*$_A$ I.def.6–I.def.11, of which only *Sph.*$_A$ I.def.10 corresponds to *Sph.* I.def.6 (Kunitzsch and Lorch 2010, 344–345). We will take up a discussion of *Sph.*$_A$ I.def.11 in the Commentary to *Sph.* II.21, below.

Sph. I.def.1, I.def.2

The first definition is modeled exactly on *Elem.* I.def.15, the definition of a circle. Moreover, it is different from *Elem.* XI.def.14, the definition of a sphere given in the *Elements*, which would not be as mathematically useful in the *Spherics*,[6] because it would first have to be demonstrated that all of the radii of a sphere are equal.[7] These definitions should be compared with *Def.* 76 in the Heronic *Definitions* (Heiberg 1912, 52), which seems to combine elements from each.

It might be argued that in the *Spherics* the assertion that the center be *inside* the sphere is strictly unnecessary, but this is still a useful way of denoting how the sphere divides any points which may be introduced into the configuration – namely into points which are *inside* the sphere and can be joined to the center by a line that does not intersect the surface of the sphere, and points which are *outside* the sphere, and cannot be so joined.[8]

Usage: *Sph.* I.1, I.4, I.6, I.7, I.16, I.23(∗),[9] II.13 (by implication), III.5, III.7, III.13.

---

[6] As mentioned in n. 14, below, however, Euclid's definition of the sphere is used directly in a number of the problems in *Elements* XII and XIII, and it may have been introduced for this purpose.

[7] Such a demonstration would be trivial, but it would still constitute a separate proposition in the Greek style.

[8] See Sidoli (2018b, 423–426) for a discussion of a similar role for the concept of *inside* in the plane definition of a circle.

[9] *Sph.* I.23 is spurious, as discussed in the Commentary to that proposition, below.

Sph. I.def.3

This definition agrees with *Elem.* XI.def.15 in defining the *axis of a sphere*
as a line about which the sphere rotates. In the *Elements*, however, the
rotation of the sphere is essential to its definition, *Elem.* XI.def.14, whereas
rotation is not explicitly discussed elsewhere in the *Spherics*. Moreover,
the term *axis* is not used in the treatise after *Sph.* I.def.4, which makes it
doubtful that this definition is authentic. *Sph.* I.def.3 is similar to *Def.* 78
in the Heronic *Definitions* (Heiberg 1912, 54).

One could argue that this definition is used in an alternative reading,
which defines *diameter*, but this is also unnecessary, because once *Sph.* I.6
shows that a great circle passes through the center of the sphere, we can
simply infer that a diameter of a great circle and a diameter of the sphere
are equal, without any need for this definition.

From the perspective of pure geometry, there is no good reason to believe that this definition is not a later interpolation. At the very least, it
adds nothing to our understanding of the mathematics expounded by the
text. On the other hand, from the perspective of spherical astronomy, this
definition can be taken to define the *celestial axis*, about which the celestial sphere rotates. Since this text was probably often taught with the use
of a solid globe or armillary sphere, this definition may have been used to
instantiate the purely geometric arguments of the text in a tangible object.
Indeed, because the *Spherics* was taken to be mathematically necessary
for reading the *Moving Sphere* and *Phenomena* when the *Little Astronomy*
was being edited, this definition was probably inserted to define the objects
introduced in those treatises (Acerbi 2014, 146).

Usage: Unused in the *Spherics*.[10]

Textual Comments: In **A**, the oldest Greek manuscript, this is the definition of a *diameter* (διάμετρος) of a sphere (Heiberg 1927, 2; Czinczenheim
2000, 52). Although Heiberg preferred to take this as διάμετρος, Czinczenheim went against the reading of **A**, contrary to her usual tendency. Since
the next definition follows immediately on this and uses the notion of axis,
which would be otherwise unintroduced, we have followed Czinczenheim in
reading ἄξων.

Comparison to Arabic: Thābit's version reads محور, indicating that the
Greek manuscript from which the source translation was made also read
ἄξων.

Comparison with Latin: Gerard's Latin simply transliterates محور as
meguar (Kunitzsch and Lorch 2010, 12–13).

---

[10] *Sph.* I.17 and II.6 might be held to rely on the alternative text of this definition,
but as mentioned above, following *Sph.* I.6 an appeal to the definition is unnecessary.

Sph. I.def.4

This definition is not used anywhere in the *Spherics*, nor, indeed, are the *poles of the sphere* themselves ever mentioned elsewhere in the text (Mogenet 1947, 238). Again, from the perspective of the geometry of the treatise, there is no reason that this definition should have been advanced in the text as it currently stands.

From the perspective of spherical astronomy, however, this definition belongs with *Sph.* I.def.3 and can be taken as describing the *celestial poles*. Perhaps it is a relic from some earlier work in spherics, or, more likely, it was included in the *Spherics* because it is related to the subject matter of the *Moving Sphere* and the *Phenomena*, for which the *Spherics* was later understood to provide a foundation (Acerbi 2014, 146).

Usage: Unused in the *Spherics*.

Sph. I.def.5

From the perspective of usage, this is the most important definition in the text. It establishes the pole of a circle on the sphere as an analog to the center of a circle in the plane, and introduces the concept, although not the terminology, of a *polar radius* ("polar distance" in the Translation) – that is, a linear distance between a pole of a circle in the sphere and its circumference. For small circles there will be two polar radii, but for practical purposes one may simply work with one of them, usually the lesser, and in the *Spherics* this is usually treated as a single object.

This definition allows the geometer to work with the polar radius in much the same way as a plane radius – that is, as a way to introduce equal objects by construction and to compare the magnitudes of certain objects in the sphere. It is similar to *Def.* 79 in the Heronic *Definitions* (Heiberg 1912, 54).

This definition also introduces the concept of a *circle in a sphere* – perhaps the most important object in the treatise. In the *Spherics*, a circle in a sphere is both the plane disk of the circle, as well as the curved line that bounds it, which is similar to the ambiguous way in which the concept of a circle is used in the *Elements*. Hence, when the term *circle* is used, it may refer to either the plane disk, the curved line, or both. When the term *circumference*, περιφέρεια, is used, only the curved line is meant, whether in part or in whole.

Usage: *Sph.* I.8, I.9, I.10, I.16, I.17, I.20, I.21, II.5, II.6, II.10, II.13, II.15, II.17, II.22, II.23, III.3, III.4, III.5.

Textual Comments: Czinczenheim (2000, 52) takes the verb of this definition to be λέγεται, following **K** (88v) and some of the other manuscripts, whereas Heiberg (1927, 2) follows **A** (1r) and reads as ἐστί – which is contrary to their usual tendency.

Comparison to Arabic: Thābit's version uses the root قول, indicating that the Greek manuscript from which the source translation was made also used λέγεται.

Comparison with Latin: Gerard's Latin follows the Arabic (Kunitzsch and Lorch 2010, 12–13).

Sph. I.def.6

Although this definition could be regarded as a combination of *Elem.* XI.def.6(∗) and XI.def.7(∗), those definitions, along with *Elem.* XI.def.5(∗), are thought to be late interpolations into the Euclidean text (as denoted by (∗)), and are not found in the Arabic versions of the *Elements* (Vitrac 1990–2001, 4.77–79). Hence, since something like *Sph.* I.def.6 is required twice in *Sph.* II.21, it is tempting to read it as authentic, but there are two features of the text that weigh against this interpretation. In the first place, when this definition is used at the end of *Sph.* II.21, it is repeated in full in the context of a parenthetical remark that probably entered the text as a gloss (see p. 147, n. a). Furthermore, the definition is introduced at the beginning of *Spherics* I, but never used in that book, whereas *Sph.* II.def.1, which is also only used in *Spherics* II, is introduced at the beginning of that book. Hence, there is little reason to believe that this definition is genuine and, indeed, it is rejected as an interpolation by both Heiberg (1927, 2) and Czinczenheim (2000, 52, 212–216).

As with *Sph.* I.def.3 and I.def.4, this definition was probably inserted into the *Spherics* when this text was put together with the other treatises of the *Little Astronomy*, in late antiquity (Acerbi 2014, 146). Notice, for example, that *Sph.* I.def.6 is used in *Mov. Sph.* 7 (Mogenet 1947, 239).

Usage: *Sph.* II.21.[11]

Constructive assumptions and procedures

Some mention should be made of the sorts of constructions that are applied in the *Spherics*, both in *theorems* and in *problems*. Since no postulates, or constructive axioms, are advanced in the text itself, an initial assumption might be that we should look to the *Elements* to see what sorts of constructive procedures will be used, and a reading of the text bears this

---

[11] The second application of this definition in *Sph.* II.21 is unusual, for which see the Commentary below.

out. In fact, the constructive methods of the *Spherics* are essentially the same as those of *Elements* XI–XIII, with the addition of a construction of a circle on the surface of the sphere about a pole, and the assumption that distances can be transferred from one place in the configuration to another. The details of these differences will be discussed momentarily. Although there are a number of open questions about the constructive methods of *Elements* XI–XIII, and a detailed study is warranted, here we simply sketch the main similarities between the ways that constructions are performed in the *Elements* and in the *Spherics*.

The first type of construction that is employed is the production of straight lines in space, which can be justified either by a problem in the *Elements*, or by a direct application of *Elem.* I.post.1 – "to produce a straight line from any point to any point." Indeed, the solid problems in the *Elements* themselves depend on either applications of previously established plane problems of the *Elements* in an assumed, or demonstrated, plane, or direct applications of *Elem.* I.post.1. For example, in the proof of *Sph.* I.1 a perpendicular line is dropped from a point to a plane, an application of *Elem.* XI.11, which demonstrates such a construction. *Elem.* XI.11, in turn, relies on *Elem.* I.11 and I.12, which demonstrate the plane construction of a perpendicular to a line. When *Elem.* I.11 and I.12 are used in *Elem.* XI.11 the constructions are sometimes carried out in planes that are neither explicitly assumed, nor produced, but which are implicitly taken to be there because a point and a line determine a plane – a result which follows immediately from *Elem.* XI.2, showing that two lines or a triangle determine a plane. In the *Spherics*, as well, plane constructions are often carried out, using problems from the *Elements*, in planes that are implicitly taken to exist, say by *Elem.* XI.2, but which are neither explicitly assumed nor produced in the text.

The second type of construction involves using a plane for the purpose of introducing its intersection either with a spherical surface or with another plane.[12] This type of construction is based on methods that derive from *Elem.* XI.2 and XI.3, which establish that a plane is determined by two intersecting lines or a triangle – and hence, also a line and a noncollinear point, or three noncollinear points – and that two intersecting planes determine a line.[13] Because *Elem.* XI.2 and XI.3 show the existence, or rather uniqueness, of these objects, they serve a role in the solid books similar to that played by the postulates in the plane books – that is, they allow the ge-

---

[12] A "plane surface," ἐπίπεδος ἐπιφάνεια, not a "plane," is defined in *Elem.* I.def.7, however, whenever a plane is dealt with in *Elements* XI it is simply called a "plane," ἐπίπεδον. According to Proklos, this usage goes back to sources prior to Euclid (Friedlein 1873, 116; Morrow 1970, 94).

[13] Although it has been pointed out that there may be problems with the proof of *Elem.* XI.2 (Heath 1908, 275), for our purposes here, we will simply assume that Greek mathematicians, including Theodosios, accepted this proposition as valid.

ometer to introduce new objects in the *construction*. Indeed, in *Elem.* XI.5 and XI.13 the plane through two lines is "extended," from ἐϰβάλλειν, or "produced through," from διάγειν, so as to make a straight line as an intersection with another plane (Mugler 1959, 166–167, 126–127). In *Elem.* XI.7 and XI.18 a line is "produced through" or "extended" through a single line, again in order to make a straight line as an intersection with another plane. In both of these later cases, however, we can regard the plane as actually being produced through a line and a noncollinear point: in *Elem.* XI.7, as part of an indirect argument, the plane is produced through a line that is assumed to have a point that is noncoplanar with the original parallel lines; and in *Elem.* XI.18 the plane is named with two points noncollinear with the line through which it passes, either one of which can be assumed to be a third arbitrary point determining the plane. In the same way, we must consider that *Sph.* I.1 – an intersecting sphere and plane determine a circle – also plays a constructive role in the *Spherics*.[14] This proposition shows that a plane can be used to introduce a circle in the sphere, and moreover, from the definition of the circle itself, *Elem.* I.def.15, also implies that a circle drawn on the surface of the sphere determines a plane, which can then be used to produce a straight line as an intersection with another plane. All of these sorts of constructions, or constructive assumptions, are used in the *Spherics*. In fact, these sorts of assumptions are not constructions in the normal sense, because there are many cases in which surfaces intersect producing lines or circles that are not named or brought into the discussion, because they are not needed in the *demonstration*.

While the previous two types of constructive procedures allow the introduction of the types of objects that we have called solid, the final type of procedure used in the *Spherics* allows for the production of objects – small and great circles – directly on the surface of the sphere. Although such a construction is not used in the *Elements*, it is used in the *Moving Sphere*, and, perhaps, implicitly in the *Phenomena*.[15] The basic operation, introduced in *Sph.* I.19 is that of drawing a circle about some point as pole and passing through another point – that is, a circle with its distance, or radius,

---

[14] A general intersection of a sphere and a plane is not used in the *Elements* to introduce a circle. In *Elem.* XII.17, however, planes are extended through the semicircle that defines a sphere and through two lines to produce great circles, involving an argument that depends directly on the definition of a sphere in the *Elements*, *Elem.* XI.def.14, based on the rotation of the semicircle. As Heath (1908, 269) remarks, the problems of *Elements* XIII also depend on this definition.

[15] In Euclid's *Phenomena*, in contrast to his *Elements*, little or no attention is paid to how objects are constructed. Hence, it is possible to argue that the objects introduced in the *constructions* of propositions in that treatise are not meant to be understood as produced by an effective procedure.

determined by two points in the solid configuration.[16] This operation is clearly analogous to the way that *Elem.* I.post.3 is used in the *Elements*, but it operates here on the surface of the sphere, not in the plane. In the notes justifying the steps of the argument and in this Commentary, we call this operation

Elem. I.post.3 (spherical).

This operation is conceptually the same as the use of *Elem.* I.post.3 in the *Elements*, with the only difference being that the distance is not in the same surface as the object that is produced.[17]

While it is well known that, in the plane books of the *Elements*, the details of the constructions in *Elem.* I.2 and I.3 show that *Elem.* I.post.3 need be used to transfer a distance from one place to another, it becomes clear that in the *Spherics* an operation involving transferring distances is crucial to the problems, and is also used in the theorems. In fact, however the actual wording of *Elem.* I.post.3 – "to draw a circle with any center and distance" – tells us nothing about whether or not we can take the distance to be found anywhere in the given configuration, or only with one of its endpoints coinciding with the center of the circle. It is the mechanics of the constructions in *Elem.* I.2 and I.3 that have led to the conclusion that Euclid is treating the circle construction postulate as an abstraction of a collapsing compass – an object which does not exist – or more precisely, as the mathematical production of a circle with two given points, as opposed to three (Heath 1908, 246; Sidoli 2018b, 418–419). In the plane, *Elem.* I.2 is used to show that the conception of *Elcm.* I.post.3 as operating on two points can be extended, with *Elem.* I.1, to allow a given length anywhere in the figure to be translated so as to have one endpoint on any given point. In this way, any length in the figure can be translated to any point, and hence, a circle can be drawn with any center and a length found anywhere in the plane figure.[18] For a distance that is noncoplanar with the circle to be drawn, however, the situation is not quite so simple. Of course, it might be possible to argue that *Elem.* I.2 is used to transfer a distance in a different plane to a point in the same plane and then a sphere of radius equal to the transferred distance is made around the point, thus producing the same distance in any other plane passing through the center point. But

---

[16] This operation is applied already in *Mon Sph.* 6, but because that takes place in the *exposition* of the proposition, it is not clear that it is an actual construction.

[17] Although, in principle, it might be possible that the distance could be taken on the spherical surface itself, it is clear in a number of places in the text – starting already with *Sph.* I.def.5 – that Theodosios means the solid, linear distance between the pole and the circumference.

[18] In fact, however, the *Elements* does not make this construction explicit, which casts doubt on the claim that Euclid intended *Elem.* I.post.3, I.2 and I.3 as abstractions of the actual operations of a compass.

nothing like this is discussed in the text. In fact, it is simpler to assume that a distance found anywhere in the configuration can be transferred to any other place in the configuration – as can be done with a real compass. Indeed, in the *Spherics*, the linear distance, or interval, between two points on a plane or a surface is transferred from one position in the configuration to another, and arcs of circles on the sphere are cut off equal to other arcs of equal circles on the sphere. In the notes justifying the steps of the argument and in this Commentary, we call this operation

Elem. I.post.3 (transfer).

These operations can all be considered to be abstractions of potential manipulations of a real compass.[19]

In fact, it is possible that the uses of *Elem.* I.post.3 in the *Spherics*, and in the other texts in ancient spherics, all go back to an older conception of these constructive operations, when they were simply understood to be anything that can be done with an actual compass[20] – that is, before Euclid produced the more abstract, and purely mathematical, interpretation of *Elem.* I.post.3 implied by *Elem.* I.2 and I.3. Nevertheless, because the *Spherics* was composed after, and appears to have been strongly influenced by, the *Elements*, we have separated these operations into one that is closely analogous to Euclid's use of *Elem.* I.post.3, but on the surface of the sphere, and one that involves the transfer of a solid distance from place to place.[21]

---

[19] It should be noted that Apollonios's approach to angle transfer – that is, his proof of *Elem.* I.23, as reported by Proklos – appears to depend on a series of operations that can actually be carried out with a real compass (Friedlein 1873, 335–336). (Note that "a length" should be excised from Morrow's (1970, 262 l.30) translation.)

[20] That is, we could refer to all of these operations as *Elem.* I.post.3 (solid) – indicating that they involve interpreting the "any center" of *Elem.* I.post.3 to mean any point in a solid configuration, and the "distance" to be the span made up by any other two points in the configuration.

This simplified reading of the operation is clearly intended by Heron in *Meas.* III.17 and III.23, where he uses an expression along the lines of "with a pole, *A*, and a distance equal to *BC*, let a circle be drawn on the surface of sphere *DE*," in which the distance used to draw the circle is a line given elsewhere in the configuration (Acerbi and Vitrac 2014, 344, 358). Also, notice that the diagram for *Meas.* III.23 in the manuscript is mathematically suspect. Following the pattern of the diagram for *Meas.* III.17, which is correct, point M should appear in the middle of the circle, and the sphere should be labeled with N, not M (Bruins 1964, 1.216; Acerbi and Vitrac 2014, 359 n. 219)

[21] This separation of the two postulates also follows the way that medieval readers appear to have understood the text. For example, we find an explicit postulate that is equivalent to *Elem.* I.post.3 (spherical) but not *Elem.* I.post.3 (transfer) in both al-Ṭūsī's version and the Latin version of Campanus (Hyderabad 1939/40, 3; Sidoli and Kusuba 2008, 15; Vat.lat. 3380, f.1r).

Constructive propositions (*Sph.* I.1, I.2)

The two opening propositions, showing that an intersecting sphere and plane make a circle and how to find the center of a sphere using an intersecting plane, are closely related to one another. *Sph.* I.1 is used in *Sph.* I.2, while *Sph.* I.2 might appear to be used in *Sph.* I.1. In these propositions, a plane, a curved line, and a circle are named by three points, *A*, *B*, and *G*. This way of naming the objects may be significant, because *Elem.* XI.2 shows that every triangle – that is, every three points – determines a plane, so that these three points may stand in for any three points that determine a plane. Although at this point we have no practical way of actually drawing the circle through these three points on the sphere itself, such a practical construction can eventually be obtained, as we will see below – although it seems to have played no role in the treatise itself.

Sph. I.1 (Theorem)

This proposition, in two cases, uses elementary solid geometry. Case 1 follows immediately from the definitions of the sphere and the circle (*Sph.* I.def.1, I.def.2, *Elem.* I.def.15), while Case 2 follows from a fairly straightforward application of *Elem.* I.47 – the so-called Pythagorean theorem – to solid geometry.

In the proof of Case 2, we let some point, *D*, be the center of the sphere. Initially, it might seem that this assumes *Sph.* I.2, which shows us how to find the center of the sphere. But Theodosios avoids this circularity by simply asserting the existence of the center – which we know must exist by definition. Here, we simply assume that the center exists. That is, we take any point and assume that it is the center of the sphere. From this assumption alone, the argument follows.

In Greek mathematical texts, the word γραμμή with no qualification means any line – that is, a straight or curved line. Hence, we could have translated it more precisely with "⟨straight or curved⟩ line," but in the interest of brevity we simply use "⟨curved⟩ line," since it is shown to be a circle.

As discussed above, this theorem has a constructive implication, since it shows that an intersecting sphere and plane determine a circle. Hence, just as with *Elem.* XI.2 and XI.3, it can be used in the *construction* of a proposition to introduce a new object into the configuration – in this case a circle.

In the course of the *demonstration*, it is asserted that $S(AE) = S(EB) \Rightarrow AE = EB$, which depends on the general claim that $S(a) = S(b) \Rightarrow a = b$. This obvious assertion follows immediately from *Elem.* VI.20.cor., but such an inference is also made in *Elem.* I.48, well before *Elements* VI, and can, in fact, be shown from the methods developed in *Elements* I. Starting with equal lines, $a = b$, a is extended with another line $b' = b$, squares $\mathbf{S}(a)$, $\mathbf{S}(b)$ and $\mathbf{S}(b')$ are constructed on all the lines, and then $\mathbf{S}(b') = \mathbf{S}(b)$, can be shown by producing a diagonal and showing that the triangles are equal pairwise, using *Elem.* I.4. Then, $\mathbf{S}(a)$ and $\mathbf{S}(b')$ can be shown to be in the same parallels, *Elem.* I.27, so that $\mathbf{S}(a) = \mathbf{S}(b') = \mathbf{S}(b)$, *Elem.* I.36, I.c.n.1. That is, $a = b \Rightarrow \mathbf{S}(a) = \mathbf{S}(b)$. Then, the converse can be shown indirectly by starting with $\mathbf{S}(a) = \mathbf{S}(b)$ and assuming that, say, $a > b$. In this situation, $b' = b$ is cut off of a, and constructing $\mathbf{S}(b')$, it can be shown that $\mathbf{S}(b') = \mathbf{S}(a)$, which is impossible, *Elem.* I.c.n.5. Hence, it is also the case that $\mathbf{S}(a) = \mathbf{S}(b) \Rightarrow a = b$.[22]

In regard to the practice of doing spherical astronomy on a solid globe or armillary sphere, this proposition has the following implications: If one wants to produce a stand, or plinth, on which to rest a globe or armillary sphere that is level to the ground, the opening in the mouth of the stand should be a circle, consist of three pegs, or of more than three pegs that lie on a circle.

Basis: *Elem.* I.def.15, I.post.2, I.c.n.1, I.c.n.4, I.47, VI.20.cor., XI.11, *Sph.* I.def.1, I.def.2.

Usage: *Sph.* I.2, I.2 (cor.), I.3, I.4, I.6 (cor.), I.8 (cor.), I.13 (cor.), I.17, I.19, I.22(∗).[23] A proposition like this was used in the introduction to the *Phenomena*, and in *Mov. Sph.* 1 and 2.

Diagram: The diagram for this proposition is a planar representation of local objects, using both intrinsic and extrinsic objects. In this case, there may appear to be a visual appearance of depth, but because the initial circle is depicted with a circle, we can think of this as representing all of the objects simply flattened out in a single plane. The axis of circle ABG is represented by an extended radius that has a false crossing with the circle itself.

Textual Comments: The manuscript evidence for the diagram(s) accompanying this proposition is inconsistent. In **A**, there is no diagram at all. In **B** (2v), and **F** (31r), there are three diagrams, but it is not clear how they represent different mathematical cases, as opposed to simply different visual perspectives, and there seems to be a problem with the last of the three diagrams (Czinczenheim 2000, 681). We include a diagram close to that found as the second image in **F** and the only image in **K** (89r).[24]

---

[22] The fact that none of this is detailed in the *Elements* is an indication that already by the end of *Elements* I, Euclid is looking ahead to the more abstract and operational concept of area that is used in the theory of the application of areas, and which is at play in the later propositions of *Elements* II.

[23] *Sph.* I.22 is spurious as discussed in the Commentary to that proposition.

[24] In **K**, the letter-names A and B are switched, but this does not change the argument in any way.

Comparison with Arabic: The diagram in **I** (21a), of Thābit's version, is a single figure, like that in **K**, but rotated 90°. Since this may indicate that the manuscript from which the Arabic was translated also had such a diagram, we have been influenced by this in deciding on the diagram for the translation.

Sph. I.2 (Problem)

This proposition presents an analogy to *Elem.* III.1, which it uses, such that the sphere corresponds to a circle in the plane, and its center corresponds to the center of the circle. The procedure of *Sph.* I.2 also follows that of *Elem.* III.1. A plane is passed through the sphere, producing a circle. The center of the circle is taken, a perpendicular is erected in both directions to the sphere, and bisected. The proof, which is indirect, involves the assumption that there is some other center.

The only two applications of this proposition in the *Spherics* as a *problem* are in *Sph.* I.3 and I.6, but it is not clear how essential these applications are to the proofs of those propositions.

This proposition has the full potential structure of a problem as discussed in the introduction (p. 61, above; Sidoli 2018b, 410–419). The problem itself, before the corollary, ends with the phrase "which it was required to show" (ὅπερ ἔδει δεῖξαι). Such QED phrases are rare in this text,[25] and this one is especially odd because it concludes a problem, whereas in the *Elements* a problem usually ends with the expression "which it was required to do" (ὅπερ ἔδει ποιῆσαι).

The first step of the *construction* is the assumption that there is some plane cutting the sphere. There is no postulate for the assumption of a cutting plane, in either the *Elements* or the *Spherics*. As mentioned above, however, the fact that the plane is named by three points may help us understand the situation. Since, we can assume that there are some three points on the sphere, then *Elem.* XI.2 assures us that there is a plane passing through these points. That is, as long as we can assume three points, we can assert the existence of the plane without a postulate. Then, *Sph.* I.1 shows that this plane produces a circle in the sphere.

In the proof, the counterfactual assumption of another center – when we are in the process of showing how to find the actual center – helps us understand the assumption of a center in the proof of *Sph.* I.1 as well. This assumption is not a real construction – in both cases, we can simply assume that any point is the center, and proceed to argue from that assumption.

In the *demonstration*, it is also possible that point Q, assumed as the center of the sphere, might fall on line ZE, which would constitute another

---

[25] See also, *Sph.* III.4. Nikolantonakis (2016, 168) claims, incorrectly, that such phrases never occur in the *Spherics*.

case (πτῶσις), as Proklos defines the term (Friedlein 1873, 222–223). In such a case, however, the proof would be simpler than that given here, since there would be fewer steps in both the *proof-construction* and the *demonstration* (see p. 61, above), with the demonstration simply pointing out that the same segment would be bisected in two different points. Such simple cases were generally omitted in Greek mathematical texts, as can be seen by the many cases provided in ancient and medieval commentaries, and by the fact that a number of the extra cases in the current text of the *Elements* have been shown to be later additions (Saito 2006, 84–90; Saito 2012).

Basis: *Elem.* I.def.15, I.def.16, I.post.2, I.10, III.1, XI.11, XI.12, *Sph.* I.1.cor.

Usage: *Sph.* I.3, I.6, I.10 (cor.), I.12 (cor.), I.16 (by implication), I.19 (by implication).

Diagram: The diagram for this proposition is a planar representation of local objects, using both intrinsic and extrinsic objects. The axis of circle *ABG* is represented by an extended diameter that has two false crossings with the circle.

Comparison with Arabic: Thābit's version adds a proof sketch of the trivial case in which the arbitrary plane passes through the center of the sphere between the *construction* and the *demonstration*. The Latin follows the Arabic and adds one further, vaguely expository, sentence (Kunitzsch and Lorch 2010, 18–21).

Plane tangent to a sphere (*Sph.* I.3–I.5)

This group of theorems develops the concept of the tangency of a plane with a sphere analogously to the way that the tangency of a line with a circle is handled in *Elements* III. These propositions are nowhere else used in the *Spherics*, so Theodosios must have thought that this analogy would be inherently interesting to his readers.

One explanation for this presumed interest might be found in gnomonics as set out in the basic configuration of the analemma described by Vitruvius in *Arch.* IX.7. In this passage, Vitruvius describes the analemma model of the celestial sphere as resting on the horizon. In such a configuration, *Sph.* I.3–5 shows that the model can be taken to be tangent to the horizon at a point, and that the line perpendicular to the horizon known as the *gnomon* will join the point of tangency with the center of the spherical model.

Sph. I.3 (Theorem)

By considering the sphere as corresponding to a circle, and the tangent plane to a tangent line, this proposition is related to *Elem.* III.def.2 – the definition of a line tangent to a circle. It is demonstrated in a similar way to *Elem.* III.13, which deals with two circles, by using *Elem.* III.2. We do not, however, find a proposition that is a direct analog of this proposition in the *Elements*, because such a claim would follow immediately from *Elem.* III.def.2.

The structure of this proposition is rather loose, or non-standard. There is no clear *exposition* or *specification*, and the *enunciation* is followed immediately by the beginning of the indirect argument. Furthermore, this is one of the few propositions in this text that includes a general *conclusion* – that is, a restatement of the *enunciation* in general terms. In this case, the verb is in a different tense.

In this proposition, the center of the sphere is asserted as "taken," which should be understood as an application of *Sph.* I.2, because the same pair of verbs are used as in the *Elements* for *Elem.* III.1 and its applications. It is not clear, however, why we could not use the same approach as in *Sph.* I.1 and I.2 and simply assume that there is some point that is the center.

Basis: *Elem.* I.post.1, III.2, XI.def.2, XI.2, XI.3, *Sph.* I.1, I.2.

Usage: *Sph.* I.4.

Diagram: The diagram for this proposition is a planar representation of global objects, containing both intrinsic and extrinsic objects. It is categorized as global because circle *ABD* must be a great circle, despite the fact that this will not be shown until *Sph.* I.6. The diagram is unique in the *Spherics*. It appears to be modeled on the digram of *Elem.* III.13 (Saito 2008, 62). All of the objects in the diagram lie in the cutting plane, so that the diagram appears to show a cross section of the sphere and the tangent plane. Because the diagram shows an impossible situation, the sphere is depicted with a lune.

Textual Comments: In the Greek manuscripts, the center of the sphere, Γ, nearly falls on its surface towards Δ (for example, **A**, 2r; **B**, 4r; **F**, 31r; **K**, 90r). Our diagram follows that in Thābit's version (**I**, 21b).

Sph. I.4 (Theorem)

This proposition is analogous to *Elem.* III.18, which it uses. It is a partial converse with the following proposition, and proceeds by passing two arbitrary planes through a line joining the point of tangency and the center of the sphere.

Here, and in *Sph.* I.6, *radius* translates the literal expression "the ⟨distance⟩ from the center," ἡ ἐκ τοῦ κέντρου. This was the standard terminology for the radius of a circle that is already drawn in the plane (Fowler and Taisbak 1999; Sidoli 2004a).

As discussed above (p. 199), there is no postulate for extending an arbitrary plane through any line. Hence, this construction is similar to that in *Elem.* XI.7 and XI.18. In this proposition, however, no other point is named; so we must simply assume that the plane is any of those which can pass through the line. Probably such an assumption was unpostulated because it would be analogous to the plane construction of passing a line through a given point, which was also unpostulated because it does not return a definite result.

Basis: Production of a plane through a line, *Elem.* III.def.2, III.18, XI.3, XI.4, *Sph.* I.def.1, I.def.2, I.1, I.3.

Usage: *Sph.* I.5.

Diagram: The diagram for this proposition is a planar representation of global objects, containing both intrinsic and extrinsic objects. All of the objects are simply drawn in the plane of the diagram as though the two cutting planes have been folded together into a single plane.

Textual Comments: There is no letter-name H in the text or the diagram although the following letter Θ is used, which is non-standard. Probably, the circle AΘ was originally called AHΘ, with the H placed opposite Δ in the diagram. Then, at some point in the transmission the H was omitted from both the text and the diagram, perhaps in stages.

The claim that an original H has dropped out of the transmission is supported by the Commentary to *Sph.* I.1 and I.2, above, which argued that, unless the pole is known, a circle on a sphere is properly specified by at least three points.

Comparison with Arabic: Following the Greek, Thābit's version also uses ظ (Θ, Q) although there is no ح (H). In the diagram for **I** (21b), د (Δ, D) has been moved over to mark the other intersection of the two circles, although this is not required by the text.

Comparison with Latin: The Latin follows the Arabic (Kunitzsch and Lorch 2010, 24–25).

Sph. I.5 (Theorem)

This proposition is analogous to *Elem.* III.19, and it is demonstrated in a similar way. It is a partial converse with the preceding proposition, which it uses in an indirect argument.

Basis: *Elem.* I.post.1, XI.13, *Sph.* I.4.

Usage: Unused in the *Spherics*.

Diagram: The figure for this proposition is a standard solid diagram, similar to some of the early diagrams of *Elements* XI (Saito 2014, 72, 74). There is no attempt to depict either the sphere or the non-cutting plane. As usual, only objects that are named in the *exposition* and *construction* appear in the diagram.

Textual Comments: In some of the manuscripts a circle is drawn around AB as a diameter (**B**, 5r; **D**, 1v). Since no such circle is named in the text, this circle would have to be interpreted as representing the sphere itself. There are, however, no diagrams in the *Spherics* – nor indeed in *Moving Sphere* or *Phenomena* – in which the sphere itself is depicted. Hence, we have followed the diagram in **A** (2v), which does not depict the sphere.

Comparison with Arabic: In Thābit's version, the diagram is similar to that in **A**, except labels *B* and *G* are interchanged, and the image has been reflected about the vertical axis (**I**, 22a).

Comparison with Latin: The diagram in Gerard's Latin follows that in the Arabic (Kunitzsch and Lorch 2010, 26–27).

Comparison of the magnitudes of circles in the sphere (*Sph.* I.6)

This proposition stands alone in dealing with the relative sizes of circles in a sphere. It shows that (a) the largest circle in a sphere – a great circle – passes through the center of the sphere and that (b) circles that are equally distant from the center are equal, while (c) the farther a circle is from the center of the sphere the smaller it is. This proposition is the most important of the early propositions and one of the more used propositions in the whole treatise.

Sph. I.6 (Theorem)

This theorem is a extension to the sphere of both *Elem.* III.14 and III.15. It is proved through solid geometry using an application of *Elem.* I.47 in a way that is similar to the method of *Sph.* I.1.

This proposition includes a full *conclusion*, summarizing the different parts of the proposition as a nearly verbatim restatement of the *enunciation*.

As in *Sph.* I.3, the center of the sphere is "taken." Again, however, it is not clear how this is mathematically different from simply assuming that there is some point that is the center, as was done in *Sph.* I.1 and the proof of I.2.

In terms of spherical astronomy, this proposition makes the claim that any circle that passes through the earth, or the center of the model – such as the horizon, meridian, vertical, equator, ecliptic and colures – are great circles; and that any small solid circles that are equidistant from the earth

– such as certain δ-circles – are equal to one another, while any small circles that are farther from the earth – such as other δ-circles – are smaller. As discussed above, however, before we have the theories of parallelism and tangency that are developed in *Spherics* II, we are not in a position to state fully how the various δ-circles satisfy the conditions of this proposition.

Basis: *Elem.* I.def.15, I.post.1, I.17, I.18, III.def.1, III.def.4 (generalized to the sphere), III.def.5 (generalized to the sphere), VI.20.cor., XI.def.3, XI.11, *Sph.* I.def.1, I.def.2, I.1.cor., I.2.

Usage: *Sph.* I.12, I.13, I.14, I.16, I.17, I.19, I.20, II.6, II.13 (by implication), II.17, III.5, III.6, III.7, III.8, III.13. A proposition like this was used in *Mov. Sph.* 6 (by implication), and 12.

Diagram: The diagram is a planar representation of global objects, containing both intrinsic and extrinsic objects. In this case, all of the objects have simply been flattened out into the plane of the figure. The axes of circles *AB* and *EZ* are depicted with extended radii of these circles that make false crossings with the circles themselves. It is not necessary that lines *QH* and *HK* appear in a straight line. There is no way of knowing whether they were originally so depicted, or whether this resulted from the tendency of copyists to produce symmetrical figures (see the discussion of such practices above, p. 78). This is a situation where the symmetry and overspecification of the text might mislead a modern reader, who hopes to take this type of information directly from the diagram. An ancient or medieval reader, however, was apparently not confused by such a figure. As is often the case in the *Elements*, this proposition depicts two cases with a single diagram (Saito 2012).

Textual Comments: In some of the later Greek manuscripts this proposition is followed by a converse that is neither in the earliest Greek manuscript, **A**, nor in the Arabic versions.[26]

Axis of a circle in a sphere (*Sph.* I.7–I.10)

This group of four propositions can be taken together to constitute a theory of the properties of the axis of a circle in a sphere. These theorems develop an analogy with *Elem.* III.3 in which a diameter of a plane circle corresponds to a diameter of the sphere, and a chord, or diameter, in the plane circle corresponds to a solid circle in the sphere. The analogy is somewhat complicated by the fact that the poles of a circle in a sphere are defined independently of the endpoints of the diameter of the sphere

---

[26] For a translation of the converse, see Ver Eecke (1927, 10–11). This spurious *Sph.* I.6 (converse) is applied in *Sph.* I.11 in a phrase which is, itself, thought to be an interpolation.

through the center of the circle – *Sph.* I.def.5. This means that a number of the propositions in the *Spherics* have no counterpart in the *Elements*.

We consider a line satisfying the following conditions with respect to the configuration of a solid circle in a sphere:

(a) passing through the center of the sphere,

(b) passing through the center of the circle,

(c) perpendicular to the circle,

(d) passing through both of the poles of the circle, and

(e) passing through one of the poles of the circle.

Then, *Sph.* I.7–I.10 show that if some of these conditions hold for a line, then one or more of the other conditions also hold for the same line. Namely, *Sph.* I.7 shows that (a) and (b) $\Rightarrow$ (c), *Sph.* I.8 shows that (a) and (c) $\Rightarrow$ (d), *Sph.* I.9 shows that (e) and (c) $\Rightarrow$ (b) and (d), and *Sph.* I.10 shows that (d) and (c) $\Rightarrow$ (a) and (b).

In terms of spherical astronomy, this group of theorems can be taken to explicate the relationship between the celestial axis and the circles that have it as their axis, the equator and the δ-circles. In particular, the celestial axis is perpendicular to the celestial and terrestrial equator and to all of the δ-circles at their centers, and passes through the celestial poles, which are the poles of all these celestial circles.

Sph. I.7 (Theorem)

This proposition, which is a partial converse to *Sph.* I.2.cor., is analogous to the first half of *Elem.* III.3. It is demonstrated directly through the definitions of the sphere and its center, the triangle congruence of *Elem.* I.8, and solid geometry.

The function of the term *base*, βάσις, in the *demonstration* is to alert the reader to the fact that we are dealing with congruent triangles through *Elem.* I.4 or I.8, which both contain the term in their *enunciations* (Heath 1908, 248–249). As Proklos points out, the *base* is either the side of a triangle that is level with the text, if no sides have been mentioned, or the third side, if two other sides have already been mentioned (Friedlein 1873, 236; Morrow 1970, 184).

The *construction* calls for the production of lines *AZG* and *BZD* through *Z*. Although such a construction does not appear to have been used in the *Elements*, it is clear why Theodosios would have felt free to introduce it without postulation. It simply involves passing an arbitrary line through a given point. If we like, we can think of it as reduced to taking another arbitrary point on a certain object – something which is often done in the *Elements* – and then applying *Elem.* I.post.1 and I.post.2. The specification of lines *AZG* and *BZD as diameters* is handled through the convention of giving them letter-names that include the center of the circle, *Z*.[27]

Basis: *Elem.* I.def.10, I.def.15, I.post.1, I.8, XI.4, *Sph.* I.def.1, I.def.2.

Usage: *Sph.* I.14. A theorem equivalent to this is applied in *Mov. Sph.* 12.

Diagram: The diagram is a planar representation of local objects, containing both intrinsic and extrinsic objects. As with the diagram for *Sph.* I.1, some viewers may regard this image as conveying a sense of depth, but it can also simply be regarded as depicting all of the objects folded into the plane. The axis of a circle is depicted with an extended radius that has a false crossing with the circle.

Textual Comments: The step in the *demonstration* that relies on *Elem.* I.def.10 is cited by an explicit quotation of the first part of that definition (p. 98, n. 8). Since a number of steps in the argument up to this point can be justified by appeal to the *Elements*, one might wonder why this place requires such an explicit reference.

Sph. I.8 (Theorem)

This proposition, which has no counterpart in the *Elements*, is the first that handles the concept of the poles of a circle in a sphere. Along with *Sph.* I.1.cor., it is demonstrated through solid geometry.

This proposition is one of the few that includes a proper *conclusion*, ending in an abbreviated form often found in the *Elements*.

Basis: *Elem.* I.def.15, I.post.1, I.4, XI.def.3, *Sph.* I.def.5, I.1.cor.

Usage: *Sph.* I.13, I.14, I.16, II.1, II.7 (by implication). A theorem equivalent to this is applied in *Mov. Sph.* 1.

Diagram: The diagram is a planar representation of local objects, containing both intrinsic and extrinsic objects. The axis of a circle is depicted with an extended diameter that has two false crossings with the circle.

Comparison with Arabic: Following this proposition, Thābit's version contains an additional proposition, *Sph.*$_A$ I.9, followed by a sketch of the proof. The complete text of *Sph.*$_A$ I.9 reads as follows, "If a circle is in a sphere, and it is joined by a straight line between one of its poles and the center, the line is an upright on the circle. The proof of this proposition is similar to the proof of the proposition that precedes it" (Kunitzsch and Lorch 2010, 42). That is, using the abbreviations introduced above to

---

[27] Netz (1999a, 20–21) and Acerbi (2020a, 51–60) provide discussions of various ways in which naming practices produce specificity in Greek geometric texts.

summarize the propositions of this section, this additional proposition states that (b) and (d) ⇒ (c).

Comparison with Latin: Gerard's Latin follows the Arabic (Kunitzsch and Lorch 2010, 43).

Sph. I.9 (Theorem)

This proposition, which also has no direct counterpart in the *Elements*, is demonstrated using the definition of a pole, *Sph.* I.def.5, and elementary solid geometry.

Although AE and EB are articulated separately in the *construction*, the analogy with same construction in *Sph.* I.7, I.8 and I.10 makes it clear that the construction again involves the production of an arbitrary line through a given point, E, such that the line meets the circle at two points, say A and B.[28]

Basis: *Elem.* I.def.15, I.post.1, I.c.n.1, I.c.n.3, I.4, I.47, VI.20.cor., XI.def.3, *Sph.* I.def.5.

Usage: *Sph.* I.10.

Diagram: The diagram is a planar representation of local objects, containing both intrinsic and extrinsic objects. The axis of a circle is depicted with an extended diameter that has two false crossings with the circle. Because lines AE and EB are constructed separately from point E, there is no reason why they need to be depicted in a straight line. Nevertheless, such a representation is not necessarily mathematically incorrect; hence, as with the diagram for *Sph.* I.6, we cannot know whether this resulted from the tendency of the copyists to produce symmetry, or indicates a different visual sensibility on the part of the ancient author.

Comparison with Arabic: In the Arabic versions this proposition is *Sph.*$_A$ I.10.

Sph. I.10 (Theorem)

This theorem, which has no direct counterpart in the *Elements*, is demonstrated through the corollary to *Sph.* I.2 and solid geometry.

Basis: *Elem.* I.def.10, I.post.1, I.c.n.3, I.4, I.8, XI.4, *Sph.* I.def.5, I.2.cor.

Usage: *Sph.* I.15, II.1, II.2, II.6, II.17, III.11.

Diagram: The diagram is a planar representation of local objects, containing both intrinsic and extrinsic objects. The axis of a circle is depicted with an extended diameter that has two false crossings with the circle.

---

[28] See Czinczenheim's (2000, 61–64) critical apparatus for the various ways that these letter-names appear in the manuscripts.

Comparison with Arabic: In the Arabic versions this proposition is
Sph.$_A$ I.11.

Intersecting great circles (*Sph.* I.11, I.12)

This pair of converses establishes the fundamental analogy between great
circles in a sphere and diameters in a circle – namely, two great circles,
just as two diameters, bisect one another. These theorems are not built on
analogy with any theorem in *Elements* III, because such a theorem would
follow almost immediately from the properties of a diameter of a circle,
Elem. I.def.15, I.def.16, and I.def.17. Nevertheless, these theorems are again
closely related to *Elem.* III.3 and, especially, III.4, which is the plane counterpart to the same situation for small circles. The analogy between great
circles and diameters introduced in *Sph.* I.11 and I.12 will be further developed in the following group of propositions.

For spherical astronomy, these propositions imply that the principal
great circles of the celestial sphere – the horizon, equator, ecliptic, meridian, vertical, and colures – all bisect one another pairwise.

Sph. I.11 (Theorem)

This proposition follows almost immediately from the properties of circles,
and especially of great circles, in a sphere.

In the *demonstration* it is asserted that the planes of two great circles
intersect in a straight line. For this claim to follow, however, it is necessary
to assume that each circle has at least one point that is noncoplanar with
the other circle. In fact, the naming convention in the *exposition* can be understood to provide this assumption. That is, when **gC**(*AB*) and **gC**(*GD*)
are introduced, *A*, *B*, *G* and *D* are not specific points, but simply parts of
the names of those circles. Nevertheless, we can understand them to be any
arbitrary points on those two great circles that are not in common, so that
when points *E* and *Z* are introduced as common points, each of the circles
will have at least three points which are noncoplanar with the other circle.
Hence, they determine two separate planes, *Elem.* XI.2, which intersect in
a line, *Elem.* XI.3.

Basis: *Elem.* I.def.17, I.def.18, I.post.1, III.1, XI.3, *Sph.* I.6 (converse, ∗).[29]

Usage: *Sph.* II.5, II.6, II.17, II.19, II.21, III.6 (by implication). A theorem
equivalent to this was used in *Phen.* 6, and in *Mov. Sph.* 1 (by implication).

---

[29] This application of the converse to *Sph.* I.6, found as an interpolation in some late
manuscripts, is probably itself also an interpolation. The claim that it supports is not
needed in the argument for *Sph.* I.11.

Diagram: The diagram is a planar representation of global objects, containing both intrinsic and extrinsic objects – namely, intersecting great circles and their shared diameter. The circles in the diagram represent circles in the surface of the sphere, and the line represents an internal straight line. In this diagram there are no false crossings. Hence, the diagram represents the topology of the configuration. The labeling of the diagram helps us understand the distortion that will be produced in surface representations of great circles. In particular, since line *EZ* bisects each great circle, the outer arcs will be equal to the inner arcs, so that arcs *AE, EB, BZ, ZA, GE, ED, DZ*, and *ZG* all represent quadrants of the two great circles.

Comparison with Arabic: In the Arabic versions this proposition is *Sph.$_A$* I.12.

Sph. I.12 (Theorem)

This proposition, although the converse of the former, relies on *Sph.* I.2.cor. and I.6, as well as solid geometry. This theorem is not used in the *Spherics*; so it is probably introduced for mathematical completeness – so as to show, with *Sph.* I.11, that mutual bisection is both necessary and sufficient for a pair of great circles.

The structure of this proposition appears to be non-standard, insofar as there are two "I say" (λέγω) statements. The first is a normal *specification* articulated using letter-names that have already been assigned to objects. The second is stated in general terms and seems to be an explanation of how the proof proceeds.[30] This proposition contains a general *conclusion*.

Basis: *Elem.* I.def.15, I.def.16, I.def.17, I.post.1, I.10, XI.12, *Sph.* I.2.cor., I.6.

Usage: Unused in the *Spherics*. A claim equivalent to this theorem is asserted in the introduction to the *Phenomena*.

Diagram: The diagram is a planar representation of global objects, containing both intrinsic and extrinsic objects. It is the same as the diagram for *Sph.* I.11, with the addition of further straight lines. The straight lines drawn from the center of the sphere have false crossings with the great circles.

Comparison with Arabic: In Thābit's version this proposition is *Sph.$_A$* I.13. The argument has been somewhat fleshed out – in particular, line *EZ* (ز ه) is introduced explicitly as the common section of the two circles (Kunitzsch and Lorch 2010, 52–54). There is also an unexpected "I say" statement in the claim that point *H* (ح) is the center of the sphere.

---

[30] Czinczenheim (2000, 67) bracketed the second statement as an interpolation.

There is no general *conclusion*, but, as usual, a QED statement has been added.

Comparison with Latin: The Latin follows the Arabic (Czinczenheim 2000, 915–916; Kunitzsch and Lorch 2010, 53–55).

Great circle through the poles of a small circle (*Sph.* I.13–I.15)

This group of propositions provides a direct analogy to *Elem.* III.3, in which, now, a diameter of a plane circle corresponds to a surface great circle, and a chord corresponds to a surface small circle. Because the poles of a circle on a sphere are defined independently, *Sph.* I.def.5, in the *Spherics* we have three properties instead of the two handled in *Elem.* III.3. We consider a surface great circle involved in the following conditions with respect to another surface circle in a sphere:

(a) bisecting a small (or any) circle,

(b) perpendicular to a small (or any) circle, and

(c) passing through the poles of a small (or any) circle.

Then, these propositions show that of if one of these configurations holds for a great and small (or any) circle, then the two others also obtain. Namely, *Sph.* I.13 shows (b) ⇒ (a) and (c), *Sph.* I.14 shows (a) ⇒ (b) and (c), and *Sph.* I.15 shows (c) ⇒ (a) and (b). In fact, *Sph.* I.14 is shown for small circles, while *Sph.* I.13 and I.15 are shown for any circles. In the *Spherics*, although *Sph.* I.13 and I.14 are both used, the most important theorem is *Sph.* I.15, which functions as a sort of goal of this section.

From the perspective of spherical astronomy, these three propositions can be taken to show that a great circle through the poles of another great circle will bisect at right angles any small circles about the same poles. Indeed, this configuration could then be used to determine coordinates for any of the principal systems – local, equatorial or ecliptic. In terms of the circles in the model of the celestial sphere that underlies this treatise, these propositions can be taken to show that the colures, in particular, the equinoctial and solstitial colures, pass through the celestial poles and bisect the δ-circles, such as the tropics and the visibility circles, at right angles. Again, we will have to wait until the theory of parallel circles developed in *Spherics* II to see the details of this configuration fleshed out.

Sph. I.13 (Theorem)

Although this theorem applies some elementary theorems of solid geometry, the core of the argument depends on a number of propositions developed in earlier sections – namely, *Sph.* I.1.cor., I.6, and I.8.

At the end of the *demonstration* it is shown that "circle *ABGD* cuts circle *EBZD* through the poles." From this point on in the treatise, "to cut through the poles," τέμνειν διὰ τῶν πόλων, or simply to *be* "through the poles," as in the following proposition, will be used as a technical expression for one circle passing through the poles of another.

This proposition contains two "I say" statements – the first introducing both parts, and the second reintroducing the second part. Both are articulated as general claims, with no use of letter-names, which is non-standard for a *specification*.

In the *construction*, *BD* is introduced as the common section of the two planes introduced in the *exposition*, those of **C**(*ABGD*) and **C**(*EBZD*), using the expression "let it be joined," ἐπεζεύχθω, which is usually used when joining two points. Hence, one may be tempted to read this as an application of *Elem.* I.post.1. Starting from *Sph.* I.17 and following, however, the common section of two planes will simply be introduced by asserting its existence.[31] Indeed, as discussed above, since *Elem.* XI.3 shows that the common section of two planes is a straight line, we do not need a construction postulate, or problem, to produce this line – it is already there.

As mentioned above, however, the assertion of the existence of *BD* as the common section of **C**(*ABGD*) and **C**(*EBZD*) is not a construction in the normal sense – as the introduction of a new object – but is rather a sort of constructive assumption. That is, the *construction* simply names some part of a line that is already there, and draws it into the diagram – in effect, bringing it into the mathematical discourse. This should be contrasted with *Sph.* I.15, and many other propositions, in which the common section of two planes is a straight line as always, *Elem.* XI.3, but is neither named nor drawn into the diagram, because it plays no role in the argument.

When *A* and *G* are introduced in the letter-name of **gC**(*ABGD*) they are initially unspecified, and are simply parts of the letter-name. At the end of the *construction*, however, they are completely specified as the intersections of *HQ* with **gC**(*ABGD*) on the surface of the sphere.[32]

Although the proof given for this theorem depends on **C**(*EBZD*) being a small circle, the theorem also holds when **C**(*EBZD*) is great. In fact, the proof for this case would follow immediately from *Sph.* I.11 and I.8. As

---

[31] See, in particular, the expression used in *Sph.* II.3, which makes the assertion of existence explicit.

[32] Other examples of this common type of naming practice are adduced by Acerbi (2020a, 51–60).

often the case in a Greek geometrical proposition, the more trivial case is omitted.

Basis: *Elem.* I.def.18, I.post.2, I.12, III.1, XI.3, XI.4, *Sph.* I.1.cor., I.6, I.8.

Usage: *Sph.* I.15, I.21, III.8 (indirect). A proposition like this was used in *Phen.* 13.

Diagram: The diagram is the first solid representation in the text, depicting global objects, and containing both intrinsic and extrinsic objects. The axis of a circle is depicted with an extended diameter that has a false crossing with the circle and two true intersections with a great circle. This diagram uses a lens-shaped figure to depict a foreshortened circle (but see the "Textual Comments," below).

Textual Comments: There are a number of mathematical issues with the diagrams of all three propositions in the Greek manuscripts, and they do not follow the visual pattern of the other diagrams in the first part of the treatise, which otherwise do not employ lens-shaped figures for circles before *Sph.* II.17 (Malpangotto 2010, 80 n. 22). The diagram for this proposition in the Greek manuscripts depicts circle EBZΔ as a lens-shaped figure about line BΔ, which is drawn through the center of circle ABΓΔ, so as to appear to be a diameter of the sphere. Hence, point Θ appears to be the center of the sphere and point H appears to be the intersection of circle EBZΔ with line AΓ, which it should not be (**A**, 6r; **B**, 12r; **F**, 32r; **K**, 96v). In **D** (13r), another smaller circle EBZΔ has been drawn below the first, in another ink.

Comparison with Arabic: In Thābit's version this proposition is *Sph.*$_A$ I.14. In this version, and the medieval traditions that are based on it, the small circle in the diagram appears as a circle; it is not drawn through the center of the sphere; and the overall configuration is fairly clear (**I**, 25a; Kunitzsch and Lorch 2010, 55). In **I**, the diagram is rotated 90°. Hence, in Thābit's text, these diagrams, like all other diagrams in the treatise, show a circle that intersects the base circle as partly inside and partly outside of the base circle. A diagram in this style which could accompany our translation, would appear as Figure 3.

This diagram is both easier to draw, and more consistent with the principles of the other diagrams in the treatise, hence, although it is possible that Thābit copied such a diagram from his source, it is also possible that he produced this diagram himself as a sort of correction. Moreover, although we have not checked all of the manuscripts systematically, this diagram appears to have been highly stable and appears in all versions of the medieval Arabic and Latin traditions of the treatise that we have seen. It is also found in the earliest modern printings, prior to Maurolico's (1558) edition.

Comparison with Latin: Gerard's Latin follows the Arabic (Kunitzsch and Lorch 2010, 55–57).

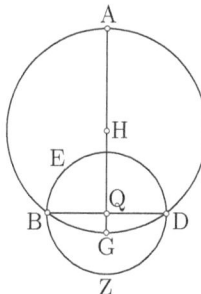

Figure 3: Diagram for *Sph.* I.13 in the Arabic and Latin traditions.

Sph. I.14 (Theorem)

As in the previous proposition, this depends on elementary solid geometry and *Sph.* I.6 and I.8.

As in *Sph.* I.13, there are two "I say" statements, for the two parts, both in general terms.

Once again, the *construction* includes the constructive assumption of line *BD*. See the Commentary to *Sph.* I.13 for a discussion of how this functions. As in the previous proposition, *A* and *G* are initially unspecified in the *exposition*, and are then fully specified at the end of the *construction* as the intersections of *HQ* with **g**C(*ABGD*).

Basis: *Elem.* I.def.16, I.def.17, I.def.18, I.post.1, I.post.2, I.10, III.1, XI.3, XI.18, *Sph.* I.6, I.7, I.8.

Usage: *Sph.* I.21.

Diagram: The diagram is a solid representation of global objects, containing both intrinsic and extrinsic objects. It is the same diagram as for the previous proposition (but see the "Textual Comments," just below).

Textual Comments: The diagram in the Greek manuscripts has all of the same issues as the previous diagram, with the added problem that in this proposition EBZΔ must not be a great circle, which it certainly appears to be (**A**, 6r; **B**, 13r; **D**, 13r; **F**, 32v; **K**, 96v).

Comparison with Arabic: In the Arabic versions this proposition is *Sph.*$_A$ I.15. The diagram for this proposition is the same as that for *Sph.*$_A$ I.14 (Kunitzsch and Lorch 2010, 59). Again, this style of diagram was used in all of the copies of the text that we have seen in the Arabic and Latin traditions up until the early modern period.

Comparison with Latin: The diagram in Gerard's Latin version follows that of Thābit (Kunitzsch and Lorch 2010, 59).

Sph. I.15 (Theorem)

This proposition depends on *Sph.* I.10 and I.13. In terms of usage, this is the most important theorem in *Spherics* I for the mathematical development of the rest of the treatise.

The *construction* in this proposition includes the introduction of the poles of $\mathbf{C}(EBZD)$, *A* and *G*. It is not clear why these points are introduced and named in the *construction*, as opposed to the *exposition*. Since the poles are mentioned in the *enunciation*, standard practice would have been to introduce them and assign them letter-names in the *exposition*. At any rate, despite the location of this passage, this is not a real construction, which is why there is no issue with introducing and naming these poles before it is shown how to find the poles of a circle in *Sph.* I.21. Because of this, *A* and *G* are again unspecified when they are introduced in the *exposition* as included in the letter-name of $\mathbf{gC}(ABGD)$. Then they are later specified as the poles of $\mathbf{C}(EBZD)$ in the *construction*.

Basis: *Elem.* I.post.1, XI.18, *Sph.* I.10, I.13.

Usage: *Sph.* I.17, I.21, II.3, II.6 (by implication), II.9, II.10, II.13, II.15, II.17, II.19, II.21, II.22, II.23, III.3, III.4, III.5, III.6, III.7, III.8, III.11, III.12 (by implication). A theorem equivalent to this was used in *Phen.* 13, and in *Mov. Sph.* 5, 6, 7 and 10.

Diagram: The diagram is a solid representation of global objects, containing both intrinsic and extrinsic objects. It is closely related to that for *Sph.* I.13 and I.14, but the intersection of the two circles is not depicted, because it is not mentioned in the text (but see the "Textual Comments," below).

Textual Comments: In the Greek manuscripts there are a number of different ways of depicting this figure, none of them convincing. In **A** (7r), arc BZΔ (*BZD*) is nearly overlapping with arc BΓΔ (*BGD*). In **B** (13r), the two arcs BZΔ and BΓΔ coincide. In **D** (13r), arc BZΔ has been drawn in with a different ink. In **F** (33r), Z and Γ seem to label the same point and another unlabeled arc joins points B and Δ below the circle. In **K** (96v), another arc has been drawn directly below arc BΔ, similar to it, with no label for either Z or Γ.

Comparison with Arabic: In Thābit's version this proposition is *Sph.*$_A$ I.16. The diagram in the Arabic text includes line د ب (*BD*) and point ط (*Q*), despite the fact that such objects are neither constructed nor named in the text, nor used in the mathematical argument (**I**, 25b; Kunitzsch and Lorch 2010, 61). In **I**, the diagram is again rotated 90°. A diagram in this style which could accompany our translation, would appear as Figure 4.

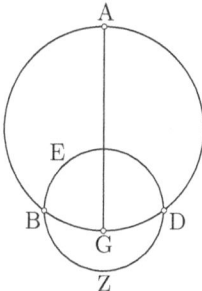

Figure 4: Diagram for *Sph.* I.15 in the Arabic and Latin traditions.

A diagram of this general style was used in all copies of the text that we have seen in the Arabic and Latin traditions up until the early modern period.

Comparison with Latin: Gerard's Latin version follows that of Thābit (Kunitzsch and Lorch 2010, 61).

Polar radius (distance) of great circles (*Sph.* I.16, I.17)

This pair of converses establishes the basic metrical property of great circles – namely, that the polar radius of a great circle is equal to the side of a square inscribed in it. These propositions are fairly crucial to the mathematical development of the text, being used in a number of key propositions, particularly problems.

Sph. I.16 (Theorem)

This proposition is developed on the basis of the fact that a great circle shares a center with the sphere, *Sph.* I.6, along with *Sph.* I.8 and the properties of a square inscribed in a circle, which are developed in the *Elements* in the context of a problem, *Elem.* IV.6.

In the *enunciation* of *Sph.* I.16, we are introduced to the terminology that will be used in this treatise for the length between a pole and the circumference of a circle in the sphere – "the ⟨distance⟩ from the pole," ἡ ἐκ τοῦ πόλου. This expression is formed on analogy with "the ⟨distance⟩ from the center," ἡ ἐκ τοῦ κέντρου (Fowler and Taisbak 1999; Sidoli 2004a), and could well be translated as *polar radius* (see Commentary to *Sph.* I.6). Since, however, when a circle is drawn on a sphere the text explicitly calls this length a "distance," διάστημα, we have chosen to translate the phrase ἡ ἐκ τοῦ πόλου as *the polar distance*, for consistency, although we will often use *polar radius* in the Commentary.

This proposition also introduces the technical terminology of "the side of the square inscribed in the great circle," ἡ τοῦ τετραγώνου πλευρὰ τοῦ εἰς τὸν μέγιστον κύκλον ἐγγραφομένου, for the polar radius of a great circle. In *Sph.* II.15, through ellipsis, this sometimes becomes "the side of the square," ἡ τοῦ τετραγώνου πλευρά, or simply "⟨the side⟩ of a square," τετραγώνου. In our Commentary we will call a square inscribed in a great circle a *great square*.

This proposition has two *specifications*. The first states in general terms what is to be shown. The second follows the *construction*, in which two lines – one the side of the great square, and another the polar radius of the great circle – have been drawn. The second *specification* is stated using the letter-names of these two lines, which will be used in the argument as well.

When point *E* is first encountered in the *exposition* it has not been introduced and is completely unspecified, but we can see that it must be the intersection of *AG* and *BD*, both from the diagram, and from its later role in the argument, in which it must be the center of **C**(*ABGD*)

The *construction* begins by stating the production of a pair of perpendicular diameters in a circle. In terms of the constructive methods of the *Elements* and this text, this can be fleshed out by finding the center of the circle, *Elem.* III.1, producing an arbitrary line through this point as done in, say, *Sph.* I.7, and producing a perpendicular at the center, *Elem.* I.11. In the next step of the *construction*, point *E* is introduced as unspecified.

Basis: *Elem.* I.def.22, I.post.1, I.4, I.11 (by implication), III.1 (by implication), IV.6, XI.def.3, XI.12, *Sph.* I.def.1, I.def.2, I.def.5, I.2.cor., I.8.

Usage: I.17, I.20, I.21, II.15, II.21, II.22.

Diagram: The diagram is a planar representation of global objects, containing both intrinsic and extrinsic objects. The axis of a circle is depicted with an extended radius that has a false crossing with the circle. It is fairly similar to the diagram for *Sph.* I.1.

Comparison with Arabic: In Thābit's version this proposition is *Sph.*$_A$ I.17.

Sph. I.17 (Theorem)

Although this short proposition is the converse to *Sph.* I.16, it shown directly through a number of other key propositions in this book, namely *Sph.* I.1, I.6, and I.15. Along with *Sph.* I.16, it is crucial to the problems developed in the next section, as well as those in *Spherics* II.

The *construction* begins by extending a plane through a line and a point. Since *Elem.* XI.2 shows that any triangle, or three points, is found in a single plane, this proposition can be used to justify this construction with no postulate. Along the same lines, as in *Sph.* I.13, a common section of two planes is invoked in the *construction*, but now the verb is different. Here, the common section is simply asserted to exist, and a letter-name is assigned to it. This is the same expression that will eventually be used in the rest of the treatise, starting in *Sph.* II.3.

Basis: *Elem.* I.def.17, I.def.18, I.post.1, III.31, XI.3, XI.12, *Sph.* I.def.3 (alternate text), I.def.5, I.1, I.6, I.15, I.16.

Usage: *Sph.* I.20, I.21, II.14, II.15.

Diagram: The diagram is a planar representation of global objects, containing both intrinsic and extrinsic objects. It depicts two great circles that have both been flattened into a plane passing through their intersection and the pole of one of them. In this diagram there are no false crossings, so that it accurately depicts the topology of the configuration.

Comparison with Arabic: In Thābit's version this proposition is *Sph.*$_A$ I.18.

Problems treating diameters, great circles, and poles (*Sph.* I.18–I.21)

This final group of propositions composes another major aim of *Spherics* I, namely the completion of problems that appear to have two goals – providing constructions that will be applied in the following propositions, and developing methods for doing actual drawings on a solid sphere (Schmidt 1943, 13–14; Sidoli and Saito 2009).

   To these different ends, the constructions in these propositions function on multiple levels. In the first place, they provide constructions that will be used later in the text, both in *theorems* and in *problems*. We can regard this usage as theoretical, in the sense that once the problems have been solved, the objects that they produce can simply be called into the mathematical discourse, usually, but not always, using the same verb as that used in the original problem. In this regard, they function like plane problems in Euclid's *Elements* (Sidoli 2018b). Secondly, in the constructions that solve the problem, but not necessarily those used in the proof, any objects produced on the surface of the sphere itself involve only the usage of *Elem.* I.post.3 (spherical, transfer), which, as discussed above, can be considered to be an abstraction of the operations of an actual compass, while applications of *Elem.* I.post.1 are made in only a single plane which is outside the sphere, so as to function as an abstraction of the operations of an actual straightedge on a drawing surface. This has the effect that objects on the surface of the sphere are produced using only constructions on the surface itself, and objects in a plane are produced in only a single plane. Finally, the proofs that the objects so produced satisfy the conditions set out in the *enunciation* are not so constrained – they may invoke planes passing through objects, objects produced inside the sphere, lines in multiple planes, and so on. Insofar as the objects used in the proofs are not subject to the same kinds of restraints as objects that solve the problem, the problems in the *Spherics*, again, follow the pattern established in the plane problems of the *Elements* (Sidoli 2018b).

Sph. I.18 (Problem)

The goal of this proposition is "to set out," ἐκθέσθαι, the diameter of a given circle in a sphere. The verb used for the constructive operation is the same as that used in *Elem.* XIII.18, which shows how to set out the edges of the five regular convex solids in a plane. Hence, in this proposition, as well, the construction involves producing a line equal to a line that is originally internal to a sphere – in this case, a diameter of a circle in the sphere. The primary use of this proposition is as a lemma for *Sph.* I.19, but as we will see, it can also be used in *Sph.* I.21, to show how that problem might be understood to provide a construction on the surface of the sphere. *Sph.* I.18 uses no previous propositions of the *Spherics* and hence is a proposition in elementary solid geometry.

The structure of this problem is like that of *Elem.* I.12, in which both the constructions that solve the problem and those that serve the proof are completed together in a unified *construction* prior to the *demonstration*, which immediately follows (Sidoli 2018b, 416). There is no *problem-conclusion* following the argument.

In the *exposition*, A, B and G are initially introduced as unspecified elements of the name of $\mathbf{C}(ABG)$ – at this point they do not denote points. Then, in the first step of the *construction*, arbitrary points are taken on $\mathbf{C}(ABG)$, and are assigned the names A, B and G. From here on A, B and G are the letter-names of specific points on $\mathbf{C}(ABG)$. Although they are chosen arbitrarily, once chosen they are assumed to be determined.[33]

In this problem, as in the problems in the *Elements*, some constructions are used to solve the problem while others are, in fact, part of the proof (Sidoli 2018b). In the perspective diagram provided here, we see the given circle as a bold line, straight lines which solve the problem as solid lines, and straight lines which serve the proof as dashed lines. The sphere itself is shown in gray. Those lines which solve the problem can be produced both with the postulates and problems of the *Elements*, supplemented with operations mentioned in the discussion of constructive assumptions above, but also with an actual compass and straightedge, whereas those lines that are part of the proof can only be produced by postulates and problems of the *Elements* – and are, hence, purely theoretical. Because the procedure

---

[33] These points are now *given* in the sense of (**G2**) (Sidoli 2018a).

for this proposition is somewhat unusual, it may be helpful to go through it in detail (Sidoli and Saito 2009, 589–591).

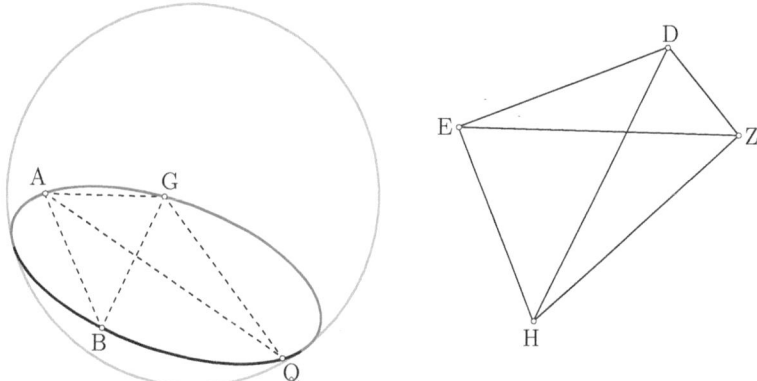

Figure 5: Perspective and plane diagrams, *Sph.* I.18.

The proposition begins with a given circle, **C**(*ABG*), seen here in bold – it has the letter-name *ABG*, but at this point *A*, *B* and *G* are simply parts of the name of the circle, they are not specified points. From a practical perspective, it may seem odd that a circle could simply be there, as *given*, with no points of the circle nor its pole given as well. Nevertheless, this sort of assumption is consistent with the practice of *Elements* and the *Data* – in which objects appear at the beginning of a problem as simply *given*, with no further assumptions about how they could have been introduced.[34] The solution to the problem then begins by taking three arbitrary points on the circle, *A*, *B* and *G*, which now serves to specify these letter-names by associating them with actual points.[35] Then a triangle, **T**(*DEZ*), is said to be constructed with *DE* set equal to the straight line, or the distance, from *A* to *B*, *DZ* to that from *A* to *G*, and *EZ* to that from *B* to *G* – making it clear that lines *AB*, *AG*, and *BG* are not yet assumed to be actually constructed. In terms of the problems of the *Elements*, the construction of **T**(*DEZ*) appears to be an application of *Elem.* I.22, using three given lines. Hence, in order to produce such lines on the plane, we suppose three uses of the strong form of *Elem.* I.post.3 (transfer), discussed above, in which the distance that is used to draw the circle can simply be any distance that appears in the assumed objects – here the distances between the arbitrary points. From a practical perspective, we simply place the compass tips on the assumed points *A*, *B*, and *G*, and transfer the lengths between them to the plane. Then, the quadrilateral with two right angles, **Quad**(*DEHZ*) can be constructed straightforwardly, using either postulates and problems

---

[34] This is the concept of *given* as (**G1**) (Sidoli 2018a).
[35] In fact, these points are *given* in the sense of (**G2**) (Sidoli 2018a).

of the *Elements*, or a compass and straightedge in the plane, such that *DH* is the sought line.

Then, for the sake of proving that *DH* is, in fact, the sought line, *AQ* is constructed as the diameter of the given circle, and lines *AB*, *BG*, *GA* and *GQ* are produced. All of these lines can be produced straightforwardly, using the postulates and problems of the *Elements* as noted in the justifications to the translation, but none of them can be drawn using an actual straightedge in a solid sphere. Moreover, the lines *AB*, *BG* and *GA* are introduced after lengths equal to them have already been produced outside the sphere. Finally, line *AQ*, produced here, is the diameter of the given circle, but it is not the object to be furnished by this problem. These considerations make it clear that the goal of this problem is not to find a diameter of a circle, such as *AQ*, but rather to produce a line outside the sphere, in a plane, using constructions that are all straightforward abstractions of operations that can be performed with an actual compass and straightedge (Schmidt 1943, 13–14). When it comes to the proof, however, further constructions, not bound by this constraint, are introduced, using the sort of constructive methods found in the solid books of the *Elements*.

The *demonstration* is elementary. It uses the properties of congruent triangles to show that the components of **Quad**(*DEHZ*), as constructed in the plane, and of **Quad**(*ABGQ*), as constructed inside the sphere, are respectively equal, so that *DH* = *AQ*.

Basis: *Elem.* I.post.1, I.post.2, I.post.4, I.c.n.1, I.8, I.12, I.22, I.26, III.1, III.21, III.22.

Usage: *Sph.* I.19, I.21 (by implication).

Diagram: The first diagram, of objects in a sphere, is a planar representation of local objects, containing both intrinsic and extrinsic objects. The second diagram is a standard planar diagram, like those in the plane books of the *Elements*. The spherical diagram shows a circle in a sphere and with internal lines that lie in the same plane, and the standard plane diagram shows congruent rectilinear objects in a different plane. In the diagram of the spherical surface there are no false crossings. Hence, the diagram represents the topology of the configuration on the sphere. In both of the diagrams, the triangles in question appear to be the isosceles triangle of a half square – whereas our inclination would be to depict these triangles as scalene. If the original diagrams had depicted these triangles as scalene, the type of overspecification that we now find in the manuscripts is just the sort that would have arisen from the tendency to produce symmetry in the course of transmission (see p. 78, above). On the other hand, since there is no mathematical error in depicting any triangle as a right isosceles triangle, and since ancient authors may have had a different view about the function of visual images in conveying the generality involved, we have decided to maintain the visual style found in the medieval manuscripts.

Comparison with Arabic: In Thābit's version this proposition is *Sph.*$_A$ I.19.

Sph. I.19 (Problem)

The goal of this problem, like that of *Sph.* I.18, is to "set out" the diameter of a given sphere. That is, once again we are producing in a plane a line that is equal to a line that would originally be found inside the sphere, namely as one of its diameters. Although this proposition is nowhere explicitly used in the rest of the *Spherics*, it is crucial for any practical constructions that involve drawing objects on the surface of a given sphere, as we will argue below. Moreover, it supplies a construction that was assumed without justification in *Elements* XIII.

The structure in this problem is a bit unusual, but is generally modeled on the full structure for a problem, such as that found in *Sph.* I.2. In this case, there is no normal *exposition* – presumably because there are no points or objects initially given on the sphere; so the assignment of letter-names, which is the primary function of the *exposition*, is unnecessary. Hence, there is also no *problem-specification* – that is, no assertion of what is to be done using letter-names. There is, however, a *proof-specification*, or rather, a statement of what is to be shown in the *demonstration*. There is also a *proof-construction*, which merely serves the *demonstration*, and which is, in this case, clearly distinguished from the construction that completes the problem. Finally, there is no *problem-conclusion* following the argument.

When $\mathbf{C}(BGD)$ is drawn in the *problem-construction*, D is initially unspecified and is simply an element in the name of the circle. Later, however, in the *proof-construction*, diameter AK is "imagined," and through it a plane which cuts $\mathbf{gC}(ABD)$ in the sphere. Presumably both the plane and the circle are also imagined. The use of D as part of the letter-name of $\mathbf{gC}(ABD)$ now specifies this point as the intersection of $\mathbf{gC}(ABD)$ and $\mathbf{C}(BGD)$. Hence, since this is now a fully specified point with a name, in the next step it can be used to join lines, AD and BD.

As in the previous proposition, some objects are produced in the construction in such a way that they could also be drawn with an actual compass or straightedge and some objects are introduced only for the sake of the *demonstration*. In fact, in this case the structure of the proposition makes these distinctions clearer, as will be discussed below. In the perspective diagram provided here, objects that are produced to solve the problem, and which can actually be drawn, are depicted with solid lines, while those which are constructed merely for their use in the *demonstration* are shown as dashed lines. The sphere itself is gray. Once again, we will go through the details of the construction (Sidoli and Saito 2009, 591–592).

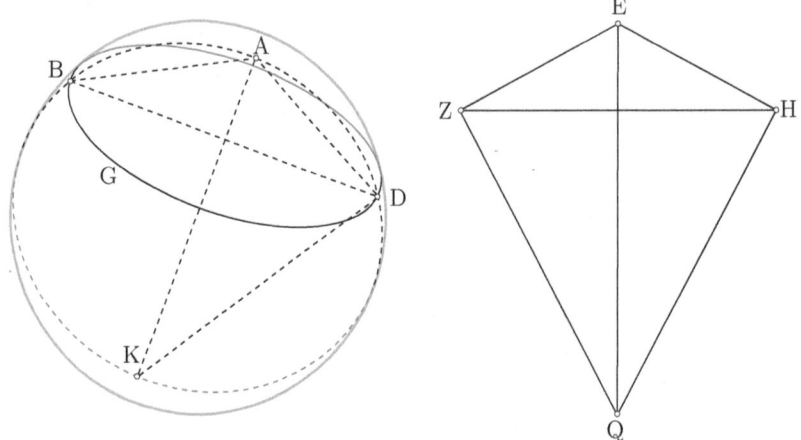

Figure 6: Perspective and plane diagrams, *Sph.* I.19.

The text begins by simply asserting the sphere to be "imagined," νενοήσθω. There are various uses of this verb in Greek geometric texts, but here, it probably simply indicates that the sphere is a solid object and none of its elements are currently being named.[36] Then two arbitrary points, *A* and *B* are taken on the surface of the sphere, again following the practice of the *Elements* and the *Data* of taking arbitrary points on given objects (Figure 6). Next, a circle, **C**(*BGD*), is drawn with *A* as a pole and *AB* as its distance – using the same terminology with which *Elem.* I.post.3 is asserted and used in the *Elements*. Here, however, as discussed above, the object is produced on the surface of the sphere. Since this can be done with two points assumed on the sphere, in a precise analogy to the way that *Elem.* I.post.3 is used in the plane, as discussed above we call this an application of *Elem.* I.post.3 (spherical). From a practical perspective, we can simply use an actual compass, setting one tip at *A* and the other at *B*.[37] At this point, *D* is unspecified and is simply part of the name of the circle. Since this circle can actually be produced on the surface of the sphere, it appears as a solid line in the diagram.

___

[36] This is similar to the way that two assumed spheres are introduced in *Elem.* XII.17, in which only the common center is named. Netz (2009) provides a discussion of the various ways that the verb νοεῖν is used in geometrical texts. See also Mugler (1959, 297) on this verb.

[37] For practical applications on a sphere the compass should be large relative to the size of the sphere and have tips that can bend in, as is still often the case with higher quality compasses.

The next stage of the construction simply points out that since a circle is now drawn on the sphere, "it is possible to set out" its diameter. Although this is a non-standard way of asserting the application of a problem, it is clear that this is a reference to *Sph.* I.18, in which it was shown how this can be done. Indeed, this use of *Sph.* I.18 follows the standard practice of using problems in the *Elements* – the object that the problem produces simply appears in the configuration, without having to go through the problem again (Sidoli 2018b). From a practical perspective, however, we would have to go through the actual steps of *Sph.* I.18 again: marking three arbitrary points on **C**(*BGD*), using the compass to transfer the lengths between the points to the plane, and using the compass and straightedge to draw the quadrilateral whose diagonal is equal to the diameter of **C**(*BGD*).

Following this, the plane construction of *Sph.* I.19 begins. Using a line equal to the set-out diameter, say *ZH*, and two lines equal to the distance from *A* to *B*, say *EZ* and *EH*, an isosceles triangle is constructed. Just as in *Sph.* I.18, the distance between *A* and *B* can be transferred to the plane using *Elem.* I.post.3 (transfer), then the triangle can be constructed using *Elem.* I.22. With right angles at *Z* and *H*, **Quad**(*ZEHQ*) is completed and line *QE* joined. This completes the constructions necessary to solve the problem – that is, the diameter of the sphere, *EQ*, has been set out. It should be clear that all these objects can actually be drawn in the plane using a compass and straightedge.

Following a *proof-specification* that states what is now to be shown, a new series of constructions is carried out. These constructions can also all be done with with postulates and problems of the *Elements*, but the fact that this sequence begins by asserting that the diameter of the sphere, *AK*, is to be "imagined," alerts us to the fact that we are now dealing with objects of a different status. In this case, the use of this terminology is probably meant to highlight the fact that this is a solid construction. Indeed, although the text states that *AK* is simply "imagined," we can use the postulates and problems of the *Elements* and the *Spherics* to flesh out the production of this line in a way that is consistent with the way that solid problems are handled in the *Elements*. That is, using *Sph.* I.2 we find the center of the sphere, and then we join this with *A* using *Elem.* I.post.1 and extend the line to the other side of the sphere at *K* with *Elem.* I.post.2. Then, a plane is imagined through *AK*. Again, this plane is not named at this point, and it may be any plane that passes through *AK*. In fact, we can simply assume that it is the plane passing through the three points *A*, *K* and *B*, by *Elem.* XI.2. Since, by *Sph.* I.1 and I.6, this plane must make a great circle in the surface of the sphere, say **gC**(*ABD*), it must also meet **C**(*BGD*), at some other point, say *D*, by *Sph.* I.15. As discussed above, it is the naming convention that indicates that *D* is the intersection of **gC**(*ABD*) and **C**(*BGD*), so that point *D* is now fully specified. Finally, just as in *Sph.* I.18, the lines joining these points are produced and the quadrilateral inside the sphere is completed.

The *demonstration* follows the model of *Sph.* I.18, simply using plane geometry to show that the elements of the two quadrilaterals are congruent.

From a practical perspective, this proposition furnishes a number of corollaries that will be of use in working with solid spheres. That is, it leads to a number of methods for carrying out constructions using an actual compass and straightedge.

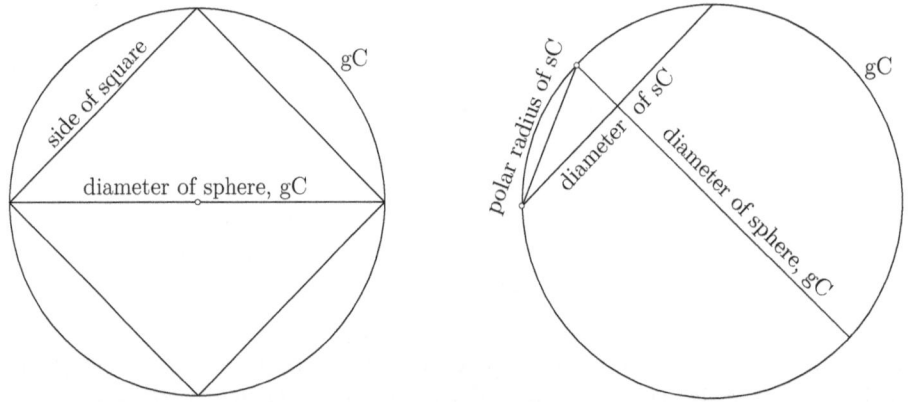

Figure 7: Practical corollaries to *Sph.* I.19, polar radius of a great circle and of a given circle.

For example, we can readily set out a great circle, and from that its polar radius, the side of the square inscribed in it, *Sph.* I.16 (Figure 7 (left)). First, the diameter of the sphere is set out, *Sph.* I.19, and then bisected, *Elem.* I.10, so that a circle, *gC*, can be drawn through its endpoints, *Elem.* I.post.3. Then, a square can be inscribed in *gC*, *Elem.* VI.6, so that its side will furnish the length we seek. By transferring this length back to the sphere, *Elem.* I.post.3 (transfer), we can actually draw a great circle about a given point as pole, *Elem.* I.post.3 (spherical).

Furthermore, combining *Sph.* I.18 and I.19, it is possible to set out the polar radius of a given small circle, *sC*, or rather of the circle that passes through a set of three given points (Figure 7 (right)), although there is no evidence from the text that such a corollary was intended by Theodosios or needed in the *Spherics*.[38] In such a situation, we would set out the diameter of the sphere, and draw the great circle, *gC*, as before. Then we would use *Sph.* I.18, to set out the diameter of the small circle, *sC*. Then, we would fit this length into *gC*, *Elem.* IV.1, as the diameter of *sC* and draw the

---

[38] Here we recall that in *Sph.* I.18 the circle was *given* as **(G1)**, while the three points were *given* as **(G2)**, so that from a practical perspective it is arbitrary which we take to be primary (see footnotes 34 and 35, above).

diameter of gC that is the perpendicular bisector of sC, using *Elem.* I.10 and I.11. Finally, joining the endpoints of the lines representing the diameters of sC and gC will give the polar radius of the given small circle.[39] Once, again, this length can be transferred back to the sphere.

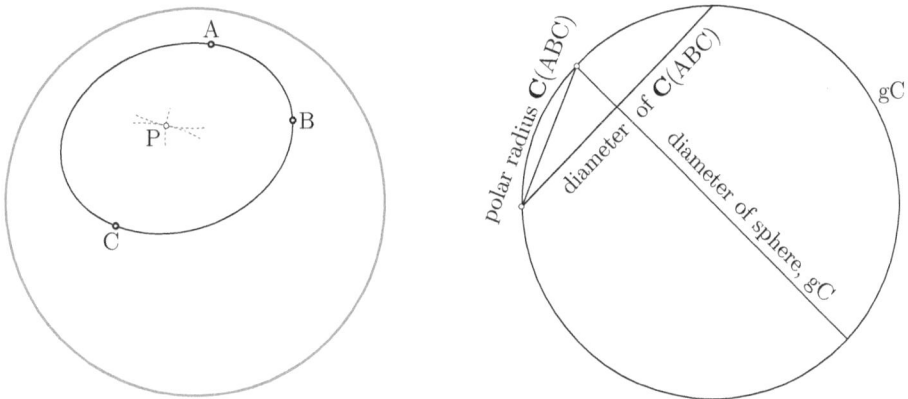

Figure 8: Practical construction of the circle passing through three arbitrary points on a sphere.

Finally, these methods can be used to actually draw the circle that passes through three arbitrary points on a sphere – for which, again, there is no evidence in the text. We begin with three points, say A, B and C, but no circle (Figure 8 (left)). Then, using the methods of *Sph.* I.18, which, as noted above, is actually carried out on three given points, we can transfer to the plane a line equal to the diameter of the circle $\mathbf{C}(ABC)$ that will be drawn passing through A, B and C (Figure 8 (right)). Then, using other arbitrarily chosen points – not shown in the diagram – we produce the diameter of the sphere in the plane, *Sph.* I.19, and following the method described in the previous paragraph we determine the polar radius of $\mathbf{C}(ABC)$ in the plane. Then, using *Elem.* I.post.3 (transfer), we can transfer this distance back to the sphere, and, using *Elem.* I.post.3 (spherical), we can draw the circles with poles A, B and C, and distance equal to the polar radius of $\mathbf{C}(ABC)$ – shown here as dashed gray arcs – which will all meet at the pole, P, of the circle that passes through A, B and C, *Sph.* I.def.5 (Figure 8 (left)).[40] Finally, with P as pole and the polar radius of $\mathbf{C}(ABC)$ as distance, an actual circle can be drawn about P passing through A, B and C – shown as a solid curved line in the diagram. Since all of these constructions can

---

[39] Again, there are actually two polar radii, but for practical purposes, we will work with the lesser of the two.

[40] In fact, any two such circles will suffice to find P, but one can see that from a practical perspective it may not always be clear where the intersection lies with just two circles.

actually be carried out with a compass and straightedge on a solid sphere and a separate plane, it is clear that if we start with three arbitrary points on a sphere, we can actually draw the circle that passes through them.

Basis: *Elem.* I.post.1, I.post.2, I.post.3 (spherical, transfer), I.post.4, I.8, I.12, I.22, I.26, III.21, III.22, XI.2 (by implication), *Sph.* I.1, I.2, I.6, I.18.

Usage: *Sph.* I.20 (by implication), I.21 (by implication), II.14 (by implication), II.15 (by implication), III.9 (by implication), III.10 (by implication). A construction such as this is assumed a number of times, without justification, in *Elements* XIII.

Diagram: The first diagram, depicting objects in a sphere, is a planar representation of global objects, containing both intrinsic and extrinsic objects. The second diagram is a standard planar diagram. The spherical diagram shows two circles in a sphere both flattened into the same plane, namely the plane that passes through their intersection and the pole of the lesser circle. There are also some internal lines that lie in the same plane. The second diagram shows congruent rectilinear objects in a different plane. Depending on the size of circle *BDG* relative to circle *ABD*, line *DK* may have a false crossing with circle *BDG*.

Comparison with Arabic: In Thābit's version this proposition is *Sph.*$_A$ I.20.

Sph. I.20 (Problem)

This problem, or rather its Case 2, can be understood as an analog to the production of a line between two points in a plane, *Elem.* I.post.1, and is related to the extension of a straight line as well, *Elem.* I.post.2.[41] The goal is to draw a great circle through any two given points on a sphere. If we consider an analogy between a plane and a spherical surface in which points correspond to points and straight lines correspond to great circles, then this proposition is an analogy to *Elem.* I.post.1 and I.post.2.

In terms of usage, *Sph.* I.20 is the single most applied proposition in the *Spherics*. Indeed, a construction of this sort was also used in Euclid's *Phenomena* and Autolykos's *Moving Sphere*, so that the execution of this *problem* may have been an important goal of *Spherics* I.

In this proposition, there is no clear division between the *construction* and the *demonstration*, since deductive steps are used in the course of the

---

[41] We will see in the Commentary to *Sph.* III.3 and III.4 that a certain construction in these propositions can be understood as an application of *Sph.* I.20 to complete a circle, in analogy with *Elem.* I.post.2.

It is well known that *Elem.* I.post.2 is sometimes associated with the so-called Archimedean postulate, asserting than a line can be made longer than any assumed line. As stated, however, *Elem.* I.post.2 only concerns the possibility of extending a line indefinitely, but does not say anything about its total length.

construction, and the final deductive step that shows that the constructed object has the properties that the problem calls for is an immediate consequence of the construction. Hence, after a normal *enunciation, exposition,* and *specification,* following the dismissal of a degenerate case, the *problem-construction, proof-construction* and *demonstration* are developed with no clear divisions between them. Finally, this problem is the first in the treatise that ends with a *problem-conclusion.*

Once again, those constructions that actually solve the problem are produced on the surface of the sphere itself, whereas the internal lines, *AE* and *EB*, are only used in the *demonstration,* and hence have a different status from the constructions that complete the *problem.* Again, we will go through the details of the construction (Sidoli and Saito 2009, 593–594).

The problem is handled in two cases: (Case 1) the given points lie on a single diameter of the sphere, or (Case 2) they do not.

Case 1 is summarily dismissed with the statement that if the given points *A* and *B* lay on a diameter, "unlimited," ἄπειρος, great circles will be drawn through them. There are a number of possibilities for why this case is not dealt with in any detail (Sidoli and Saito 2009, 594), but probably the most important is that there is no finite set of solutions to this problem – setting it on the level of producing a line through a single point, a circle through two points, or a plane through a single line – so that it does not constitute what a Greek geometer would have thought of as a well-defined problem. That is, in Case 1 this proposition no longer acts as an analog to *Elem.* I.post.1 and I.post.2, and hence ceases to be of mathematical interest. Furthermore, the method for actually drawing any one of these infinite great circles will be obvious as soon as we read Case 2. Namely, we set a compass to the distance of the side of the great square, as described in the Commentary to *Sph.* I.19, and, placing a tip on either *A* or *B*, draw a great circle, such that *A* and *B* are the poles of this great circle. Then, setting the tip of the compass *anywhere* on this great circle, we may draw a second great circle passing through both *A* and *B*.

Case 2, in which *A* and *B* are not the endpoints of a single diameter of the sphere, is the only case actually treated in the text (Figure 9). The construction involves using both *A* and *B* as poles and drawing two great circles, using the side of the great square as distance, *Elem.* I.post.3 (spherical). Then, one of the intersections of these great circles is joined by two lines with *A* and *B*, *Elem.* I.post.1, and it is pointed out that these lines must also be sides of the great square. It should be noted that these internal lines are not part of the solution to the problem, but are used, as before, for the proof that the solution is sound. In this case, however, they are introduced in the middle of the construction of the problem itself so that it can be shown that the final great circle that is drawn will, indeed, pass through *A* and *B*. This is, again, done by setting an intersection, say *E*, as a pole, and

pointing out that a circle drawn about it with the side of the great square
as distance, *Elem.* I.post.3 (spherical), will be a great circle passing through
A and *B*.

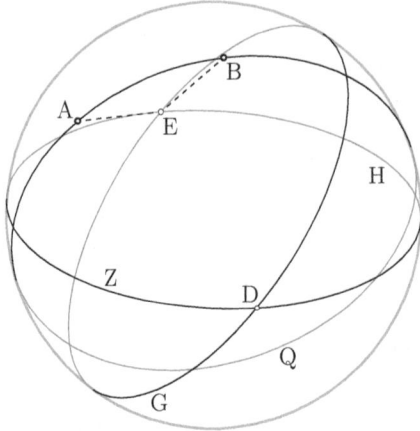

Figure 9: Perspective diagram, *Sph.* I.20, Case 2.

The construction for Case 2 also gives us a practical way of determining
whether or not the original given points, *A* and *B*, lay on the same diameter
of the sphere. Namely, when there are great circles about *A* and *B* as poles,
they will either intersect at two points, *D* and *E*, or they will coincide. If
they intersect, the construction for Case 2 can be carried through, if they
coincide, it cannot and any point on the single great circle so produced
will be a pole of another great circle passing through *A* and *B*, reducing to
Case 1.

Once again, taking into consideration the construction of the side of the
great square explained in the Commentary to *Sph.* I.19, it is clear that all
of the operations that are performed on the surface of the sphere itself in
Sph. I.20 can actually be drawn with real instruments. The internal lines,
AE and *EB*, however, which cannot be so drawn, merely serve to show that
the sought object has been produced.

The construction in this proposition hints at a practical corollary using
a construction that is analogous to that in *Elem.* I.1 – namely, to draw a
circle through two given points with a given line as polar radius, where the
given line is not greater than the side of the great square and not less than
the chord subtending half of the great-arc separation between the two given
points. Although such a construction is not directly used in the treatise, it
can be used with *Sph.* II.9 to explain some constructions carried out later
in the *Spherics*.

Specifically, with two given points, *A* and *B*, and a given line *c*, meeting
the stated conditions, two circles are drawn with polar radius *c* about poles

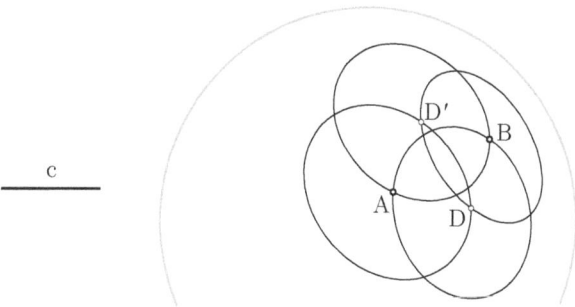

Figure 10: Constructive corollary to *Sph.* I.20.

A and *B* (Figure 10). Then, these circles will intersect in two points *D* and *D'* – or at a single point *D* when *c* is equal to the chord subtending half of the great-arc separation between *A* and *B*. Then, with either *D* or *D'* as a pole another circle is drawn with the same polar radius *c*, which will pass through both *A* and *B*, by an argument analogous to that in *Elem.* I.1.

Basis: *Elem.* I.def.17, I.post.1, I.post.3 (plane, spherical, transfer), I.c.n.1, I.10 (by implication) IV.6 (by implication), *Sph.* I.6, I.16, I.17, I.19 (by implication).

Usage: *Sph.* I.21, II.4 (counterfactual), II.5 (normal, counterfactual), II.6, II.8, II.13, II.14, II.15, II.16, II.17, II.19, II.20, II.21, II.22 II.23, III.3, III.4, III.5, III.7, III.8, III.9, III.10, III.12. The construction treated in this problem had been carried out in *Phen.* 2, and in *Mov. Sph.* 6, 7 and 10.

Diagram: The diagram is a planar representation of global objects, containing both intrinsic and extrinsic objects. In this case, we see three great circles all flattened into the plane of the diagram, as well as the lines joining the poles of one of them to the poles of the other two. In this diagram there are no false crossings, so that it well represents the topology of the configuration.

Textual Comments: We have been guided by the diagram of Thābit's version for the placement of the letters in the diagram (**I**, 27a).

Comparison with Arabic: In Thābit's version this proposition is *Sph.$_A$* I.21.

Sph. I.21 (Problem)

This problem is a second analog to *Elem.* III.1, in which, now, the circle on the sphere corresponds to a circle in the plane, and a pole of the circle in the sphere corresponds to the center of the plane circle. Moreover, it presents the same issues of interpretation that have been raised against *Elem.* III.1 – namely, if we take a strictly constructive interpretation of the objects that are introduced in the text, it is unclear how a circle could ever have been produced without its center, on the plane, or its pole, on the sphere, having been used to generate it (Sidoli 2018b, 411). Of course, in the spherical situation, since a circle can also be generated by passing a plane through a sphere, *Sph.* I.1, there is clearly a theoretical need to be able to find the pole of such a circle, but we should note that the construction provided in *Sph.* I.21 does not proceed by passing planes thorough the sphere, which would be the simplest way to solve this problem. Hence, it appears that the problem was introduced from a perceived theoretical need to provide the pole of a circle that is assumed, and there was a desire that this should be carried out through the production of objects on the surface of the sphere. Like the other problems, however, this proposition was probably meant to function on two levels, and following a detailed discussion of the operations from a theoretical perspective, we will point out an application of this problem to practical configurations that would have often arisen in working with a solid globe.

The structure in this proposition is similar to that in *Sph.* I.20, with the exception that here there is a division into two cases, which must be handled separately, and there is no *problem-conclusion*. Otherwise, however, in each case the *problem-construction* is not separated from the *demonstration*, and no *proof-construction* is required.

The proposition begins with the assumption of a given circle in a sphere, $\mathbf{C}(ABG)$, of which no points are initially assumed as given. The *construction* itself begins with the assumption of a random point, A, which follows the practice of the *Elements* and the *Data* of assuming arbitrary points on given objects, and then lays off two equal arcs, $\mathbf{Arc}(DA) = \mathbf{Arc}(AE)$, on either side of A. This can be fleshed out by simply assuming another arbitrary point, say D, and then laying off $\mathbf{Arc}(AE) = \mathbf{Arc}(DA)$, using *Elem.* I.post.3 (spherical). Next, the other arc, $\mathbf{Arc}(DE)$, is bisected at Z, so as to locate the other point on the same diameter as A.

It is not explained how to do this construction, but there are various options (Czinczenheim 2000, 924–925). If the side of the great square has already been produced, we can carry out a construction analogous to *Elem.* I.10, or rather I.1, but the proof that this is valid will require *Sph.* II.9, which has not yet been shown.[42] Using propositions that have already been elaborated, however, the construction can be made as follows. Using the three points so far determined on the given circle, A, D, and E, an equal circle can be set out in the plane by setting out its diameter, as detailed in *Sph.* I.18. Then, $\mathbf{Arc}(DE)$ is transferred from the sphere to this plane circle,

---

[42] See the Commentary to *Sph.* II.9.

using *Elem.* I.post.3 (transfer), bisected using *Elem.* III.30, and the half arc is transferred back to the sphere, again using *Elem.* I.post.3 (transfer). The proof that this is valid follows almost immediately from the construction itself.

The *construction* and *demonstration* are then developed for two cases: Case 1, in which the given circle is small, and Case 2, in which the given circle is great.

In Case 1, a great circle is drawn through the diametrically opposite points, A and Z, *Sph.* I.20, and then the arc of this great circle is bisected. Again, this bisection can be done by setting out the great circle, first setting out the diameter, *Sph.* I.19. Of course, in order to draw the great circle through A and Z, we will have already had to set out the great circle, as described in the Commentary to *Sph.* I.20. Notice that G is unspecified in Case 1.

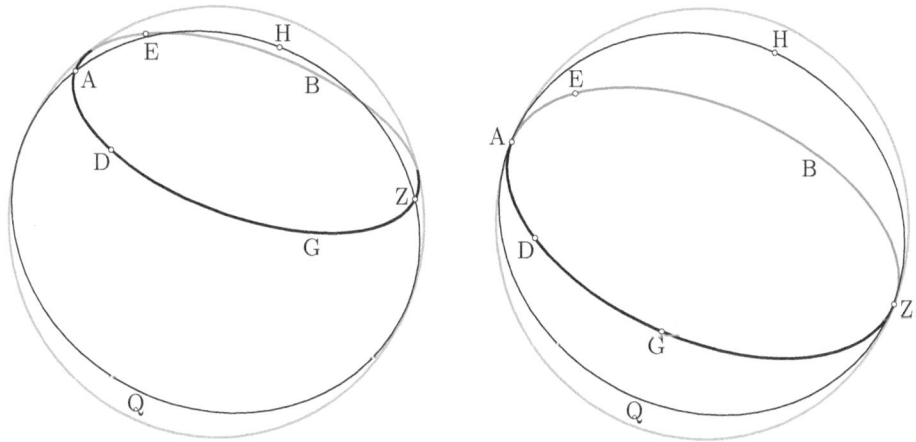

Figure 11: Perspective diagrams, *Sph.* I.21: (left) Case 1, (right) Case 2.

In Case 2, $\mathbf{C}(ABG)$ is assumed to be great, point Z is determined in the same way as in Case 1, and it is shown that $\mathbf{Arc}(ADZ) = \mathbf{Arc}(AEZ)$. Then, point G is determined as the midpoint of $\mathbf{Arc}(ADZ)$. Again, this could be done on the surface of the sphere, using an analogy to *Elem.* I.10, or I.1, but there will be no proof that this construction is correct until later in the treatise. As in Case 1, however, we can construct this point using the methods developed so far in this treatise by drawing a great circle in the plane, using the methods described in the Commentary to *Sph.* I.19 and then transferring the requisite arcs back and forth, as in Case 1. In fact, if we have already set out the great square of this sphere, we can simply use that. Furthermore, it should be noticed that this construction also gives us a constructive way to check whether or not $\mathbf{C}(ABG)$ is great. Namely, we determine point G as the midpoint of $\mathbf{Arc}(ADZ)$, using the methods described in the Commentary to Case 1, and then draw a circle about *G* passing through *Z*, using *Elem.* I.post.3 (solid). If this circle passes through point *A*, then **C**(*ABG*) is great, otherwise it is small.

For Case 1, the proof follows from *Sph.* I.14 and the definition of a pole, *Sph.* I.def.5, so that there is little distinction between the *construction* and the *demonstration*. If the conditions of Case 2 apply, then a great circle passing through *A* and *Z* that is perpendicular to the given great circle is drawn about *G* as a pole, so that with one of its arcs bisected, the pole of the original great circle will have been located, by *Sph.* I.13 and *Sph.* I.def.5. Once again, the steps of the *construction* and the *demonstration* are interspersed.

Since these constructions would not be applicable to a practical construction in which the given circle has actually been drawn around some assumed pole, we may well ask what purpose this proposition serves.[43] In fact, some construction like this is theoretically required later in the text, since it is applied in a fair number of the following propositions, both in *problems* and *theorems*. In this way, *Sph.* I.21 acts like *Elem.* III.1, insofar as it allows the introduction of the pole of a circle into the diagram and domain of discourse, when a proposition begins with a circle that is simply assumed. That is, on analogy with the usage of *Elem.* III.1 in the *Elements* and the *Data*, the assumption of a circle in a proposition of the *Spherics* must not be taken to also imply the assumption that it has been constructed around its pole.[44] Hence, if the pole of some assumed circle is needed, it must be found.

Furthermore, it should be noted that this problem, like the preceding three problems, shows a preference for using operations that can actually be performed on the surface of a sphere, without entering the sphere, or assuming cutting planes. This can be seen from the fact that a simpler, purely theoretical, solid construction could be provided by passing two different planes through the center of the given circle, *Elem.* III.1, and the center of the sphere, *Sph.* I.2. The fact that Theodosios prefers to operate entirely

---

[43] One can also observe that the construction provided in *Sph.* I.21 might point towards a practical construction for producing a circle passing through three given points (Thomas 2018a, 13–14). For such a construction we would first produce the points opposite two of the given points in the plane, using *Sph.* I.18, and then transfer these points back to the sphere using *Elem.* I.post.3 (transfer). If two great circles are drawn through the given points and their opposites, they will meet in the pole of the circle that would pass through the originally given points, about which the circle can now be drawn.

[44] Indeed, any strictly constructive interpretation of the *Elements* that holds that every object that enters the discourse should have first been constructed through a postulate or *problem*, must contend with the fact that in *Elem.* III.1 a circle is simply *given*, despite the fact that its center is not known and must be found – and furthermore that *Elem.* III.1 is used in other propositions to find the centers of other circles, which are assumed to have been introduced without their centers being known, or having been used for their construction.

in the spherical surface, indicates that providing constructions that operate within such constraints was one of the goals of these problems.

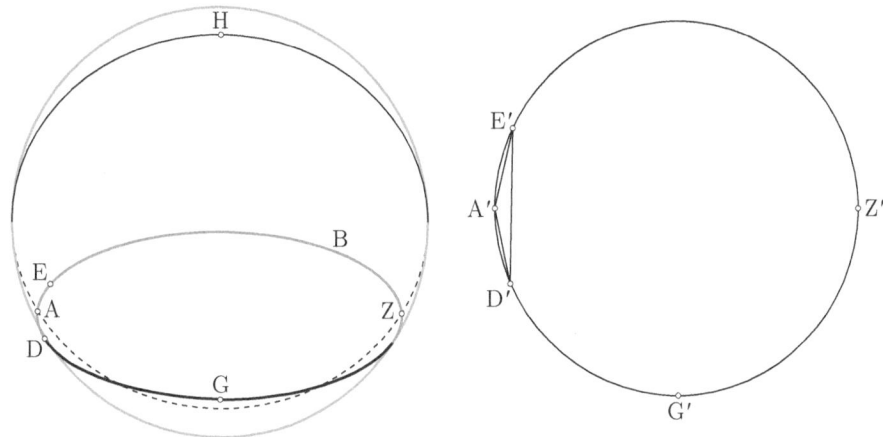

Figure 12: Practical application of *Sph.* I.21 to a solid globe.

Finally, in working with the sorts of solid globes that were used to depict the celestial sphere, and which we have argued were probably used for teaching and studying spherical astronomy, this proposition has a fairly clear application. Since the globes that were used for doing spherical astronomy would probably be set on a stand, or plinth, such that they could be arranged so as to configure the horizon of the sphere parallel to the local horizon (Figures 8 and 9), this problem could be used to find the zenith of the globe. Namely, if we consider that a solid globe will usually be set on a stand with a mouth equal to a great or small circle of the globe (Figure 12),[45] then it is clear that all of the operations described in the constructions above can actually be carried out with a real compass on the globe and a straightedge in an auxiliary plane, and the zenith for any particular configuration of the globe can thus be marked on the globe itself. Furthermore, in actually making a globe, or a spherical sundial, if any circles are produced on the solid surface by submerging it in a liquid, or spinning it on a lathe, the methods of this proposition can then be used to locate the poles of such a circle.

Basis: *Elem.* I.def.17, I.def.18, I.post.3 (normal, transfer), I.post.4, I.c.n.2, I.10 (by implication), III.29, III.30 (by implication), *Sph.* I.def.5, I.6, I.13, I.14, I.15, I.16, I.17, I.18 (by implication), I.20.

---

[45] Indeed, the stand may simply be three points that determine a circle parallel to the local horizon.

Usage: *Sph.* II.1, II.6, II.13, II.14, II.15, II.17, II.22, II.23, III.7, III.8 (by implication), III.12, III.14 (by implication). A construction demonstrated by this problem had been carried out in *Mov. Sph.* 7.[46]

Diagram: This diagram is the first that is a planar representation of global objects, containing only intrinsic objects. In this case, we see two intersecting great circles, following the same encoding as was used in the diagrams for *Sph.* I.11, I.12, I.17, and I.20. This proposition depicts two cases in a single diagram.

Comparison with Arabic: In Thābit's version this proposition is *Sph.*$_A$ I.22.

Stragglers (*Sph.* I.22, I.23)

The final two propositions in the Greek manuscripts are clearly spurious (Heiberg 1927, 40; Czinczenheim 2000, 80–81). Neither of these propositions is found in the Arabic versions of the treatise, so that they were either added to a Greek manuscript when it was copied in the 9th century, or they were removed by Thābit, who would have recognized them as mathematically pointless.

Sph. I.22 (Theorem, spurious)

This proposition shows the obvious fact that *Elem.* III.3, which has been a guiding analogy for a number of the theorems in this book, holds on a plane passing through the center of a sphere. This claim has already been implicit in a number of propositions – particularly, *Sph.* I.7–I.10, and I.13–I.15.

The style of this proposition is non-standard: it has no letter-names, no division into parts, and no diagram. Furthermore, it handles a pair of converse statements, which would usually be treated as two separate propositions in the *Spherics*.

Basis: *Elem.* III.3, *Sph.* I.1.

Usage: Unused in the *Spherics*.

Diagram: There is no diagram.

Comparison with Arabic: This proposition is not found in the Arabic versions.

---

[46] Schmidt (1943, 6) held that this proposition was also used in *Mov. Sph.* 6, but the assertion of the visible pole in that proposition is not the construction of the pole of any given circle and, hence, is probably simply the exposition and naming of the visible pole that is assumed in the *enunciation*.

Another possible usage of *Sph.* I.21 could be in II.20, but we argue against this. See the Commentary to *Sph.* II.20.

Sph. I.23 (Theorem, spurious)

This proposition shows the same mathematical fact as was shown in *Sph.* I.7, although the expression is somewhat different. The proof is slightly different, and more compact, but the basis is very nearly the same.

Basis: *Elem.* I.def.10, I.def.15, I.post.1, I.8, XI.def.3, *Sph.* I.def.1, I.def.2.

Usage: Unused in the *Spherics*.

Diagram: The diagram is a planar representation of local objects, including both intrinsic and extrinsic objects, but it does not follow the encoding used elsewhere in the rest of the text.

Comparison with Arabic: This proposition is not found in the Arabic versions.

Spherics II

With this book, the focus of the treatise begins to shift away from purely geometric considerations of the sphere itself towards propositions concerning circles on the surface of a sphere that are readily understood as circles on the celestial sphere. Although the underlying goals of *Spherics* II appear to be increasingly astronomical, the approach is, nevertheless, still geometrical in its articulation, insofar as almost no astronomical terminology is introduced, and motion is not directly mentioned. Although most of the propositions in *Spherics* II do not have direct analogs in the *Elements*, those few that do continue to develop the analogy established in the final problems of *Spherics* I, and again establish analogs to propositions of Euclid's theory of the circle, *Elements* III. These can be summarized as follows:

Spherics II	*Elements* III	Analogies
II.3–II.5	III.11, III.12(∗)	surface great circle ↔ any line
II.14	III.16	surface small circle ↔ any circle
II.15	III.17	

The early propositions, *Sph.* II.1–II.6, continue to develop theories about objects that could be considered as purely geometrical – namely, parallel and tangent circles. In the next section, *Sph.* II.10–II.16, these objects are joined together to form the two fundamental elements of the ancient field of spherics – namely, a bundle of parallel circles and a pencil of great circles intersecting them, which are either through their poles or tangent to the same pair of parallels. Internal to this section, we also have some important lemmas dealing with comparisons of arcs, and problems for drawing tangent great circles. The next group of propositions, *Sph.* II.17–II.20, treats various relations that obtain for a bundle of parallel circles and a single great circle that is oblique to the parallels. The final group of theorems, *Sph.* II.21–II.23, for which *Sph.* II.21 is a lemma, deal with a bundle of parallel circles and two great circles that are inclined on them, or rather one inclined great circle and a pencil of great circles tangent to another of the parallel circles. That is, *Spherics* II begins by developing the theories of parallelism and tangency for circles on the sphere, introduces a bundle of parallel circles and a pencil of great circles through the poles or tangent to two of the parallel circles, and then a bundle of parallel circles intersected by a single, inclined great circle, and finally combines the later two topics into a treatment of a bundle of parallel circles with a pencil of great circles tangent to two of the parallel circles and another single great circle inclined on all of the other circles.

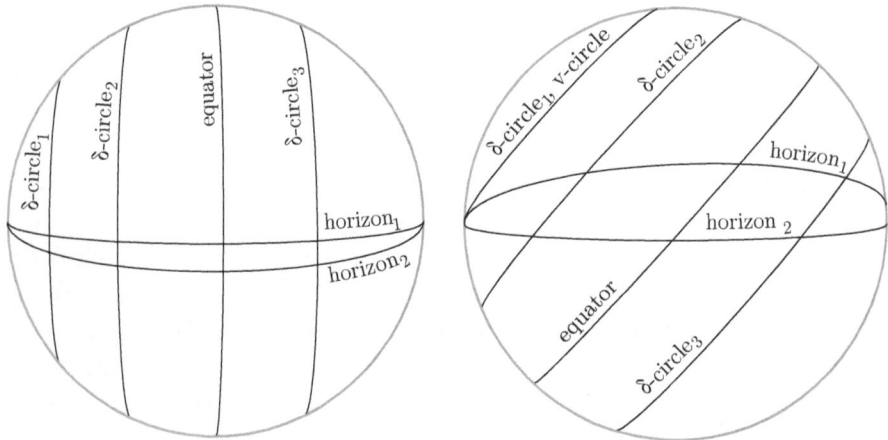

Figure 13: Diagrams for the astronomical implications of *Sph.* II.10–II.16.

As in *Spherics* I, this book exhibits shifts from solid considerations to surface considerations, but now these take place not gradually throughout the whole book, but rather topic by topic. Each of the four groups of propositions discussed above begins with solid concepts and methods, like those developed in the *Elements* and *Spherics* I, and then proceeds by asserting properties that can be established directly on the surface of the sphere. The propositions in each group that introduce solid methods, sometimes relying only on the solid geometry of the *Elements*, often act as lemmas to the later propositions. The final theorems in each group use purely surface methods and considerations.

Moreover, just as *Spherics* II is no longer built on direct analogies with the propositions of the *Elements*, so the methods of establishing propositions in *Spherics* II are somewhat different from those of *Spherics* I. In the first book of this treatise, each step of the *demonstration* of each proposition can be justified on the basis of earlier theorems of the *Spherics* itself, or the *Elements*. In *Spherics* II, however, a number of propositions – *Sph.* II.6,

II.11, II.12, II.21, II.22 – require justifications that are more involved than some two or three theorems of the *Elements*, and some of which are shown in marginal scholia in the extant Greek manuscripts. In almost all cases, these lemmas can be supplied by considering the geometry of a circle in a cutting plane, and using the plane propositions of the *Elements*. Whether or not lemmas like these were shown in some other ancient treatises of spherics is unknown, but the fact that such lemmas are required in this book, whereas they were not needed in *Spherics* I, is another indication that the subject matter is moving away from pure geometry – analogous to and justified by the *Elements* – into more specifically astronomical topics – treating objects of special interest in spherical astronomy and using moderately more specialized methods.

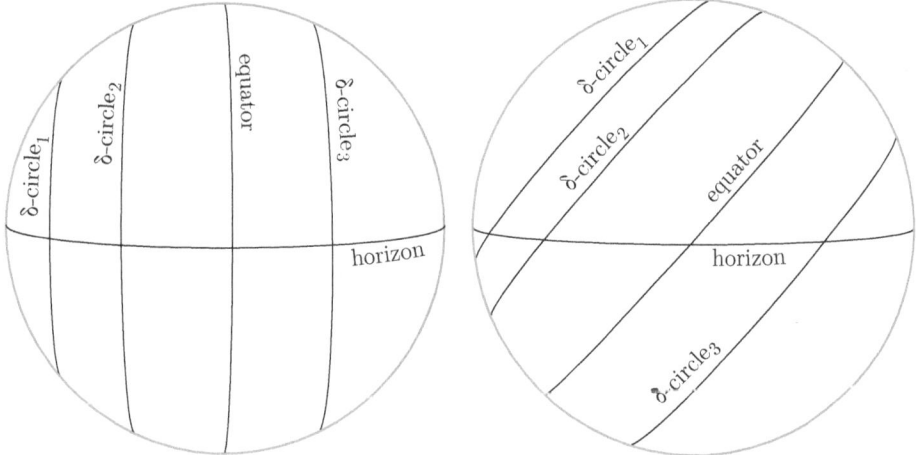

Figure 14: Diagrams for the astronomical implications of *Sph.* II.17–II.20.

As mentioned, the primary transition that takes place over the whole course of *Spherics* II is from treating the geometry of a sphere generally to treating the celestial sphere. Whereas the first group of propositions mentioned above can be understood as purely geometrical theorems of use in the rest of the treatise, the second group are rather naturally understood as treating the relationships that hold between the bundle of δ-circles, including the celestial equator, and different configurations of the horizon. *Sph.* II.10 treats horizons at locations on the terrestrial equator, known as the upright sphere, while *Sph.* II.13 treats the horizons at locations between the terrestrial poles and the equator, known as the inclined sphere, and *Sph.* II.16 is a partial converse to both (Figure 13).

The next group of theorems, *Sph.* II.17–II.20, concerns the relative sizes of δ-circles with respect to horizons on the upright and inclined spheres and the ways in which the arcs of the δ-circles are cut off above and below the horizon (Figure 14).

The final group of theorems deals with various possible configurations of the ecliptic relative to the horizon on the inclined sphere (Figure 15). The details of the claims made in all these propositions, and their implications for spherical astronomy will be taken up below.

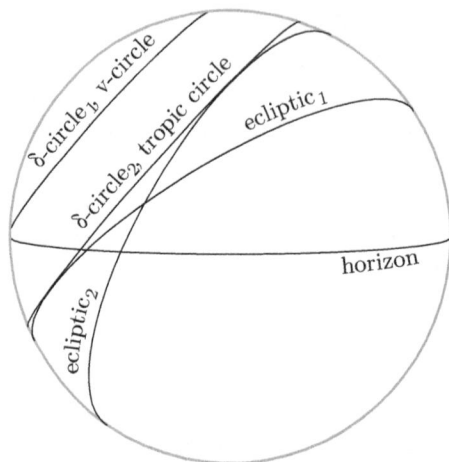

Figure 15: Diagram for the astronomical implications of *Sph.* II.21–II.23.

For now it is sufficient to point out that we can see from the diagrams illustrating these topics of spherical astronomy that the underlying, and tacit, subject matter of this book is the motion of the celestial sphere relative to the horizon – that is, the diurnal motion of the sphere – whether this is regarded as the motion of the actual cosmos, or that of a solid globe or armillary sphere. In this text, however, motion is treated in a completely abstract, geometrical way. Indeed, strictly speaking, *Spherics* II does not deal with motion at all. Rather, it treats geometrical configurations that could result from motion – that is, the geometrical relationships that obtain between the instantaneous positions that the geometric objects could assume as the result of a certain motion. If we consider a solid globe or an armillary sphere, as we believe was done when spherics was studied in antiquity, the implications of *Spherics* II for spherical astronomy concern statements that can be made about the δ-circles and the ecliptic at one configuration of the sphere compared to another, following a rotation about the celestial axis. The rate of the rotation, and even its direction, however, are irrelevant. Although, the text itself simply deals with different configurations, or the results of transformations, an informed ancient reader would have known that these changed configurations result from a daily, uniform motion of the sphere. Hence, the astronomical topic of *Spherics* II is the daily motion of the principal objects in the celestial sphere – the ecliptic and fixed objects moving on δ-circles – relative to the horizon, taken as a stable frame of reference.

The only explicit indications that we are dealing with topics of spherical astronomy come at the end of *Spherics* II, starting from *Sph.* II.19, in which we encounter terminology that makes it clear that we are dealing with a horizon, and that the horizon serves as the frame of reference. Although these terminological markers are somewhat subtle, they provide a clear sense of astronomical orientation.

Orientation

As discussed in the Introduction (p. 62), from the final propositions of *Spherics* II, the text begins to describe the objects in the sphere using language that strongly suggests a privileged orientation. In particular, there is a *visible pole* and an *invisible pole*, separated by a base circle. Furthermore, certain objects are *higher* or *lower*, and more or less *upright*, relative to this base circle. All of this implies an astronomical context, in which the base circle is taken as the local horizon. Although the claims made in reference to this sense of orientation can be understood in terms of either the celestial sphere around us or a solid or armillary sphere before us, they are certainly easier to visualize on a model.

Definition

There is only one new definition introduced in *Spherics* II, that for tangent circles in a sphere.

Sph. II.def.1

Theodosios's definition of the tangency of two circles in a sphere is fully solid in its conception. Each of the two circles determines a plane, *Elem.* I.def.15, which then together determine a straight line as their common section, *Elem.* XI.3. When this common section is tangent to each circle, the two circles will be tangent to one another. It has been noted that this definition is apparently different from the concept of tangency on a sphere that was used by Autolykos, and perhaps Euclid (Mogenet 1947, 240; Schmidt 1943, 11–12; Berggren 1991, 242). Since it is only used in *Sph.* II.3, which, as we will see below, transforms the concept of tangency on a sphere from one based on solid considerations to a surface specification, it is not unlikely that this definition was set out so as to satisfy the needs of that theorem.

Usage: *Sph.* II.3.

Parallel circles and their poles (*Sph.* II.1, II.2)

This pair of converses shows that circles in a sphere are parallel if and only if they share the same poles. This introduces us to one of the principal objects of investigation for the rest of *Spherics* II, namely, a bundle of parallel circles about a pair of poles.

An immediate corollary of these propositions taken together is that, for circles in a sphere, the property of parallelism is transitive. Although it is obvious, the transitivity of parallelism is not shown for planes in the *Elements*, despite being shown for lines in both plane and solid configurations, *Elem.* I.30 and XI.9 respectively.[47] Hence, we will treat *Sph.* II.1 and II.2 together as the justification, in the *Spherics*, for a claim that the parallelism of circles in a sphere is transitive.

In terms of spherical astronomy, these propositions imply that the δ-circles, such as the tropics and the visibility circles, form a bundle of parallel circles about the celestial poles. Hence, taken together with *Sph.* I.7–I.10, we can now state the full properties of the δ-circles as a bundle of parallel circles that share a single axis, the line through the celestial poles.

Sph. II.1 (Theorem)

This theorem is essentially solid geometry in the sphere, since it is based on *Sph.* I.8 and I.10, both from the group of propositions treating the properties of the axis of a circle in a sphere, and an immediate corollary to *Elem.* XI.16, namely the claim that if a line is perpendicular to one of two parallel planes, it is also perpendicular to the other, which is the converse of *Elem.* XI.14.[48]

Basis: *Elem.* I.post.1, XI.16 (by implication), *Sph.* I.8, I.10, I.21.

Usage: *Sph.* II.7 (with II.2), II.8 (with II.2).

Diagram: The diagram is a planar representation of local objects, containing both intrinsic and extrinsic objects. A pair of parallel circles on the surface of the sphere is represented with concentric circles in the plane of

---

[47] Transitivity of parallel planes can be shown, in general, using propositions from the *Elements*, as follows. If we have three planes, of which two pairs are pairwise parallel, then we can show that the remaining pair are also parallel by considering a pair of intersecting planes that also intersect the original planes such that the line of intersection of the new planes falls on the original three planes at any angle. Then, by *Elem.* XI.16, the intersections of these newly constructed planes with the original planes are pairwise parallel lines, and hence, by *Elem.* XI.9, all the lines are parallel. Moreover, each pair must intersect on the intersection of the constructed planes, so that, by *Elem.* XI.15, all the planes, taken in any pair, are parallel (Heath 1908, 301).

[48] The claim can also be shown by an indirect argument from *Elem.* XI.14, assuming the transitivity of parallelism for planes. But since showing this transitivity also requires *Elem.* XI.16, as discussed above, it is probably simplest just to read this as an application of *Elem.* XI.16 as well (Heath 1908, 301).

the figure. The axis of both circles is depicted with an extended radius that has two false crossings with each circle.

Sph. II.2 (Theorem)

This theorem is the converse of the foregoing proposition, and is demonstrated directly using *Sph.* I.10 and the converse of one of the core claims used in *Sph.* II.1, namely *Elem.* XI.14.

Basis: *Elem.* I.post.1, XI.def.8, XI.14, *Sph.* I.10.

Usage: *Sph.* II.4, II.6, II.7 (with II.1), II.8 (independently and with II.1), II.22, II.23. A proposition like this had been used in *Mov. Sph.* 1, 6 and 8.

Diagram: The diagram is the same diagram as that of the previous proposition.

Tangent circles and the great circle through their poles (*Sph.* II.3–II.5)

These three theorems present Theodosios's theory of the tangency of two circles in a sphere. They should be compared with *Elem.* III.11, which deals with the tangency of two circles in the plane. (*Elem.* III.12(∗), which treats the same subject matter, is understood to be a late interpolation.) *Sph.* II.3–II.5, taken together, establish that two circles on the surface of the sphere are tangent if and only if a single great circle passes through their four poles and the point of tangency. Although *Sph.* II.def.1 describes a solid configuration, *Sph.* II.3 transitions the criteria for tangency to purely surface conditions, while *Sph.* II.4 and II.5 show what an assumption of tangency implies.

Because of the transition between solid and surface considerations, the logical implications of these theorems involve some partial converses. That is, we consider the following conditions with respect to two circles on a sphere:

(a) a great circle passes through the poles of both,

(b) a great circle passes through the poles of one,

(c) the great circle passes through their point of contact, and

(d) the two circles are tangent.

Then, *Sph.* II.3 shows (a) and (c) ⇒ (d), *Sph.* II.4 shows (a) and (d) ⇒ (c), and *Sph.* II.5 shows (b), (c), and (d) ⇒ (a). Notice that all three of these propositions can be applied with essentially the same argument whether the smaller circle is depicted as inside or outside the larger circle, which on a sphere is simply a matter of perspective. If we consider these three theorems as treating the configuration of a great circle passing through the poles of two circles that meet in a sphere, then we can regard that topic as being extended in *Sph.* II.9. That is, as a group these propositions deal with a great circle passing through the poles of two circles that meet in one point, *Sph.* II.3–II.5, or in two points, *Sph.* II.9.

For spherical astronomy, these theorems show that certain great circles passing through the celestial poles also pass through the points of tangency of δ-circles with the ecliptic and the horizon. That is, the solstitial colure passes through the points of tangency of the ecliptic with the tropics, and the meridian passes through the points of tangency of the visibility circles with the horizon.

Sph. II.3 (Theorem)

As discussed above, the force of this theorem is to introduce a condition for the tangency of two circles on a sphere that involves purely surface considerations. That is, this proposition shows that a certain configuration of two circles and a great circle through their poles implies the solid configuration specified in *Sph.* II.def.1, following which that definition is never again used (Schmidt 1943, 11). This proposition is essentially solid geometry, since it is based mostly on *Elements* XI, along with *Sph.* I.15, which is itself from that part of *Spherics* I that deals with solid considerations. Following this proposition, however, tangency on a sphere can be dealt with through purely surface considerations, so that this proposition has a similar effect to the whole of *Spherics* I, namely shifting the focus from solid objects to objects on the surface of the sphere.

In the *construction*, *Elem.* XI.3 is used to introduce three lines as pairwise intersections of three planes, using wording that is different from that used in *Sph.* I.13 and I.14, but similar to that used in I.17. Here again, this is not strictly speaking a construction, since those lines exist as common sections whether or not they are required for the argument, but nevertheless, it is a sort of constructive assumption, insofar as the *construction* is that part of the proposition in which new objects are introduced and named.

Basis: *Elem.* I.def.17, III.16.cor., XI.def.3, XI.def.4, XI.3, XI.19, *Sph.* I.15, II.def.1.

Usage: *Sph.* II.4, II.6, II.8, II.14, II.15.

Diagram: The diagram is a planar representation of local objects, containing both intrinsic and extrinsic objects. Although a great circle appears in the diagram, the whole circle is not represented (but see the "Comparison with Arabic," below). Straight lines, which occur in two different planes have been drawn into the plane of the diagram. Two circles are shown as tangent in the surface of the sphere by representing them as touching in the plane of the diagram. In this diagram, the circles are shown as both externally tangent, but if one were internally tangent to the other, the proof could still be carried through with no changes. In this diagram there are no false crossings. Hence, the diagram represents the topology of the configuration.

Comparison with Arabic: In the diagrams of Thābit's Arabic version, circle ا ح ه (AGE) is represented by a full circle (**I**, 28b; Kunitzsch and Lorch 2010, 332). This would change our categorization of the diagram from local to global.

Sph. II.4 (Theorem)

As mentioned above, this is a partial converse to *Sph.* II.3, showing that if two circles are tangent, now interpreted through the result of *Sph.* II.3, then the great circle through their poles must pass through their point of tangency. The objects considered, and thus the arguments, now take place within the surface of the sphere. Hence, this theorem is analogous to *Elem.* III.11 and III.12(∗), of which the latter is regarded as a late interpolation, because it was not known to Heron (Besthorn and Heiberg 1897–1905, 46–47; Lo Bello 2009, 95).

The crux of the indirect argument – namely, the assertion that $\mathbf{C}(ABG)$ and $\mathbf{C}(BKQ)$ are intersecting, despite having being shown to be tangent – appears to be drawn directly from the diagram, since nothing in the text up to now gives one warrant to make this claim. It is possible to reconstruct various arguments for this, such as by pointing out that since B is assumed to lie somewhere on $\mathbf{C}(ABG)$ farther away from H than G, if a great circle were joined through H and G it would meet $\mathbf{C}(BKQ)$ somewhere inside circle $\mathbf{C}(ABG)$, so that those two circles must intersect one another. But no argument of any kind is made in the text.

Basis: *Elem.* I.post.3 (spherical), III.def.3, *Sph.* I.20 (counterfactual), II.2, II.3.

Usage: *Sph.* II.5.

Diagram: The diagram is a planar representation of local objects, containing only intrinsic objects. Although it is used for an indirect argument, and hence represents an impossible configuration, there are no false crossings. Hence, it represents the assumed topology of the objects.

Sph. II.5 (Theorem)

This proposition is a partial converse to *Sph.* II.3 and II.4, again using only surface considerations.

Basis: *Elem.* I.def.17, I.def.18, I.post.1, *Sph.* I.def.5, I.6, I.11, I.20 (normal, counterfactual), II.4.

Usage: *Sph.* II.6, II.13, II.22, II.23. A proposition like this had been used in *Phen.* 2, and in *Mov. Sph.* 10.

Diagram: The diagram is a planar representation of local objects, containing both intrinsic and extrinsic objects. Again, the great circles are represented with arcs in the diagram. As in the previous proposition, it is used for an indirect argument, and hence represents an impossible configuration. Nevertheless, there are no false crossings, so that it represents the assumed topology of the objects. In this diagram, the two circles are drawn as externally tangent, but if the diagram depicted one of them as internally tangent to the other, the same argument could be carried through with only slight changes to the letter-names.

Textual Comments: When the great circle passing through poles Z and H is introduced in the *construction*, it is called ZΓH in the oldest manuscript, but ZH in many of the other manuscripts (Czinczenheim 2000, 85). Since, at this point in the argument, it has not yet been stated that $\mathbf{gC}(ZH)$ passes through point G, the shorter name may make better sense.

Comparison with Arabic: In Thābit's version, the question of whether to call the circle ZH or ZΓH is resolved by simply not giving the circle a name until it has already been noted that it passes through ج (Γ) (Kunitzsch and Lorch 2010, 98).

Parallel circles tangent to a great circle (*Sph.* II.6–II.8)

This group of theorems introduces one of the most important configurations in the text, which constitutes a cornerstone of Theodosios's approach to spherics – namely, a bundle of parallel circles and an oblique great circle that is tangent to an equal pair of the parallels. Taken together these theorems show that a pair of circles in a sphere are equal and parallel if and only if there is a great circle that is tangent to both of them.

Because *Sph.* II.6 acts as a constructive proposition, allowing the introduction of new objects – along the lines of *Elem.* XI.3 or *Sph.* I.1 – these theorems do not make up a bundle of partial converses in the normal sense. Nevertheless, their content can be summarized if we consider the following claims:

(a) a great circle is tangent to a small circle,

(b) a great circle is oblique to some circle (of a bundle of parallels),

(c) there are two equal and parallel small circles,

(d) the great circle is tangent to another equal and parallel small circle,

(e) the great circle is tangent to the other small circle, and

(f) the great circle is tangent to both parallel small circles.

Then, *Sph.* II.6 shows (a) ⇒ (d), and hence also (c), *Sph.* II.7 shows (a) and (c) ⇒ (e), and *Sph.* II.8 shows (b) ⇒ (c) and (f).

From the perspective of spherical astronomy, this group of theorems implies the configuration of the tropics and the visibility circles with respect to the ecliptic and the horizon. Namely, the tropics and visibility circles are two pairs of equal and parallel δ-circles tangent to the great circles of the ecliptic and horizon, respectively, which each cuts at least one circle parallel to the δ-circles, namely the equator. Indeed, these theorems are related to the contents of *Mov. Sph.* 6, which shows that if the horizon is inclined to the celestial axis, then it will be tangental to the two visibility circles.

Sph. II.6 (Theorem)

This proposition, although simple, relies for its proof on a number of different concepts developed so far in the text. Because it is proved by construction it relies on *Sph.* I.20 and I.21, as well as a number of constructive steps that can be understood as solid applications of *Elem.* I.post.3 (transfer, spherical). Furthermore, the proof depends on both the solid and surface aspects of the theory of parallelism, *Sph.* II.2, and tangency, *Sph.* II.3 and II.5, as well as the concept of the axis of a circle, *Sph.* I.9.

The function of this proposition is essentially constructive, since it seems to have been written so that it can be used, in the following two propositions, to introduce a new small circle that will play a role in the *demonstration*. Indeed, both the construction and the argument of this proposition could easily have been reformulated in the form of a *problem* – namely, *on the surface of a sphere, draw a small circle equal and parallel to a given small circle, both tangent to the same great circle*.[49] Instead, this proposition is expressed as what is essentially an existence claim, but not merely about the other circle – for some equal and parallel circle will surely exist – but rather about the overall configuration, namely two equal and parallel circles, tangent to a great circle, all on the surface of a sphere. At any rate, like *Elem.* XI.3 and *Sph.* I.1, this theorem will be used to introduce a new object

---

[49] Notice that *Sph.* II.14, which would also be required, depends on nothing later than *Sph.* II.3.

– namely, a small circle not mentioned in the *enunciation* for use in the *demonstration*.[50]

In the *construction*, an arc of a great circle, BZ is cut off equal to the great-circle arc GE. Although this construction is analogous to that in *Elem.* I.3, in the plane, considering the constructive methods that have been developed already in the text, we can simply regard this as *Elem.* I.post.3 (transfer, spherical), or an operation of a normal compass.

In the *demonstration*, Theodosios claims that, since $\mathbf{gCArc}(GE) = \mathbf{gCArc}(BZ)$, the small circles about poles E and Z and passing through G and B, respectively, are equal. Although this claim is fairly straight-forward, it is not any more so than many other things that are shown in this treatise. We can furnish a lemma to this effect along the lines of the analemma considerations that we introduced in the Commentary to *Spherics* I (see also Ver Eecke 1927, 39 n. 2).

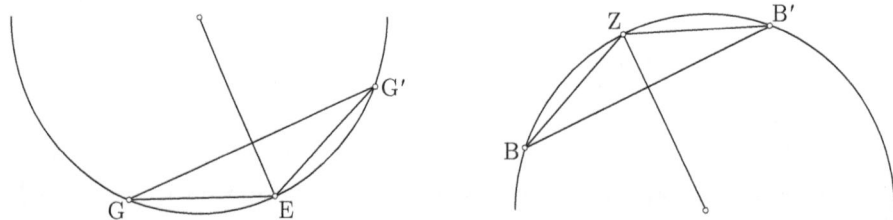

Figure 16: Lemma for *Sph.* II.6, small circles with equal great-circle arcs through their poles and circumferences are equal.

In general, if great circles pass through the poles, E and Z, and points, G and B, on the circumferences of any two small circles, then we can represent the small circles as chords on the planes of the great circles passing through these pairs of points and the center of the sphere, which will be equal to the diameters of the small circles, *Sph.* I.15 (Figure 16). Then, naming the other ends of the diameters G' and B', and joining EG, EG', ZB and ZB', they will all be equal, *Elem.* III.29 and *Sph.* I.def.5. Then, since the circles are equal, being great, they will cut off equal angles, so that $\mathbf{Ang}(EGG') = \mathbf{Ang}(EG'G) = \mathbf{Ang}(ZBB') = \mathbf{Ang}(ZB'B)$, *Elem.* III.27. Hence, $\mathbf{Ang}(GEG') = \mathbf{Ang}(BZB')$, *Elem.* I.32, so that $GG' = BB'$, *Elem.* III.26, III.29. Therefore, the two small circles of which these lines are diameters are equal, *Elem.* III.def.1.

---

[50] Indeed, it is possible that the reason for asserting this proposition as a *theorem*, as opposed to a *problem*, is because the two times it is used are in the *demonstrations* of *Sph.* II.7 and II.8. Although it is not an absolute rule, there is a strong tendency in Greek geometrical texts to justify steps of the *construction* with *problems* and steps of the *demonstration* with *theorems*.

Basis: *Elem.* I.post.3 (spherical, transfer), I.c.n.1, I.c.n.2, I.26 (by implication), I.32 (by implication), III.def.1 (by implication), III.27 (by implication), *Sph.* I.def.5, I.6, I.10, I.11, I.20, I.21, II.2, II.3, II.5.

Usage: *Sph.* II.7 (counterfactual), II.8. Something like this proposition seems to have been required at the end of the first *demonstration* in *Mov. Sph.* 6. A proposition similar to this had been used in *Phen.* 2.[51]

Diagram: The diagram is a planar representation of global objects, containing only intrinsic objects. Circles that are tangent on the surface of the sphere are encoded as touching in the plane of the figure. A circle is depicted as dividing the surface of the sphere into two by depicting the objects in one part inside it and the objects in the other part outside of it. There are no false crossings in the diagram, so that it represents the topology of the configuration.

Sph. II.7 (Theorem)

This proposition is a partial converse to the previous proposition, and it is proven indirectly on that basis.

As mentioned above, this proposition introduces another parallel circle as part of its counterfactual indirect argument. The justification for introducing this new object is clearly *Sph.* II.6, which is reiterated. Hence, *Sph.* II.6 is here used constructively.

In this proposition there is no distinction between the *construction* and *demonstration*. Indeed, if there is a construction step at all it is the introduction of $sC(EZ)$, just mentioned, which occurs in the middle of the *demonstration*.

The crux of the indirect argument hinges on the claim that there cannot be more than two equal and parallel circles in a sphere. While this is obvious, the reasons for this claim can be fleshed out.[52] Since parallel circles share a pole, *Sph.* II.1, if we assume that there are more than two parallel circles in a sphere, then they will all have the same axis, which will pass through the center of the sphere and the centers of the circles, *Sph.* I.8. Hence, if they are all equal, their centers must be different points on the same line, their axis, that are equidistant from the same point, the center of the sphere, *Sph.* I.6; which is impossible for more than two points. Hence, no more than two such parallel circles can be equal.

---

[51] Schmidt (1943, 89) also notes this proposition as used in *Phen.* 14, but there the two visibility circles are assumed in the *enunciation*, and the tangent circle is drawn as tangent to one and then asserted to be tangent to the other. Hence, *Sph.* II.7 is a better fit.

[52] In fact, if *Sph.* II.6 had been developed as a *problem*, and especially if an analysis had been included, this claim would be implicit in the construction itself and would not need to be shown. It may be that Theodosios was considering the situation from such a constructive perspective, but it remains the case that he did not make this explicit.

Basis: *Elem.* I.c.n.1, *Sph.* I.6 (by implication), I.8 (by implication), II.1, II.2, II.6 (counterfactual).

Usage: Unused in the *Spherics*. A proposition like this had been used in *Phen.* 14 (see n. 51, above).

Diagram: The diagram is a planar representation of global objects, containing only intrinsic objects. It uses the same encoding as the previous diagram, but now, because it is used for an indirect argument, it also presents an impossible situation. Nevertheless, there are no false crossings, so that it represents the assumed topology of the objects.

Sph. II.8 (Theorem)

This proposition, like *Sph.* II.6, is essentially constructive in its significance, and like that proposition, it demonstrates a configuration that establishes the conditions of *Sph.* II.7, and hence leads to its conclusions. We will see, below, that the only time this proposition is used in the treatise, in *Sph.* II.16, it is used in a way that is essentially constructive.

Like *Sph.* II.7, this proposition uses *Sph.* II.6 constructively in the middle of the *demonstration* for the purpose of introducing an equal and parallel circle.

From the point of view of spherical astronomy, this proposition has some implications that distinguish it from the other two in this group – namely, that the ecliptic, being oblique to the equator, must be tangent to two equal δ-circles, the tropics. Moreover, the horizon will be tangent to the visibility circles, when it is oblique to the equator. When, however, the horizon is at locations on the terrestrial equator or at the terrestrial poles, there will be no visibility circles.

Basis: *Elem.* I.post.3 (spherical), *Sph.* I.13 (indirect), I.20, I.21, II.1, II.2, II.4, II.6.

Usage: *Sph.* II.16.[53]

Diagram: The diagram is a planar representation of global objects, containing only intrinsic objects. It uses the now standard encoding for intersections, parallel circles, tangent circles and separation of the spherical surface into two parts. There are no false crossings; so it accurately depicts the topology of the objects.

---

[53] Schmidt (1943, 8) notes this proposition as being used in *Phen.* 7, but we cannot find this requirement. In *Phen.* 7, the tangent parallels, the tropics, are assumed from the beginning and the oblique parallels are constructed in the course of the argument, so there is no need to apply something like *Sph.* II.8, which proceeds the other way.

Great circle through the poles of intersecting circles (*Sph.* II.9)

This proposition stands alone in establishing the relationship between a pair of intersecting circles on a sphere and a great circle that, passing through their poles, intersects each of the original circles. Namely, such a great circle bisects all four arcs of the two circles on each side of their intersection. It should be noted that the original intersecting circles may be either small or great circles in the sphere. That said, as mentioned above, if we regard this proposition as dealing with a great circle that passes through the poles of two meeting circles, then *Sph.* II.9 belongs together with *Sph.* II.3–II.5, because it extends the configuration of those propositions from the situation of two circles that meet at one point to that in which they meet at two points.

Sph. II.9 (Theorem)

As well as some basic theorems on the circle, the key to this proposition involves solid geometry and *Sph.* I.15, which is one of the propositions from *Spherics* I that deals with the surface properties of a great circle through the poles of small circles.

This proposition is an important lemma for a number of theorems in *Spherics* II, particularly *Sph.* II.13, II.22, and II.23, all of which deal with topics important to spherical astronomy. Although this proposition could have been shown directly after *Sph.* II.3–II.5 – where from a geometrical perspective it perhaps more naturally belongs – it is probably introduced here as a lemma to the group consisting of *Sph.* II.10, II.13 and II.16.

In the *construction, Elem.* XI.3 is used to introduce two straight lines as the intersections of two pairs of planes. We look back to *Sph.* I.13 and II.3 to understand the force of this constructive assumption, and, to *Sph.* II.3 especially, as a model for the translation.

A constructive corollary that follows from this theorem could show how to bisect a given arc of a circle, using the non-attested corollary to *Sph.* I.20 discussed in the Commentary to that proposition (see p. 234, above). That is, with some given arc **Arc**(AB) of **C**(ABC), we use the constructive corollary to *Sph.* I.20 to draw, with arbitrary but suitable polar radius, two equal circles that also pass through A and B, say **C**(ABD) = **C**(ABF) about poles E and G (Figure 17). Then, we join E and G with a great circle, *Sph.* I.20, intersecting **C**(ABC) at H and **C**(ABD) at J, which, by *Sph.* II.9, is the midpoint of the **Arc**(AB) that is a part of **C**(ABD). Then, this circle must also pass through the pole of the original circle **C**(ABC), say point K. For, if not, we take the pole of **C**(ABC), say K' (not shown), and drawing a great circle passing through K' and E, the pole of **C**(ABD), it will pass through point J, by *Sph.* II.9, such that two different great circles will pass through points J and E, which is impossible unless these points are diametrically opposite, by *Sph.* I.20. Finally, *Sph.* II.9 assures us that the great circle passing through E and K bisects the given **Arc**(AB) at point H.

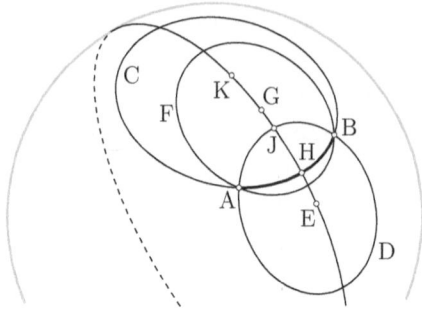

Figure 17: Constructive corollary to *Sph.* II.9.

We could also establish this constructive corollary through a fact that is mathematically related to *Sph.* II.9 – namely, that if two circles on the surface of the sphere intersect then the midpoints of the arcs cut off between them are joined by a single great circle that passes through their poles.[54] The fact that none of this is made explicit in the *Spherics*, however, makes it doubtful that Theodosios was thinking of such a constructive procedure. On the other hand, since arcs were already bisected in *Sph.* I.21, we can also understand the bisection of arcs on the sphere to always be carried out by the method discussed in the Commentary to that proposition.

In terms of spherical astronomy, this proposition shows that, wherever $\varphi \neq 90°$, the meridian, in passing though the poles of both the horizon and the equator, will bisect the arcs of those great circles between their intersections at the east and west points. Moreover, again where $\varphi \neq 90°$, for all δ-circles that are intersected by the horizon, the meridian bisects both the parts above and the parts below the horizon. In particular, where the tropics intersect the horizon, $\varphi < 90° - \varepsilon$, then the meridian bisects the arcs of the tropics and all of the day-circles, both those above and those below the horizon.

Basis: *Elem.* I.def.17, I.post.3, I.4, III.3, III.29, XI.def.3, XI.3, XI.19, *Sph.* I.15.

Usage: *Sph.* II.13, II.15, II.17, II.22, II.23. A proposition like this had been used in *Phen.* 2.

---

[54] Clavius (1586, 37–38) provides a proof of this claim, which he says is added "*in alia versione.*" This apparently refers to an addition found in the Campanus version (Milan Ambr. C 241 inf., f.139v; Vat.lat. 3380, f.8v).

Diagram: The diagram is a planar representation of local objects, containing both intrinsic and extrinsic objects. If great circle *AGBD* were drawn as a full circle the categorization would change to global. Just as in the diagram for *Sph.* II.3, objects that occur in the surface of the sphere appear as circles and circular arcs in the plane of the figure, while straight lines that lie in two different planes have been drawn into the plane of the diagram. The straight lines have three false crossings with the circles.

Textual Comments: The naming-sequence used in this proposition – Z, A, E, B, Γ, Δ, H – deviates from the alphabetic order, which is unusual in a Greek mathematical text, and has not previously happened in the *Spherics*. Hence, it is possible that this proposition was adopted with little alteration from a previous treatise in which this sort of ordering was not unusual, or made contextual sense (Sidoli and Saito 2009, 585 n. 15). Another possibility is that the text originally introduced the two initial circles as AB and ΓΔ, and then a later editor inserted the Z and E into these letter-names. If this is what happened, then the original naming sequence would have been alphabetical.

Comparison with Arabic: In Thābit's version, the naming sequence does not follow the *abjad* order, but rather that of the Greek text – ز, ا, ه, ب, ج, د, ح (Kunitzsch and Lorch 2010, 106–108).

Great circles intersecting a bundle of parallel circles (*Sph.* II.10, II.13, II.16)

This group of three theorems deals with great circles intersecting a bundle of parallel circles. In particular, they show that the great circles cut off similar arcs of the parallel circles if and only if the great circles either pass through the poles of the parallels or are tangent to the same pair of parallels, and, moreover, that the arcs of the great circles cut off between two parallels are equal.

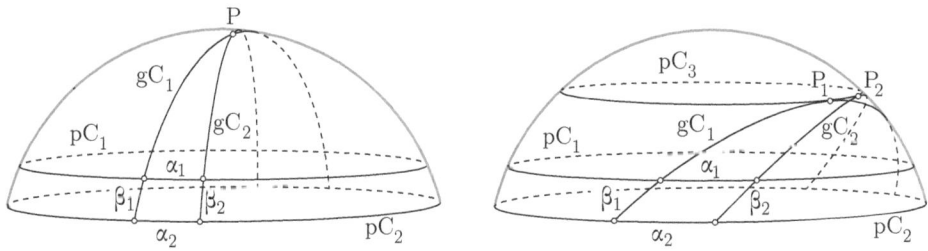

Figure 18: Perspective diagrams, *Sph.* II.10, II.13, and II.16.

In particular, *Sph.* II.10 Part 1 shows that if gC_1 and gC_2 pass through a pole, P,[55] of parallels pC_1 and pC_2, then the arcs cut off of the parallel circles will be similar, $\alpha_1 \sim \alpha_2$; while II.10 Part 2 shows that the parallel circles will cut off equal arcs of the great circles, $\beta_1 = \beta_2$ (Figure 18 (left)). *Sph.* II.13 Part 1 shows that if gC_1 and gC_2 are tangent to one of the parallel circles, pC_3, then, considering arcs of gC_1 and gC_2 taken in the same direction from points P_1 and P_2, once again, the arcs cut off of pC_1 and pC_2 will be similar, $\alpha_1 \sim \alpha_2$; and II.13 Part 2 shows that those arcs cut off of gC_1 and gC_2 will be equal, $\beta_1 = \beta_2$ (Figure 18 (right)). *Sph.* II.16 gives the converses to the first part of each of these, showing that where a pair of great circles cuts off arcs $\alpha_1 \sim \alpha_2$ from a pair of parallel circles, then the great circles either pass through the poles of the parallel circles, or are both tangent to another circle that is parallel to the first two (Figure 18 (left and right)).

That these propositions belong together is clear from the mathematical contents, but it is also made explicit by the way that the letter-names are introduced in the text and applied to the diagram. In *Sph.* II.13 and II.16, the letter-names are not introduced in alphabetical order, which would be standard, but such that the same letter-names will indicate the same, or analogous, objects in the diagrams of all three theorems. In a Greek mathematical text, this is as good an indication as any that we are meant to understand these theorems as treating the same subject matter.

In terms of the overall structure, these theorems are interrupted by two pairs of propositions that can be understood as lemmas to II.13 and to II.16, despite the fact that they are also used elsewhere in the treatise. In this way, the whole run of propositions *Sph.* II.10–II.16 should be understood as a single group with two auxiliary subgroups *Sph.* II.11, II.12, and II.14, II.15. The function of these subgroups will be treated separately, below.

These propositions have a number of important implications for spherical astronomy in terms of the rising times of arcs of δ-circles, as well as the use of the arc of the horizon known as the ortive amplitude, η, as a characteristic arc for a δ-circle (Schmidt 1943, 15–17). The details of these implications will be taken up in the individual Commentaries to *Sph.* II.10 and II.13, below.

Sph. II.10 (Theorem)

This proposition is essentially based on solid geometry, although it also uses some basic properties of the circle.

Although in this treatise objects are generally called parallel using a form of the adjective παράλληλος, when *Elem.* XI.10 (p. 124, n. 3) is applied,

---

[55] Of course, if they pass through one pole, they will also pass through the other, but here, for the sake of simplicity, we discuss only a single pole.

lines are designated as parallel using the preposition παρά, following the terminology of *Elem.* XI.10 itself.

There is a normal *specification* for the whole proposition, and then the second part has another "I say" statement, which acts as a general specification, setting out what will be shown in that part. Each part has a proper *conclusion*, in general terms.

The letter-names L, and M are so completely unspecified as to indicate that there is probably some textual problem. See the "Textual Comments," below. The inclusion of K within the names of circles ZKQ and EKH in the *construction* is in alphabetic order and specifies this point as the intersection of those two circles.

An alternative proof for this proposition is found in the scholia, Scholium 183 (Czinczenheim 2000, 402).

For spherical astronomy, this theorem states that arcs of δ-circles – such as the visibility circles, the tropics and the equator – that are cut off between a pair of colures are similar, and that any pair of δ-circles cut off equal arcs from the colures. Furthermore, for horizons at the terrestrial equator, the horizon itself will cut off similar arcs of the δ-circles. In the case of a solid globe or armillary sphere, this proposition describes the situation in which the celestial poles are set on the horizon. Then, when we rotate the instrument, following the rotation of the celestial sphere, we see that arcs of the δ-circles rise in the same time as similar arcs of the equator.

Basis: *Elem.* I.def.17, I.c.n.3, III.def.11, III.28, XI.3, XI.10, XI.16, *Sph.* I.def.5, I.15.

Usage: Part 1: II.20, II.22, II.23, III.12, III.13. A proposition like this had been used in *Mov. Sph.* 2. Part 2: *Sph.* II.15(∗), III.5(∗), III.6(∗) – see "Textual Comments."

Diagram: The diagram represents local objects that are both intrinsic and extrinsic, but it could be classified as either planar or solid. As in the diagrams for *Sph.* I.1 and I.7, some viewers may find that this image conveys a sense of depth, but we can also simply consider it as depicting all of the objects flattened into the plane of the diagram. As usual, circles and circular arcs represent objects in the surface of the sphere and straight lines represent internal lines in two different planes, all drawn into the plane of the figure. The straight lines have false crossings, both among themselves and with the circles.

Textual Comments: It is likely that the second part of this proposition was not included in the original composition. As will be discussed in the Commentaries to those propositions below, when the geometric fact that is established in *Sph.* II.10 Part 2 needs to be used in *Sph.* II.15 and III.5, it is not simply stated, as would be standard, but is re-established. Although it appears to be simply stated in *Sph.* III.6, this could be an appeal to what was just shown in III.5 (Czinczenheim 2000, 936).

There are some issues with the letter-names in this proposition. A number of crucial points – Λ, and M – are not introduced as letter-names in the normal way. This is not just the usual situation of a point initially being an unspecified part of an object's name and then becoming specified later in the course of the proposition. For example, Λ and M are each *shown* to be the center of a circle before any point, or letter-name, Λ or M has been introduced. Taken together with the fact that the arrangement of the letter-names of this proposition has been made to agree with that of *Sph.* II.13 and II.16 and the addition of Part 2, as discussed above, this situation may be further indication that we are dealing with some revision of this proposition from an earlier version, or some later editorial process that was not carefully done.

Comparison of the magnitudes of arcs (*Sph.* II.11, II.12)

This important pair of converses is used to show congruencies among certain arcs on the sphere. In fact, these are propositions of elementary solid geometry, and they do not mention the sphere at all, although they are, of course, applicable to the sphere and will be so used in the *Spherics*.

These theorems show that if two equal segments, $Seg_1 = Seg_2$, are perpendicular on the diameters of two equal circles, C_1 and C_2, and equal arcs, $\beta_1 = \beta_2 \neq Seg/2$, are cut off from the segments, then lines d_1 and d_2 joining the endpoints of these arcs to the circles are equal if and only if they cut off equal arcs, α_1 and α_2, on the circumference of the circles (Figure 19). That is, where $\beta_1 = \beta_2 \neq Seg/2$, these propositions show that

$$d_1 = d_2 \Leftrightarrow \alpha_1 = \alpha_2.$$

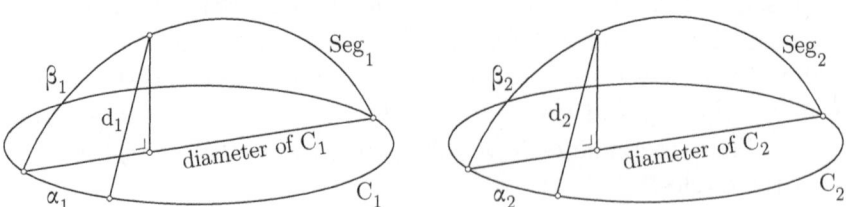

Figure 19: Perspective diagram, *Sph.* II.11 and II.12.

It should be noted that, with the same configuration of an upright segment, it is not always the case that $d_1 = d_2$ and $\alpha_1 = \alpha_2$ imply that $\beta_1 = \beta_2$.

This is because, as shown in *Sph.* III.1, the least line d_n dropped from the endpoints of α_n onto Seg_n will be that which intersects the diameter of Seg_n joined through the foot of the perpendicular dropped on Seg_n from the endpoint of α_n (not shown), and the d_ns will monotonically increase when taken further and further away on either side of this least line. Hence, there will be situations in which $d_1 = d_2$, joined from the endpoints of $\alpha_1 = \alpha_2$, can fall on each of Seg_1 and Seg_2 at two different places cutting off β_1 and β_2 of different lengths.

In the *enunciations* and *expositions* of these two propositions it is explicitly asserted that $\beta_1 = \beta_2 < Seg/2$, but it should be pointed out these propositions remain true also when $\beta_1 = \beta_2 > Seg/2$ (Ver Eecke 1927, 45 n. 2, 47 n. 2), as follows immediately from the theorems themselves. Indeed, since there is no restriction on the size of $\alpha_1 = \alpha_2$, and since $Seg_1 - \beta_1 = Seg_2 - \beta_2$ and $180° - \alpha_1 = 180° - \alpha_2$, if one started with a situation in which $\beta_1 = \beta_2 > Seg/2$, one would simply consider the other ends of the diameters in which $Seg_1 - \beta_1 = Seg_2 - \beta_2 < Seg/2$, in which case *Sph.* II.11 and II.12 show that

$$d_1 = d_2 \Leftrightarrow 180° - \alpha_1 = 180° - \alpha_2 \text{ or } \alpha_1 = \alpha_2.$$

These theorems are apparently introduced at this place in the text because *Sph.* II.11 is used in II.13, and they come together as a pair of converses. Hence, their position in the overall treatise indicates that they serve as lemmas, although they are used in more propositions than just those that immediately follow. This positioning of these theorems as lemmas, despite the fact that they play a similar function in Theodosios's work as the congruence theorems will play in Menelaos's approach to spherical geometry (Schmidt 1943, 17–20), is an indication of how completely different Theodosios's conception of the geometry of the sphere is from the later reorientation of the field that was made by Menelaos.

Because of the way that *Sph.* II.12 is applied in some of the later propositions, *Sph.* II.15 and II.22, it is worth pointing out that *Sph.* II.11 and II.12 hold for any *Seg*, whether less than, equal to, or greater than a semicircle. Indeed, although the diagrams of *Sph.* II.11 and II.12 depict the case in which *Seg* < 180°, the proofs will hold exactly as written, for *Seg* = 180°, and the diagrams need only modest changes in the case that *Seg* > 180°. In the case where *Seg* > 180°, we need to consider the possibility that the perpendicular from the endpoint of β can also fall outside of *C* (Figure 20). Nevertheless, the propositions hold in this case as well.

The *demonstrations* of *Sph.* II.11 and II.12 rely on the claim that if perpendiculars are dropped from the endpoints of equal arcs on equal segments, then they are equal and they divide the bases of the segments into equal parts. The formulaic way in which this fact is expressed in the text suggests

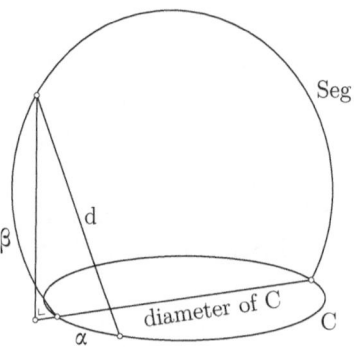

Figure 20: Perspective diagram, *Sph.* II.11 and II.12, *Seg* > 180°.

that it was well known and that we are being referred to some previous proof of this fact, but we do not find such a theorem in the *Elements*.[56] Nevertheless, such a lemma can be readily furnished from the propositions of the *Elements* – namely, *Elem.* III.21 and I.26 – and we find a lemma to this effect as a scholium in some of the medieval manuscripts. Scholium 190 reads as follows (Czinczenheim 2000, 404; Figure 21):[57]

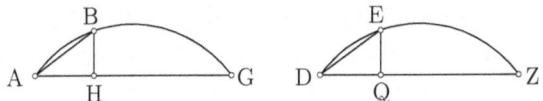

Figure 21: Scholium 190 to *Sph.* II.11.

> For let there be two equal segments of circles, *ABG* and *DEZ*, and let there be cut off equal arcs, *AB* and *DE*, and from *B* and *E* perpendiculars *BH* and *EQ*.
>
> I say that *BH* is equal to *EQ*, and *AH* to *DQ*.
>
> For, *ABG* is equal to *DEZ*, of which *AB* ⟨is equal⟩ to *DE*, therefore circumference *BG*, as a remainder, is equal to circumference *EZ*, as a remainder. Let *AB* and *DE* be joined. Therefore, angle *BAG* is equal to angle *EDZ*, since also circumference *BG* is equal to circumference *EZ*, and ⟨angle⟩ *AHB*, as upright, is equal to ⟨angle⟩ *DQE*, as upright. Then, there are two triangles, *BAH* and *EDQ*, having the two angles at *A* and *H* equal to the two angles at *D* and *Q* and one side to one side, *BA* to *DE*, equal.

---

[56] The same formulaic expression is used again in *Sph.* III.3 (see p. 162, n. 11).

[57] We have not attempted to preserve the visual characteristics of the manuscript diagrams.

And, the remaining sides will be equal to the remaining sides, *BH* to *EQ* and *AH* to *DQ*.

Furthermore, this lemma also holds when **Seg**(*ABG*) = **Seg**(*DEZ*) ≥ 180° (Figure 22). In both of these cases – that is, **Seg**(*A′BG′*) = **Seg**(*D′EZ′*) = 180°, and **Seg**(*A″BG″*) = **Seg**(*D″EZ″*) > 180° – it is clear that once the diagram is produced according to the changed conditions, the argument articulated in the scholium still works exactly as expressed. Indeed, such cases as only require a new diagram but no change in the wording of the *demonstration* are often neglected in Greek geometrical texts, outside the commentary tradition. Since this lemma is one of the key steps in *Sph.* II.11 and II.12, it is clear that the argument of these propositions will also hold, as written, for the case in which the perpendicular falls outside the circle (Figure 20).

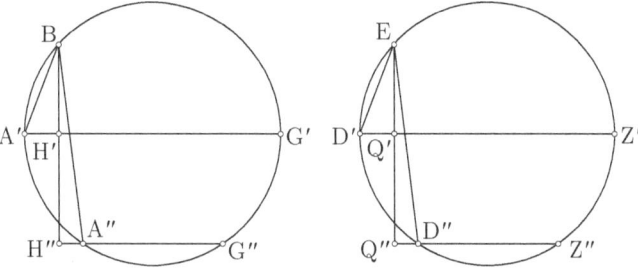

Figure 22: Reconstructed diagram for Scholium 190 to *Sph.* II.11, **Seg**(*ABG*) = **Seg**(*DEG*) ≥ 180°.

Although Scholium 190 is mathematically quite simple it is not the sort of thing that is generally passed over in silence in Greek mathematical texts. Perhaps in Theodosios's time it was shown in another text of mathematical astronomy that he could have assumed his readers would know.

Sph. II.11 (Theorem)

This is a proposition in elementary solid geometry and follows, primarily, from propositions of *Elements* XI, as well as *Elements* I and III.

After producing perpendiculars, *HK* and *QL*, the *construction* includes the claim that these perpendiculars will fall on the common sections of the two perpendicular planes. A general proposition to this effect was included following *Elem.* XI.37 by Peyrard (1814–1818, 3.108) in his edition of the *Elements*, but this is now generally accepted as a late addition to the text (Heiberg and Stamatis 1969–1977; Heath 1908, 3.360; Vitrac 1990–2001, 4.229–230). Indeed, an argument that is verbatim the same as the core of

Peyrard's *Elem.* XI.38 is found in Scholium 186 to the *Spherics* (Czinczen-heim 2000, 403), which is an indication that the author of this scholium did not find such a proposition in any manuscripts of the *Elements* that were consulted at that time. In fact, however, if we read *Elem.* XI.def.4 as stating the necessary and sufficient conditions for two planes being perpen-dicular, then it is implicit in this definition that if, as here, two planes are assumed to be perpendicular, then any line in one plane that is produced perpendicular to the other plane will fall on their common section. Hence, it seems likely that Theodosios simply took this claim for granted as an immediate corollary of *Elem.* XI.def.4, as is done sometimes in Greek math-ematical texts for results that follow immediately from a stated principle or proposition.

As mentioned in the Commentary to both *Sph.* II.11 and II.12, it is specif-ically asserted in the *construction* that perpendiculars *HK* and *QL* will fall on the common sections *AG* and *DZ*. In fact, however, the proposition also holds when the perpendiculars fall onto the extensions of the common sections, beyond segments *AG* and *DZ*. With this minor adjustment, the argument as written in *Sph.* II.11 works for all cases.

Basis: *Elem.* I.def.15, I.def.16, I.post.1, I.c.n.1, I.c.n.3, I.8, I.47, III.1, III.26, VI.20.cor., XI.def.3, XI.def.4 (by implication), XI.11.

Usage: *Sph.* II.13, III.12 (by implication).

Diagram: This is a standard solid diagram, of the sort found in the solid books of the *Elements*. As in the previous diagram, the image may convey a sense of depth, but it can also be understood as a representation of the objects flattened into the plane of the diagram. The straight lines have a false crossing among themselves.

Textual Comments: Since Pappos considered the contents of what is now *Sph.* II.13 to be the twelfth proposition of *Spherics* II (Hultsch 1876–1878, 612, 616), a 4th-century version of the text apparently treated *Sph.* II.11 and II.12 together as a single proposition.

Comparison with Arabic: In Thābit's version, *Sph.* II.11 and II.12 appear together as a single proposition, *Sph.*₄ II.11. Hence, the Arabic translation may have been made on the basis of a Greek text earlier than that in **A**, or, at least in this respect, closer to the text read by Pappos.

Comparison with Latin: The text of Gerard's Latin version follows that of Thābit's Arabic (Kunitzsch and Lorch 2010, 116–125). In a number of the later manuscripts, there are two different supplementary notes, dealing with the case in which the perpendicular segments are greater than a semicircle, as discussed in the introductory Commentary to both *Sph.* II.11 and II.12, above (Lorch 1996, 177–174; Kunitzsch and Lorch 2010, 323–327). Since this material is not found in the late 12th-century manuscript of the Gerard tradition, and since related material is found directly in the text of the Latin version of Campanus (for example, Milan Ambr. C 241 inf., f.140r; Vat.lat. 3380, f.9r), it is possible that it was adopted for the Gerard version from the Campanus version in the 13th century.

Sph. II.12 (Theorem)

This proposition is a converse to the foregoing, and, like that theorem, it is proved directly through solid geometry and some basic properties of circles. It also requires the same lemma discussed in the Commentary to both propositions, above, and, once again, the argument as written will work for all cases, once it is recognized that perpendiculars *HK* and *QL* will sometimes fall on the extensions of common sections *AG* and *DZ*.

Basis: *Elem.* I.def.15, I.def.16, I.post.1, I.c.n.3, I.4, III.1, III.27, XI.def.3, XI.def.4, XI.11.

Usage: *Sph.* II.15, II.17, II.22, II.23.

Diagram: The diagram for this proposition is the same as that for the previous proposition.

Comparison with Arabic: In Thābit's version, this proposition appears with the previous proposition as *Sph.*$_A$ II.11.

Comparison with Latin: Gerard's translation follows Thābit's Arabic (Kunitzsch and Lorch 2010, 116–125).

Sph. II.13 (Theorem)

As discussed in the introduction to this group of theorems, this proposition is a sort of extension of *Sph.* II.10. The argument now, however, is more complicated, and it relies on a number of core theorems and problems of the *Spherics* as well as the lemma *Sph.* II.11. It should be noted that the qualification that the great circles should be "cutting the remaining ⟨parallel circles⟩" is somewhat loosely expressed since, in most cases, there will also be ever-visible and ever-invisible circles that are not so cut. Hence, what is really meant is that all of the parallel circles between the two equal and parallel tangent circles are cut.

This proposition introduces the technical terminology *non-intersecting*, referring to semicircles, which is also used in the texts on spherics by Autolykos and Euclid (Schmidt 1943, 15–16; Neugebauer 1975, 758–759; Berggren 1991, 243). In its current state, the text makes this introduction in a rather confused way, but, as we discuss below, it is likely that originally this term was simply used with no discussion, under the assumption that its meaning would be well known to readers. The Greek term is ἀσύμπτωτος, but it has a different meaning, in this context, from the way that we now usually employ the term *asymptotic*, which follows Apollonios's use of the same Greek word.

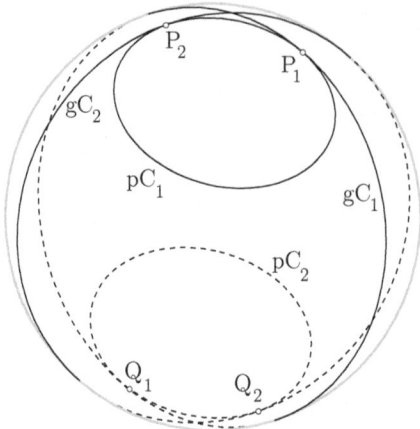

Figure 23: Explanation of non-intersecting semicircles.

If we consider any two great circles, gC_1 and gC_2, that are tangent to a pair of equal and parallel small circles, pC_1 and pC_2, at points P_1, P_2 and the diametrically opposite points Q_1, Q_2, then it is clear that the pairs of opposite points of tangency, P_1, Q_1 and P_2, Q_2, divide each great circle into a pair of semicircles. Moreover, of the four semicircles, each pairwise with the semicircles from the other great circle, two pairs are intersecting and two are non-intersecting. In particular, the pair of semicircles going out from P_1, P_2 or Q_1, Q_2, in the same direction – that is, both clockwise or both counterclockwise – is non-intersecting. The semicircles from P_1 counterclockwise and P_2 clockwise meet near those points, while those from P_2 counterclockwise and P_1 clockwise meet before they reach Q_2 and Q_1. Indeed, as was done in the introduction to this group of propositions, it would be simpler to refer to these non-intersecting semicircles as orientated in the same direction with respect to an assumed tangent small circle.

Once we have stripped out the long digression about non-intersecting semicircles from the *specification* – for which excision see the "Textual Comments," below – the structure of this theorem is essentially the same as that of *Sph.* II.10. There is a normal *specification* for the whole proposition, and then the second part has another "I say" statement, which acts as a general *specification*, stating what will be shown in that part. Once again, each part has a proper *conclusion*, in general terms.

In terms of astronomy, this proposition describes the orientation of the celestial sphere on all horizons other than those at the terrestrial poles or on the terrestrial equator – that is, in the open interval $0° < \varphi < 90°$. That is, the great circles touching a pair of equal and parallel circles in this proposition model the changing orientations of the local horizon, with respect to a bundle of δ-circles, as the celestial sphere rotates, and the equal and parallel circles are the visibility circles. In this way, the theorem asserts that, relative to a given pair of visibility circles, two different orientations of the horizon, taken such that they are both the western or eastern horizons, cut off similar arcs of the δ-circles – the tropics and the equator – and, moreover, any pair of δ-circles – the visibility circles, tropics and equator – cut off equal arcs of the horizon. Technically, this means that, for any given horizon, any given δ-circle, and in particular a day-circle, determines a constant ortive amplitude, η. Hence, this proposition, along with *Sph.* II.17 and II.18, shows us why the ortive amplitude can be treated as a characteristic of a given day-circle at a given terrestrial latitude. Moreover, since the horizon is tangent to the visibility circles, and, usually, cuts the parallel day-circle, this proposition begins to show us why the seasonal hours will be different from the equinoctial hours, and different for different positions of the day-circle. This proposition, along with *Sph.* II.19 and II.20, helps us understand, qualitatively, how the lengths of daytime and nighttime vary with changes in positions of the day-circles between the equator and the tropics, and at different terrestrial latitudes.

Finally, on a solid globe or armillary sphere this proposition deals with all situations in which the celestial axis is neither in the plane of, nor normal to, the horizon – that is, terrestrial latitudes such that $0° < \varphi < 90°$. Once again, when we rotate the instrument, modeling the daily rotation of the celestial sphere, we see that arcs of the δ-circles rise in the same times as similar arcs of the equator; or in the case of the visibility circles, that the point of tangency rotates about the circle at the same rate as the rotation of the sphere. Moreover, since the orientation of the δ-circles is assumed to be fixed relative to the horizon, as we rotate the instrument, we see that any pair of δ-circles always cuts off the same arc on the horizon, and in particular, for a given terrestrial latitude, the ortive amplitude of any two pairs of δ-circles, η, is fixed.

Basis: *Elem.* I.c.n.1, I.c.n.2, I.c.n.3, III.def.1 (by implication), III.def.11, III.27, III.28, III.29, V.9, *Sph.* I.def.1 (by implication), I.def.2 (by implication), I.def.5, I.6 (by implication), I.15, I.20, I.21, II.5, II.9, II.10, II.11.

Usage: Part 1: *Sph.* II.16, III.12, III.13, III.14. A proposition like this had been used in *Phen.* 2, 4, 5, 9(∗),[58] and 12, and in *Mov. Sph.* 8. Part 2: *Sph.* II.19, III.7, III.8. Something like this had been used in *Phen.* 7 and 14.

Diagram: The diagram for this proposition is a planar representation of global objects, using both intrinsic and extrinsic objects. As we will argue

---

[58] Although this usage is not asserted by Menge (1916, 46–47) or Schmidt (1943, 8), something like *Sph.* II.13 is needed to assert that "*M* is on a diameter with *N*." This occurs in the so-called **b** version of the text.

below, the parts of great circles *AKG* and *BLD* beyond circle *ABGD* are not necessary for the proof and could be left out of the diagram. If that were done, the diagram would be a planar representation of local objects. In either case, the diagram encodes information about intersections, parallel circles, tangent circles, intersecting great circles, and the division of the spherical surface into two parts, using the conventions developed so far. There are no false crossings; so it accurately depicts the topology of the objects.

Textual Comments: There are serious textual issues with this proposition (Czinczenheim 2000, 97–98). The letter-names are introduced in a non-standard, and indeed non-sensical, way, and it seems that a passage that was once a gloss explaining, or discussing, the concept of non-intersecting semicircles has entered the text itself. If, indeed, this is what happened, then the gloss was inserted into the main text before the 9th century, since it is found in all the Greek manuscripts as well as Thābit's version (Kunitzsch and Lorch 2010, 128–129).

The letter-names used in this auxiliary discussion were probably added to the diagram when the gloss was included in the text – and then were presumably shifted around the diagram in the course of transmission. More-over, it seems that a few of these labels were added into later passages of the text as part of the letter-names for objects in which they are not needed – for example, Φ and Τ in the circle ΑΕΚΗΓΦΤ, which would have originally been called simply ΑΕΚΗΓ.

Based on the assumption that the letter-names were originally intro-duced in alphabetic order, as is standard, and the observation that if the concept of non-intersecting semicircles was assumed to be known to the reader, such that the proof would follow without any need of the objects with letter-names following Π, we have proposed some relabeling of the orig-inal great circles, and agree with Czinczenheim in excising a large part of the *specification*.

As usual, we have not actually changed the text or the diagram in our translation, nevertheless, by following our notes, the reader can see what the original text – at least in outline – was likely to have been. If these suggestions are followed, all of the objects are introduced in alphabetic order, with the exception of Ξ before Ν, and none of the letter-names beyond Π are needed. In the diagram, the letter-names Ρ, Σ, Τ, Υ, Φ and their points can be removed – they are not used in any essential way, and other than Τ they do not appear in any of the parts of the proposition that can reasonably be considered genuine. In fact, although Τ appears in the letter-names of two great circles in the *demonstration*, it is not needed for the argument, and can be regarded as a later addition to these letter-names. Hence, it may be removed as well.

In most of the Greek manuscripts, the three letters T, Υ, and Φ are all clustered around the same point (see Figure 12, p. 69; **B**, 29r; **F**, 36r). In our diagram, we have followed the placement of these letters in the late recension **K** (110r), and in **I** (33a) of Thābit's Arabic version, both of which accurately portray the endpoints of the non-intersecting semicircles.

There is also some possibility that the second part of this proposition may have been a later addition, both to the *enunciation* and the *demonstration*. We do not stress this possibility, however, because while the appeal to the claim of *Sph.* II.13 Part 2 in *Sph.* II.19 may be a later addition, those in *Sph.* III.7 and III.8 appear to be genuine.

Problems treating great circles tangent to a small circle (*Sph.* II.14, II.15)

These two problems continue to develop the analogy between straight lines and circles on the plane and great circles and small circles on the sphere that was established at the end of *Spherics* I. In particular, they show how to draw a great circle tangent to a given small circle passing through a given point.

Sph. II.14 (Problem)

This proposition corresponds to *Elem.* III.16.cor., although it is different insofar as that corollary simply suggests a method for constructing a tangent to a point on a circle in the plane, which will be used, for example, in *Elements* IV, whereas this proposition is a standard *problem* that furnishes a complete construction for a tangent great circle on the sphere.

This *problem* has the same simplified structure of those of *Elements* I – namely, a single *specification* is followed by a clearly delineated *construction* and then *demonstration*. The proposition ends with a *problem-conclusion*.

As with the problems at the end of *Spherics* I, here a great circle corresponds to a line and small circle corresponds to a circle, and all of the operations can be performed on the surface of the sphere. Moreover, in this case, the proof requires no objects that are not produced in the *problem-construction* itself. As with the problems in *Spherics* I, it may be useful to go through the construction in detail (Sidoli and Saito 2009, 596).

With **sC**(AB) as given, and B a given point on it, the pole of **sC**(AB), say point G, is found using the methods set out in *Sph.* I.21. Then **gC**(GB) is drawn through G and B, *Sph.* I.20, and the arc that subtends the great square is cut off of this great circle from B, as **Arc**(BD). Although the details of this construction are not spelled out, they can easily be supplied from the problems established in the text (Sidoli and Saito 2009, 593 n. 44). The details of this construction are fleshed out above, in the Commentary to

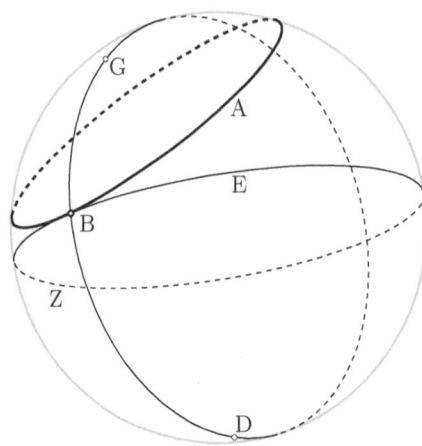

Figure 24: Perspective diagram, *Sph.* II.14.

Sph. I.20. Then, with *D* as a pole, and the distance *BD*, equal to the side of
the great square, **gC**(*EBZ*) is drawn passing through point *B*, *Elem.* I.post.3
(spherical). Following the explanations in the Commentaries to *Sph.* I.18,
I.19, it is clear that all of these operations can also be carried out with a
compass on the surface of a solid globe.

It should also be noted that the method of construction is general. If,
in place of the side of the great square, a different length is used for the
polar radius of **C**(*BEZ*), a small circle with an arbitrary polar radius can
be drawn tangent to **sC**(*AB*) at point *B*.

The *construction* in *Sph.* II.14 can be used to illustrate an important
feature of the *problems* produced by mathematicians such as Euclid and
Theodosios – namely, that they are idealizations and do not generally use
a procedure that would be carried out by a human geometer. In particular,
when a new object is introduced in the *construction* using a previously
established *problem*, it simply appears in the configuration, but none of the
auxiliary objects that were used in the previous *problem* are available for use
(Sidoli 2018b, 432–442). Indeed, if an auxiliary object that would have been
used to produce the new object is called for in a later step of the *construction*,
it must be produced separately – whereas, in a construction produced by a
human geometer that auxiliary object would already be there in the figure.
In particular, in *Sph.* II.14, the construction requires a great circle going
through the given point, *B*, and the pole of **sC**(*AB*), say *G*. Here, this
is effected by taking pole *G* in the first step of the *construction*, which
assuming the methods of the *Spherics* must be done using *Sph.* I.21, working
back through *Sph.* I.20–I.18, and then drawing a great circle through *B* and
G using *Sph.* I.20. But when we look into the details of the *construction*
in *Sph.* I.21, we see that this involves producing a great circle through an

arbitrary point on the given $\mathbf{sC}(AB)$, which may be taken to be B. Hence, a human geometer, who was interested in carrying out this construction on a solid sphere using a compass, would simply produce $\mathbf{gC}(GBD)$ directly, without first finding G. The *construction* provided in *Sph.* II.14, however, prefers to apply previously established *problems* and does not try to find the most effective procedure using basic operations, or manipulations of a mechanical device such as a compass.

The steps of the *demonstration* follow immediately from the *construction*, based on the properties of the polar radius of a great circle, *Sph.* I.17, and the surface characteristics of the tangency of two circles on a sphere, *Sph.* II.3.

Basis: *Elem.* I.post.3 (by implication, transfer, spherical), I.10 (by implication), IV.6 (by implication), *Sph.* I.17, I.19 (by implication), I.20, I.21, II.3.

Usage: *Sph.* II.16 (counterfactual).

Diagram: The diagram is a planar representation of local objects, containing only intrinsic objects. If circle EBZ were complete, the diagram would be global – in some of the medieval manuscripts this is the case. The diagram encodes tangent circles on the sphere with tangent circles in the plane of the figure. It makes no difference whether the small circle is depicted inside or outside the great circle.[59] There are no false crossings; so it represents the topology of the objects.

Comparison with Arabic: In Thābit's version this proposition, $Sph._A$ II.12, is found before *Sph.* II.13, which separates it from the other problem dealing with tangent great circles, *Sph.* II.15 ($Sph._A$ II.14). It is possible that Thābit thought that this problem should appear before *Sph.* II.13 so that it could be used to draw the tangent great circles that are introduced in the *exposition* of that proposition. If this was the case, it would place Thābit in that group of scholars who believe that every object that is introduced in the *enunciation* and *exposition* of a proposition should have already been produced in a problem. In fact, however, the objects mentioned at the beginning of a proposition in a Greek geometrical text are simply assumed, and in many propositions the objects so assumed have not been previously produced in a problem.

Sph. II.15 (Problem)

This proposition corresponds to *Elem.* III.17, and is established by an analogous procedure (Bjørnbo 1902, 47; Schmidt 1943, 12–13). In fact, although

---

[59] For example, a manuscript of the Latin version that was probably composed by Campanus depicts the small circle on the inside (Vat.lat. 3380 9v).

the analogy between lines and circles on the plane and great circles and small circles on the sphere was established at the end of *Spherics* I, this is the only proposition in Theodosios's *Spherics* in which a plane proposition, *Elem.* III.17, is modeled directly on the sphere. That is, this is the only proposition in the entire treatise that undertakes an intrinsic geometry of spherical figures of the sort that Menelaos later developed.

In this *problem*, there is no separate *specification* for the *demonstration*, and indeed, as in *Sph.* I.20, there is no clear division between the *construction* and the *demonstration* – deductive steps are interspersed through the construction, and new constructions are introduced when needed. Moreover, the inferences that show that the problem has been satisfied follow almost immediately from the constructions themselves. Indeed, the argument depends on the same key propositions as *Sph.* II.14. There is a clearly stated *problem-conclusion*, which completes the original presentation of the proposition.

Although there are discussions of at least one further case in all of the manuscripts, as discussed below, these are clearly interpolations. It is likely that the original version of this proposition only contained one case, as is often the situation in Greek mathematical texts.

As with the *problems* at the end of *Spherics* I, the objects that produce the solution to the problem in this proposition can all be carried out on the surface of the sphere, whereas auxiliary objects that are introduced for use in the *demonstration* can simply be imagined and need not actually be drawn. Once again, we will go through the details of the construction (Sidoli and Saito 2009, 597–598).

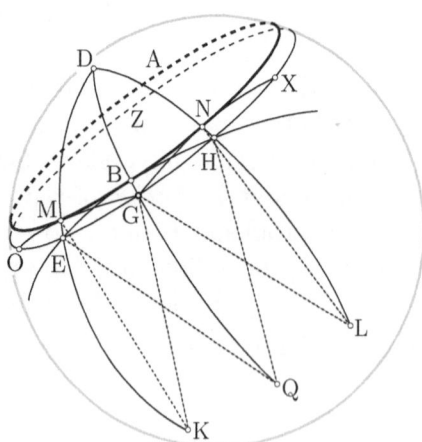

Figure 25: Perspective diagram, *Sph.* II.15.

With $\mathbf{sC}(AB)$ as given, and G a given point between $\mathbf{sC}(AB)$ and the circle equal and parallel to it, the pole of $\mathbf{sC}(AB)$, say point D, is found, using the methods set out in *Sph.* I.21. Then, about pole *D*, and with distance *DG*, a circle parallel to **sC**(*AB*) is drawn, say **C**(*GZ*), *Elem.* I.post.3 (spherical), and through points *D* and *G*, **gC**(*DGQ*) is drawn, *Sph.* I.20. Next, using the same operation as in *Sph.* I.14, the arc that subtends the side of the great square is cut off of **gC**(*DGQ*) from *B*, as **Arc**(*BQ*). Then, with *Q* as pole and the linear distance from *B* to *Q* as the distance, **gC**(*EBH*) is drawn passing through *B* and intersecting **C**(*GZ*) at points *E* and *H*, *Elem.* I.post.3 (spherical). Next, **gC**(*DEK*) and **gC**(*DHL*) are drawn through points *D* and *E*, as well as *D* and *H*, respectively, *Sph.* I.20. Then, **Arc**(*EK*) = **Arc**(*GQ*) is cut off from **gC**(*DEK*), and **Arc**(*HL*) = **Arc**(*GQ*) from **gC**(*DHL*), *Elem.* I.post.3 (transfer), and it is shown that **Arc**(*MK*) = **Arc**(*NL*) = **Arc**(*BQ*) – that is, they are all arcs subtending the side of the great square. As in *Sph.* I.20, in the middle of the surface constructions that solve the problem, internal lines – *LN*, *LG*, and *QE* – are produced, which cannot be drawn in a solid sphere, and which merely serve the *demonstration*. It is then shown that *GL* = *LN*, so that the great circle drawn about pole *L* and with distance *GL* will pass through *N*, and hence be tangent to **sC**(*AB*) at that point. A similar construction and argument is then used for the great circle drawn about pole *K* and with distance *GK* = *GL*.[60] Once again, it is clear that all of the operations necessary for producing the tangent great circles can also be carried out with a compass on the surface of a solid globe.

In the course of the argument, the text shows that the arcs of great circles passing through the poles of a pair of parallel circles that are cut off between those parallel circles are equal. In this proposition, the argument that this is the case is stated in detail, despite the fact that this was already shown in *Sph.* II.10 Part 2 (Czinczenheim 2000, 936). Since it is not standard practice in a Greek mathematical text to reiterate an argument that has already been established in a previous proposition, it is likely that this passage originates in a version of the treatise in which *Sph.* II.10 Part 2 was not shown. (See the "Textual Comments" to *Sph.* II.10.)

Another indication that the procedure of *Sph.* II.15 may have been developed independently from *Sph.* II.10 Part 2 can be obtained from considering a simpler, intrinsic version of this problem that may be derived directly from the insight provided by that claim. Using the same letter-names as the translation, such that the given circle is **sC**(*AB*) and the given point is *G* (Figure 26), the pole *D* of **sC**(*AB*) is found, *Sph.* I.21, and a great circle, **gC**(*DBG*) is drawn through points *D* and *G*, *Sph.* I.20. Then, **Arc**(*BQ*) is cut off **gC**(*DBG*) such that it subtends the side of the great square, *Elem.* I.post.3 (transfer), and a circle **pC**(*KQL*) is drawn about pole *D* through point *Q*, *Elem.* I.post.3 (spherical), such that it is paral-

[60] For summaries of the constructions in the other two cases, which are later interpolations, see Sidoli and Saito (2009, 597–598).

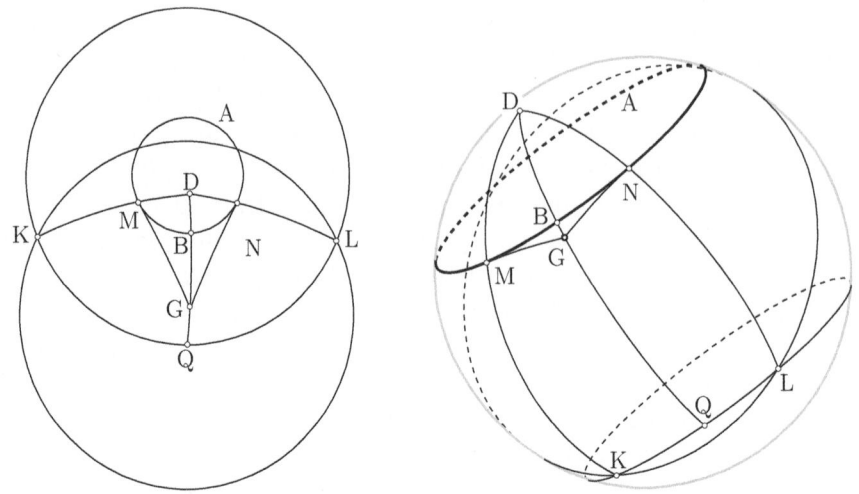

Figure 26: Diagrams for an unattested alternate approach to *Sph.* II.15: (left) ancient and medieval style diagram, (right) perspective diagram.

lel to **sC**(*AB*), *Sph.* II.2. Next, with *G* as a pole and a distance equal to the side of the great square, **gC**(*KL*) is drawn intersecting **pC**(*KQL*) at points *K* and *L*. Now, joining points *D*, *L* and *D*, *K* with **gC**(*DK*) and **gC**(*DL*), intersecting **pC**(*AB*) at *M* and *N* respectively, it must be that **Arc**(*BQ*) = **Arc**(*MK*) = **Arc**(*NL*), by *Sph.* II.10 Part 2. Hence, great circles about poles *K* and *L* will pass through *G*, *M* and *G*, *N*, respectively. Let **gC**(*GM*) and **gC**(*GN*) be drawn, *Elem.* I.post.3 (spherical). Finally, each of **gC**(*GM*) and **gC**(*GN*) must be tangent to **sC**(*AB*), by *Sph.* II.3.

The construction by Theodosios, on the other hand, both makes no use of *Sph.* II.10 Part 2, and also follows the construction of *Elem.* III.17 closely. Hence, we see again that Theodosios was guided by direct analogies with *Elements* III in producing his *Spherics*, and did not seek to develop an intrinsic geometry of the surface of the sphere.

Basis: *Elem.* I.post.1, I.post.3 (by implication, transfer, spherical), I.c.n.1, I.c.n.2, I.c.n.3, I.10 (by implication), III.28, IV.6 (by implication), *Sph.* I.def.5, I.15, I.16, I.17, I.19 (by implication), I.20, I.21, II.3, II.9, II.10,[61] II.12.

Usage: *Sph.* II.16 (counterfactual), III.7, III.12, III.14. A construction justifiable by this problem was used in *Phen.* 2, 4, 5, and 12.

Diagram: The diagram is a planar representation of local objects, containing both intrinsic and extrinsic objects. As usual, circles and circular arcs represent objects in the surface of the sphere and straight lines represent internal lines in a number of different planes, all drawn into the plane of

---

[61] See the remarks about this above.

the figure. Intersections on the spherical surface are encoded with inter-sections between circles and circular arcs in the diagram, parallel circles on the sphere are encoded with concentric circles, tangent circles with tan-gent circles, different regions of the spherical surface are depicted by the regions inside and outside of the various circles. The straight lines, which are only introduced for the sake of the *demonstration*, have false crossings both among themselves and with the circles.

Textual Comments: The texts established by Heiberg (1927, 68–76) and Czinczenheim (2000, 102–105) for this proposition are substantially dif-ferent. The text that Heiberg produced follows **K**, which is his tendency. Indeed, this manuscript is preferable from a mathematical perspective, since it presents three cases, clearly laid out and treated separately, and the lan-guage is closer to the standard mathematical prose that we expect from texts like the *Elements*, with fewer personal expressions. Nevertheless, **K** is a 14th-century Byzantine recension of the *Spherics*, not an accurate copy of an ancient source (Czinczenheim 2000, 239–258; Acerbi 2016, 162–163). Hence, it is clear that the text that Czinczenheim furnished, following **A**, is historically preferable – it is that found in the older Greek manuscripts, as well as in Thābit's Arabic version. Furthermore, as noted by both Heiberg and Czinczenheim, the final paragraph, with its personal expressions and demonstration of the case in which **Arc**$(BG) = 90°$, is also clearly an in-terpolation, which must have been made before the 9th century, since it is found both in the 9th-century Greek manuscript **A**, and in Thābit's Arabic version (Kunitzsch and Lorch 2010, 144–146).

Comparison with Arabic: In the Arabic versions this proposition is *Sph.*$_A$ II.14. The overall structure of Thābit's version agrees with the text found in the 9th-century **A** as opposed to that found in the 14th-century **K** (Kunitzsch and Lorch 2010, 136–146).

Sph. II.16 (Theorem)

As discussed above, this proposition is a partial converse to *Sph.* II.10 and II.13, and relies on those theorems, as well as three problems from the *Spherics*, and *Sph.* II.8, which is here also used constructively. It is unused in the text; so it was probably introduced for the sake of mathematical completeness, as the converse to the claims of *Sph.* II.10 Part 1 and II.13 Part 1.

The structure of this proposition is somewhat convoluted. Mathemati-cally, the assumed objects divide into three possible orientations, which are treated in Case 1, Case 2.1 and Case 2.2, each by an indirect argument. In Case 1, one of the great circles is assumed to pass through the poles of two unequal parallel circles, in Case 2.1 it is assumed to be tangent to the lesser of the parallel circles, and in Case 2.2 it is assumed to be oblique to the lesser of the parallels. There is an individual *specification* for each case and no general *conclusion*.

The way in which *Sph.* II.8 is applied in Case 2.2 warrants some comment. The first time that *Sph.* II.8 is used, in the *enunciation* for Case 2.2, it functions purely as a theorem, to affirm that there will exist two equal and parallel tangent circles. In the opening statement of the indirect argument, however, we assume that $\mathbf{gC}(BZQD)$ is tangent to a newly introduced circle, $\mathbf{sC}(LMX)$. This assumption and the introduction of the new object $\mathbf{sC}(LMX)$ also depends on *Sph.* II.8, now used constructively to introduce a new object satisfying certain conditions. Hence, taken together, we can see that the primary purpose of relying on *Sph.* II.8 in this proposition is to introduce a new object, so that its function is essentially constructive.

Basis: *Elem.* I.c.n.5, III.def.11, III.26, *Sph.* I.20, II.8, II.10, II.13, II.14 (counterfactual), II.15 (counterfactual).

Usage: Unused in the *Spherics*.

Diagram: The diagram is a planar representation of local objects, containing only intrinsic objects. It uses the now standard planar encoding for intersections, parallel circles, tangent circles and division of the spherical surface into two parts. There are no false crossings; so it accurately depicts the topology of the objects.

Textual Comments: There are problems with the diagram, the lettering of the diagram, and the assignment of the letter-names for Case 2.1. The first issue is that although, in the *exposition* of all three cases, circle AHΓ is the circle that is assumed to be either through the poles or not, such a circle is not found in the second diagram – neither in the manuscripts that we examined, nor in the editions of Heiberg (1927, 79) and Czinczenheim (2000, 696). For example, in the diagram for **A** (18r), there is no circle called AHΓ and a point H, along with another point Γ that belongs to circle AE, as well as another point Z have been added in a different ink.

Moreover, in the course of the argument for Case 2.1, a different point Γ, local to this argument, is introduced as part of the letter-name of circle ZΓ,[62] which is assumed to be tangent to circle EZΘ. Hence, there were either two Γs, or more likely, this second Γ was originally called something else – such as K – which was then later changed by someone who did not understand the argument.

Ver Eecke (1927, 57 n. 2) recognized that there was a problem with the diagram and provided what he considered to be a corrected figure, following Nizze. But this figure is also incorrect for two reasons. In the first place, Θ floats freely on circle EZΘ, whereas, according to the overall *exposition* for

---

[62] We call this ZG' in the translation.

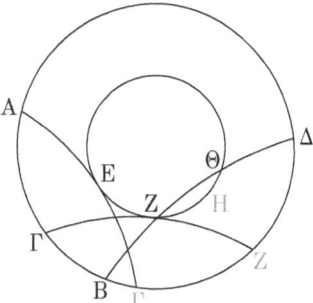

Figure 27: Manuscript diagram, *Sph.* II.16, Case 2.1 (**A**, 18r). Gray labels have been added in a different ink.

the whole proposition, it must also belong to circle BZΘΔ. More significantly, the two Γs are collapsed into a single point, which is both different from the manuscript diagrams, and does not provide the generality that the theorem requires. Although each step of the argument is valid when the two Γs coincide, in order for the argument to be fully sound, the Γ of circle ZΓ – *G'* in our translation – must meet circle ABΔ wherever it happens to fall when circle ZΓ is drawn tangent to circle EZΘ. Hence, it will not generally coincide with the Γ of circle AHΓ.

For these reasons, we have proposed the following changes, which will preserve the topology of the manuscript diagrams and make minimal changes to the text itself. We have transliterated the local Γ with *G'*, both in the diagram and in the text. Moreover, we have added a *G* and an *H* to the diagram, following the requirements of the text. With these changes, the text and the diagram are both mathematically correct and better represent the generality required by the theorem.

Comparison with Arabic: In the Arabic versions this proposition is *Sph.*_A II.15. Thābit's version was either based on a better Greek manuscript, or, more likely, was changed so as to resolve the problem with Case 2.1 in the Greek, since both the diagram and the text are slightly different from the Greek (Kunitzsch and Lorch 2010, 150). Probably, Thābit, or someone else in the transmission, realized that there was a problem and changed the text accordingly.[63]

---

[63] In Kunitzsch and Lorch's (2010, 151, ll.32–34) edition of Gerard's Latin, here the letter *L* is a typographic error for *K*.

A great circle cutting parallel circles: Magnitudes of the parallel circles (*Sph.* II.17, II.18)

This pair of converses provides a comparison of the magnitudes of small parallel circles by measuring with respect to the arc-distance along any great circle intersecting the greatest of the parallels – that is, along any other great circle. Considering a bundle of parallel circles, pC_1, gpC and pC_2, and a great circle gC intersecting the parallel circles such that the small parallel circles cut off arcs, η_1 and η_2, along gC between each of pC_1, pC_2 and gpC (Figure 28), these two theorems show that (1) the two parallel circles are equal if and only if their great-arc separation from the greatest of the parallels is equal, $pC_1 = pC_2 \Leftrightarrow \eta_1 = \eta_2$, and (2) one is less than the other if and only if its great-arc separation is greater than that of the other, $pC_1 < pC_2 \Leftrightarrow \eta_1 > \eta_2$.

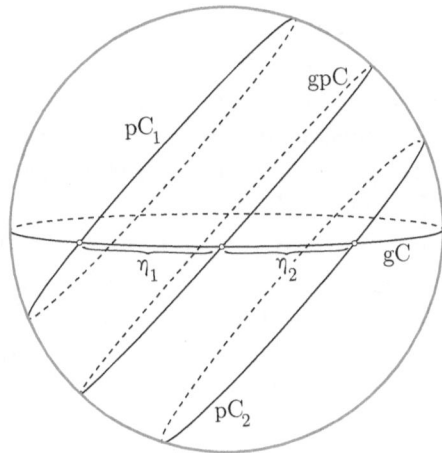

Figure 28: Perspective diagram, *Sph.* II.17 and II.18.

The mathematical import of these theorems is closely related to that of *Sph.* I.6. Now, however, all of the objects and arguments involved take place on the surface of the sphere, whereas *Sph.* I.6 was framed in terms of solid objects. Moreover, *Sph.* II.17 and II.18 are expressed in terms of a bundle of parallel circles, which is one of the key objects of *Spherics* II. Nevertheless, it is easy to see that *Sph.* II.17 and II.18 are true if *Sph.* I.6 is true.

From the view of spherical astronomy, these propositions complete the approach of *Sph.* II.13 in establishing the ortive amplitude, η, as a characteristic measure of a δ-circle. That is, where we take the oblique great circle, gC, as the horizon and the greatest of the parallels, gpC, as the equator, then these propositions assert that the ortive amplitudes of a pair of δ-circles – such as the visibility circles, the tropics, or a pair of day-circles

– are equal if and only if the δ-circles are equal, and that the sizes of the δ-circles vary inversely with the ortive amplitudes.

Sph. II.17 (Theorem)

This proposition is based on basic properties of the circle, a bit of solid geometry and a number of core propositions of the *Spherics*, but the most essential tool used for the proof is *Sph.* II.12, which determines the magnitude of the polar radii of the parallel circles in question.

The proposition has two parts – Part 1 treating equalities, Part 2 treating inequalities. Part 1 divides into two cases depending on whether or not the cutting great circle passes through the pole of the parallel circles. Part 2 is shown immediately on the basis of Part 1.

When the letter-names H, Q, K, L and M are initially introduced, they are unspecified parts of geometric objects that only become specified as the discussion continues. Moreover, as the text currently stands the letter-names are introduced somewhat haphazardly and out of alphabetic order. See the "Textual Comments," below.

The application of *Sph.* II.12 in this proposition might be somewhat unexpected, because that theorem is articulated in terms of a pair of circles, whereas here it is applied in the two hemispheres produced by a single great circle. Nevertheless, the full articulation of the conditions of *Sph.* II.12 before stating the result is a clear indication that we must understand this step in the argument as an application of that theorem (see p. 140, n. 1).

In the final step of the *demonstration*, Part 1, Theodosios asserts that circles that have equal polar radii are equal. Although there is no justification for this in the text (Ver Eecke 1927, 61 n. 9), it easily follows from propositions of the *Elements*, or from *Sph.* I.6. This claim was essentially already used in *Sph.* II.6. See the Commentary to that proposition (p. 252, above).

Basis: *Elem.* I.def.17, I.post.3 (spherical, transfer), I.c.n.1, I.c.n.2, I.c.n.3, I.5 (by implication), I.11, I.27, III.def.1, III.14 (by implication), III.26, III.28 (by implication), III.29, V.15, XI.3, XI.16, *Sph.* I.6, I.10, I.11, I.15, I.20, I.21, II.9, II.12.

Usage: *Sph.* II.18, III.13. A proposition like this was used in *Phen.* 6, 8, 12, and 14.

Diagram: The diagram is a solid representation of global objects, containing both intrinsic and extrinsic objects. In this case, the solid articulated style presents a fairly vivid visual impression of the sphere in something approaching the perspective style. The diagram is built around a base circle that represents a great circle in the sphere. A bundle of parallel circles is encoded with a set of arcs of concentric circles that meet the base cir-

cle in half-lenses. The diameters of these parallel circles are also depicted,
which gives the figure its articulated appearance. The base circle divides the
sphere into two hemispheres which are encoded as inside and outside of the
base circle. Indeed, point X which is on the other hemisphere from that facing us is displayed on a continuation of arc *LNHQRKM*, following the usual
pattern for planar diagrams. The straight lines inside the sphere have false
crossings with a great circle on the surface of the sphere, which is depicted
by a circular arc meeting the base circle with a half-lens configuration.

Textual Comments: Notice that in the current state of the text H is not
introduced until after M, as opposed to after Z, which would be standard.
This could be corrected if, in the *construction*, the circles that have common
sections with ABΔΓ were named AHB, EΘZ and ΓKΔ, successively. These
names would agree with what the latter two objects are called immediately
below, when they are discussed as planes. Moreover, when H is finally
introduced it is in the letter-name for circle AHB, which is the same object
that was called circle AB in the *construction*. In either case, the letter-names
H, Θ, and K are unspecified by the text when they are introduced and it is
only when great circle ΛΘKMΞ is introduced that Θ and K become specified.
The letter-name H remains completely unspecified by the text and it is only
by looking at the diagram that we can see that it is the intersection of one
of the parallel circles with a great circle through its poles.

Comparison with Arabic: In the Arabic versions this proposition is
Sph.$_A$ II.16.

Sph. II.18 (Theorem)

This proposition is a converse to the preceding proposition and is proved on
that basis by an indirect argument.

Basis: *Sph.* II.17.

Usage: *Sph.* II.19, III.13. A proposition like this was used in *Phen.* 8 and
12.

Diagram: The diagram is a solid representation of global objects, containing only intrinsic objects. It produces the visual impression of the sphere.
The diagram shows a great circle with the base circle and a bundle of parallels that intersect the great circle as concentric half-lenses that meet the
base circle. There are no false crossings, so the diagram depicts the topology
of the configuration.

Comparison with Arabic: In the Arabic versions this proposition is
Sph.$_A$ II.17.

A great circle inclined on parallel circles: Comparison of segments of the parallel circles (*Sph.* II.19, II.20)

These theorems provide us with two different ways of comparing the sizes of the arcs of parallel circles that are cut off by a great circle falling on them obliquely.

In *Sph.* II.19 the segments of the parallel circles that are to be compared are divided into the four regions produced by the geometric intersection of the oblique great circle, gC, with the greatest of the parallel circles, gpC (Figure 29 (left)). These regions can be labeled as follows: Π_1, the set-intersection of both upper hemispheres produced by the great circles, which contains the visible pole, P_v; Π_2, the set-intersection of the upper hemisphere of the parallel circles with the lower hemisphere produced by the oblique great circle; Π_3, the set-intersection of both lower hemispheres produced by the great circles, which contains the invisible pole, P_i; and Π_4, the set-intersection of the lower hemisphere of the parallel circles with the upper hemisphere produced by the oblique great circle. Then, *Sph.* II.19 shows that gC will cut all of the small parallel circles unequally and that the segment of any pC_1 in Π_1 will be greater than the segment of the same circle in Π_2, and moreover the alternate segments of the equal and parallel circle pC_2 will be equal. That is, where $\eta_1 = \eta_2$, such that $pC_1 = pC_2$, by *Sph.* II.17, then the segment of pC_1 in Π_1 will be equal to that of pC_2 in Π_3, and that of pC_1 in Π_2 will be equal to that of pC_2 in Π_4.

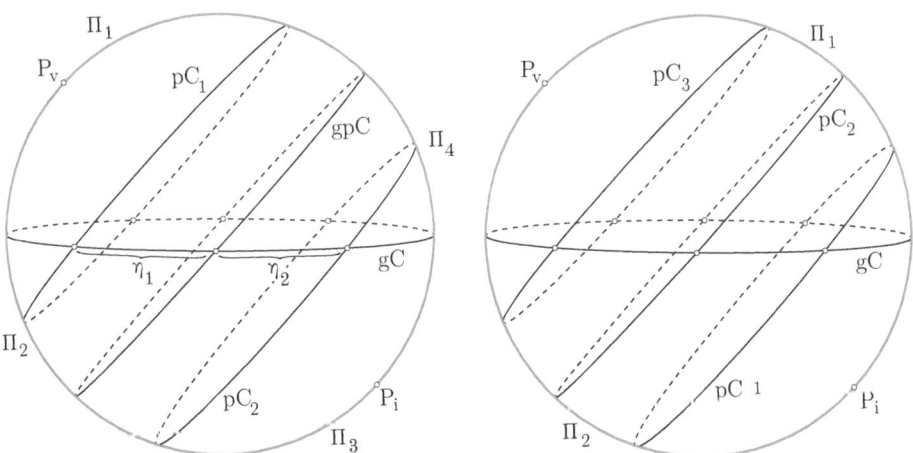

Figure 29: Perspective diagrams, (left) *Sph.* II.19 and (right) II.20.

In *Sph.* II.20 the arcs to be compared are divided into two hemispheres – namely, Π_1, the hemisphere produced by gC that contains the visible pole, P_v, and Π_2, the opposite hemisphere, containing the invisible pole, P_i (Figure 29 (right)). With this distinction, *Sph.* II.20 shows that in Π_1 the angular measures of the arcs cut off of any three parallel circles, pC_1, pC_2, and pC_3, increase monotonically from pC_1 to pC_2 to pC_3 as the parallel circles are taken closer to P_v. Although it is not mentioned in the text, taken together with the previous proposition this makes it clear that the same situation holds for the arcs in Π_2 as they are taken closer to P_i.

One of the astronomical implications of *Sph.* II.20 is that if two stars rise at the same time, that which is nearer to the visible pole will set later (Schmidt 1943, 22–23). Furthermore, together these propositions fill out the implications of *Sph.* II.13 for understanding the length of the daytime and nighttime, and hence of the seasonal hours, for inclined latitudes.[64] That is, since *Sph.* II.13 shows that similar arcs of all the δ-circles are cut off between two configurations of the horizon as the celestial sphere rotates, if we consider the greatest of the parallel circles to be the celestial equator, and the inclined great circle to be a horizon of the inclined sphere, then we can use the differences of the arcs of the day-circles cut off by the horizon to compare the lengths of time that the sun will be above and below the horizon. Then, *Sph.* II.19 implies that day-circles in the direction of the visible pole will produce a greater daytime than nighttime, and that two day-circles equally separated from the celestial equator will produce symmetrically opposite equal daytimes and nighttimes. Moreover, with the same assumptions, *Sph.* II.20 implies that the lengths of the daytimes will monotonically increase as the day-circles are taken closer to the visible pole, and, inversely, that the nighttimes will monotonically decrease. Finally, since the seasonal hours are simply the daytimes and the nighttimes divided into six parts, the implications of these theorems for the seasonal hours are clear.

Sph. II.19 (Theorem)

This proposition is based on some basic properties of circles, some of the core propositions of *Spherics* I, and *Sph.* II.13 and II.18. This broad basis, along with the fact that it is not used in this treatise, indicates that this theorem was regarded as an inherently worthwhile part of the subject matter treated in *Spherics* II.

Here, we are introduced to the expressions *visible pole*, φανερὸς πόλος, and *non-visible pole*, ἀφανὴς πόλος for the two poles of a bundle of parallel circles on opposite sides of an assumed great circle. These expressions almost certainly originate in the context of spherical astronomy, in which they mean the celestial poles above and below the horizon – indeed, they are found already so used in Autolykos's *Moving Sphere* and Euclid's *Phe-*

---

[64] Schmidt (1943, 21–24) discusses *Sph.* I.20 along with *Sph.* III.14, which we will discuss below.

nomena. In this text, however, they are some of the few expressions that have an explicitly astronomical referent, so that one may wonder why these astronomical terms are used while others seem to be avoided. One explanation for their continued use in the *Spherics* is that they may simply serve as a way to denote the two hemispheres created by an assumed great circle. In this sense, we can take these as technical terms that have directionality, but not a strong astronomical meaning. Of course, following the general orientation of directionality discussed above, the visible pole is considered to be higher, while the invisible pole is lower. Hence, the astronomical interpretation is always there, just below the literal expression of the text.

The first *specification*, for the whole proposition, is somewhat unusual, insofar as it is stated mostly in general terms with the exception of *EZ*, which is the only letter-name used. The second *specification*, for Part 2, is expressed normally.

The *construction* of this proposition warrants some comment. At the beginning of the *construction* the visible pole of the parallel circles, *H*, is simply asserted to be there, but is not said to be taken. This can be compared with the first steps in the *construction*s of *Sph.* II.14 and II.15. It is unclear, in *Sph.* II.20, whether this is simply the act of naming the pole that had been mentioned in the *enunciation*, or is an application of *Sph.* I.21. Either interpretation is defensible, but we tend to the former, because the pole is one of the objects mentioned in the *enunciation*, so that, in normal practice, it would be named in the *exposition* and not introduced as a new object in the *construction*.

Furthermore, in the *construction*, there are some repeated constructions involving **gC**(*HNZK*) and **pC**(*BG*). The first time that **gC**(*HNZK*) is drawn through *E* and *H*, using *Sph.* I.20, it is apparently not completed, because it is then shown that this circle "filled out," προσαναπληρούμενος will pass through *Z*, and then it is so constructed. The construction using *Sph.* I.20, however produces this entire circle, so that one might think it sufficient to simply argue that it will pass through *Z*. Furthermore, when points *Q* and *K* need to be introduced this is done by letting **gC**(*BG*) be "filled out," despite the fact that this circle, being assumed in the *enunciation* and set out and named in the *exposition*, should be assumed to be already drawn. In both cases, forms of the verbs προσαναπληροῦν and ἔρχεσθαι, which are not usually used to draw circles, are employed for these operations (Mugler 1959, 199, 363). This seems to be another indication that, in introducing new objects, the *construction* serves a number of different roles that are are not all strictly mathematical. Here, as much attention seems to be paid to producing a practical diagram that can be used in the argument as to generating objects from postulates and previously established problems.

Basis: *Elem.* I.def.18, I.post.3 (spherical), I.c.n.2, I.c.n.3, I.c.n.5, III.28, III.29, *Sph.* I.11, I.15, I.20, II.13, II.18.

Usage: Unused in the *Spherics*. A proposition like this had been used in *Phen.* 2 and 6.

Diagram: The diagram is a solid representation of global objects, containing only intrinsic objects. Like the diagram for *Sph.* II.17, however, the base circle divides the sphere into two hemispheres, which are encoded as inside and outside of the base circle, following the convention for planar representations. That is, points Q and K, which are on the other hemisphere from the pole of the parallels are drawn outside of the base circle. The diagram shows a great circle as the base circle, a bundle of parallels intersecting it as concentric half-lenses that meet the base circle, and another great circle that passes through the pole of the parallels as a half-lens with a different orientation. In this diagram, however, there are no false crossings, so that it represents the topology of the configuration.

Textual Comments: There is a minor irregularity in setting out the letter-names, insofar as there are points named M and N, but not Λ.

Comparison with Arabic: In the Arabic versions this proposition is *Sph.$_A$* II.18.

Sph. II.20 (Theorem)

Although thematically this proposition belongs with the previous one, its primary basis is *Sph.* II.10, to which it is almost a corollary. The mathematics of this proposition is closely related to the claims about the relative risings and settings of points made in *Mov. Sph.* 9. Indeed, Autolykos's theorem depends directly on something like *Sph.* II.20, and since this proposition is not used elsewhere in the *Spherics*, it was probably included in the text because of its astronomical significance.

   The core of the theorem relies on an intuitive notion of betweenness that is introduced through the construction and encoded in the diagram. In particular, since the pole of the parallels, H, is either assumed or constructed in a region of the surface of the sphere between $\mathbf{gC}(GABD)$ and $\mathbf{pC}(GD)$, while the two circles intersect at points G and D, then when $\mathbf{gC}(HG)$ and $\mathbf{gC}(HD)$ are drawn, they must each lie between the pair $\mathbf{Arc}(GABD)$ and $\mathbf{Arc}(GD)$, on either side. Hence, since $\mathbf{pC}(AB)$ is closer to pole H than $\mathbf{pC}(GD)$, it must cut $\mathbf{gC}(HG)$ and $\mathbf{gC}(HD)$ inside the figure contained by $\mathbf{Arc}(GABD)$ and $\mathbf{Arc}(GD)$. Hence, *Sph.* II.10 can be applied. The concept of betweenness that is involved here is similar to that at work in *Elem.* I.6 and I.7.

   This proposition introduces the first usage in the *Spherics* of the terminology *greater than similar to*, μείζων ἢ ὁμοία, which had already been used in Autolykos's *Moving Sphere* and Euclid's *Phenomena.*[65] Although this relation is not defined in the *Elements*, the idea is intuitively clear, and, based on *Elem.* III.def.11, we can say that $\alpha_1 \succ \alpha_2$ when the angle contained in, or subtended by, the segment bounded by α_1 is greater than the corresponding angle in the segment bounded by α_2. In this proposition, the method for showing this inequality involves showing that one arc is similar to an arc that is a part of the arc that must be shown to be *greater than similar to* the first. This relation will be used again in *Sph.* III.6, III.8, and III.14. In each case, this is a way of expressing an inequality in the angular measure of an arc.

As in the previous proposition, at the beginning of the *construction* the visible pole of the parallel circles, H, is simply asserted to be there, but is not said to be found. See the Commentary to *Sph.* II.19 for a discussion of this.

Basis: *Elem.* I.c.n.5, III.def.11, *Sph.* I.20, II.10.

Usage: Unused in the *Spherics*. A proposition like this had been used in *Mov. Sph.* 9,[66] and in *Phen.* 9 and 14.

Diagram: The diagram is a solid representation of global objects, containing only intrinsic objects. The diagram shows a great circle as the base circle, a bundle of parallels intersecting it as concentric half-lenses that meet the base circle, and two other great circles that meet at the pole of the parallels as circular arcs. This diagram has no false crossings. Hence, it represents the topology of the configuration.

Textual Comments: The diagram in most of the manuscripts is mathematically incorrect in placing H on circle ABΓ, where it does not lie (**B**, 38r; **D**, 9r; **F**, 38v). That in **A** (20r), which probably originally looked like those in most of the other early manuscripts, has been corrected in a different ink. In **K** (117v), although H correctly does not lie on circle ABΓ, the two great circles ΓΘH and HKΔ appear as a single arc passing through all five points. The diagram in our translation follows the corrected diagram in **A**.

Comparison with Arabic: In Thābit's version this proposition is *Sph.*_A II.19. The diagram in the Arabic is somewhat different. As in the

---

[65] Notice that Bruin and Vondjidis's (1971, 16–17) translation of this phrase with "larger than, or the same, as" does not make mathematical sense.

[66] Schmidt (1943, 8) also notes this proposition as being used in the introduction to the *Phenomena*. The argument there, however, is ostensibly empirical. The author of the introduction argues that we see certain arcs as longer and shorter based on the durations of the motions of various stars, and on this basis, among others, claims that the cosmos is spherical (Menge 1916, 4; Berggren and Thomas 1996, 44–45). Hence, the claim that the arcs are of different sizes cannot be based on *Sph.* II.20, an application of which must *assume* a spherical cosmos.

previous proposition, the great circles through the poles are extended down to meet the lower equal and parallel circle, at ς and ظ. Hence, the letter-names in the text are also somewhat different (**I**, 37a; Kunitzsch and Lorch 2010, 335, 385 n. 2). The diagram in **I**, like that in some of the Greek manuscripts, incorrectly depicts the two great circles ح م د ς and ح ج ط with a single arc, but this is corrected in the critical edition (Kunitzsch and Lorch 2010, 170).

Comparison of the inclinations of great circles upon one another (*Sph.* II.21)

This proposition stands alone in establishing a determining characteristic for the inclination of one great circle upon another. Namely, the inclination of a pair of great circles on one another is greater than that of another pair of great circles when the perpendicular dropped from the pole of one circle of the first pair onto the plane of the other is greater than the corresponding perpendicular in the other pair. Likewise, the inclination of one pair of great circles is equal to that of another pair when the same perpendiculars are equal.

Sph. II.21 (Theorem)

This theorem, which is a lemma to the following theorem, is the first propo-sition in the treatise that mentions two or more spheres.[67] In fact, however, when it is applied, just as with *Sph.* II.11 and II.12, it will be used for two configurations of great circles in a single sphere. It relies on elementary solid geometry as well as some core propositions of *Spherics* I. Indeed, from a geometrical perspective, this theorem is a proposition of solid geometry involving circles and semicircles, but it is expressed in terms of spheres so that some of the core propositions of *Spherics* I can be applied directly.

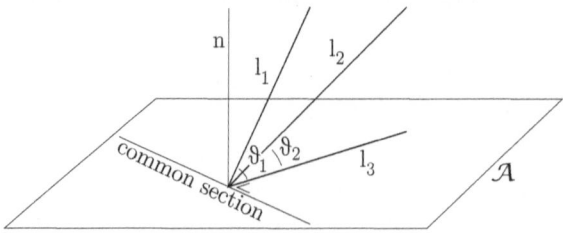

Figure 30: Explanation of *more* and *less inclined.*

---

[67] *Sph.* II.11 and II.12 deal with two configurations that can, of course, be found in one or more spheres, but they do not actually mention the spheres.

As well as establishing a way of comparing the *inclination* of two great circles, this proposition also introduces technical terminology for comparing inclinations more generally, which, in translation, may strike some readers as non-intuitive. According to *Elem.* I.def.8, *a plane angle* is the inclination of two lines to one another, where inclination is apparently taken as a primitive concept. Such an equivalence between angles and inclinations is also at work in this proposition. In fact, in Part 1.1 of the *demonstration*, in treating the inclination of two planes, Theodosios carefully argues that a certain line in one plane is perpendicular to the common section of both planes, and then asserts that the angle between this line and the line in the base plane that meets it *is* the inclination between the planes. This seems to be a direct appeal to *Elem.* XI.def.6(∗), but this definition is generally believed to be a much later addition to the text of the *Elements*. Whatever the case, the argument in *Sph.* II.21 clearly implies that the inclination between two planes can be compared by treating the angles between these lines. That is, if l_1 and l_2 are two lines lying in planes that meet some other plane, $\mathcal{A}$, at the same common section, such that they meet the common section at right angles at the same point as l_3, perpendicular to the common section, in plane $\mathcal{A}$, then the inclinations of the planes with $\mathcal{A}$ are angles $\vartheta_1(l_1, l_3)$ and $\vartheta_2(l_2, l_3)$, respectively (Figure 30).

What becomes clear in *Sph.* II.21 and II.22, however, is that as the angle of inclination becomes less, the line, or the plane, is regarded as being *more inclined*. That is, since $\vartheta_1 > \vartheta_2$, l_2 is more inclined on the plane than l_1. While this may not agree with standard mathematical usage in English, it follows from the fact that a perpendicular line, n, was regarded as upright, so that a line in the plane at the base of a perpendicular, l_3, could be regarded as fully inclined. Hence, where the normal n is upright, l_1 can be called *more upright* while l_2 can be called *more inclined*. Indeed, the fact that *more inclined* was regarded as the opposite of *more upright* is made explicit in *Sph.* II.22 (p.151, above). Although readers of Greek probably would have found these distinctions obvious from the meaning of the common language expressions, this was apparently not the case for mathematical scholars working in Arabic, because someone added a special definition explaining this situation. The final definition of Thābit's version, *Sph.*$_A$ I.def.11, in discussing the inclination of planes, states that "... those whose angles are less are more inclined" (Kunitzsch and Lorch 2010, 14).

The structure of *Sph.* II.21 is non-standard, partly because the proposition is in two parts, but there are other complications. The condition of Part 1 is introduced in the *exposition* for that part, followed by an initial *specification* for Part 1, and a *construction* that will serve for both parts. There are two *specifications* in Part 1, one for the general concept of inclination, and another for the specific angles that have been argued to determine the inclinations in question. A brief *exposition* sets out the conditions for

Part 2, and then there are again two *specifications*, repeating the pattern of
Part 1.[68] As usual, there are no general *conclusions*, simply the conclusions
to the *demonstrations*.

The argument for Part 1 requires the claim that a perpendicular dropped
from a point on the surface of a sphere to the plane of a great circle is greater
when the point is closer to the pole of the great circle. If we pass a plane
through the elevated point, the pole of the great circle, and the center of
the sphere, thus producing another great circle that is perpendicular to
the initial great circle, this claim follows from *Elem.* III.15, since the most
elevated point on the great circle of the cutting plane is the pole on the
perpendicular diameter, whereas all other perpendiculars are chords, and
those that are progressively farther from the center of both sphere and the
cutting great circle will be progressively lesser, by *Elem.* III.15.

The *demonstrations* of both parts rely on a pair of lemmas that are
closely related to the lemmas used in *Sph.* II.11 and II.12 – for which see the
introductory Commentary to those theorems. Here, the required claims are
that, in equal segments of circles, equal perpendiculars raised on the bases
will cut off equal arcs of the segments, and greater perpendiculars will cut
off greater arcs. The first of these claims is the converse of the lemma used
in in *Sph.* II.11 and II.12, and the second is an obvious extension to this
pair of converses. A related marginal gloss, Scholium 274, reads as follows
(Czinczenheim 2000, 412; Figure 31):[69]

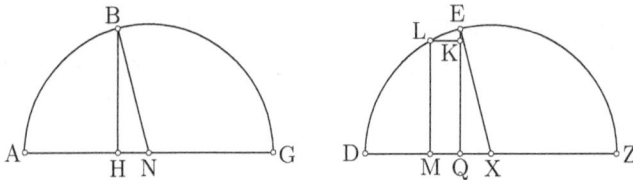

Figure 31: Scholium 247 to *Sph.* II.21.

For, let there be two semicircles, *ABG* and *DEZ*, two points on
them, *B* and *E*, and perpendiculars to them, *BH* and *EQ*, and,
first, let them be equal.

I say that circumference *AB* is also equal to circumference *DE*.

For, let there be the centers of the circles, *N* and *X*. Let it be
constructed.

---

[68] Notice that the final clause of the first *specification* for Part 2, which is redundant
with the second *specification* for Part 2, is probably an interpolation (see p. 146, n. a,
above).

[69] The diagram in **A** (21r) is rather poorly executed and has T in place of K, but that
provided by Czinczenheim (2000, 717) exaggerates the errors even further. We have not
attempted to preserve the visual characteristics of the manuscript diagrams.

Since the square on *BN* is equal to the square on *EX*, and the squares on *BHN* are equal to the squares on *EQX*,[70] of which *BH* is equal to *EQ*, therefore, *HN* is equal to *QX*, as a remainder. So, two, *QXE*, are equal to two, *HNB*,[71] and *BH*, as base, is equal to *EQ*, as base. Therefore, angle *X* is equal to angle *N*, so that circumference *AB* is also equal to circumference *DE*.

So, with that being already shown and with the same assumptions, let perpendicular *EQ* be greater than *BH*.

I say that circumference, *DLE*, is also greater than circumference, *AB*.

For, let *QK* be cut off equal ⟨to *BH*⟩, and *KL* through *K* parallel to *QM*, and *LM* through *L* parallel to *KQ*.

Therefore, *LM* is equal to *KQ*, but *KQ* is equal to *BH*, therefore *BH* is also equal to *LM*. And, both *BH* and *LM* are perpendiculars. Therefore, circumference *AB* is equal to circumference *DL*, so that *DLE* is greater than circumference *AB*.

This scholium, in fact, shows that the two claims hold under the assumption that the segments are semicircles. For their use in *Sph.* II.21, this restriction is acceptable, but does not agree with what the text actually asserts. The more general situation, involving equal segments, not necessarily semicircles, can however, be shown fairly straightforwardly. The converse to the lemma established in the introductory Commentary to *Sph.* II.11 and II.12, Scholium 190, can be established from that theorem by an indirect argument. Once this pair of converses has been demonstrated, the extension to unequal arcs and perpendiculars is straightforward.

Basis: *Elem.* I.c.n.2 (normal, inequalities), I.c.n.3 (normal, inequalities), III.15, III.27, III.28, XI.def.3, XI.def.4, XI.def.6(∗), XI.def.7(∗), XI.3[72] XI.19, *Sph.* I.def.6,[73] I.11, I.15, I.16, I.20.

Usage: *Sph.* II.22.

Diagram: This is a standard solid diagram of the sort found in the *Sph.* II.11 and II.12. It is similar to the diagrams for *Sph.* I.1, I.7, and II.10–II.12 in visually conveying a sense of depth. As usual, circles and circular arcs represent objects in the surface of the sphere and straight lines are internal to the sphere, so that they have false crossings with the surface objects.

---

[70] That is, $\mathbf{S}(BH) + \mathbf{S}(HN) = \mathbf{S}(EQ) + \mathbf{S}(QX)$.

[71] That is, $QX = HN$ and $XE = NB$, so that $QX + XE = HN + NB$.

[72] *Elem.* XI.def.6(∗) and XI.def.7(∗) are believed to be late interpolations to the *Elements*.

[73] The second usage of *Sph.* I.6 in this proposition is probably a later addition. See the Commentary to *Sph.* I.def.6 (p. 198, above), and the "Textual Comments" to this proposition, immediately below.

Textual Comments: The proposition ends with what appears to have been a gloss that has entered the text. Since this passage is found in both the early Greek manuscripts and in Thābit's version (Kunitzsch and Lorch 2010, 182), it must have entered the text before the 9th century. It states a definition of similarly inclined planes that "we learned," which is a nearly verbatim quotation of the current *Sph.* I.def.6 – both of which are a combination of *Elem.* XI.def.6(∗) and XI.def.7(∗), now believed to be late additions to the *Elements*. This similarity of expression indicates that both the gloss and definition are probably interpolations (Czinczenheim 2000, 52, 122), which may have originated in an educational context.

Comparison with Arabic: In Thābit's version this proposition is *Sph.$_A$* II.20.

Comparison of the inclination between a pencil of great circles that are equally inclined on a bundle of parallel circles and a great circle oblique to the parallel circles (*Sph.* II.22, II.23)

The final two propositions of *Spherics* II deal with comparisons of various angles between great circles that are oblique to a bundle of parallel circles. The stipulated configuration is that of a base great circle, gC, that is tangent to one of the parallel circles, pC_1, and cuts into greater and lesser parts a larger parallel circle, pC_2, to which, in turn, a pencil of great circles, gC_n, are tangent (Figures 32 and 33). Then, under the condition that the pole of gC, say point Z, lies between pC_1 and pC_2, *Sph.* II.22 makes the following claims about the angles, ϑ_n, between gC and various positions of gC_n:

(a) all of the positions of gC_n are inclined on gC – that is, $\vartheta_n \neq 0°$ or $90°$,

(b) that gC_n tangent at the midpoint of the greater part of pC_2, say gC_1, has the greatest angle – that is, ϑ_1 is greatest,

(c) that gC_n tangent at the midpoint of the lesser part of pC_2, say gC_2, has the least angle – that is, ϑ_2 is least,

(d) two gC_ns whose points of tangency are equidistant from either midpoint, say gC_3 and gC_4, have the same angle – that is, $\vartheta_3 = \vartheta_4$,

(e) the angles, ϑ_n, monotonically decrease as the points of tangency are taken farther from the midpoint of the greater part of pC_2 – that is, $\vartheta_3 > \vartheta_5$, and

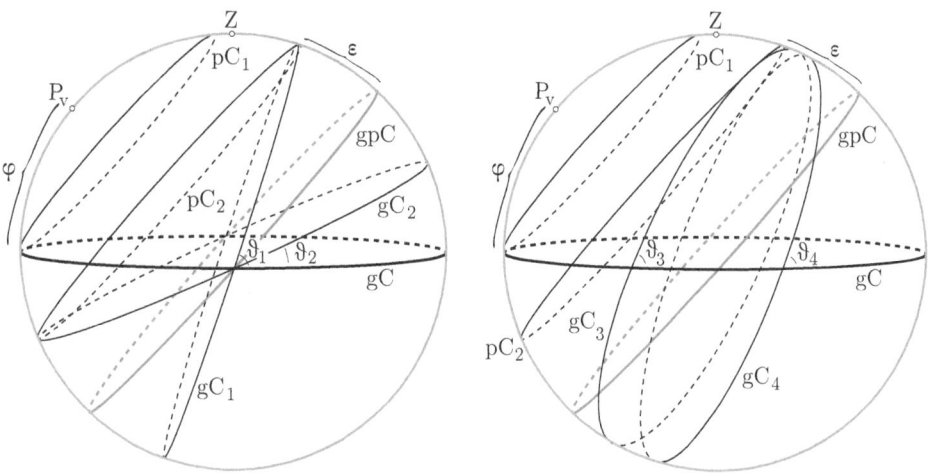

Figure 32: Perspective diagrams, *Sph.* II.22, (left) claims (b) and (c), and (right) claim (d).

(f) the poles of all of the gC_ns lie on another pair of parallel circles (of which one is say pC_3, such that $pC_3 < pC_1$).[74]

Then, *Sph.* II.23 shows that, with the same overall configuration, if the arcs of any two gC_ns between their points of tangency with pC_2 and their intersections with gC are equal, then those gC_ns have the same angles of inclination with gC. As becomes clear in the course of the proof, this proposition provides another condition for the claim made in (d), in the list above.

In *Sph.* II.22 it is stated as a premise that one of the poles of gC, point Z, be between the two parallel circles, pC_1 and pC_2. Hence, we may take this as the stipulation of the bounds, at which at least one of the claims made in the theorem ceases to be true. In order to understand what these bounds mean, it will be helpful to use an analemma diagram (Schmidt 1943, 24–26, fig. 11).

In such a diagram, we set the base great circle, gC, as the base diameter of the analemma and we depict, in gray, the mapping of the greatest of the parallel circles, gpC, and, perpendicular to this, the axis of the parallels, although neither of these plays a role in the text or the diagram of the proposition itself (Figure 34). The orientation of the parallel circles to the base circle gC can described by the angle between the axis of the parallels and gC, say φ. Then, the two parallel circles, pC_1 and pC_2, will be depicted as parallel to gpC, and perpendicular to the axis – such that pC_1 always meets gC at the analemma circle and pC_2 lies between pC_1 and gpC. In such a configuration, the diameters of the great circles that are tangent to pC_2

---

[74] The necessity for the claim that $pC_3 < pC_1$ is explained below (p. 293).

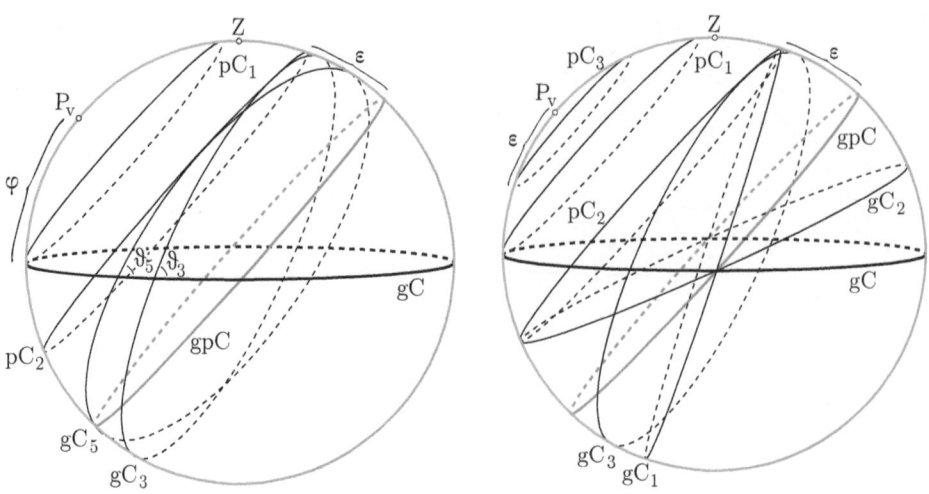

Figure 33: Perspective diagrams, *Sph.* II.22, (left) claim (e), and (right) claim (f).

at the midpoints of its greater and lesser segments, will be two diameters making the same angle, say ε, with gC, namely gC_1 and gC_2. Hence, the angle ε can be taken as determining the position of pC_2, since it measures the great-arc distance between gpC and pC_2. Although such a parallel circle is not mentioned in the text, for the purposes of illustrating the bounds, we may introduce another parallel circle, pC_4, passing through the zenith Z. Then, on such an analemma diagram, the claim that the pole Z of base circle gC must be between pC_1 and pC_2 is equivalent to the statement that pC_4 must be between pC_1 and pC_2. That is,

$$\varphi < 90° - \varphi < 90° - \varepsilon.$$

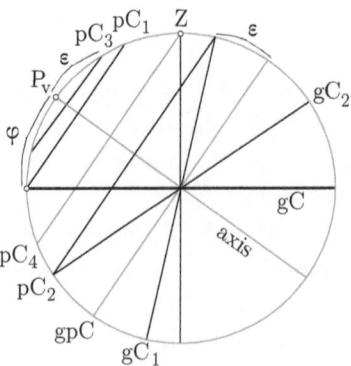

Figure 34: Analemma figure for the configuration of *Sph.* II.22 and II.23.

Since the overall interval of φ is $(0°, 90°)$, however, we see that $\varphi < 90° - \varphi$ implies that $\varphi < 45°$. Moreover, since ε is also in the interval $(0°, 90°)$, it is clear that $90° - \varphi < 90° - \varepsilon$ implies $\varepsilon < \varphi$. Indeed, a geometric argument for a configuration that is mathematically equivalent to this claim is made in the *construction* of Sph. II.22. Hence, the claims of Sph. II.22 are asserted to hold when φ is in the open interval

$$\varepsilon < \varphi < 90° - \varphi, \text{ or } \varepsilon < \varphi < 45°.$$

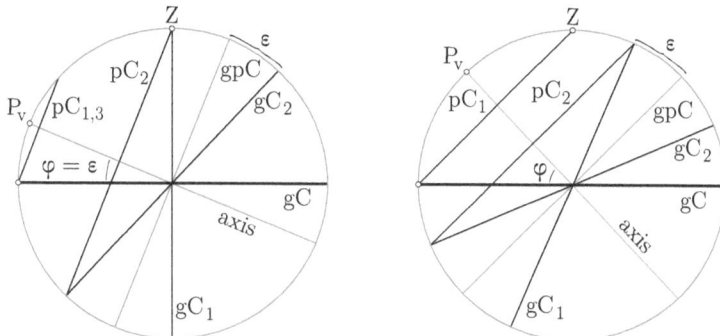

Figure 35: Bounds for *Sph.* II.22 and II.23, analemma diagrams: (left) lower bound $\varphi = \varepsilon$, (right) upper bound of the text $\varphi = 90° - \varphi = 45°$.

On the analemma, it can be seen that the lower bound occurs when φ is set such that Z is an endpoint of pC_1 and the upper bound occurs when φ is set such that Z is an endpoint of pC_2 (Figure 35 (left) and (right)). Indeed, as seen in the diagram, at the lower bound, when $\varphi = \varepsilon$, the angle between gC_1 and gC is 90°, so that the great circles are no longer inclined (Figure 35 (left)). At the upper bound assumed by the text, however, when $\varphi = 90° - \varphi = 45°$, the claims (a)–(f) remain true, and they continue to be true until $\varphi = 90° - \varepsilon$ (Schmidt 1943, 25–26). Indeed, when $\varphi = 90° - \varepsilon$, point Z is the endpoint of a diameter of a parallel circle, say pC_3, that is the same great-arc distance from the pole of the parallels as pC_1 and pC_2 are from gpC, namely ε (Figure 36). In such a configuration, gC_2 coincides with gC, and the angle between them is 0°, so that there is no inclination. In the analemma diagram, we can see that the claims of the theorem hold true as long as $pC_3 < pC_1 < pC_2$ (Figure 34). That is, although the proposition is true as it is stated, geometrically, claims (a)–(f) are also true in the larger open interval

$$\varepsilon < \varphi < 90° - \varepsilon.$$

Hence, for the claims asserted in *Sph.* II.22, the geometrical bounds for angle φ can be stated as a lower bound of $\varphi = \varepsilon$ and an upper bound of $\varphi = 90° - \varepsilon$.

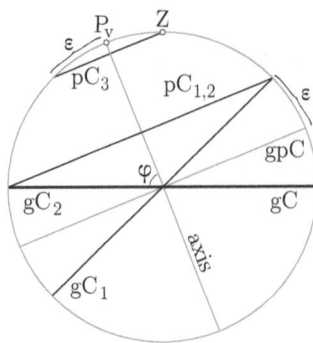

Figure 36: Bounds for *Sph.* II.22 and II.23, analemma diagram: geometrical upper bound $\varphi = 90° - \varepsilon$.

These bounds also help us understand the claim made in (f), above, that the poles of the various positions of gC_n lie on a parallel circle that is smaller than pC_1, say pC_3 (Figure 33 (right); Figure 34; and Figure 36). Since the gC_ns are tangent to pC_2, these great circles will form an angle with the greatest of the parallels, gpC, that can be characterized by the great-arc distance between pC_2 and gpC, namely ε. Hence, the pole of the parallels, P_v, and the poles of the gC_ns – being perpendicular to the planes of the great circles at the same point, the center of the sphere – will be separated by the same the great-arc distance, again ε. Then, as can be seen from the analemma diagram, in the interval treated by *Sph.* II.22, when φ is in the interval $\varepsilon < \varphi < 90° - \varepsilon$, the small circle of the poles, pC_3, will always be smaller than pC_1, because at the lower bound of $\varphi = \varepsilon$ it coincides with pC_1, and when $\varepsilon < \varphi$, it is clear that $pC_3 < pC_1$.

Although *Sph.* II.22 and II.23 are expressed in purely geometric terms, their primary purpose was almost certainly astronomical. Indeed, based on the use of the diagrams presented here to articulate these propositions, the reader will have already surmised how they can be interpreted from the perspective of spherical astronomy. If we set gC as the horizon, Z as the zenith, P_v as the visible pole determined by terrestrial latitude φ, gpC as the celestial equator, pC_1 as the v-circle, and pC_2 as the upper tropic determined by the obliquity of the ecliptic, ε, then the midpoints of the arcs of pC_2 above and below the horizon lie on the meridian, and the gC_ns can be understood to be various instantaneous positions of the ecliptic as the celestial sphere rotates throughout a day (Schmidt 1943, 24–26). In particular, gC_1 is the position of the ecliptic at midday, gC_2 its position at midnight, gC_3 and gC_4 are two of its positions at equal intervals before and after midday, and gC_3 and gC_5 are two of its positions such that gC_5 is farther from midday than gC_3. Under such an interpretation, *Sph.* II.22 and II.23 concern the changing angles that the ecliptic makes with the horizon throughout the course of a day. In particular, under the assumption that the terrestrial latitude is greater than the obliquity of the ecliptic and less than 45°, then *Sph.* II.22 shows that,

(a) the ecliptic is always inclined on the horizon,

(b) the angle of inclination is greatest at midday,

(c) the angle of inclination is least at midnight,

(d) at equal intervals from midday or midnight, the angle of inclination is equal,

(e) angles of inclination monotonically decrease as the ecliptic is taken farther from midday, and

(f) the pole of the ecliptic traces out a δ-circle that is smaller than the v-circle, namely the $v\varepsilon$-circle.

In fact, as shown above, these claims all continue to hold so long as the terrestrial latitude is greater than ε and less than $90° - \varepsilon$.

In this vein, *Sph.* II.23 shows that if the lengths of the arcs of ecliptic at two positions between the upper tropic and the horizon are equal, then those positions of the ecliptic stand on the horizon at equal angles of inclination.

Sph. II.22 (Theorem)

This theorem appears to be a significant goal of *Spherics* II, since it is based on a number of the key theorems in the book and, although used in the following theorem, does not play any role in *Spherics* III. Hence, the purpose for developing this theorem must have been its relevance to spherical astronomy, as discussed above.

The bounds for this proposition and the next, as discussed above in the introductory Commentary to both of these theorems, are handled in this proposition by placing the upper pole of $\mathbf{gpC}(ABG)$ – that is, point K – on the arc of the great circle, $\mathbf{gC}(AL)$, that passes through the upper pole of the parallels, point L, such that K falls between $\mathbf{pC}(AD)$ and $\mathbf{pC}(EZHQ)$. This is done as part of the *construction*, before $\mathbf{pC}(FCUW)$ is introduced. This has the effect of setting the upper bound as $\varphi = 90° - \varphi$, as discussed above, because then point K would coincide with point Z on $\mathbf{pC}(EZHQ)$ and be beyond the range treated by the theorem. The *construction* then proceeds with an argument that if we cut off a quadrant from Z on $\mathbf{gC}(AL)$ in the direction of K, it will be less than $\mathbf{Arc}(ZA)$. This has the effect of specifying the other bound, namely $\varepsilon < \varphi$, because when $\varphi = \varepsilon$ point F would coincide with point A on $\mathbf{pC}(AD)$, and hence be beyond the range allowed by the theorem. That is, from an astronomical perspective, the argument in the extended *construction* has the effect of bounding the terrestrial latitude in the open interval $\varepsilon < \varphi < 45°$, which is the same bound as found in *Phen.* 14. This astronomical background may provide a historical explanation for the stated bound shown in the text. Indeed, even much later, in the 2nd century CE, the standard list of latitudes that Ptolemy used included only two that are at or above the limit of $\varphi = 45°$, and in earlier schemes – such as those given in P.Hebeh 27 or P.Louvre 2388 Ro + P.Louvre 2329 Ro (formerly P.Paris 1) – the northernmost latitude is well below this bound (Neugebauer 1975, 706, 725–727). Hence, if Theodosios's intention was to address the inclinations of the ecliptic on horizons at which people in his general vicinity lived, these could all be handled by the single configuration and proposition that he provides.

Nevertheless, we can also think of some more mathematical reasons why Theodosios may have chosen to express the situation in the way that he did. Given the methods of ancient geometry, in order to explicitly handle a bound of $\varepsilon < \varphi < 90° - \varepsilon$ one might separate the current theorem into two, one merely a lemma to the other, and both based on essentially the same diagram. First, it would have to be shown that the poles of the great circles tangent to $\mathbf{pC}(EZHQ)$ fall on circle $\mathbf{pC}(FCUW)$, which is parallel to $\mathbf{pC}(AD)$, independently of the placement of the pole of $\mathbf{gpC}(ABG)$. This would be a separate theorem showing Part 1 of the current theorem, in a somewhat different way. Then, having shown this, in a separate proposition, the pole of $\mathbf{gpC}(ABG)$, point K, could be placed in between the two parallels $\mathbf{pC}(EZHQ)$ and $\mathbf{pC}(FCUW)$, and the rest of the theorem would be more or less the same. In the conventions of Greek geometry, however, each of these theorems would have to be fully enunciated and set out with a complete diagram, and such a revision would result in a significantly longer text than simply splitting the current theorem into its two parts.

Another possibility is that Theodosios chose to simply show the most difficult case, assuming that the other cases would be clear to the reader, as is often done in Greek geometrical texts. According to this interpretation of the situation, we would take the primary condition to be that point K is inside $\mathbf{pC}(EZHQ)$, which is equivalent to $90° - \varphi < 90° - \varepsilon$, which, where φ and ε are each in the range $(0°, 90°)$, implies that $\varepsilon < \varphi$. Under such an assumption, we have three possibilities: (1) point K falls outside $\mathbf{pC}(AD)$, or $\varphi < 90° - \varphi < 90° - \varepsilon$, (2) point K coincides with point D, or $\varphi = 90° - \varphi = 45° < 90° - \varepsilon$, and (3) point K falls inside $\mathbf{pC}(AD)$, or $90° - \varphi < \varphi < 90° - \varepsilon$. The first part of the *demonstation*, Part 1, shows that if case (1) is assumed, then $\mathbf{pC}(FV) < \mathbf{pC}(AD)$, or $\varepsilon < \varphi$. For the other two cases we may reason as follows. For (2), if K coincides with D, then $\mathbf{Arc}(ALD)$ is a quadrant, hence a quadrant from Z to F will fall inside $\mathbf{pC}(AD)$, so that $\mathbf{pC}(FV) < \mathbf{pC}(AD)$. In the case of (3), if K is inside **pC**(*AD*) then **Arc**(*ALD*) is greater than a quadrant so that, since **Arc**(*ZDF*) is a quadrant, again **pC**(*FV*) < **pC**(*AD*). That is, so long as *K* is inside **pC**(*EZHQ*), it can be shown that **pC**(*FV*) < **pC**(*AD*).

Because this theorem involves so many assertions, its structure is more complicated than usual. In the *enunciation*, great circle **gC**(*ABG*) is set out tangent to parallel circle **pC**(*AD*), and intersecting a larger parallel circle **pC**(*EZHQ*); then in the *specification* the following claims are made (Figure 32 and 33): (a) great circles **gC**(*MNX*), **gC**(*BZG*), **gC**(*OPR*), **gC**(*ST*), **gC**(*UQ*) that are tangent to **pC**(*EZHQ*) will all be inclined on **gC**(*ABG*), (b) the tangent great circle at the midpoint of the greater arc of **pC**(*EZHQ*), namely **gC**(*BZG*), will be the most upright, (c) the tangent great circle at the midpoint of the lesser arc of **pC**(*EZHQ*), namely **gC**(*UQ*), will be least upright, (d) those that are tangent at points equally distant from the midpoints, namely **gC**(*MNX*) and **gC**(*OPR*), will be similarly inclined, (e) those whose point of tangency are progressively farther from the midpoint of the greater arc, such as **gC**(*ST*) farther than **gC**(*OPR*), will be monotonically more and more inclined, and (f) the poles of all of the tangent great circles lie on a single parallel circle, later constructed as **pC**(*FCUW*), and such that **pC**(*FCUW*) < **pC**(*AD*). The *demonstrations* of these claims, however, are given in a different order. Indeed, the *construction* introduces some new objects – in particular, the great circle through the midpoints, **gC**(*AL*), and small circle **pC**(*FCUW*), as mentioned. Moreover, the *construction* contains a deductive argument that **pC**(*FCUW*) < **pC**(*AD*), which is made before the circle is actually drawn. We can understand the placement of this argument in the *construction* to indicate that Theodosios understands the relative size of this new circle to be one of the overall premises of the theorem, although it is not independent of the others. Following this there is a *demonstration* that shows the first part of (f), which we call Part 1 in the Translation. Then, a secondary *specification* sets out, once again, the remaining claims. After the *demonstration* of (b), another auxiliary *specification* sets out claim (c). Finally, (a) is not demonstrated separately, but follows from the fact that in the other arguments each tangent great circle is shown to be inclined to **gC**(*BZG*), so that (a) simply results from all of the arguments taken together, and is asserted in the conclusion to all of the *demonstrations* that corresponds to the first *specification*.

As in *Sph.* II.19, the *construction* in this proposition involves some repeated constructions. Initially, **gC**(*AL*) is drawn through points *A* and *L*, after which it is shown that this circle "filled out," προσαναπληρούμενος, will pass through *K* and then again through *Z* and *Q* and these further constructions are again made by using one of the verbs used in *Sph.* II.19 (ἔρχεσθαι). As before, the text gives a sense of the practical aspect of drawing the diagram, because from a mathematical perspective it would suffice to simply show that points K, Z, and Q lie on $\mathbf{gC}(AL)$ and assign them names.

Near the beginning of Part 2.1, the text asserts that $\mathbf{Arc}(IV) = \mathbf{Arc}(FW)$, because they are vertical arcs. This is based on the general claim that two great circles through the pole of a small circle cut the small circle into two vertically opposite pairs of equal arcs. The argument for this can be fleshed out a little. Since $\mathbf{gC}(ILW)$ and $\mathbf{gC}(VLF)$ both pass through pole L of $\mathbf{sC}(FWVI)$, by *Sph.* I.15, they will both bisect it perpendicularly. Hence, lines FV and WI – not drawn in the diagram, will pass through the center of $\mathbf{sC}(FWVI)$ and make vertical angles equal, *Elem.* I.15. Hence, by *Elem.* III.26, $\mathbf{Arc}(IV) = \mathbf{Arc}(FW)$. A somewhat longer, but geometrically related argument, is given in Scholium 294 (Czinczenheim 2000, 415). In fact, in *Mov. Sph.* 10 we find the same geometrical configuration, and from it the same inference.

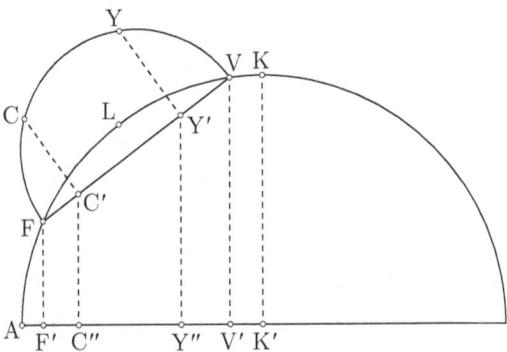

Figure 37: Analemma diagram, lemmas to *Sph.* II.22.

In Parts 2.1, 2.2, and 2.4, the argument makes claims about the relative magnitudes of perpendiculars dropped from different points of $\mathbf{pC}(FCYVW)$ onto the plane of $\mathbf{gC}(ABG)$. None of these claims are justified in the text, but they all follow from considerations that can be based on a single analemma diagram (Figure 37).[75] The text asserts that, of perpendiculars dropped to the plane of $\mathbf{gC}(ABG)$, (Part 2.1) that from point C is greater than that from point F, (Part 2.2) that from point V is greater than that from point Y, and (Part 2.4) that from point Y is greater than that from point C. If we take the plane of $\mathbf{gC}(AFLVK)$, through pole L of $\mathbf{pC}(FCYVW)$, as the analemma plane, then the points of $\mathbf{pC}(FCYVW)$ can be represented in the usual way by folding this circle into the analemma plane. Then, as in the solid diagram, point C will be between points F and

---

[75] Figure 37 is drawn within the geometrical upper bound of $\varphi = 90° - \varepsilon$. For the texts upper bound of $\varphi = 90° - \varphi = 45°$, the slope of line FV will be more pronounced and the inequalities discussed here will be greater, and in the same direction.

Y, and point *Y* will be between points *C* and *V*. The locations of these points on the solid sphere are found by mapping these points to their orthogonal projections on the analemma plane, such that $F \mapsto F$, $C \mapsto C'$, $Y \mapsto Y'$, and $V \mapsto V$.[76] Then the perpendicular from *F* is *FF'*, that from *C* is *C'C''*, that from *Y* is *Y'Y''*, and that from *V* is *VV'*. Then, since line *FV* lies in the quadrant *AK*, an indirect argument can be used to show that $FF' < C'C'' < Y'Y'' < VV'$, since otherwise, at least one of points *F* or *V* must fall beyond *K*.

Basis: *Elem.* I.post.3 (spherical, transfer), I.c.n.1, I.c.n.2, I.c.n.5, I.15 (by implication), I.33, III.def.11, III.26, III.27, XI.16, *Sph.* I.def.5, I.15, I.16, I.20, I.21, II.2, II.5, II.9, II.10, II.12, II.21.

Usage: *Sph.* II.23. A proposition like this was used in *Phen.* 7.[77]

Diagram: The diagram is a planar representation of global objects, containing only intrinsic objects – and one of the best of its kind in the *Spherics*. It uses all of the conventions developed so far – intersections on the spherical surface are encoded with intersections between circles and circular arcs in the diagram, parallel circles on the sphere are encoded with concentric circles, tangent circles on the sphere with tangent circles in the plane, different hemispheres of the spherical surface are depicted by the regions inside and outside a base circle in the figure, and so on. There are no false crossings in the diagram, so that it almost completely displays the topology of this rather involved configuration. The exception is that the circular arcs of circle *QU* are not shown meeting points *B* and *G* in opposite directions, as they in fact do.

Textual Comments: In Part 2.1, the point I ($\varnothing$) is used in two separate clauses. This letter-name is not commonly used in Greek mathematical texts,[78] and the point itself serves no role in the argument presented here. Hence, as noted by both Heiberg (1927, 104) and Czinczenheim (2000, 127–128) the two phrases that contain I should be struck from the text, and point I ($\varnothing$) should be removed from the diagram.

Comparison with Arabic: In Thābit's version this proposition is *Sph.*$_A$ II.21. The presence of I ($\varnothing$) in the Greek manuscripts from which

---

[76] For a discussion of these constructions on the analemma, see Sidoli (2020b, 36–40).

[77] Berggren and Thomas (1996, 25–24) note that the conditions of this proposition are closely related to the conditions of a proposition, or lemma, that is required in the two principal versions of *Phen.* 14, which they call the "Inclined Arcs Theorem." Indeed, a proposition like *Sph.* II.22 was used in an alternative proof for *Phen.* 14 (Menge 1916, 122–125). Schmidt (1943, 8) notes this proposition as used in "E 22" (denoting a *Phen.* 22), which must be a typographic error.

[78] I is sometimes, but not always, used in Archimedes' writings (Heiberg and Stamatis 1969–1977). There, however, when it is used, it is introduced alphabetically, between Θ and K, whereas in *Sph.* II.22 its introduction follows that of ᛌ.

that translation was made seems to have caused some confusion in the Arabic transmission, so that the argument is somewhat different, and, in particular, a claim made about the elevation of point X (*C*) in the Greek, is, instead, made about point ﺱ (I, Ø) in the Arabic (Kunitzsch and Lorch 2010, 194–195, 392 n. 8). Although the claim made in the Arabic text is not incorrect, it is also not necessary. Hence, as argued above, we believe that all references to point Ø (I, ﺱ) should be removed, and that the claims about the elevation of the various points can all be justified by analemma considerations, as explained above.

Sph. II.23 (Theorem)

This theorem provides another condition for part of the previous theorem and is demonstrated using a simpler version of the same diagram. In fact, it shows another condition that will lead to the configuration treated in *Sph.* II.22 Part 2.3, or what we called (d) in discussing the claims of that proposition. This proposition applies some basic properties of the circle and some of the core propositions of the *Spherics* to show that the conditions of this proposition imply the conditions of of *Sph.* II.22 Part 2.3, so that the result follows directly. Hence, *Sph.* II.23 can be regarded as an addendum to *Sph.* II.22.

As with *Sph.* II.10, II.13, and II.16, these two propositions are marked as belonging together by the distribution and organization of the letter-names in both the text and the diagram. Letter-names that mark the same points in both propositions remain the same, while those that mark points in *Sph.* II.22 that are unused in *Sph.* II.23 are repurposed to name different points.

The structure of this proposition is shortened because of its relationship with the previous proposition. Both the *enunciation* and the *exposition* are highly abbreviated and simply assume the same objects, with the same letter-names, as set out in *Sph.* II.22.

Basis: *Elem.* I.post.3 (spherical), I.c.n.3, III.def.11, III.27, III.28, III.29, V.15, *Sph.* I.def.5, I.15, I.20, I.21, II.2, II.5, II.9, II.10, II.12, II.22.

Usage: Unused in the *Spherics*.

Diagram: The diagram is a planar representation of global objects, containing only intrinsic·objects. It is a slight modification and simplification of the diagram for the preceding proposition.

Textual Comments: As noted above, the letter-names are introduced non-alphabetically, such that most of them refer to the same objects in this proposition as they did in *Sph.* II.22. The letters Ξ, O, Σ, T, and Υ name completely different points, whereas B and Γ name the same circle, but different points on it. As usual, this can be taken as an indication that these two theorems form a unified treatment of a single subject.

The word in the *enunciation* and *exposition* that we have translated by "junction" is σύνδεσμος, which by Ptolemy's time had become the technical term in mathematical astronomy for a node – that is, one of the intersections of the ecliptic with the plane of the primary circle of the lunar, or a planetary, model. Indeed, both Ver Eecke (1927, 78 n. 1) and Czinczenheim (2000, 803) have taken the term, in a sort of generalization of this technical sense, as the intersection of two great circles. Since, however, in the *exposition*, points N and P are set as the letter-names of the "junctions," this creates problems for their reading, because N and P are not the intersections of two great circles. Hence, Ver Eecke (1927, 78) added the expression "jusqu'aux points de contact" and again "jusqu'aux" with no justification in the text, and Czinczenheim (2000, 130) simply removed the naming of points N and P as an interpolation. It should be noted that Heiberg (1927, 109) simply took the text as received and translated with "a punctis coniunctionis" and "a punctis coniunctionis N, Π," with no implication that we must read the expression to mean node in the technical sense.

Indeed, it is not clear that σύνδεσμος had the technical sense of "node" in early Greek writing on astronomical subjects and we have not found it so used in any writing prior to the so-called *Art of Eudoxos* papyrus (P.Louvre 2388 Ro + P.Louvre 2329 Ro, formerly P.Paris 1),[79] dated to around the mid-2nd century BCE (Blass 1887, 23; Tannery 1893, 292). Furthermore, we have not found any instance in which it is used in the here requisite sense of a general intersection of two great circles. Moreover, the fact that σύνδεσμος is a direct translation of the technical term *kiṣ(a)ru*, "knot," used for lunar nodes in the Babylonian astronomical procedure texts (Ossendrijver 2012, 596), indicates that it probably did not derive from a Greek geometrical tradition, and hence never had the sense of a general intersection of two great circles. For all of these reasons, we have understood Theodosios's use of σύνδεσμος in the non-technical sense, not as a node, but simply a junction. We believe that Heiberg understood the term in the same way.

Comparison with Arabic: In Thābit's version this proposition is *Sph.~A~* II.22. The word σύνδεσμος was translated into Arabic by using the normal technical term for node, عقدة, but within the context of an explanation, not found in the Greek, that makes it clear that here we must understand the word in a different sense (Kunitzsch and Lorch 2010, 198–200). The passage in question reads, "The arc that is produced in what is between the places of the node, namely in what is between the locations of the touching of the circles and the intersection of the circles." It is likely

---

[79] We thank Alexander Jones for pointing out this usage in the papyrus.

that Thābit was concerned that the reader would take this term in its later technical sense and wanted to warn against this tendency.

Comparison with Latin: Gerard's Latin translation follows the Arabic, translating عقدة with *nodus* (Kunitzsch and Lorch 2010, 199–201).

Spherics III

The final book of the treatise continues to develop topics that have an underlying astronomical theme; however, there is no astronomical terminology and motion is, again, never mentioned. Although a number of propositions, *Sph.* III.1–III.4, III.10, do not have a clear astronomical interpretation, for the most part they can be understood as lemmas to the theorems that follow. A possible exception to this is *Sph.* III.3, which is mathematically equivalent to a special case of one of Menelaos's later congruence theorems, but which is used by Theodosios simply as a lemma, without developing any of its deeper implications. On the other hand, the geometric claims established in *Sph.* III.4 and III.10 are so specific that it is unlikely that they would have been explored at all except for their usefulness in the theorems that follow.

Following the opening lemmatic theorems, *Sph.* III.5–III.8 (Figure 38 (left)) deal with the arcs cut off from both the greatest circle of a bundle of parallels and another great circle that is oblique to the parallels by the great circles that are treated in *Sph.* II.10 and II.13 – namely, those that are either through the poles of the bundle of parallel circles or tangent to the same parallel. *Sph.* III.9 is an extension of the claim made in *Sph.* III.6, but is proved using a new method that is also used in *Sph.* III.10. The next group of theorems, *Sph.* III.10–III.12, of which III.10 is a lemma, deal with the same arcs of the greatest of the parallels and the oblique great circle, but now stating ratio inequalities that hold between these arcs and the diameters of the sphere and the parallels tangent to the oblique great circle. *Sph.* III.13 also deals with these arcs, now taken on either side of the intersection of the greatest of the parallels and the oblique great circle. Finally, *Sph.* III.14 treats the relative magnitudes of arcs of the parallel circles cut off between two great circles that are tangent to two different pairs of the parallels (Figure 38 (right)).

Spherics III, like the previous book, involves some steps that cannot be directly justified by the material in Euclid's *Elements* or previous propositions in the *Spherics*. A number of these are shown in scholia that accompany the medieval manuscripts, and some of which probably originate in ancient sources. Two of these lemmas are of particular interest. The first is a sort of *problem* that shows, in a general way, the possibility of constructing a magnitude that satisfies certain conditions and which is needed

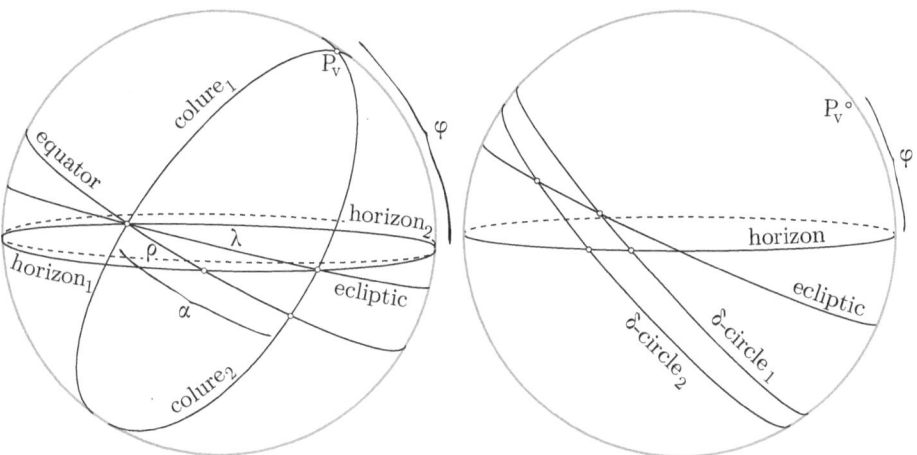

Figure 38: Diagrams for the spherical astronomy of *Spherics* III: (left) *Sph.* III.5–13, (right) *Sph.* III.14.

in a two-stage argument that involves proving a relation between two magnitudes, first under the assumption that they are commensurable, and then extending this to the case in which they are not. The second establishes a ratio inequality between two sides of a right triangle and two of its angles. The mathematical methods related to these lemmas are also found in works by Euclid, Aristarchos, and Archimedes, dealing with astronomy and the exact sciences. Hence, this book exhibits the use of mathematical methods that were not treated in the *Elements*, but which nevertheless appear to have been commonly employed in the exact sciences, at least by the early Hellenistic period. This is one of a number of indications that by the time this book begins, Theodosios assumes that his readers will understand that the subject matter is directly applicable to spherical astronomy.

Another indication comes from the way that certain mathematical objects are introduced into the discussion. In *Spherics* II, a bundle of parallel circles and their poles are introduced using indefinite forms, as though these might be one of any number of such bundles that are mathematically possible. In *Spherics* III, however, we consistently encounter *the* parallel circles and their pole, which implies that there is only one such bundle of parallel circles with their pole that is of any interest to the discussion – namely, the δ-circles and the visible pole. This is the same sort of terminology that was used in the *Phenomena*.

In this regard, the propositions *Sph.* III.5–III.13 can all be understood to deal with the relationships between, on the one hand, an arc of the ecliptic, λ, cut off between two colures, *colure₁* and *colure₂*, and two configurations of the horizon, *horizon₁* and *horizon₂*, and, on the other hand, the arc of the celestial equator cut off between the colures, α, and that cut off between the two configurations of the horizon, ϱ (Figure 38 (left)). That is, α and ϱ can be understood as determined by selecting some λ and then projecting this onto the equator by the colures and the positions of the local horizon, $\alpha(\lambda)$ and $\varrho(\lambda, \varphi)$. Hence, these propositions can all be understood as dealing with the rising and setting times of arcs of the ecliptic on both the upright and the inclined sphere. Finally, *Sph.* III.14 compares arcs of two δ-circles, δ-*circle*₁ and δ-*circle*₂, cut off between the horizon and the ecliptic, which is related to *Phen.* 4 and deals with the successive risings and settings of points on the ecliptic (Figure 38 (right)). The details of the claims made in these propositions will be discussed below.

Here we will merely remark that the overall topic of this book is clearly the rising and setting of the ecliptic, considered both arc-by-arc and point-by-point. Again, however, no explicit mention is made of any of these topics and the concept of motion is once again handled by considering different orientations of the great circle that can be interpreted as the horizon, or comparing different arcs of the circles that can be understood as the δ-circles.

Comparison of lines produced from a segment standing on a circle to the circumference of the circle (*Sph.* III.1, III.2)

The opening pair of theorems act as lemmas for the first block of propositions that treat a topic of spherical astronomy. *Sph.* III.1, III.2, each consisting of two cases, compare the lengths of certain lines joined from a point above a circle to its circumference, and, hence, will give a means of comparing the great-circle arcs that subtend such lines.

In particular, *Sph.* III.1 Case 1 shows that where a segment of a circle, *Seg* ≤ 180°, stands perpendicular on the chord of a base circle, C, dividing C into two arcs, $\alpha_1 > \alpha_2$ (not labeled), and some point P divides *Seg* into two arcs, $\beta_1 \neq \beta_2$, and a perpendicular is dropped from P onto the chord, and the diameter of the circle through the foot of this perpendicular is produced, and P is joined to the circumference of the circle with line b_1 subtending β_1, b_2 subtending β_2, b_3 meeting the far endpoint of the diameter, b_4 between b_1 and b_3, and b_5 between b_3 and b_2, respectively, then, of the lines b_n joining P to points on the larger arc of circle C, α_1 (not labeled), the following claims hold (Figure 39 (left)):

(a) b_1 is least,

(b) b_4 monotonically increases from b_1 towards b_3,

(c) b_3 is greatest,

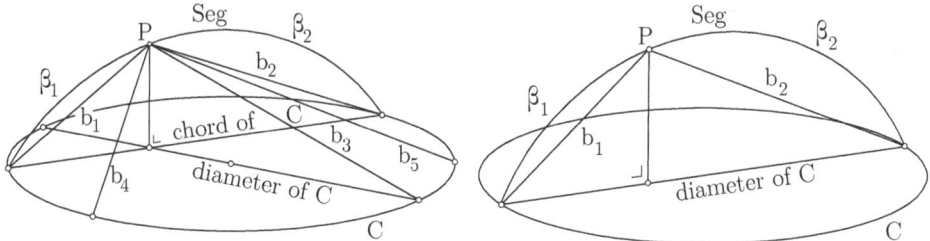

Figure 39: Perspective diagram, *Sph.* III.1: (left) Case 1, (right) Case 2.

(d) b_2 is locally least, $b_2 < b_5 < b_3$, and

(e) b_5 monotonically decreases from b_3 towards b_2.

Furthermore, *Sph.* III.1 Case 2 shows that if *Seg* is perpendicular on a diameter of *C*, so that $\alpha_1 = \alpha_2$ (not labeled), and, again, *P* is taken such that $\beta_1 \neq \beta_2$ (Figure 39 (right)), then

(f) b_1 is least, and

(g) b_2 is greatest.

Sph. III.2 Case 1 deals with the same configuration with the exception that now *Seg* $\leq 180°$ is inclined on circle *C* in the direction of the lesser arc cut off from *C* (Figure 40 (left)). Once again, where lines b_n join *P* with points on the greater arc of *C*, $\alpha_1 > \alpha_2$ (not labeled), the theorem shows that same statements (a)–(e) hold.

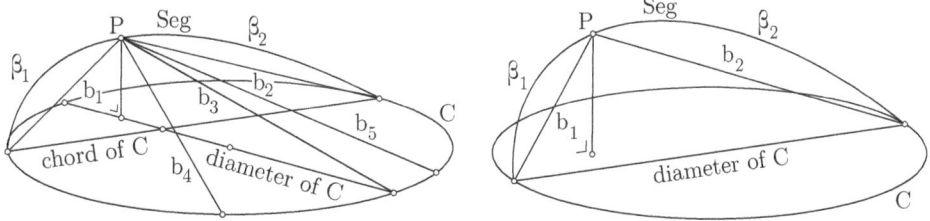

Figure 40: Perspective diagram, *Sph.* III.2: (left) Case 1, (right) Case 2.

It is important, however, to recognize, that, in the situation in which *Seg* stands on a chord but is inclined in the other direction, towards the greater segment of circle *C*, then the statements (a)–(e) may not hold, and, in particular, b_1 may not be the least line. This is because *Seg* may be close enough to *C* that some other least line joins point *P* to the greater arc of *C*, $\alpha_1 > \alpha_2$. This will become important when *Sph.* III.2 is used in

Sph. III.8, with respect to understanding the restrictions to the claims made in *Sph.* III.8 – particularly regarding its implications for spherical astronomy.

Finally, *Sph.* III.2 Case 2 shows that if *Seg* is inclined on a diameter of C, $\alpha_1 = \alpha_2$ (not labeled), and P is taken such that $\beta_1 \neq \beta_2$, then statement (f) holds (Figure 40 (right)).

Sph. III.1 (Theorem)

This is a theorem of solid geometry, which follows from an argument involving the solid geometry of *Elements* XI, and uses propositions of Euclid's plane geometry in the usual way. It makes no use of any previous propositions of the *Spherics*. This proposition functions as a lemma to the first block of theorems of *Spherics* III that treat a topic of spherical astronomy.

Because there are some discrepancies between the details of what is stated in the *enunciation* and shown in the proof, the structure of this proposition is somewhat involved. In particular, the assertions of the *enunciation* are more limited than what is actually shown. In terms of the statements in the introductory Commentary to both *Sph.* III.1 and III.2, above, the *enunciation* simply claims (a) for Case 1, and both (f) and (g) for Case 2. The *demonstration* of Case 1, however, shows (a), (b), (c), and then (d), from the proof of which (e) is asserted almost as a corollary. Then, Case 2 shows both (f) and (g) at the same time. Moreover, these differences cannot be explained by the way that the theorem is used in this treatise. In fact, this theorem is used three times in the rest of the book as follows: *Sph.* III.1 Case 1 (a) is used in both *Sph.* III.5 and III.6, whereas *Sph.* III.1 Case 1 (a) and (b) are used in *Sph.* III.7. There are no uses of the other claims of this theorem in the *Spherics*.

The *demonstration* for Case 2 requires, as a lemma, the claim that a perpendicular dropped from a point on a segment to its base divides the base unequally – or, in particular, **Arc**$(EZ) >$ **Arc**$(EB) \Rightarrow EZ > EB$. A marginal gloss in some of the manuscripts, Scholium 332, shows how this follows from elementary plane considerations (Czinczenheim 2000, 419). That is, since the arcs are unequal, the subtending lines are unequal, *Elem.* III.29, so that, say, **S**$(ED) >$ **S**(EB), so that by *Elem.* I.47, **S**$(EZ) +$ **S**$(ZD) >$ **S**$(EZ) +$ **S**(ZB), and since **S**(EZ) is common, **S**$(ZD) >$ **S**(ZB), so that $ZD > ZB$, by *Elem.* VI.20.cor.

In the course of the *construction*, a line is produced from point E to a particular arc of the base circle, *ABG*. We can regard this as an extension of *Elem.* I.post.1, where an arbitrary point, say L or M, is first taken on the arc in question and then *Elem.* I.post.1 is applied.

Basis: *Elem.* I.def.17, I.post.1, I.post.2, I.c.n.2 (inequalities), I.47, III.1, III.7, VI.20.cor., XI.def.3, XI.def.4, XI.11.

Usage: *Sph.* III.5, III.6, III.7. A proposition like this had been used in *Mov. Sph.* 7.

Diagram: This is a standard solid diagram, of the same sort as those accompanying *Sph.* II.11, II.12, and II.21. Despite the fact that the initial circles are displayed as full circles, the diagram conveys a sense of depth. As usual for this type of diagram, straight lines have false crossings among themselves and with one of the circular arcs.

Sph. III.2 (Theorem)

Once again, this is a proposition of solid geometry that relies on no previous propositions of the *Spherics*. Moreover, it is also a lemma to the first group of theorems with implications for spherical astronomy.

The difference between the claims made in the *enunciation* and those shown in the proof are even greater here than was the case in *Sph.* III.1. In terms of the statements introduced in the Commentary to both *Sph.* III.1 and III.2, the *enunciation* of *Sph.* III.2 makes a single statement combining both (a) and (f). Nevertheless, the proof Case 1, as in *Sph.* III.1, shows (a), (b), (c) and (d), the arguments for which also implies (e), which is simply asserted. The final sentence is a proof sketch of Case 2, asserting (f).

It seems that the *enunciation* of this proposition is modeled on that of the previous proposition. In fact, however, if this proposition were asserted to be taking place in a sphere, as, indeed, it will be used in the rest of the treatise, a number of issues would be resolved. These issues can be explained as follows.

In the *construction* it is asserted that perpendicular EZ, dropped from the point Z to the plane of the base circle, will fall such that foot Z will be within **Seg**(ADG). Now, since the plane of **Seg**(AEG) is inclined towards **C**(ABG), it is clear that Z must, indeed, be on the side of AG towards which **Seg**(AEG) is inclined, but it does not follow from the conditions for the segments, as asserted in the *enunciation*, that it will necessarily fall within **Seg**(ADG). That is, since both **Seg**$(ADG) \leq 180°$ and **Seg**$(AEG) \leq 180°$, but there is no necessary relationship between **Seg**(ADG) and **Seg**(AEG) themselves, it is possible that Z might also fall outside of **C**(ABG) but still in the direction of **Arc**(ADG). Nevertheless, with some minor changes, the argument in the proposition will still hold and all of the claims of this theorem are still demonstrably true. That said, since the *construction* actually asserts that foot Z will fall inside **Seg**(ADG), the argument seems to assume that point E will be somewhere inside a right cylinder passing through **C**(ABG), as it would be if the whole configuration were regarded as taking place in a sphere. Indeed, since this proposition is actually a lemma for configurations on the sphere, whenever it is it applied it will be the case that foot Z will fall within **Seg**(ADG).

Indeed, if we restrict the location of E to the surface of a sphere containing $\mathbf{C}(ABG)$ and $\mathbf{Seg}(AEG)$, then there is a proof for this theorem that relies directly on *Sph.* III.1, and which some readers may regard as simpler.

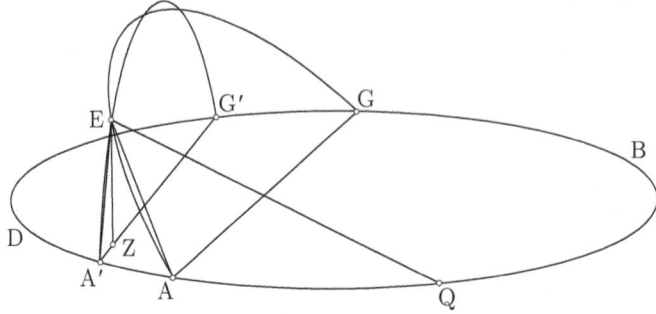

Figure 41: Diagram for an alternative proof of *Sph.* III.2, using III.1.

If $\mathbf{Seg}(AEG)$ inclines towards $\mathbf{Arc}(ADG)$, such that foot Z of perpendicular EZ falls inside $\mathbf{Seg}(ADG)$, as stated in the *construction*, then a parallel $A'G' \parallel AG$ can be drawn through Z in the plane of $\mathbf{C}(ABG)$ (Figure 41). In this way, a segment $\mathbf{Seg}(A'EG')$ that passes through points A', E, and G', in the plane passing through lines EZ and $A'ZG'$ will be perpendicular to $\mathbf{C}(ABG)$. Then, *Sph.* III.1 can be applied directly to show that $EA' < EA < EQ$, and all of the other claims made in the proof of *Sph.* III.2.

Basis: *Elem.* I.post.1, I.post.2, I.c.n.2 (inequalities), I.47, III.1, III.7, VI.20.cor., XI.def.3.

Usage: *Sph.* III.8.

Diagram: The diagram is the same as that for the previous proposition.

Textual Comments: Czinczenheim (2000, 140–141) regards the end of this proposition, starting from Case 1, Part 2, as spurious. Indeed, the final paragraph uses the 3rd person, future, passive form of δειχνύναι, the only time that this expression is used in the treatise. In any case, as it stands, the structure and content of this proposition is similar to that in the previous proposition, which reflects the mathematical situation.

Congruence in pairs of intersecting equal great-circle arcs (*Sph.* III.3)

This congruence theorem shows that if equal pairs of arcs, $\alpha_1 = \alpha_2$, $\beta_1 = \beta_2$, are cut off of two great circles, gC_1, gC_2, from one of their points of intersection, and if straight lines, d_1, d_2, are joined between the endpoints of the pairwise equal arcs, then the straight lines are equal. That is, in the stated configuration $\alpha_1 = \alpha_2, \beta_1 = \beta_2 \Rightarrow d_1 = d_2$ (Figure 42).

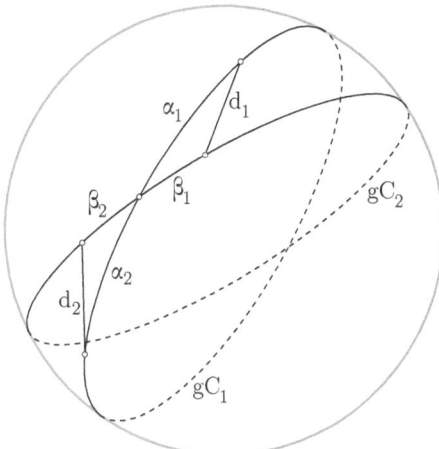

Figure 42: Perspective diagram, *Sph.* III.3.

From this it immediately follows that the great-circle arcs subtending d_1 and d_2 are also equal, *Elem.* III.28. Hence, *Sph.* III.3 can be understood as a sort of congruence theorem for great-circle arcs (Schmidt 1943, 17). In particular, this is, mathematically, a special case of side-angle-side congruence for spherical triangles, as in the conception of spherical figures later developed by Menelaos in his *Spherics*. Nevertheless, the underlying insight in Theodosios's text is different, because the idea of a spherical figure, as bounded by great circles, is never articulated and the congruence treated in this proposition is never shown or used in the general sense, but only in the special case of pairwise equal arcs about a pair of vertical angles on the sphere.

Sph. III.3 (Theorem)

This proposition, which is proved by construction in two cases, is primarily based on solid geometry, but also requires some of the core concepts from *Spherics* I. Although it might be viewed as an important theorem of spherical geometry, in this text it simply serves as a lemma to propositions that treat topics of interest to spherical astronomy.

The *demonstration* is divided into two cases, based on whether or not **Arc**(AE) = **Arc**(EB) are equal to **Arc**(GE) = **Arc**(ED). The first possibility is treated in Case 1, while the second possibility is treated in Case 2, by assuming that one pair, say **Arc**(AE) = **Arc**(EB), is the greater. This division into cases is handled in the text through the construction, and then instantiated in the diagram, insofar as **C**(AB), which is drawn about pole

E with distance EA, will either pass through G and D, or not. In Case 1 it is assumed to pass through G, while in Case 2, it is assumed to pass through a point H such that $\mathbf{Arc}(EH) > \mathbf{Arc}(EG)$. It should be noted that there is no loss of generality in this way of proceeding, because it is simply a constructive way of instantiating the assumption that one of the pairs of arcs is greater than the other pair.

In the second part of the *demonstration* for Case 2, the argument requires the same lemma as was required by *Sph.* II.11 and II.12, but now applied on opposite sides of the same segment, as opposed to two equal segments (Scholium 190; p. 262, above). Despite the fact that the wording of the justification is the same in all three propositions, a scholiast included a special lemma for this proposition, which is more complicated than is necessary, and only proves half of what is needed (Scholium 350; Czinczenheim 2000, 422).

Basis: *Elem.* I.def.15, I.def.16, I.def.17, I.post.1, I.post.3 (spherical), I.c.n.3, I.4, III.28, XI.3, XI.11, *Sph.* I.def.5, I.15, I.20.

Usage: *Sph.* III.6, III.8, III.13. A proposition like this was used in *Phen.* 12.[80]

Diagram: The diagram, like that of *Sph.* II.10, represents local objects that are both intrinsic and extrinsic, but it could be classified as either planar or solid. It is similar to the diagrams for *Sph.* I.1, I.7, and II.10 in visually conveying a sense of depth, but all of the objects could also be regarded as folded into the plane of the figure. There are number of false crossings between surface and solid objects.

Comparison of the arcs that certain parallel circles cut off from two great circles (*Sph.* III.4)

The configuration handled by *Sph.* III.4 is rather specific. Essentially, this proposition compares the magnitudes of the arcs that three parallel circles passing through the endpoints of two equal and contiguous great-circle arcs cut off from another great circle. In particular, the theorem deals with the situation in which there are two intersecting great circles, gC_1, gC_2, and equal arcs, $\alpha_1 = \alpha_2$, are cut off from say gC_1 on opposite sides of one of the points of intersection, and parallel circles, pC_1, pC_2, pC_3, are passed through the endpoints of α_1 and α_2, such that the plane of one of them, say pC_1, meets the intersection of the planes of gC_1 and gC_2 outside the sphere, say at P, and such that pC_1, pC_2 and pC_3 cut off two contiguous arcs, β_1,

---

[80] Schmidt (1943, 8) indicates that this theorem was also required by *Mov. Sph.* 10, but see the Commentary to *Sph.* II.22, above.

β_2, from gC_2, where β_1 is in the direction of the intersected plane of pC_1. Under these circumstances, *Sph.* III.4 shows that (Figure 43),

$$\alpha_1 = \alpha_2 > \beta_1, \beta_2 \Rightarrow \beta_1 < \beta_2.$$

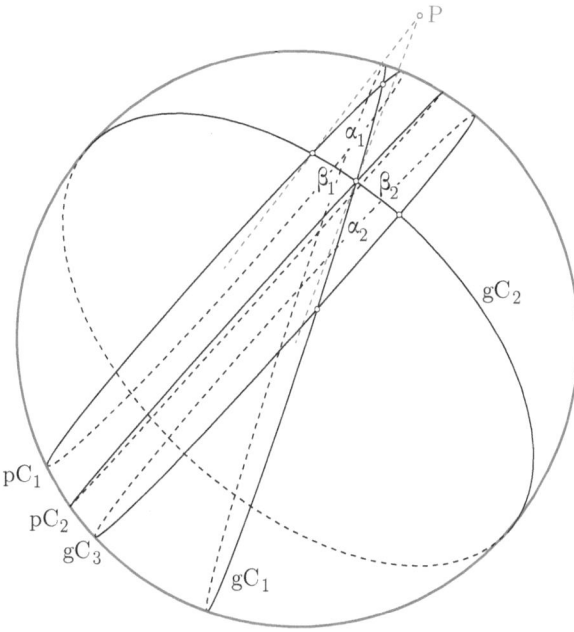

Figure 43: Perspective diagram, *Sph.* III.4.

If the pole of the parallel circle lies on gC_2, as it will when this proposition is applied in the *Spherics*, the condition that the intersection of the planes of gC_1 and gC_2 should meet the plane of one of the parallel circles, say pC_1, outside the sphere is geometrically equivalent to the assertion that (1) gC_1 and gC_2 should not be perpendicular to one another, and that (2) they should not meet one another at a rotation about the axis of the parallels of 90° from such a position – that is, the intersection of gC_1 and gC_2 should not lie in the plane of the greatest of the parallel circles, gpC (not shown). In the first excluded situation, the intersection of gC_1 and gC_2 would meet the plane of a parallel circle tangent to gC_1, say pC_4 (not shown), at the surface of the sphere itself, because pC_4 would be one of the two parallels tangent to gC_1, and only one parallel, say pC_5 (not shown) would pass through both of the endpoints of $\alpha_1 = \alpha_2$, so that there would be only one arc β cut off from gC_2, and the intersection of gC_1 and gC_2 would meet pC_5 inside the sphere. In the second excluded situation, the parallel circle drawn through the intersection would be gpC itself (that is, $pC_2 := gpC$), so that the intersection of gC_1 and gC_2 would lie in the plane of gpC and it would not meet the plane of any of the small parallel circles, such as pC_1 or pC_3. Indeed, in this situation, the two parallel circles drawn through the endpoints of $\alpha_1 = \alpha_2$ on gC_1 would be equal parallels and the two arcs cut off from gC_2 would be equal, $\beta_1 = \beta_2$.

Stated in such general terms, as is done in the *Spherics* itself, it may be difficult to understand the significance of this theorem. Hence, as usual, it can be useful to think of the implications that this proposition would have for spherical astronomy. If we regard gC_1 as representing the ecliptic, gpC as the equator, gC_2 as an oblique colure,[81] and pC_1, pC_2, and pC_3 as three δ-circles that cut off equal and contiguous arcs of the ecliptic, $\alpha_1 := \lambda_1$, $\alpha_2 := \lambda_2$, then *Sph.* III.4 asserts that these δ-circles will cut off arcs of the colure, $\beta_1 := \delta_1$, $\beta_2 := \delta_2$, such that $\delta_1 < \delta_2$, where δ_1 lies towards the circle of the tropic that is on the same side of the equator as the local intersection of the ecliptic and the oblique colure.

Sph. III.4 (Theorem)

The basis of this proposition is closely related to that of the previous theorem, insofar as it relies on the solid geometry of the *Elements* and the same material from *Spherics* I. Although *Sph.* III.4 has some implications for spherical astronomy, as specified above, its primary function in the *Spherics* is as a lemma to the following group of propositions.

The structure of this proposition is fairly standard, but exceptionally, for this treatise, it ends with a QED phrase – the only other occurrence of this phrase is in *Sph.* I.2.

The condition that $\mathbf{Arc}(AE) = \mathbf{Arc}(EB) > \mathbf{Arc}(DE)$ or $\mathbf{Arc}(EG)$ – above stated as $\alpha_1 = \alpha_2 > \beta_1, \beta_2$ – is introduced in the *construction* through the introduction of $\mathbf{sC}(AHBZ)$ about pole E with polar radius $AE = EG$ such that D and G fall on the same side of the spherical surface as pole E, between Z and H.

The final stage of the proof requires a lemma to the effect that if two parallel lines meet two points on the base of a segment of a circle that are equidistant from the endpoints of the base, and if acute and obtuse angles are cut off from each arc, the arc subtending the acute angle will be the lesser. An argument showing this is found among the marginal scholia. Scholium 362 reads as follows (Czinczenheim 2000, 424; Figure 44):

> For, let there be the aforementioned semicircle, $ABGDZE$, and let equals, AK, LE, be cut off from the diameter, and let parallels, KB, LD, be produced, and let angle AKB be acute, and ELD obtuse.
>
> I say that circumference DZE is greater than circumference AB.

---

[81] That is, neither the equinoctial nor solstitial colure.

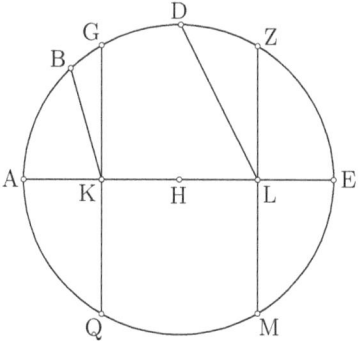

Figure 44: Scholium 362 to *Sph.* III.4.

For, let *KG* and *LZ* be produced from *K* and *L* on *AE* at an upright, and let the circle be filled in, and let *GK* and *LZ* be extended to points *Q* and *M*, and let the center of the circle, *H*, be taken. It is obvious that *GQ* and *ZM* are equally distant from the center and are equal.

Since some line through the center, *HK*, cuts some line not through the center, *GQ*, at an upright, it will also bisect it. Therefore, *GK* is equal to *KQ*. Then, similarly, also *ZL* to *LM*. But, also *GQ*, as a whole, is equal to *ZM*, as a whole, therefore *GK* is also equal to *ZL*. Therefore, the ⟨straight line⟩ from *A* to *G*, a base, is equal to the ⟨straight line⟩ from *E* to *Z*, a base, and circumference, *ZE*, is equal to circumference, *GA*. So that circumference *DZE* is greater than circumference *GA*, therefore much greater than *AB*.

It will be proven both if the aforementioned segment is less than a semicircle, and if it is greater.

Basis: *Elem.* I.def.11, I.def.12, I.def.15, I.def.16, I.def.17, I.post.2, I.post.3 (spherical), I.c.n.1, I.c.n.3, I.13, I.16, I.28, XI.def.3, XI.3, XI.16, XI.19, *Sph.* I.def.5, I.15, I.20.

Usage: *Sph.* III.5, III.7.

Diagram: The diagram is a solid representation of local objects, containing both intrinsic and extrinsic objects. It is similar to the diagrams for *Sph.* I.1, I.7, II.10, and III.1–III.3 in visually conveying a sense of depth. In fact, this diagram, along with that for *Sph.* III.11, comes fairly close to using techniques of perspective drawing, such as became common practice in the early modern period. As usual in perspective drawings, there are number of false crossings between the various objects.

Textual comments: The conjunction ἐπειδήπερ, which we translate with "because" is used only one other time in this text, in *Sph.* III.1. Although it may have been a later interpolation, and is marked as such by Heiberg (1927, 8) and Czinczenheim (2000, 57), it is contained in the oldest Greek manuscript, **A**, and was translated into Arabic (Kunitzsch and Lorch 2010, 26).

Comparison of the projections of equal and contiguous arcs of one great circle onto other great circles (*Sph.* III.5–III.8)

This group of theorems assumes (1) a bundle of parallel circles, including the greatest of them, and one of their poles, (2) a pencil of great circles that are either (2a) passing through the poles of the parallels or (2b) tangent to the same parallel, and (3) one more great circle that is oblique to the parallels, and, in the case of (2b), is tangent to a greater parallel circle than the pencil of great circles. The use of the terminology of "the parallels," and the "the pole of the parallels," with no further qualification – which is the same terminology as is used sometimes in Euclid's *Phenomena* – is a linguistic clue that we are dealing with propositions whose tacit subject matter is the celestial sphere. Nevertheless, since the language is still ostensibly geometrical, we will simulate that in this Commentary.

These theorems divide twice into two pairs. *Sph.* III.5 and III.6 deal with the configuration of (2a), while *Sph.* III.7 and III.8 deal with that of (2b). This division of the material corresponds to that between *Sph.* II.10 and II.13, and shows, again, that the configuration of a pencil of great circles tangent to a pair of equal parallel circles is understood as a geometrical extension of a pencil of great circles passing through the poles of the parallels. Furthermore, the first proposition of each of these pairs, *Sph.* III.5 and III.7, deals with arcs of these great circles, (2a), (2b), cut off between parallel circles, (1), while the second pair, *Sph.* III.6 and III.8 deals with arcs of the greatest of the parallels, (1), cut off between the great circles through the poles or tangent to the same parallel (2a), (2b). In each case, equal and contiguous arcs of the other oblique great circle, (3), are cut off by either the parallel circles, (1), or the related pencils of great circles, (2a), (2b). As is well known, these can all be regarded as projections of arcs of the oblique great circle, (3), onto either the related pencils of great circles, (2a), (2b), or the greatest of the parallels, (1) (Neugebauer 1975, 766).

In particular, in *Sph.* III.5 and III.6, we begin with the assumption of some base great circle, gC, on which lies the pole, P, of a bundle of parallel circles that are perpendicular to it, such that the greatest of the parallels, gpC is perpendicular to gC, as well as another great circle, gC_1, that is

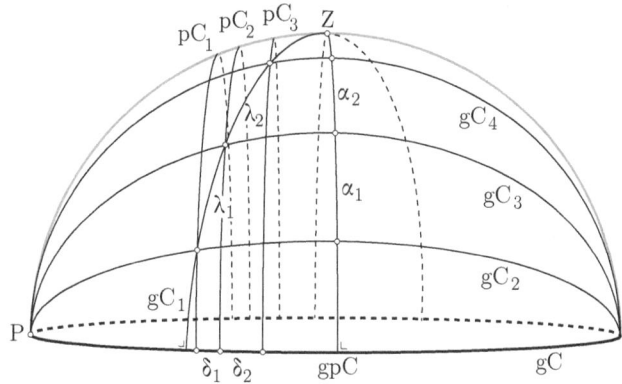

Figure 45: Perspective diagram, *Sph.* III.5 and III.6.

oblique to gpC and perpendicular to gC (Figure 45). Then, equal and contiguous arcs of gC_1, $\lambda_1 = \lambda_2$, are cut off between the greatest of the parallels and the intersection of gC and gC_1, and parallel circles, pC_1, pC_2, pC_3, are drawn through the endpoints of λ_1 and λ_2 cutting off δ_1 and δ_2 on gC, while great circles through the poles, gC_2, gC_3, gC_4, are also passed through the endpoints, cutting off α_1 and α_2 on gpC. Then, *Sph.* III.5 shows that arcs

$$\delta_1 < \delta_2$$

monotonically increase as the λs arc taken going from the point at which gC_1 meets gC towards gpC. Moreover, *Sph.* III.6 shows that arcs

$$\alpha_1 > \alpha_2$$

monotonically decrease in the same direction.

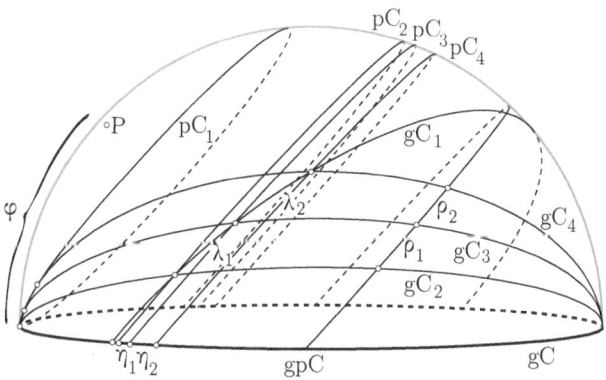

Figure 46: Perspective diagram, *Sph.* III.7 and III.8.

In *Sph.* III.7 and III.8, we begin, again, with the assumption of some base great circle, gC, but now the pole of the parallels, P, does not lie on gC, but rather gC is tangent to some parallel circle, pC_1, and the great circle gC_1 that is oblique to gpC is tangent to some other parallel circle that is greater than pC_1 (not shown), and meets gC at this point of tangency (Figure 46). Once again, equal and contiguous arcs of gC_1, $\lambda_1 = \lambda_2$, are cut off between gpC and the intersection of gC and gC_1, and parallel circles, pC_2, pC_3, pC_4, are drawn through the endpoints of λ_1 and λ_2 cutting off η_1 and η_2 on gC, while great circles tangent to pC_1, namely gC_2, gC_3, gC_4, are also passed through the endpoints, cutting off ϱ_1 and ϱ_2 on gpC. In this configuration, *Sph.* III.7 shows that arcs

$$\eta_1 < \eta_2$$

monotonically increase as the λs are taken going from the point at which gC_1 meets gC towards gpC. Again, *Sph.* III.8 shows that

$$\varrho_1 > \varrho_2$$

monotonically decrease in the same direction.

The distinction between *Sph.* III.5, III.6 and III.7, III.8 can be thought of as determined by the location of the pole, P, of the parallel circles relative to the base circle, gC. If the distance of this pole from gC along a great circle passing through the poles of gC is arc φ, then *Sph.* III.5 and III.6 deal with the situation in which $\varphi = 0°$, while *Sph.* III.7 and III.8 deal with φ in the open interval $0° < \varphi < 90° - \varepsilon$, where ε is the angle between gpC and gC_1. The upper bound is determined by the stipulation that gC_1 is tangent to a parallel circle that is greater than pC_1.[82]

As mentioned above, these propositions have a number of astronomical implications and they were probably written to demonstrate such astronomical claims (Schmidt 1943, 38–42; Neugebauer 1975, 799). As well as the terminology of the parallel circles mentioned above, the consistent use of a base circle, which is then referred to as the "initial," ἐξ ἀρχῆς, great circle is another linguistic marker that we are here dealing with an astronomical topic (Le Meur 2012, 185–188). Indeed, in each proposition we can set the base great circle, gC, as the horizon, the greatest of the parallels, gpC, as the celestial equator rotating about a pole, P, such that the parallel circles, pC_n, are δ-circles, the great circle gC_1, oblique to gpC, is the ecliptic, and the great circles through the pole or tangent to a parallel circle, gC_2, gC_3, gC_4, are other configurations of the horizon. In this way, it is clear that these propositions deal with the arcs of the horizon over which arcs of the

---

[82] This upper bound can, again, be seen directly in the analemma figure for the introductory Commentary to *Sph.* II.22 and II.23 (Figure 35 (right), p. 293).

ecliptic rise and set, and the arcs of the equator that rise and set in the same times as arcs of the ecliptic. The details of the claims demonstrated by these propositions will be explicated individually, below.

Here, however, it may be helpful to point out a number of symmetries in the assumed configurations that can be used to understand the astronomical implications of these theorems (Schmidt 1943, 40–41). If we consider the manuscript diagrams to represent a solid or armillary sphere, as was most likely the intention, we see that the horizon, $\mathbf{gC}(ABG)$, is the base circle inside which everything is drawn, the pole, A, U, or W, respectively, is located at the top of the diagram, and the celestial equator, $\mathbf{gC}(BZ)$, crosses the diagram horizontally. Furthermore, the ecliptic, $\mathbf{gC}(EZ)$, crosses the diagram diagonally, with the upper tropic point in the upper left and the lower tropic point in the lower right. Hence, the equal arcs of $\mathbf{gC}(EZ)$, $\mathbf{Arc}(HQ) = \mathbf{Arc}(QK)$ or $\mathbf{Arc}(QK) = \mathbf{Arc}(KL)$, are cut off in the upper left-hand portion of the diagrams. It must be recognized, however, that the pole, A, U, or W, can be either the north or the south pole. Hence, if the pole is taken to be the north pole, the diagrams and, indeed, the arguments of *Sph.* III.5–III.8, deal with arcs of the ecliptic in $\mathbf{Quad}(\text{Can} \rightarrow \text{Vir})$ setting over the western horizon, on the left. Whereas, if the pole is taken to be the south pole, they deal with arcs of the ecliptic in $\mathbf{Quad}(\text{Sag} \rightarrow \text{Lib})$ setting over western horizon. In discussing the astronomical implications of these theorems below, we will refer to this symmetry as the *polar symmetry*, and it holds for each proposition in *Sph.* III.5–III.8.

In *Sph.* III.5 and III.6, however, the pole of $\mathbf{gC}(BZ)$ is located on the base great circle, $\mathbf{gC}(ABG)$, which can be understood as the configuration of the upright sphere, $\varphi = 0°$. Hence, the geometry of the configuration is symmetrical on both sides of $\mathbf{gC}(BZ)$, so that the arguments apply just as well to the quadrant of the ecliptic, $\mathbf{gC}(EZ)$, in the lower right-hand portion of the figures. That is, when the pole is taken as the north pole, *Sph.* III.5 and III.6 also deal with the arcs of the ecliptic in $\mathbf{Quad}(\text{Lib} \rightarrow \text{Sag})$ setting over the western horizon, and when the pole is taken as the south pole, they also deal with $\mathbf{Quad}(\text{Vir} \rightarrow \text{Can})$ setting over the horizon. We will call this the *symmetry of the upright sphere*, and it does not hold for *Sph.* III.7 and III.8.

Sph. III.5 (Theorem)

This theorem is demonstrated by using some of the key propositions of *Spherics* I, along with *Sph.* III.1 and III.4 as lemmas. As well as having some implications for spherical astronomy, it is also applied in the following proposition, *Sph.* III.6. The treatment of the contents of this theorem was extensively revised in the 3rd century CE by Pappos, basing his approach on that of Menelaos and using spherical triangles (*Coll.* VI.1–VI.11; Malpangotto 2003, 124–130).

In the description of the configuration, in both the *enunciation* and the *exposition*, the fact that **gC**(*DZE*) is perpendicular to the base, or "initial," great circle, **gC**(*ABG*), implies that the parallel circles that are tangent to **gC**(*DZE*), not shown in the diagram, have their points of tangency where **gC**(*DZE*) and **gC**(*ABG*) meet at points *D* and *E*. Furthermore, this implies that point *Z* is the upper pole of **gC**(*ABG*).

As in *Sph.* II.15, in the course of the *demonstration* the text argues that the arcs of great circles passing through the poles of a pair of parallel circles that are cut off between those parallel circles are equal (Czinczenheim 2000, 936; p. 273, above). But this was already established in *Sph.* II.10 Part 2, so that it should not need to be demonstrated again here. Hence, as in *Sph.* II.15, the text of this *demonstration* may originate from a version of the treatise in which *Sph.* II.10 Part 2 had not been shown (see the "Textual Comments" to *Sph.* II.10, above).

In the final phase of the *demonstration* it is claimed that if there are two parallel circles, **pC**(*LM*) and **gpC**(*BG*), and the common section of **gC**(*HQK*) and **gC**(*AQR*) meets the plane of **gpC**(*BG*) at the center of the sphere, then under the assumption – not mentioned in the *demonstration*, but made explicit in the *specification* – that the common section meets the sphere at *Q*, between the two parallel planes, then the common section of **gC**(*HQK*) and **gC**(*AQR*) will meet the plane of **pC**(*LM*) outside the sphere.

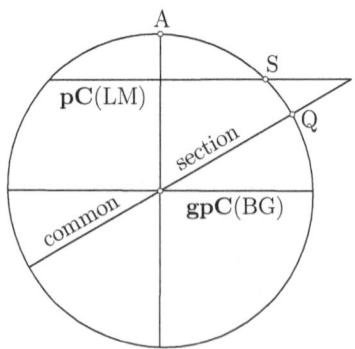

Figure 47: Analemma diagram for a lemma to *Sph.* III.5.

This can readily be seen on an analemma diagram (Figure 47). In this case, the analemma plane is that through **gC**(*AQR*), so that **gpC**(*BG*) is depicted by a diameter through the analemma circle and **pC**(*LM*) is depicted by a chord parallel to this. Thus, the common section of **gC**(*HQK*) and **gC**(*AQR*) will be depicted by a diameter of the analemma circle that meets the circle at some point *Q* between *S* on **pC**(*LM*) and the intersection of **gpC**(BG) and the analemma circle. Hence, it is clear that the common section will meet the plane of **pC**(LM) outside the analemma circle – that is, outside the sphere.

In terms of spherical astronomy, the content of this proposition is related to the first part of *Phen.* 8. If we understand **gC**(ABG) as a colure, or a horizon on the upright sphere, $\varphi = 0°$, **gC**(BZ) as the celestial equator, **gC**(DZE) as the ecliptic, and **Arc**(HQ) = **Arc**(QK), as equal and contiguous arcs of the ecliptic, then this proposition demonstrates that parallel circles passed through the endpoints of arcs of the ecliptic, $\lambda_1 = \lambda_2$, in **Quad**(Can → Vir) or, by the polar symmetry discussed above (p. 317), in **Quad**(Sag → Lib), will cut off unequal arcs of the colures, or the horizon, $\delta(\lambda_1) < \delta(\lambda_2)$, that increase as the λs are taken from the tropics towards the equator – that is, the differences of the arcs of declination of equal arcs of longitude will decrease as the arcs of the ecliptic are taken from Virgo towards Cancer, and from Libra towards Sagittarius. In particular,

$$\delta(\text{Vir}) > \delta(\text{Leo}) > \delta(\text{Can}), \text{ and}$$
$$\delta(\text{Lib}) > \delta(\text{Sco}) > \delta(\text{Sag}).$$

Moreover, because there are two great circles intersecting the same bundle of parallel circles, by the symmetries of *Sph.* II.17 and II.18, equal arcs of the ecliptic, $\lambda_m = \lambda_n$, in the symmetrical quadrants on either side of the tropic points will cut off equal arcs of the colures or horizon, $\delta(\lambda_m) = \delta(\lambda_n)$, so that

$$\delta(\text{Ari}) = \delta(\text{Vir}) = \delta(\text{Lib}) = \delta(\text{Pis}) >$$
$$\delta(\text{Tau}) = \delta(\text{Leo}) = \delta(\text{Sco}) = \delta(\text{Aqu}) >$$
$$\delta(\text{Gem}) = \delta(\text{Can}) = \delta(\text{Sag}) = \delta(\text{Cap}).$$

That is, on the upright sphere, the signs of the zodiac will rise and set over arcs of the horizon, or ortive amplitudes, of different magnitudes such that the arc of the horizon over which Aries rises is greater than that over which Taurus rises, and so on, in the same pattern as the arcs of declinations above. That is, $\eta(\lambda, 0°) = \delta(\lambda)$.

Basis: *Elem.* I.c.n.3, III.28, *Sph.* I.def.1 (by implication), I.def.2 (by implication), I.def.5, I.6 (by implication), I.15, I.20, III.1, III.4.[83]

Usage: *Sph.* III.6.

Diagram: The diagram is a solid representation of global objects, containing only intrinsic objects. It depicts a certain great circle with a base circle and a bundle of parallels that intersect the great circle as concentric half-lenses that meet the base circle. Further great circles that intersect the original great circle are shown with large half-lenses at various orientations.

---

[83] This proposition also repeats the argument of *Sph.* II.10.

There are no false crossings, so the diagram depicts the topology of the configuration.

Sph. III.6 (Theorem)

The basis of *Sph.* III.6 is similar to that of the previous proposition, which it also uses. Again, as well as having some astronomical implications, it is used in *Sph.* III.9 and III.10. Pappos also provided a treatment of the contents of this proposition using the methods introduced by Menelaos (*Coll.* VI.12–VI.27; Malpangotto 2003, 131–148).

Furthermore, the same symmetries that hold for *Sph.* III.5 are in effect here as well. Namely, the equal parallels that are tangent to $\mathbf{gC}(DZE)$ have their points of tangency at D and E, and point Z, the intersection of $\mathbf{gC}(DZE)$ and $\mathbf{gC}(BZG)$, is the upper pole of the initial great circle, $\mathbf{gC}(ABG)$.[84]

The *demonstration* of this proposition again uses the *greater than similar to* relation for arcs. See the Commentary to *Sph.* II.20.

In the course of the *demonstration*, Theodosios notes that $\mathbf{Arc}(SP) = \mathbf{Arc}(UQ)$ and $\mathbf{Arc}(PX) = \mathbf{Arc}(QF)$. This might be read as a direct usage of *Sph.* II.10 Part 2, in the current state of the text, but it may also be a reference to the argument that was just made to the same effect in *Sph.* III.5 (Czinczenheim 2000, 936).

Towards the end of the *demonstration*, Theodosios notes that $\mathbf{Arc}(HF) \succ \mathbf{Arc}(YC)$, because $\mathbf{Arc}(HF)$ is in a lesser circle but subtends a greater line. This can be seen by drawing the circle of $\mathbf{Arc}(HF)$ in the same plane as the circle of $\mathbf{Arc}(YC)$, as is often done in arguments involving solid objects in Greek geometrical texts (Figure 48).[85] In such a situation, it is clear that, since the triangles subtending $\mathbf{Arc}(YC)$ and $\mathbf{Arc}(Y'C')$ are similar, *Elem.* I.5 and I.32, while $\mathbf{Arc}(Y'C')$ is in a lesser circle, then $YC > Y'C'$, *Elem.* VI.4. Hence, the arc subtending a line equal to YC in $\mathbf{C}(Y'C')$ will be greater than similar to $\mathbf{Arc}(YC)$. Hence, even more $\mathbf{Arc}(HF) \succ \mathbf{Arc}(YC)$.

This proposition has some important implications for spherical astronomy – namely, where $\mathbf{gC}(ABG)$ is taken as a colure, or a horizon on the upright sphere, $\varphi = 0°$, $\mathbf{gC}(BZ)$ as the celestial equator, $\mathbf{gC}(DZE)$ as the ecliptic, and $\mathbf{Arc}(HQ) = \mathbf{Arc}(QK)$, as equal and contiguous arcs cut off of the ecliptic, again, because of the two symmetries of the configuration discussed above, this theorem implies that the colures, or horizons on the upright sphere, $\varphi = 0°$, drawn through the endpoints of equal and con-

---

[84] In the second sentence of the *exposition*, Ver Eecke (1927, 97 l.13) has BZΓ as a typographical error for ABΓ. He is followed in this by Nikolantonakis (2016, 161).

[85] This is done, for example, by Diodoros and Eutokios (Hogendijk 2001, 56, 70–71; Sidoli 2004b, 160–161).

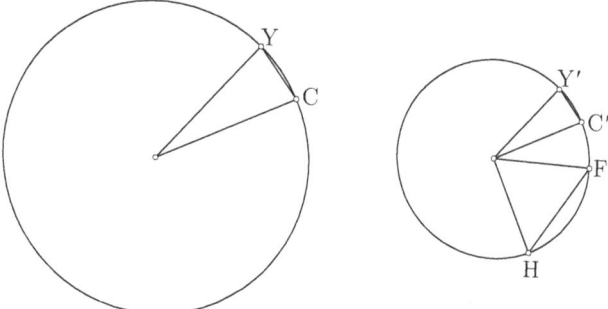

Figure 48: Diagram for a lemma to *Sph.* III.6.

tiguous arcs of the ecliptic in **Quad**(Can → Vir) and **Quad**(Lib → Sag), $\lambda_1 = \lambda_2$, will cut off unequal arcs of the celestial equator, $\alpha(\lambda_1) \neq \alpha(\lambda_2)$, such that the $\alpha(\lambda)$s increase as the λs are taken nearer to the tropic (Schmidt 1943, 38–40).

In particular, this proposition shows that, the colures or horizons through the endpoints of the signs cut off arcs of the ecliptic such that,

$$
\begin{aligned}
\alpha(\text{Can}) &= \alpha(\text{Sag}) &&> \\
\alpha(\text{Leo}) &= \alpha(\text{Sco}) &&> \\
\alpha(\text{Vir}) &= \alpha(\text{Lib}).
\end{aligned}
$$

Furthermore, because arcs of the ecliptic rise in the same times as the opposite arcs set, and, because of the symmetries of the configuration, rising times and setting times are equal on the upright sphere, we have

$$
\begin{aligned}
\alpha(\text{Can}) &= \alpha(\text{Sag}) &= \alpha(\text{Cap}) &= \alpha(\text{Gem}) &&> \\
\alpha(\text{Leo}) &= \alpha(\text{Sco}) &= \alpha(\text{Aqu}) &= \alpha(\text{Tau}) &&> \\
\alpha(\text{Vir}) &= \alpha(\text{Lib}) &= \alpha(\text{Pis}) &= \alpha(\text{Ari}).
\end{aligned}
$$

Hence, on the upright sphere the stated arcs of the ecliptic both rise and set in the same times as their projections onto the equator by the colures – for which reason these projections are called *right ascensions*.

Basis: *Elem.* I.def.17, I.post.3 (spherical, transfer), III.def.11, III.26, XI.16, *Sph.* I.6, I.11, I.15, II.10,[86] III.1, III.3, III.5.

Usage: *Sph.* III.9, III.10.[87]

Diagram: The diagram is a solid representation of global objects, containing only intrinsic objects. As in the previous diagrams, it represents a

---

[86] This may rely on the argument in *Sph.* III.5, as discussed above.

[87] This proposition is explicitly referenced towards the end of *Phen.* 12, in a passage that Menge (1916, 76) marks as a later addition. Indeed, as Berggren and Thomas (1996, 86 n. 53) point out, the entire argument to which this reference belongs is probably spurious.

certain great circle with the base circle and a bundle of parallels that intersect the great circle as concentric half-lenses that meet the base circle. Furthermore, it is the first that uses the solid representation encoding for a bundle of great circles. In particular, these great circles are shown with large half-lenses, as well as arcs that appear to stand upright on a horizontally orientated half-lens, which itself represents another great circle. This is a different visual convention than that used in *Sph.* II.10 for the same objects, which may be because *Sph.* II.10 was meant to be read as a geometrical theorem, while this proposition is meant to be understood from an astronomical interpretation (Malpangotto 2010). Because there are no false crossings, the diagram accurately depicts the topology of the configuration.

Sph. III.7 (Theorem)

This theorem depends on propositions from all three books, and is, hence, probably a major goal of the treatise. It is worth remarking, however, that it is independent of the first two theorems in this group. As well as having an astronomical implication, it is directly used in the following proposition.

The placement of the tangent points on the "initial" (ἐξ ἀρχῆς) great circle insures that $\mathbf{gC}(EZH)$ is tangent to a pair of parallel circles at the point where it intersects the initial circle $\mathbf{gC}(ABG)$, which implies that the claims made in this theorem are restricted to the quadrant of $\mathbf{gC}(EZH)$ between E and Z. However, since by *Sph.* II.17 and II.18, we know that if equal arcs of $\mathbf{gC}(EZH)$ are cut off on either side of $\mathbf{gC}(BZG)$, then the equal parallels through them will, in turn, cut off equal arcs of $\mathbf{gC}(ABG)$, this symmetry allows us to extend the results stated here to the full semicircle of $\mathbf{gC}(EZH)$ of which point E is the midpoint.

In the *enunciation* and *exposition*, we read that $\mathbf{gC}(EZH)$ is tangent to a parallel circle that is greater than $\mathbf{pC}(AD)$. As mentioned in the introduction to *Sph.* III.5–III.8, above, this serves to impose bounds on the position of $\mathbf{gC}(EZH)$, and hence provides the overall orientation of the configuration. That is, if the angle between $\mathbf{gC}(EZH)$ and $\mathbf{gC}(BZG)$ is taken to be ε, and the pole U of the parallels is assumed to be separated from the initial circle $\mathbf{gC}(ABG)$ by great-circle arc φ, this construction has the effect of setting φ in the open interval $0° < \varphi < 90° - \varepsilon$ (Figure 35 (right), p. 293).

The *demonstration* for this proposition relies on two lemmas that are not shown in the *Elements*. The first is the claim that in equal circles the greater chord subtends the greater arc. This can be shown indirectly from *Elem.* III.28. The second is the same lemma that was required in *Sph.* III.5 – namely, the claim that the intersection of the planes of $\mathbf{gC}(QKL)$ and $\mathbf{gC}(YKS)$ meets the plane of $\mathbf{pC}(MQN)$ at some point outside the sphere (see Commentary to *Sph.* III.5).

In regard to spherical astronomy, this proposition is also related to the first part of *Phen.* 8. That is, we assume that $\mathbf{gC}(ABG)$ is the base horizon on an inclined sphere of terrestrial latitude $0° < \varphi < 90° - \varepsilon$, point U is the pole, $\mathbf{gC}(BZ)$ the celestial equator, $\mathbf{gC}(DZE)$ the ecliptic, $\mathbf{Arc}(HQ) = \mathbf{Arc}(QK)$, are equal and contiguous arcs of the ecliptic, and $\mathbf{gC}(DKS)$ is the orientation of the horizon when point K is on it. Then, because of the polar symmetry discussed above (p. 317), this proposition shows that equal arcs of the ecliptic, $\lambda_1 = \lambda_2$, in $\mathbf{Quad}$(Can → Vir) and $\mathbf{Quad}$(Sag → Lib) set over unequal arcs of the horizon or ortive amplitudes, $\eta(\lambda_1, \varphi) < \eta(\lambda_2, \varphi)$, which increase as the λs are taken going from the tropics towards the equator.

In particular, the proposition directly shows that the ortive amplitudes of Cancer to Virgo and Sagittarius to Libra are such that

$$\begin{aligned} \eta(\text{Can}) &< \eta(\text{Leo}) < \eta(\text{Vir}), \text{ and} \\ \eta(\text{Sag}) &< \eta(\text{Sco}) < \eta(\text{Lib}). \end{aligned}$$

Moreover, because, as in *Sph.* III.5, two great circles intersect the same bundle of parallel circles, by the symmetries shown in *Sph.* II.17 and II.18, the same parallel circles will pass through the endpoints of the signs of the zodiac on either side of the equator, and the ortive amplitudes of all of the signs will be such that

$$\begin{aligned} \eta(\text{Can}) &= \eta(\text{Sag}) = \eta(\text{Cap}) = \eta(\text{Gem}) < \\ \eta(\text{Leo}) &= \eta(\text{Sco}) = \eta(\text{Aqu}) = \eta(\text{Tau}) < \\ \eta(\text{Vir}) &= \eta(\text{Lib}) = \eta(\text{Pis}) = \eta(\text{Ari}). \end{aligned}$$

Basis: *Elem.* III.def.1 (by implication), *Sph.* I.def.1 (by implication), I.def.2 (by implication), I.6 (by implication), I.15, I.20, I.21, II.13, II.15, III.1, III.4.

Usage: *Sph.* III.8. A proposition similar to this was used in *Phen.* 8.

Diagram: The diagram is a solid representation of global objects, containing only intrinsic objects. It is similar to the diagram for the previous proposition with the addition of two new conventions. This is the first diagram that represents a bundle of parallel circles in two different ways. Most of the parallels are encoded in the now standard way for a solid representation, using half-lenses facing in the same direction. Then, the first parallel that is entirely within the visible hemisphere produced by the base circle is depicted with a full circle that is tangent to the base circle at the top of the figure. This is probably to facilitate showing that these two circles are tangent, using the encoding developed in the diagrams using surface representation. Moreover, great circles are depicted as being tangent to this full parallel circle by being shown to meet it. Because there are no false crossings, the diagram accurately depicts the topology of the configuration, but we must remember that great circles that meet the full parallel are tangent, not intersecting.

Textual Comments: In the manuscript diagrams, circle MΘN (*MQN*) appears to be tangent to circle AΔ (*AD*), which it must not be (**A**, 31v; **B**, 67v; **D**, 13v; **F**, 43v; **K**, 135r). In some of the manuscript diagrams, the letter T appears to label this point of tangency (for example, **B**, **D**, **F**). Furthermore, T has either been erased or added in a different ink in the oldest manuscript, **A**. In the diagram in **K**, the letter T marks the intersection of circles ΥΦ (*UF*) and MN, but there is no letter Ψ (*Y*).

There is currently no letter-name T in the text, although the letters that follow it in the alphabet are used. This non-standard situation would be rectified if, at the end of the *construction*, circle ΥΚΦ (*UKF*) were instead named ΥΤΚΦ, with T labeling the intersection of circles ΥΚΦ and MN. It may be that the difficulties with the diagram led someone to remove T from the text at some point in the transmission.

Comparison with Arabic: In Thābit's version there is also no ش (T) in the text. The diagram in **I** (47b) is rotated 45° clockwise, but it clearly shows circle م ط ن (MΘN, *MQN*) as separated from circle د ا (AΔ, *AD*), and ش marking the intersection of circle ت ث (ΥΦ, *UF*) and circle م ط ن. This is probably a case in which the diagram was corrected by considering the mathematics of the situation, without reference to a better source text. There are a number of other interesting features to this diagram, including labeling with terrestrial place names, which are later additions and need not concern us here (Kunitzsch and Lorch 2010, 338). The diagram in our translation has been influenced by that in Thābit's version.

Sph. III.8 (Theorem)

As with the previous theorem, this proposition uses key results from all three books and was probably an important goal of the treatise. The key theorem upon which *Sph.* III.8 is based is *Sph.* III.2, which allows for the introduction of an inequality between great-circle arcs. Furthermore, as will be discussed below, the details of what is shown in *Sph.* III.2 and the way that that theorem is applied here provides a geometrical explanation of the difference between the symmetry found in the previous three propositions and the asymmetry implied in *Sph.* III.8. As well as having significant astronomical implications, *Sph.* III.8 is used in *Sph.* III.14.

In this theorem, the placement of the points of tangency, *E*, *G*, on the "initial" great circle, **gC**(*ABG*), performs the same function that it played in the previous proposition. Here, however, since the arcs that this theorem treats are cut off from the greatest of the parallels, not from a great circle tangent to one of the parallels, the symmetries of *Sph.* II.17 and II.18 do not apply, as will be explained below.

Again, as in *Sph.* III.7, the fact that **gC**(*EZG*) is introduced as tangent to a parallel circle that is greater than **pC**(*AD*) has the effect of imposing a bound on the great-arc distance, φ, between the pole W and the initial great circle $\mathbf{gC}(ABG)$, such that $0° < \varphi < 90° - \varepsilon$, where ε is the angle between $\mathbf{gC}(EZG)$ and $\mathbf{gC}(BZ)$.

The crux of the argument in *Sph.* III.8 is an application of *Sph.* III.2 to show that $FY < FK$ – which lines are neither constructed nor drawn in the figure. Since this application of *Sph.* III.2 can be difficult to conceptualize, it may be helpful to summarize the approach with a perspective figure, which is a way of simulating what it might have been like to discuss this proposition with reference to a globe or armillary sphere. The use of *Sph.* III.2 can be explained by considering only $\mathbf{gC}(XKO)$ and the relevant parallel circles that meet this great circle. In the perspective diagram provided here, we imagine $\mathbf{gC}(XKO)$ along with the parallel circles standing on it rotated into the plane of our visual horizon (Figure 49). If we are following this proposition on an armillary sphere, the parallel circles standing on $\mathbf{gC}(XKO)$ are the arcs of those parallels that are underneath the horizon. In fact, some three-quarters of the *demonstration* of *Sph.* III.8 involves carefully establishing that the conditions under which *Sph.* III.2 can be applied actually hold. Namely, it is first shown that each of the parallel circles, and in particular $\mathbf{pC}(CFY)$, is inclined on $\mathbf{gC}(XKO)$ towards point X – that is, that $\vartheta < 90°$. It is then shown that $\mathbf{pC}(CFY)$ stands on a chord of $\mathbf{gC}(XKO)$ that is (a) parallel to the diameter of $\mathbf{gC}(XKO)$ passing through point O and (b) between the diameter and point X. Hence, the inclination of $\mathbf{pC}(CFY)$ on $\mathbf{gC}(XKO)$ is toward the smaller segment of $\mathbf{gC}(XKO)$. Then, a laborious sentence states each of the conditions of *Sph.* III.2, from which it is deduced that the line "from F to Y is less than that from F to K." The rest of the proof follows fairly straightforwardly from this claim.

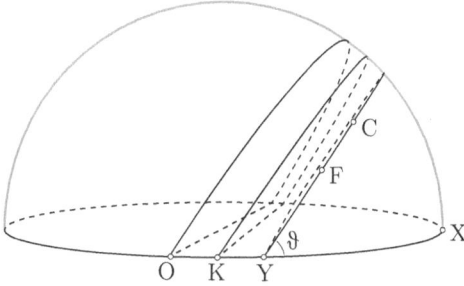

Figure 49: Diagram for the key to the proof of *Sph.* III.8.

In the *enunciation* of this theorem the description of the configuration becomes difficult to the point of being baffling without skipping ahead to the *exposition* and working with a solid sphere. In particular, in the final phrases of the *enunciation*, the claim that the arcs cut off from the greatest of the parallels will be continuously greater the closer they are to the "initial" great circle may seem like a peculiar way to describe the situation. It is important, in this regard, to remember that the points of tangency of the oblique great circle with the two equal and parallel circles, not shown in the diagram, were themselves set on the "initial" great circle. Hence, the reference to the "initial" great circle is, here, also a reference to one of the cardinal points of the oblique great circle.

Furthermore, the expression "between them," μεταξὺ αὐτῶν, is used with two different referents. In the first instance, any pair of tangent great circles, say $gC(DHL)$, $gC(MQN)$, $gC(XKO)$, cut off "between them" pairwise similar arcs from each parallel circle, including the greatest, *Sph.* II.13. That is, all of the arcs of parallel circles between, say, $gC(DHL)$ and $gC(MQN)$ are similar, and likewise all of those between, say, $gC(MQN)$ and $gC(XKO)$. Then, in the following phrase, any *other* pair of tangent great circles cut off "between them" unequal arcs of the greatest of the parallels, and indeed of all the parallels between them. That is, in this phrase the arcs of parallel circles cut off by one pair of great circles, say, $gC(DHL)$ and $gC(MQN)$, are compared, not with each other, but with the arcs of the same parallel circles cut off by some other pair of great circles, say $gC(MQN)$ and $gC(XKO)$.

In the *construction*, there is a sort of repeated construction, because, according to the methods developed in *Spherics* I, the pole must already have been assumed or found for $pC(PHR)$, $pC(SQ)$, and $pC(TUK)$ to be drawn through points H, Q, and T at the beginning of the *construction*, but then the pole, W, is later taken at the end of the *construction*. In fact, this procedure of drawing parallel circles through points on a sphere that is simply assumed to contain a bundle of parallel circles, and hence a pole, without first finding the pole, is also carried out in *constructions* in the *Phenomena* – another indication that we are dealing with a more astronomical subject matter at this point in the *Spherics*.

In terms of spherical astronomy, this proposition is related to the first part of *Phen.* 12 and 13, but it is more sophisticated because it asserts clear bounds on the geometrical configuration, and relies on the concept of the inclination of two circles on one another, as well as the application of *Sph.* III.2, as discussed above. Since the concept of the inclination of two planes on one another is developed in neither the *Elements* nor the *Phenomena*,[88] the approach of *Sph.* III.8 was likely developed after that of the *Phenomena*, although the fact that it proves appears to have been known. (See also the discussion of the relationship between these various propositions in the Introduction, p. 48.)

---

[88] The definitions treating such concepts in *Elements* XI are considered to be medieval additions to the text (Vitrac 1990–2001, 77–79), and are not used in that treatise.

We can take $\mathbf{gC}(ABG)$ as the base horizon on an inclined sphere of terrestrial latitude $0° < \varphi < 90° - \varepsilon$, point W as the pole, $\mathbf{gC}(BZ)$ as the celestial equator, $\mathbf{gC}(EZG)$ as the ecliptic, $\mathbf{Arc}(HQ) = \mathbf{Arc}(QK)$, as equal and contiguous arcs cut off of the ecliptic, and $\mathbf{gC}(DHL)$, $\mathbf{gC}(MQN)$, and $\mathbf{gC}(XKO)$ as the orientations of the horizon when points H, Q, and K are on the horizon. Then, bearing in mind the polar symmetry, *Sph.* III.8 shows that equal and contiguous arcs of the ecliptic, $\lambda_1 = \lambda_2$, in $\mathbf{Quad}(\text{Can} \to \text{Vir})$ or $\mathbf{Quad}(\text{Sag} \to \text{Lib})$, will set with unequal arcs of the celestial equator, $\bar{\varrho}(\lambda_1, \varphi) > \bar{\varrho}(\lambda_2, \varphi)$, that increase as the λs are taken from the equator to the tropic (Schmidt 1943, 38–40).

In particular, the setting times of the signs of the zodiac from Cancer to Virgo, and from Sagittarius to Libra, will be such that

$$\bar{\varrho}(\text{Can}) \quad > \quad \bar{\varrho}(\text{Leo}) \quad > \quad \bar{\varrho}(\text{Vir}), \text{ and}$$
$$\bar{\varrho}(\text{Sag}) \quad > \quad \bar{\varrho}(\text{Sco}) \quad > \quad \bar{\varrho}(\text{Lib}).$$

Moreover, the diametrically opposite arcs will have the same rising times, so that the signs from Capricorn to Pisces, and from Gemini to Aries, will be such that

$$\varrho(\text{Cap}) \quad > \quad \varrho(\text{Aqu}) \quad > \quad \varrho(\text{Pis}), \text{ and}$$
$$\varrho(\text{Gem}) \quad > \quad \varrho(\text{Tau}) \quad > \quad \varrho(\text{Ari}).$$

That is, *Sph.* III.8 can be understood as making the claim that, on an inclined sphere of latitude $0° < \varphi < 90° - \varepsilon$, equal and contiguous arcs of $\mathbf{semiC}(\text{Cap} \to \text{Gem})$ have rising times that increase and decrease regularly about the midpoint of the semicircle. Furthermore, the same is true for the setting times of $\mathbf{semiC}(\text{Can} \to \text{Sag})$.

It seems likely that already when Euclid composed the *Phenomena* it was known that these relations do not hold for the rising times of $\mathbf{semiC}(\text{Can} \to \text{Sag})$ or the setting times of $\mathbf{semiC}(\text{Cap} \to \text{Gem})$, and one can easily convince oneself of this fact intuitively by examining the rising times on a globe or armillary sphere (see p. 50, above). Nevertheless, because the mathematical methods of the *Phenomena* are rather vague, it is not clear how this could have been shown geometrically based on the material in that text. The approach of *Sph.* III.8, however, gives us some insight into this, through its application of *Sph.* III.2.

Although this argument is not found in the text, it may be useful to go through it using the style of diagram found in the text, in order to convince ourselves that this argument could have been made by an ancient geometer. In order to consider the other semicircle, we need to look at the setting times of arcs of the ecliptic of $\mathbf{Quad}(\text{Cap} \to \text{Pis})$, or, using the polar symmetry, those of $\mathbf{Quad}(\text{Gem} \to \text{Ari})$. That is, we rotate the solid or armillary sphere until $\mathbf{semiC}(\text{Cap} \to \text{Gem})$ is above the horizon, and we again consider the arcs that are setting over the western horizon. In an ancient diagram, this

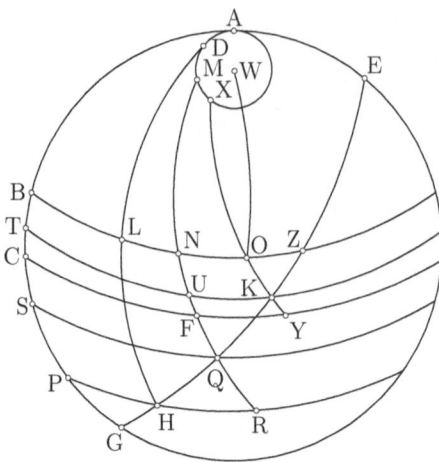

Figure 50: Diagram for a reconstructed application of the argument of *Sph.* III.8 to the other hemisphere.

would be represented by placing a pole, W, at the top of the figure, letting the greatest circle of the parallels, $\mathbf{gC}(BZ)$, cross the diagram horizontally, while the oblique great circle, $\mathbf{gC}(EZG)$ crosses the diagram from the upper right to the lower left, and taking the equal arcs of the oblique great circle, $\mathbf{Arc}(HQ) = \mathbf{Arc}(QK)$, on the side of $\mathbf{gC}(BZ)$ opposite the visible pole (Figure 50). Then the rest of the diagram can be produced as in *Sph.* III.8. Then, if the pole is taken as the north pole, the diagram depicts the arcs $\mathbf{Quad}(\text{Cap} \to \text{Pis})$ setting over the western horizon, while if the pole is the south pole, the arcs of $\mathbf{Quad}(\text{Gem} \to \text{Ari})$ are setting.

The proof would proceed in the same way by constructing $\mathbf{pC}(PHR)$, $\mathbf{pC}(SQ)$, and $\mathbf{pC}(TUK)$, such that $\mathbf{Arc}(ST) > \mathbf{Arc}(SP)$, *Sph.* III.7, while $\mathbf{Arc}(SP) = \mathbf{Arc}(QR)$ and $\mathbf{Arc}(ST) = \mathbf{Arc}(QU)$, by *Sph.* III.13. Hence, $\mathbf{Arc}(QU) > \mathbf{Arc}(QR)$. Once again, $\mathbf{Arc}(QF) = \mathbf{Arc}(QR)$ can be cut off from $\mathbf{Arc}(QU)$, *Elem.* I.post.3 (transfer), so that $HR = FK$, by *Sph.* III.3. It is at this point that one would seek to apply *Sph.* III.2, but there is a problem.

As in *Sph.* III.8, $\mathbf{gC}(WO)$ can be drawn from the pole W to point O, the point at which $\mathbf{gC}(XKY)$ intersects $\mathbf{gC}(BZ)$, *Sph.* I.20. From this, it can be shown, again, that $\mathbf{gC}(XKY)$ is inclined on $\mathbf{gC}(BZ)$, and hence on all of the parallel circles, in the direction of point X. But now $\mathbf{pC}(CFY)$ stands on $\mathbf{gC}(XKY)$, having as base a chord that is farther away from point X than the diameter through point O. To see this more clearly, we can, once again, imagine $\mathbf{gC}(XKY)$ rotated into the plane of our visual horizon with the parallel circles standing on it (Figure 51). This makes it clear that $\mathbf{Seg}(CFY)$ stands on a chord of $\mathbf{gC}(XKY)$, but is inclined towards its greater segment, namely $\mathbf{Arc}(YOX)$. Hence, *Sph.* III.2 cannot

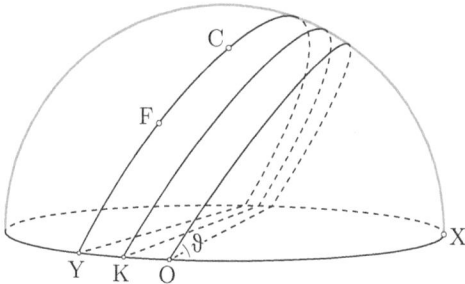

Figure 51: Diagram for the problem with applying *Sph.* III.2 in the reconstructed argument of *Sph.* III.8 to the other hemisphere.

be applied both because **Seg**(CFY) > 180°, by *Sph.* II.19, and also because it is inclined in the direction of the greater segment of the circle on which it stands, as discussed above (see p. 305, above). Thus, no general claim can be made about the relationship between FY and FK, and the rest of the proof cannot go through.

Basis: *Elem.* I.def.17, I.post.3 (spherical, transfer), III.def.11 (by implication), III.26 (by implication), XI.9, XI.16, *Sph.* I.6, I.15, I.20, I.21, II.13, III.2, III.3, III.7.

Usage: Unused in the *Spherics.* A proposition that stated some facts related to those shown in *Sph.* III.8, was used in *Phen.* 12 (see p. 48, above).

Diagram: The diagram is a solid representation of global objects, containing only intrinsic objects. It follows the same conventions as the diagram for the previous proposition. Because there are no false crossings, the diagram accurately depicts the topology of the configuration, where it must be remembered that great circles that meet the full parallel are tangent, not intersecting.

Comparison of the projections of equal and non-contiguous arcs of one great circle onto another oblique great circle (*Sph.* III.9)

The result of this proposition can be associated with the foregoing four theorems, whereas its method of proof links it with the following proposition. In terms of what is shown, *Sph.* III.9 is a generalization of *Sph.* III.6 to non-contiguous arcs, while the way that it is demonstrated is a complete departure.

In terms of what is shown, *Sph.* III.9 starts with the same assumption of a base great circle, gC, on which lies the pole, P, of a bundle of parallel circles and to which are perpendicular both the greatest of the parallels,

gpC, and a great circle oblique to it, gC_1 (Figure 52). Then, equal and non-contiguous arcs of gC_1, $\lambda_1 = \lambda_2$, are cut off between the greatest of the parallels and the intersection of gC and gC_1, and great circles through the poles, gC_2, gC_3, gC_4, gC_5, are passed through the endpoints of λ_1 and λ_2, cutting off α_1 and α_2 on gpC. Then, *Sph.* III.9 shows that arcs

$$\alpha_1 > \alpha_2$$

decrease as the λs are taken going from the point at which gC_1 and gC meet towards gpC.

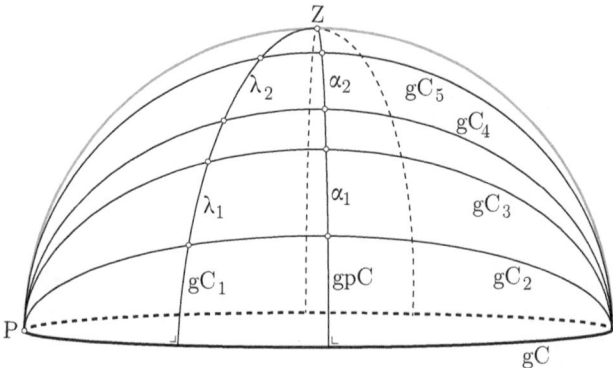

Figure 52: Perspective diagram, *Sph.* III.9.

Since the methods used in *Sph.* III.9 could be used to generalize the other theorems in the foregoing group in the same way as is done here for *Sph.* III.6, and since *Sph.* III.9 is not used in the current treatise, one might be led to wonder what role *Sph.* III.9 plays in the *Spherics*. Perhaps it is a relic from some previous arrangement of this material, or was meant as an introduction to the two-stage method of proof used in both *Sph.* III.9 and III.10, as discussed below. Another possibility is that it serves as a demonstration that the claims of *Sph.* III.5–III.8 can be extended to non-contiguous arcs, under the assumption that the reader will see that the methods of this proposition can be applied to the configurations of the other three propositions in that group as well.

The interpretation in terms of spherical astronomy is essentially similar to that for *Sph.* III.6. Once again, we can take gC as either a horizon on the upright sphere, $\varphi = 0°$, or a colure, gpC as the celestial equator, and gC_1 as the ecliptic. Then, because the celestial pole P lies on gC, while the intersections of gC_1 and gpC are the poles of gC, both of the symmetries discussed in the Commentary introducing *Sph.* III.5–III.8 apply here as well. This means that the claim of *Sph.* III.9 applies to each quadrant of the ecliptic. In particular, for both the inclined and upright spheres, this theorem implies that the colures through the endpoints of equal, non-contiguous arcs of the ecliptic, $\lambda_1 = \lambda_2$, will cut off unequal arcs of the celestial equator, $\alpha(\lambda_1) \neq \alpha(\lambda_2)$, such that the $\alpha(\lambda)$s increase as the arcs of the ecliptic are taken nearer to the tropic; and on the upright sphere, $\varphi = 0°$, the arcs of the ecliptic will rise in the same times as the $\alpha(\lambda)$s.

Sph. III.9 (Theorem)

Although in its details this proof relies on *Sph.* III.6, its most conspicuous feature is the use of a much discussed two-stage argument that was used by a number of Greek mathematicians for proving a relation between two magnitudes. Because this proposition is unused in the *Spherics*, and because, as noted, its astronomical significance seems to be only a modest extension of that of *Sph.* III.6, while the other theorems in that group are not so generalized, it is possible that it is included here to facilitate the reader's understanding of the two-stage argument by starting with the relations of equality and inequality, before moving on to ratios in *Sph.* III.10.

The two-stage argument establishes some relation involving two magnitudes, first in the case that they are commensurable; and then uses this result to extend to the case in which they are incommensurable. Aside from its use here in *Sph.* III.9 and III.10, we find this method employed in Archimedes' *Equi. Plan.* I.6 and I.7,[89] and by Pappos in his *Commentary to the Almagest* VI.7 and *Coll.* V.12,[90] as well as in his treatment of these *Spherics* propositions, *Coll.* VI.7 and VI.8. Because of its relation to the operation of repeated and alternating subtractions known as *anthyphairesis*, this material is thought to have its origin in work on magnitudes that circulated prior to the development of the theory of ratio with respect to magnitudes that we read in *Elements* V,[91] especially since many of the *anthyphairesis* arguments could be simplified through the approach of *Elements* V (Knorr 1978; Mendell 2007).[92]

---

[89] Berggren (1976) has argued that *Equi. Plan.* I.6 and I.7, which have serious textual issues, are, in fact, later additions to the text. However, others regard them as having been included by Archimedes and then corrupted in the transmission, offering various ways that the argument can be restored (Knorr 1978, 184–189, 222–227; Hayashi and Saito 2009, 225–230). Whatever the case, some Greek mathematician offered these two theorems, proved by the two-stage method, as a means of establishing the law of the lever.

[90] The text of these two examples is essentially the same. For that in the *Commentary* see Rome (1931–1943, 256–257). In the course of the argument there is a reference, whether by Pappos or added later, to a "lemma of the *Spherics*," which must refer to Scholium 416, discussed below (Rome 1931–1943, 257; Hultsch 1876–1878, 338)

[91] It may be worth pointing out that discussions of these propositions were not included in some of the more important attempts to construct an anthyphairetic ratio theory (Becker 1933; Fowler 1999, Chap. 2).

[92] Saito (2003) argues that there is little evidence for a theory of ratios prior to that found in *Elements* V, and that even that found in *Elements* V is far from general and

The orientation of the configuration is the same in this proposition as in *Sph.* III.5 and III.6. Namely, $\mathbf{gC}(DZE)$ and $\mathbf{gC}(BE)$ are perpendicular to the base great circle, $\mathbf{gC}(ABG)$, so that their intersection, point Z, is the upper pole of $\mathbf{gC}(ABG)$ and all of the objects drawn in the diagram are actually situated in the upper left-hand quadrant of the figure.

The structure of this theorem is somewhat involved, because the two-stage argument makes a division of cases, while the incommensurable case is shown by a double indirect argument, and hence, itself, involves cases. Moreover, there are two, essentially identical, *specifications* introducing each of the cases of the two-stage division.

The *construction* in Case 1 requires dividing three commensurable arcs into parts according their common measure. This requires first finding their greatest common measure, which is a constructive application of *Elem.* X.3, a general *problem* treating magnitudes. *Elem.* X.3 is general both in the sense that it describes a general *anthyphairesis* procedure for finding the common measure of two commensurable magnitudes, and insofar as it treats magnitudes generally, not any specific type of magnitude. Hence, although the general procedure is described in *Elem.* X.3, whether or not it can be reduced to an effective procedure will depend on the type of magnitude involved and the constructions that can be carried out on such magnitudes. In order to see how the procedure in *Elem.* X.3 can be applied in *Sph.* III.9, it may be useful to go through some of the details.

Elem. X.3 divides into two cases. *Elem.* X.3 Case 1 assumes that the lesser of the two magnitudes is, itself, a common measure, in which case it must be the greatest, while *Elem.* X.3 Case 2 assumes that the lesser of the two magnitudes is not a common measure, in which case, "with the lesser ever taken away in turn (ἀνθυφαιρουμένου) from the greater, that left over will at some time measure that before it, through the ⟨fact that⟩ they are not incommensurable" (Heiberg and Stamatis 1969–1977, III.5), so that whatever magnitude measures the one before it completes the process of *anthyphairesis* and is the greatest common measure. In *Elem.* X.4, this strategy is expanded to three magnitudes.

In terms of the arcs in *Sph.* III.9, since $\mathbf{Arc}(ZH) = \mathbf{Arc}(QK)$, we need only consider two arcs, say $\mathbf{Arc}(ZH)$ and $\mathbf{Arc}(HQ)$. If $\mathbf{Arc}(ZH) = \mathbf{Arc}(HQ)$, then we can simply apply *Sph.* III.6 directly to $\mathbf{Arc}(LM)$, $\mathbf{Arc}(MN)$, and $\mathbf{Arc}(NX)$, as is done in the crux of the *demonstration*. Otherwise, let one of them be greater, say $\mathbf{Arc}(ZH) > \mathbf{Arc}(HQ)$, in which case the process of *anthyphairesis* is carried out using *Elem.* post.3 (transfer), until some remaining arc, α_i, is found that measures the previously

complete. Nevertheless, even if we grant that there may not have been a general theory of ratios based on *anthyphairesis*, it seems that this was another approach to ratios with respect to magnitudes that probably originated before the theory found in *Elements* V, and certainly continued to be used for centuries afterwards.

remaining arc, α_{i-1}, which must occur at some time, because **Arc**(ZH) and **Arc**(HQ) are assumed to be commensurable, *Elem.* X.def.1. Then, α_i can be the *measure* used to divide **Arc**(ZH) = **Arc**(QK), and **Arc**(HQ) "into the measures," as Theodosios says, again using *Elem.* post.3 (transfer). That is, $m\alpha_i = $ **Arc**(HQ), $n\alpha_i = $ **Arc**(ZH) = **Arc**(QK), where $m, n \in \{1, 2, 3, ...\}$, from which the proof will follow. Likewise, if **Arc**(ZH) < **Arc**(HQ), the least common arc, α_j, can again be found, the three arcs divided by it, and the proof will follow. In each case, the generality of the proof in *Sph.* III.9 is based on the fact that the actual multitude of the *measures* in each arc, m, n, plays no role in the argument for *Elem.* X.3. That is, the overall argument for *Sph.* III.9 Case 1 consists in showing an effective procedure for reducing two non-contiguous arcs to a series of equal and contiguous arcs, so that *Sph.* III.6 can be applied to each successively.

In the *construction* for Case 2.1, another important construction is taken for granted; namely, an arc is taken between two unequal arcs, such that it is commensurable with another arc. As is well known, this construction has been shown in a scholium, dealing with magnitudes in general (Knorr 1978, 187–188; Mendell 2007, 5–7). Scholium 416 reads as follows (Czinczenheim 2000, 431; Figure 53):[93]

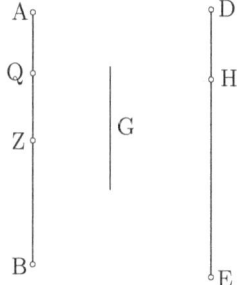

Figure 53: Scholium 416 to *Sph.* III.9.

With three magnitudes of the same kind, *AB*, *G*, *DE*, supposed, and *AB* being greater than *G*, while *DE* is arbitrary, let it be necessary to find a magnitude less than *AB*, greater than *G* and commensurable with *DE*.

Let there be an equal to *G*, *BZ*. Then, bisecting *DE* and the half of it, and ever doing this, some magnitude less than *AZ* remains. Let it be left, and let it be *DH* – less than *AZ*, being a measure of *DE*.

But, *DH* either measures *BZ*, or not.

First, let it measure it, and let there be an equal to DH, ZQ.

Since DH measures ZB, and DH is equal to ZQ, therefore DH also measures BQ, but it also measured DE. Therefore, BQ is commensurable with DE, being less than AB but greater than G.

Then, let DH not measure ZB, and let DH measuring out ZB exceed by a lesser than it, ZQ. Therefore, BQ is commensurable to DE, being less than AB but greater than G.

Since this lemma is also an argument treating general magnitudes, it may again be useful to go through the details of how this can be applied to the specific magnitudes dealt with in *Sph.* III.9, namely great-circle arcs. That is, we assume, by the hypothesis of Case 2, that $\mathbf{Arc}(HQ)$ is incommensurable with $\mathbf{Arc}(QK)$, and since, by construction, point P falls on $\mathbf{Arc}(QK)$, we have three arcs, of which $\mathbf{Arc}(QP) < \mathbf{Arc}(QK)$ and $\mathbf{Arc}(HQ)$ is arbitrary. Then, where $m, n \in \{1, 2, 3, ...\}$, by the construction described in the Commentary to *Sph.* II.9 – using *Elem.* I.post.3 (spherical) and *Sph.* I.20 – $\mathbf{Arc}(HQ)$ is repeatedly bisected until some arc remains, say $\alpha_i = \mathbf{Arc}(HQ)/2^i$, such that $\alpha_i < \mathbf{Arc}(PK) < \alpha_{i-1}$, which must result by *Elem.* X.1.cor. Then, either (a), for some m, $m\alpha_i = \mathbf{Arc}(QP)$, or (b), for all m, $m\alpha_i \neq \mathbf{Arc}(QP)$. In the case of (a), $\mathbf{Arc}(PR)$ is set out equal to α_i, using *Elem.* I.post.3 (transfer), so that $\mathbf{Arc}(QP) = (m+1)\alpha_i$, being commensurable with $\mathbf{Arc}(HQ)$, since they share the measure α_i; and $\mathbf{Arc}(QP) < \mathbf{Arc}(QR) < \mathbf{Arc}(QK)$. In the case of (b), where $m\alpha_i \neq \mathbf{Arc}(QP)$, let α_i be laid out from point Q repeatedly, using *Elem.* I.post.3 (transfer), some number of times, say $n\alpha_i = \mathbf{Arc}(QR)$, such that, once again, $\mathbf{Arc}(QR)$ and $\mathbf{Arc}(HQ)$ are both measured by α_i, and $\mathbf{Arc}(QP) < \mathbf{Arc}(QR) < \mathbf{Arc}(QK)$.

Finally, the *construction* for Case 2.2 explicitly requires the bisection of a given arc of a great circle, which can, again, be done using the problems of the *Elements* and this treatise by the method set out in the Commentary *Sph.* I.21 using *Sph.* I.19 to construct a great circle in the plane and then *Elem.* I.post.3 (transfer), III.3 and I.post.3 (transfer) (see p. 236). Alternatively, one might use the construction set out in the Commentary to *Sph.* II.9 (see p. 255).

Basis: *Elem.* I.post.3 (spherical, transfer), I.c.n.5, X.4, *Sph.* I.20, III.6.

Usage: Unused in the *Spherics*.

Diagram: The diagrams are solid representations of global objects, containing only intrinsic objects. They show two great circles intersecting the base circle with large half-lenses, and a pencil of great circles passing through a point, using the encodings developed in *Sph.* III.6. There are no false crossings, so that they depict the topology of the configuration.

Textual Comments: In the *construction* for Case 2.1, **Arc**(*KQ*), **Arc**(*QP*), and **Arc**(*HQ*) are qualified as being "of the same kind," ὁμοιογένεια – an expression not used elsewhere in the text and possibly redundant, since they are all arcs of the same great circle. This may have been a verbal indication that the reader should look to find the justification for the claim made in this passage in a treatment of general magnitudes, which makes claims about various magnitudes that are of the same kind. This qualification for magnitudes, however, is also not much used in *Elements* V and X – although it is there used in the definition of a ratio between magnitudes. On the other hand, since this qualification is found in Scholium 416, above, perhaps it was added to the text itself later to help indicate to the reader precisely which passage in the proposition the scholium is intended to justify.

Comparison of ratios, the diameter of the sphere to that of a small circle and related great-circle arcs to one another (*Sph.* III.10–III.12)

This group of theorems belongs together, because, although *Sph.* III.11 has some independent astronomical significance, *Sph.* III.10 is a lemma to III.12, and III.11 is also used in *Sph.* III.12. Hence, the primary purpose of these theorems probably has to do with the astronomical implications of *Sph.* III.12. Like the foregoing propositions, this group assumes that there is only one bundle of parallel circles.

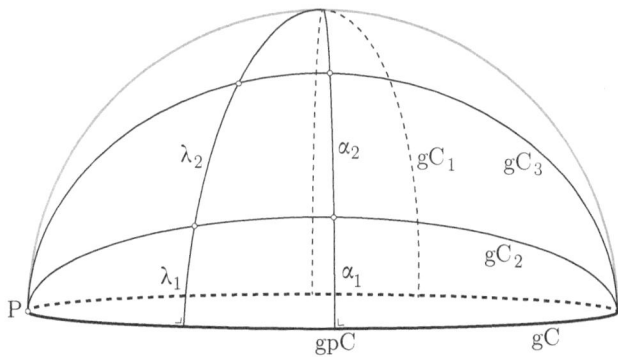

Figure 54: Perspective diagram, *Sph.* III.10.

The assumptions of *Sph.* III.10 are similar to those of III.9, namely a base great circle, gC, on which lies the pole, P, of the parallel circles, and to which are perpendicular both the greatest of the parallels, gpC, and another great circle, gC_1, that is oblique to gpC (Figure 54). Furthermore, if two arcs are cut off of gC_1 between its intersections with gC and gpC, say λ_1

and λ_2, such that λ_1 is contiguous with gC, and if great circles, gC_2, gC_3, are passed through pole P and the endpoints of λ_1 and λ_2, cutting off arcs α_1 and α_2 from gpC, then *Sph.* III.10 shows that

$$\alpha_1 : \lambda_1 = \alpha_2 : \chi, \text{ where } \chi < \lambda_2.$$

That is,

$$\alpha_1 : \lambda_1 > \alpha_2 : \lambda_2.$$

In the proof of the theorem, arc χ is introduced as an arc that satisfies the relation that is to be shown, but it is neither constructed nor named, as will be discussed in the Commentary to *Sph.* III.10, below.

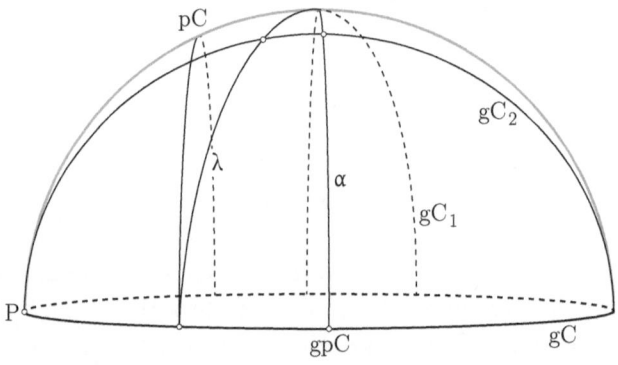

Figure 55: Perspective diagram, *Sph.* III.11.

Sph. III.11 starts with the same configuration of a base great circle, gC, on which lies the pole, P, of the parallels and to which are perpendicular both the greatest of the parallels, gpC, and an oblique great circle, gC_1, that is tangent to one of the parallels, pC (Figure 55). Moreover, if some arc, λ, is cut off of gC_1 contiguous with gC, and if another great circle, gC_2, is passed through pole P and the other endpoint of λ, cutting off arc α on gpC, then *Sph.* III.11 shows that

$$\text{diameter}_{sphere} : \text{diameter}_{pC} > \alpha : \lambda.$$

The configuration of *Sph.* III.12 is somewhat different from that of any other proposition in the text. We begin by assuming two great circles, gC_2, gC_3, tangent to some parallel circle pC_1, while great circle gC_1 is oblique to the greatest of the parallels, gpC, and tangent to a parallel circle greater than pC_1, say pC (Figure 56). Where P is one of the poles of gpC, another great circle, gC_4, is drawn through pole P and the point of tangency of gC_1 and pC, such that the intersections of gC_1 with gC_2 and gC_3 are between gpC and pC_1. Then if gC_2 and gC_3 cut off λ on gC_1 and ϱ on gpC, *Sph.* III.12

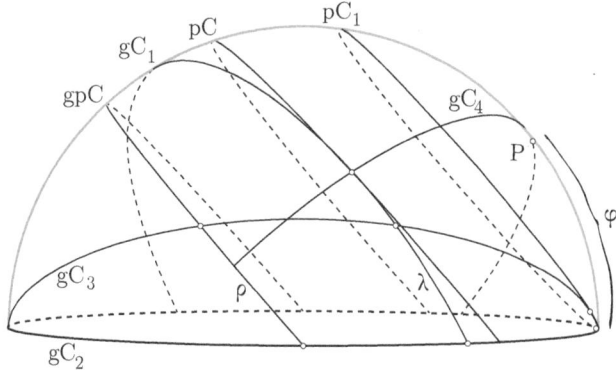

Figure 56: Perspective diagram, *Sph.* III.12.

shows that

$$2\,\mathrm{diameter}_{sphere} : \mathrm{diameter}_{pC} > \varrho : \lambda.$$

As with *Sph.* III.7 and III.8, the stipulation that gC_1 is tangent to pC, which is greater than pC_1, has the effect of setting bounds on the claim of the proposition. Namely, where φ is an arc of a great circle passing through the poles of gC_2 and cut off between P and gC_2, and ε is the inclination of gC_1 on gpC, the ratio inequality of *Sph.* III.12 holds in the open interval $0° < \varphi < 90° - \varepsilon$.

For spherical astronomy, *Sph.* III.11 and III.12 deal with the relationship between an arc of the ecliptic and its rising time (Schmidt 1943, 44–46). The details of these relations will be explained in the individual Commentaries to these propositions.

Sph. III.10 (Theorem)

This proposition is a lemma to *Sph.* III.12, and it is demonstrated using the same two-stage approach as used in *Sph.* III.9. The first case of the two-stage argument assumes that $\mathbf{Arc}(ZH)$ and $\mathbf{Arc}(DZ)$ are commensurable, and shows directly that $\mathbf{Arc}(BQ) : \mathbf{Arc}(DZ) > \mathbf{Arc}(QK) : \mathbf{Arc}(ZH)$, so that $\mathbf{Arc}(BQ) : \mathbf{Arc}(DZ)$ must be the same ratio as $\mathbf{Arc}(QK)$ to some arc that is less than $\mathbf{Arc}(ZH)$. The second case then assumes that $\mathbf{Arc}(ZH)$ and $\mathbf{Arc}(DZ)$ are incommensurable and uses the results of Case 1 in a double indirect argument to show that the theorem holds in this case as well. The constructions used in these arguments are the same as those discussed in the Commentary to *Sph.* III.9, above. The individual steps of the *demonstration* are based on claims from ratio theory, such as from *Elements* V generalized to ratio inequalities, and *Sph.* III.6.

This proposition uses both of the standard ways of expressing proportions and ratio inequalities (Heath 1912, clxxxviii; Netz 1999a, 156).

Namely, (1) *as* A *is to* B*, so* C *is to* D, ὡς [an object in the nominative] πρὸς
[an object in the accusative] οὕτως [an object in the nominative] πρὸς [an
object in the accusative], and (2) A *to* B *has a ratio greater than* C *to* D,
[an object in the nominative] πρὸς [an object in the accusative] μείζονα λόγον
ἔχει ἤπερ [an object in the nominative] πρὸς [an object in the accusative].
The use of the first expression involves Theodosios in asserting the existence
of an unconstructed and unnamed object that may stand in as one of the
terms of the proportion.

The structure of this proposition is similar to that of the previous proposition. In this theorem, there are three *specifications*, or "I say" statements –
two nearly identical assertions of what is to be shown, introducing each part
of the two-stage argument, and one shorter statement, introducing what is
to be shown in the second leg of the double indirect argument.

As well as the constructions discussed in the Commentary to *Sph.* III.9,
there is a claim made at the end of the *demonstrations* of both Case 1 and
Case 2.2 that one might be tempted to read as a construction. At the end
of Case 1, after noting that $\mathbf{Arc}(BQ) : \mathbf{Arc}(DZ) > \mathbf{Arc}(QK) : \mathbf{Arc}(ZH)$,
Theodosios states that "if we make (ποιῶμεν) as *BQ* to *DZ* so *QK* is to
some other, it will be less than *ZH*." This arc that is less than $\mathbf{Arc}(ZH)$,
however, is neither named, nor drawn into the diagram. In fact, there is
no general construction of such an arc using the methods of construction so
far developed in this treatise – namely, the abstractions of a straightedge
and actual compass. Indeed, if there were such a construction, it would
be possible to cut an arc, or an angle, in any given ratio, such as that of
1 : 3. Although Greek geometers devised a number of methods for producing
this particular construction using conic sections, special curves, and verging
lines, they believed, correctly, that such a construction is not possible using
only circles and lines.[94] There are a number indications that Theodosios is
not, in fact, intending this step as a construction in the normal sense. The
first, which carries little weight, is that this step is not developed in the
construction, but rather in the *demonstration*. More importantly, the form
of the verb is the subjunctive, 1st person, plural, and stands in the protasis
of a conditional sentence, an unlikely expression for a construction, which
would normally be indicated with an imperative, 3rd person, aorist verb.
Furthermore, the arc which is introduced in this way is neither named nor
drawn into the figure, so that it seems to have a different ontological status
from the other objects under discussion. Finally, the verb that expresses
this action, *to make*, ποιεῖν, is used by mathematical authors to express the
indirect consequences of an action (Mugler 1959, 350–351), or, more per-

---

[94] A number of the ancient methods for the trisection of an angle are given by Pappos
in his *Collection* IV (Hultsch 1876–1878, 272–288; Sefrin-Weis 2010, 146–155). Heath
(1921, 234–244) gives an overview of various solutions to this problem put forward by
Greek geometers. Knorr (1986a) also studies the ancient solutions to trisecting an angle.

tinently, for introducing a ratio or proportional term into a discussion of general magnitudes, whether or not any effective procedure has been established for doing this, for example in Euclid's *Data* 6, 8, 14, 15, 20, 21. For these reasons, it is likely that Theodosios does not intend us to understand the introduction of this unnamed arc as a construction, but simply as the hypothetical claim that if we were to make such a proportion, the fourth proportional that satisfied it would have such a property. Moreover, since this arc is less than $\mathbf{Arc}(ZH)$, which has been constructed, and since arcs were assumed to be continuous magnitudes, this amounts to an assumption of the existence of such an arc (Berggren 1991, 247).[95]

Throughout the *demonstration* we encounter a number of claims made about ratio inequalities that can be understood as generalizations of propositions demonstrated for proportions in *Elements* V, namely *Elem.* V.11, V.12, and V.16. The extension of *Elem.* V.16 concerns the ratio operation known as *alternation*, one of the classic operations used extensively by Greek geometers. The earliest extant proofs of these ratio operations applied to ratio inequalities are Pappos's *Coll.* VII.46–VII.51,[96] but, according to a scholium to Euclid's *Optics*, Heron also treated these operations for ratio inequalities (Heiberg 1895, 275). Furthermore, they are used freely by Euclid, Aristarchos, Archimedes and other mathematicians, and were apparently well-known (Dijksterhuis 1987, 53–54; Berggren and Sidoli 2007, 225–227).

The extension of *Elem.* V.11 to ratio inequalities is fairly straightforward, because it is essentially just the substitution of a ratio that is the same as one of the ratios in a ratio inequality. The generalization of *Elem.* V.12, however, may warrant some comment.

Towards the end of the *demonstration* for Case 1, Theodosios notes that since

$$\mathbf{Arc}(BX) > \mathbf{Arc}(XO) > \mathbf{Arc}(OQ) > \mathbf{Arc}(QP) > \mathbf{Arc}(PK), \text{ and}$$
$$\mathbf{Arc}(DL) = \mathbf{Arc}(LM) = \mathbf{Arc}(MZ) = \mathbf{Arc}(ZN) = \mathbf{Arc}(NH), \qquad \text{(a)}$$

while the multitude ($\pi\lambda\tilde{\eta}\vartheta o\varsigma$) of

$$\{\mathbf{Arc}(BX), \mathbf{Arc}(XO), \mathbf{Arc}(OQ)\} \text{ equals that of}$$
$$\{\mathbf{Arc}(DL), \mathbf{Arc}(LM), \mathbf{Arc}(MZ)\}, \qquad \text{(b.1)}$$

---

[95] Notice, however, that Berggren seems to equate this assumption with that of the midpoint of an arc, which is made a number of times in this text. The language of the *Spherics* indicates, however, that such midpoints are being constructed, and we showed in the Commentaries to *Sph.* I.21 and II.9 how this can actually be done.

[96] In this context it may be worth pointing out that the verb that Pappos uses in his *constructions* to introduce a magnitude as a fourth proportional is ποιεῖν, just as Euclid did in the *Data* propositions mentioned above (Hultsch 1876–1878, II.686–690; Jones 1986, 128–130).

and likewise that of

$$\{\mathbf{Arc}(QP), \mathbf{Arc}(PK)\} \text{ equals that of}$$
$$\{\mathbf{Arc}(ZN), \mathbf{Arc}(NH)\}, \tag{b.2}$$

therefore

$$\mathbf{Arc}(BQ) : \mathbf{Arc}(DZ) > \mathbf{Arc}(QK) : \mathbf{Arc}(ZH).$$

In order to flesh this out, we can note, first, that, by *Elem.* V.8, the condition (a) implies that

$$\begin{aligned}
\mathbf{Arc}(BX) : \mathbf{Arc}(DL) \quad &> \quad \mathbf{Arc}(XO) : \mathbf{Arc}(LM) \quad > \\
\mathbf{Arc}(OQ) : \mathbf{Arc}(MZ) \quad &> \quad \mathbf{Arc}(QP) : \mathbf{Arc}(ZN) \quad > \\
\mathbf{Arc}(PK) : \mathbf{Arc}(NH). &
\end{aligned} \tag{c}$$

Next, we note that *Elem.* V.12 shows that for any number of proportional magnitudes, the sum of the antecedents will be to the sum of the consequents in the same ratio. The proof of this theorem – which involves the claim that there are an equal "multitude," πλῆϑος, of antecedents and consequents (Heiberg and Stamatis 1969–1977, 20) – makes it clear, however, that the argument can be extended to show that if there is any multitude of magnitudes in decreasing ratio, the sum of the antecedents will have a ratio to the sum of the consequents that is less than the first ratio in the series, and greater than the last – that is,

$$\begin{aligned}
a_1 : a_2 \quad &> \quad b_1 : b_2 \quad &> \quad \dots \quad &> \quad n_1 : n_2 \quad \Rightarrow \\
a_1 : a_2 \quad &> \quad (a_1 + b_1 + \dots + n_1) \quad &: \quad (a_2 + b_2 + \dots + n_2) \quad &> \quad n_1 : n_2.
\end{aligned} \tag{d}$$

Then, if we separate the ratio inequalities in (c) into the two sets stated in (b.1) and (b.2), by the second ratio inequality that is concluded in (d), we have

$$\begin{aligned}
\mathbf{Arc}(BX) + \mathbf{Arc}(XO) + \mathbf{Arc}(OQ) &: \mathbf{Arc}(DL) + \mathbf{Arc}(LM) + \mathbf{Arc}(MZ) \\
&> \mathbf{Arc}(OQ) : \mathbf{Arc}(MZ),
\end{aligned} \tag{e}$$

and lesser than these, by (c), the first ratio inequality implied in (d) indicates that

$$\mathbf{Arc}(QP) : \mathbf{Arc}(ZN) > \mathbf{Arc}(QP) + \mathbf{Arc}(PK) : \mathbf{Arc}(ZN) + \mathbf{Arc}(NH). \tag{f}$$

But, looking at the diagram we see that

$$\begin{aligned}
\mathbf{Arc}(BQ) \quad &= \quad \mathbf{Arc}(BX) + \mathbf{Arc}(XO) + \mathbf{Arc}(OQ), \\
\mathbf{Arc}(DZ) \quad &= \quad \mathbf{Arc}(DL) + \mathbf{Arc}(LM) + \mathbf{Arc}(MZ), \text{ and} \\
\mathbf{Arc}(QK) \quad &= \quad \mathbf{Arc}(QP) + \mathbf{Arc}(PK), \\
\mathbf{Arc}(ZH) \quad &= \quad \mathbf{Arc}(ZN) + \mathbf{Arc}(NH),
\end{aligned} \tag{g}$$

so that, substituting the terms from (g) into (e) and (f), we have

$$\mathbf{Arc}(BQ) : \mathbf{Arc}(DZ) > \mathbf{Arc}(QK) : \mathbf{Arc}(ZH),$$

as stated in the text.

Basis: *Elem.* I.post.3 (spherical), I.c.n.5, V.def.5, V.def.7, V.8 (generalized to ratio inequalities), V.11 (for proportions and ratio inequalities), V.12, V.16 (for proportions and ratio inequalities), V.24 (generalized to ratio inequalities), *Sph.* I.20, III.6.

Usage: *Sph.* III.12.

Diagram: The diagrams are solid representations of global objects, containing only intrinsic objects. They are essentially the same as those in the previous proposition, having no false crossings, and depicting the topology of the configuration.

Sph. III.11 (Theorem)

This proposition has some astronomical implications for the upright sphere, and, perhaps more importantly, it is used in the following proposition. In terms of the *Spherics* it relies only on *Sph.* I.10 and I.15, but it also uses propositions from *Elements* I, III, V, VI, and XI.

This proposition also uses both of the expressions for proportions and ratio inequalities discussed in the Commentary to *Sph.* III.10.

The structure of this theorem is fairly standard, but there is one construction that is performed midway through the *demonstration*. This is because it must first be shown that $OR > RP$ before $RT = RP$ can be cut off from OR.

In this proposition, the normal procedure of introducing the letter-names alphabetically is disrupted, insofar as Q (Θ) is introduced in the *specification* following M, which was the last letter introduced in the *exposition*. Furthermore, two of the objects treated in the *specification*, $\mathbf{Arc}(BQ)$ and $\mathbf{Arc}(DH)$, were not mentioned at all in the *exposition*, so that they appear to be completely unspecified in the context of this proposition. In fact, however, these objects play a mathematically related role in this proposition as the two objects $\mathbf{Arc}(BQ)$ and $\mathbf{Arc}(DZ)$ had played in *Sph.* III.10 – or rather they will do so when both propositions are applied in *Sph.* III.12. This uncommon way of introducing $\mathbf{Arc}(BQ)$ and $\mathbf{Arc}(DH)$ in the *specification* of *Sph.* III.11 was likely meant to emphasize the underlying mathematical relationships, because at this point in the text, the relationships that are being discussed have become fairly complicated.

Towards the end of the *demonstration*, Theodosios articulates the conditions of a claim about a right triangle from which he asserts that $OR :$ $RT > \mathbf{Ang}(RTH) : \mathbf{Ang}(ROH)$ (Figure 57 (left)). The way that he expresses this claim suggests that he is referring to a well-known fact, and, indeed, Archimedes, in the course of his *Sand Reckoner*, gives a sort of enunciation, without proof, which asserts that in two right triangles with an equal leg, such that $\vartheta_1 > \vartheta_2$, so that $a_2 > a_1$ and $b_2 > b_1$, the following relations hold (Figure 57 (right)):

$$a_2 : a_1 < \vartheta_1 : \vartheta_2 < b_2 : b_1.$$

The right-hand inequality is the lemma that Theodosios requires.

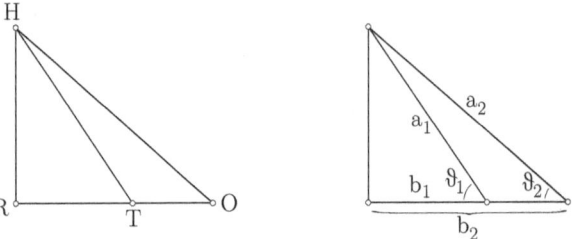

Figure 57: Diagrams for (left) a step in *Sph.* III.11, and (right) the proto-trigonometric lemmas.

The uses of these inequalities, along with their proofs in the ancient and medieval traditions, have been much studied (Schmidt 1943, 46–47; Knorr 1985; Knorr 1986b; Sidoli and Kusuba 2008, 22–27; Kunitzsch and Lorch 2010, 316–323; Acerbi 2014, 145); here we mention only the main points. Aristarchos used them – along with a related relation – without enunciation, computationally in his *Sizes and Distances of the Sun and the Moon* (Berggren and Sidoli 2007, 224–225). Archimedes enunciated the two ratio inequalities without proof, and used them computationally in his *Sand Reckoner*. These uses of the lemmas can be regarded as proto-trigonometric, in the sense that they were applied to determine numerical ratios as bounds on the sides and angles of right triangles.

The right-hand ratio inequality was proved by Euclid in *Optics* 8,[97] in a non-computational context, and used by Zenodoros and by Theodosios, here, in purely geometric contexts. Ptolemy, in *Alm.* I.10, in the course of producing a numerical approximation of the chord of 1°, demonstrated an equivalent of the left-hand ratio inequality. Finally, the right-hand inequality is demonstrated in a scholium to *Sph.* III.11, Scholium 450.

Hence, it seems that these lemmas were known by at least the early Hellenistic period. They were employed geometrically as well as computationally, and were regarded as sufficiently well known that they could be used without comment. The lemma that was transmitted with the *Spherics*

---

[97] Jones (1994) and Knorr (1994) have argued that the version of the *Optics* that Heiberg attributed to Theon's editorial work, should, in fact, be regarded as the older version of the treatise.

was probably included by Theon of Alexandria in his lost *Commentary to the Little Astronomy* (Acerbi 2014, 145).

Scholium 450 reads as follows (Czinczenheim 2000, 435; Figure 58):[98]

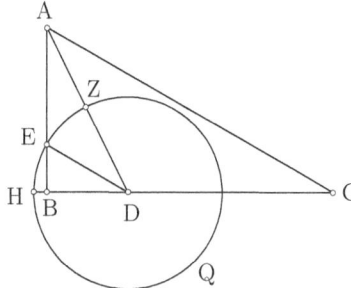

Figure 58: Scholium 450 to *Sph.* III.11. In the diagram in **A** (35r) line *AG* appears to be tangent to the circle, which it need not be.

> Let there be a right triangle, *ABG*, and let some ⟨line⟩, *AD*, be produced through ⟨it⟩.
>
> It is necessary to show that *BG* has to *BD* a greater ratio than angle *ADB* to *AGB*.
>
> For, let a parallel to *AG*, *DE*, be produced through *D*. And, since *DE* is greater than *BD* – through the ⟨fact of⟩ subtending the greater angle, for it is right – and ⟨angle⟩ *E* is acute, therefore angle *AED* is obtuse. Therefore, *AD* is greater than *ED*. Therefore, the circle drawn with center *D* and distance *DE* will cut *AD*, but fall beyond *BD*. Let it pass as *EQZ*.
>
> Therefore, triangle *AED* has to sector *EDZ* a greater ratio than triangle *EBD* to sector *EHD*. And, alternately, triangle *AED* has to triangle *EBD* a greater ratio than sector *EDZ* to sector *EHD*. But, as triangle *AED* to triangle *EBD*, so *AE* to *BE*, while as sector *EDZ* to sector *EHD*, so angle *ZDE* to angle *EDB*. And, by composition, *AB* has to *BE* a greater ratio than angle *ZDH* to angle *EDB*. But, angle *EDB* is equal to angle *AGB*, through the ⟨fact of⟩ *ED* being parallel to one of the sides, *AG*, of triangle *ABG*. Therefore, *AB* has to *BE* a greater ratio than angle *ZDB* to angle *AGB*. Therefore, *GB* has to *BD* a greater ratio than

---

[98] This translation can be compared with that by Knorr (1985, 364–365), from which it rarely differs. Note, however, that Knorr sometimes followed the mathematically superior readings of Munich Monac. 301. This manuscript, however, was copied at the end of the 16th century, and there is no reason to believe that these superior readings are not, in fact, improvements.

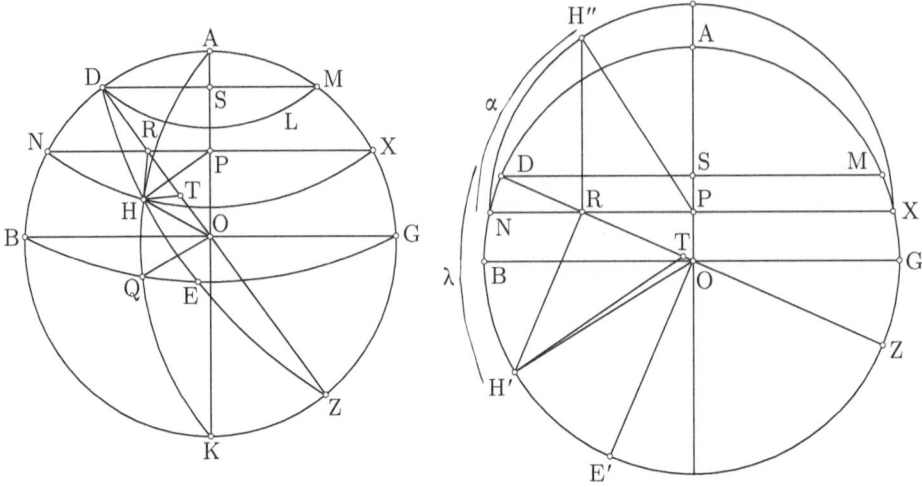

Figure 59: An analemma interpretation of *Sph.* III.11: (left) diagram from the translation, (right) reconstructed analemma diagram for *Sph.* III.11. Points marked with simple letter-names are in the plane of the base circle of *Sph.* III.11, which is also the analemma circle. Points marked with primes are rotated into the plane of the analemma.

angle *ZDB* to angle *EDB*, for *ED* cuts the sides proportionally,
and becomes as *AB* to *BE* so *BG* to *BD*.

The text of the scholium does not bother to point out that this means that $BG : BD > \mathbf{Ang}(ADB) : \mathbf{Ang}(AGB)$, as was to be shown.

The use of this lemma, and the applications of the proto-trigonometric lemmas in texts treating mathematical astronomy more generally, can be used to provide some insight into how the bound for *Sph.* III.11 was obtained.[99] The key insight, as often when reading the *Spherics*, is to model the configuration addressed in this theorem on an analemma diagram.[100] Considering the medieval diagram for this proposition (p. 181, Figure 59 (left)), the analemma circle will be the base circle, $\mathbf{gC}(ABG)$, and the two arcs that must be modeled on the analemma plane are $\mathbf{Arc}(DH)$ and $\mathbf{Arc}(NH) \sim \mathbf{Arc}(BQ)$.

[99] Schmidt (1943, 45–46) provides a discussion of this bound that will be satisfying to most modern readers, but which uses concepts for which we have no evidence in ancient or medieval sources.

[100] The use of an analemma diagram for explicating *Sph.* III.11 is suggested, both by the diagram for the proposition, and also by the similarity of the argument with that in Ptolemy's *Anal.* 6 (Sidoli 2020b, 54–59). The analemma configuration proposed here is similar to that provided by 'Id (1969) for rising times more generally with the difference that, in order to more closely model *Sph.* III.11, we take λ and α as contiguous with *D* and *B* on $\mathbf{gC}(ABG)$ as opposed to contiguous with an equinox, say *E*, as is done by 'Id.

Hence, in order to produce the analemma diagram, we set $\mathbf{gC}(ABG)$ as the analemma circle, project the solid circles $\mathbf{pC}(DLM)$, $\mathbf{pC}(NHX)$, $\mathbf{gpC}(BQG)$, and $\mathbf{gC}(DHEZ)$ onto their plane diameters as DM, NX, BG, and DOZ, respectively, and rotate the solid circles $\mathbf{gC}(DHEZ)$ and $\mathbf{pC}(NHX)$ 90° into the plane, producing $\mathbf{semiC}(NH''X)$ and $\mathbf{semiC}(DH'E'Z)$ in the analemma diagram (Figure 59 (right)).[101] The two circles $\mathbf{pC}(NHX)$ and $\mathbf{gC}(DHEZ)$ are mapped in two ways, both through orthogonal projection to lines, and through rotation to semicircles, so that $H \mapsto R$ and $E \mapsto O$, while $H \mapsto H''$ and $H \mapsto H'$, and $E \mapsto E'$. In this way, point H, the endpoint of both $\mathbf{Arc}(DH)$ and $\mathbf{Arc}(NH)$, which we wish to model, maps to the three points R, H', and H'' in the analemma, and $HR \mapsto H'R$, $HR \mapsto H''R$. Hence, $\mathbf{Arc}(DH)$ on the sphere, say λ, maps to $\mathbf{Arc}(DH')$ on the analemma, $\lambda \mapsto \mathbf{Arc}(DH')$. Likewise, if we set $\mathbf{Arc}(NH)$ as, say α, then $\alpha \mapsto \mathbf{Arc}(NH'')$ on the analemma. That is, in the analemma diagram, $\lambda := \mathbf{Ang}(DOH')$ and $\alpha := \mathbf{Ang}(NPH'')$. As always in an analemma construction, the angles and lengths at issue are conserved through the mapping.

Then, as point H is taken as an endpoint of λ at any position between the intersection of the oblique circle with the greatest of the parallels, E and E', and the tangent point of the oblique circle with one of the parallels, D, R will lie somewhere on line DO between D and O. Hence, the two angles in which we are interested, $\mathbf{Ang}(DOH')$ and $\mathbf{Ang}(NPH'')$, are two angles in right triangles, $\mathbf{rightT}(ROH')$ and $\mathbf{rightT}(RPH'')$, under the same height, $H'R = H''R$. Moreover, the bases of $\mathbf{rightT}(ROH')$ and $\mathbf{rightT}(RPH'')$, RO and RP, are themselves the sides of a right triangle, $\mathbf{rightT}(RPO)$, of which RO is the hypothenuse and RP a leg. Hence, $RO > RP$, *Elem.* I.19. This would be sufficient to apply the proto-trigonometric lemma discussed above, but in order to make the situation analogous to what is done in the solid configuration, we can construct $\mathbf{rightT}(RTH') \cong \mathbf{rightT}(RPH'')$ inside of $\mathbf{rightT}(ROH')$. Then the proto-trigonometric lemma can be applied exactly as expressed, so that $RO : RP = RO : RT > \mathbf{Ang}(DTH') : \mathbf{Ang}(DOH') = \mathbf{Ang}(NPH'') : \mathbf{Ang}(DOH') := \alpha : \lambda$.

Now, since R always lies on line DO between D and O, the geometry of the construction makes it clear that T always lies on line RO between R and O, so that, on the analemma, the least upper bound on the ratio $\mathbf{Ang}(DTH') : \mathbf{Ang}(DOH')$ is that of the ratio of the sides of $\mathbf{rightT}(RPO)$ itself. But $\mathbf{rightT}(RPO) \sim \mathbf{rightT}(DSO)$, of which the sides are the radii of the sphere and of the parallel circle tangent to $\mathbf{gC}(DEZ)$ in the sphere, so that the upper bound of the ratios of the angles in the analemma configuration is $DO : DS = DZ : DM = \text{diameter}_{sphere} : \text{diameter}_{tangent\,circle}$. Furthermore, since, if $\mathbf{Arc}(DE)$ on the sphere is divided up into some num-

---

[101] See Sidoli (2020b, 36–40) for a recent discussion of the types of operations used in analemma constructions.

ber of equal arcs, λ_n, the corresponding arcs of $\mathbf{Arc}(BE)$, say $\alpha(\lambda_n)$, will be ever greater as λ_n is taken closer to D, *Sph.* III.6. Taking λ contiguous with D will produce the greatest $\alpha(\lambda) : \lambda$, so that the configuration represented in this analemma diagram will determine the upper bound on the ratio $\alpha(\lambda) : \lambda$ in general.

In terms of spherical astronomy, if we take $\mathbf{gC}(ABG)$ as the solstitial colure, $\mathbf{gC}(BEG)$ as the celestial equator of which A is a pole, $\mathbf{gC}(DEZ)$ as the ecliptic with $\mathbf{pC}(DLM)$ as a tropic circle, then *Sph.* III.11 shows that the colure, or horizon, $\mathbf{gC}(AHK)$, that cuts off an arc of the ecliptic, λ, contiguous with the tropic circle will also cut off an arc of the equator, $\alpha(\lambda)$, such that

$$\text{diameter}_{sphere} : \text{diameter}_{tropic\ circle} > \alpha(\lambda) : \lambda.$$

That is, on the upright sphere, $\varphi = 0°$, the ratio of the setting time of an arc of the ecliptic contiguous with the solstitial colure to the corresponding arc of the ecliptic itself is less than the ratio of the diameter of the sphere to the diameter of the tropic circles (Schmidt 1943, 44–46). Moreover, because this is a relation on the upright sphere, it has all of the symmetries discussed with regard to *Sph.* III.5 and III.6, above. Because the diameter of the tropic circles depends on the magnitude of the obliquity of the ecliptic, ε, this proposition can be regarded as stating a bound on the ratio between an arc of the ecliptic and its right ascension that is itself determined by ε.

Basis: *Elem.* I.def.17, I.post.1, I.post.3 (spherical), I.post.4, I.c.n.1, I.c.n.2, I.3, I.4, I.12, I.19, V.15, VI.2, VI.33, XI.def.3, XI.10, XI.16, XI.19, *Sph.* I.10, I.15.

Usage: *Sph.* III.12.

Diagram: The diagram is a solid representation of global objects, containing both intrinsic and extrinsic objects, in the same style that was seen before in the diagram for *Sph.* II.17. It is a good example of a diagram that seems to be in linear perspective, with the only exception being that the circles that meet the base circle are dawn as half-lenses, not as sections of a smooth curve. As discussed in the Introduction, this should be understood as a deliberate choice, not a failure to understand how circles appear in perspective (see p. 85, above). The diagram uses the convention of arcs facing in the same direction to denote parallel circles, and everything in the diagram represents objects in one hemisphere, appearing inside the base circle. Because it is essentially a perspective diagram, there are many false crossings among the various objects.

Textual Comments: The letter Θ (Q) is introduced out of the standard alphabetical order. If in the *exposition*, the great circle through the poles had been called AHΘK ($AHQK$) instead of AHK, this would be corrected.

As it is, Θ is introduced, seemly out of the blue, as part of the letter-name of arc BΘ. Probably there has been an error in the transmission. In the diagrams of most of the medieval manuscripts, the letter-name Λ (ﻝ, *L*) appears to mark the false crossing of circle ΔΛM (*DLM*) and line AK, which is not a point (for example, **A**, 36r; **B**, 79v; **D**, 15r; **F**, 45r; **I**, 51a). Moreover, in the text itself Λ is not used in the letter-name of any object besides circle ΔΛM, so we have taken it to be part of the name of **pC**(*DLM*), and not as designating any specific point. Hence, in our diagram, we place *L* free-floating on **pC**(*DLM*).

Sph. III.12 (Theorem)

This proposition appears to be an important goal of the treatise. It depends on the problems *Sph.* I.20, I.21, II.15, and key theorems from *Spherics* II, such as *Sph.* II.10–II.13, as well as both of the two previous theorems, *Sph.* III.10 and III.11.

In the *enunciation* and *exposition*, bounds on the position of **gC**(*EZ*) are imposed in two ways. It is stated, firstly, that **gC**(*EZ*) is tangent to a circle greater than **pC**(*AG*), and, secondly, that **gC**(*EZ*) cuts **gC**(*AB*) and **gC**(*GD*) between **gpC**(*BZ*) and **pC**(*AG*). Both of these claims imply that if the angle between **gC**(*EZ*) and **gC**(*BZ*) is ε and the angular separation of pole *L* from **gC**(*AB*) or **gC**(*GD*) is φ, then $0° < \varphi < 90° - \varepsilon$. Moreover, despite the fact that these claims are both articulated as assumptions, they are not independent and either could be used to show the other.

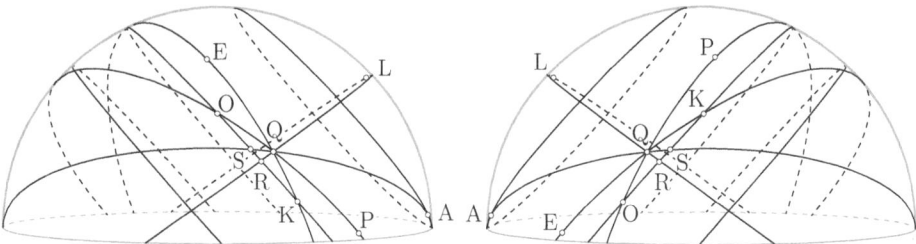

Figure 60: Perspective diagrams comparing the orientation of *Sph.* III.12: (left) the orientation presented in the medieval manuscripts, (right) the opposite orientation, similar to that found in the other diagrams for this book.

The great circles that are introduced first and then referred to later as the "initial" circles in this proposition play the same role that such "initial" circles have played in the earlier propositions of this book. Since, however, the base circle of the diagram, **gC**(*LEM*), is not one of these "initial" circles, the relationship between the diagram and the geometric objects that it depicts is different in this proposition, and we must understand the diagram as different from those of *Sph.* III.5–III.11. Although there is a new arrangement of the principal circles in this proposition, neither the *enunciation* nor *exposition* make this clear. It can, however, be seen in the diagram, and becomes clear as soon as one models the diagram on a solid or armillary sphere. In order to clarify this situation, we can consider a perspective diagram, as a way of emulating the use of a solid sphere (Figure 60).

In the diagram for *Sph.* III.12, the points, A and G, at which the "initial" circles, $\mathbf{gC}(AB)$ and $\mathbf{gC}(GD)$, meet one of the parallels, $\mathbf{pC}(AG)$, are depicted on the right-hand side of the diagram, whereas in the diagrams for *Sph.* III.7 and III.8 the analogous points are depicted on the left-hand side of the diagram (Figure on p. 182). The effect of this is that, in *Sph.* III.12, when $\mathbf{gC}(AB)$ and $\mathbf{gC}(GD)$ are projected through Q and K onto one of the parallels they will meet the parallel at a point that is in the direction of the tangent point, E, of the oblique circle, $\mathbf{gC}(EZ)$. That is, if great circles are drawn through the pole L and each of Q and K, they will meet the parallels in points that are farther away from point E than the points where $\mathbf{gC}(AB)$ and $\mathbf{gC}(GD)$ meet the parallels. Thus, in the figure for *Sph.* III.12, point N is farther from E than point B, and point X is farther from E than point D (see Figure on p. 182). Indeed, in the manuscript diagrams to *Sph.* III.12, we see that point S – the projection of point Q by $\mathbf{gC}(AB)$ onto $\mathbf{pC}(OK)$ – is in the direction of E with respect to R – the projection of Q by $\mathbf{gC}(LQ)$ onto $\mathbf{pC}(OK)$. That is, S lies on $\mathbf{Arc}(OR)$, not on $\mathbf{Arc}(RK)$ (Figure on p. 182 and Figure 60 (left)). In fact, however, the argument presented in this proposition will hold for either orientation, since the key relation is $\mathbf{Arc}(SK) < 2\mathbf{Arc}(RK)$, which holds in either orientation (Figure 60 (left) and (right)).

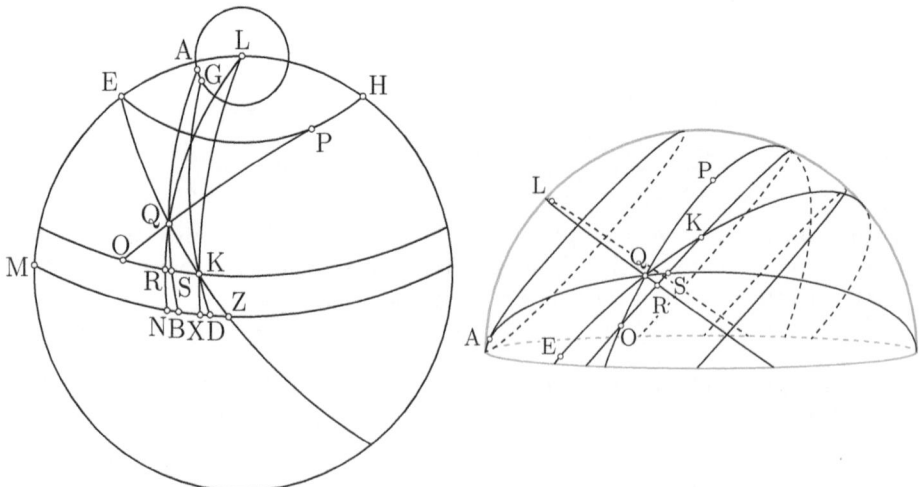

Figure 61: Diagrams for the other orientation of *Sph.* III.12: (left) a medieval style diagram, (right) perspective diagram of the key arcs.

These considerations can also help us understand how the bound asserted in this proposition was obtained, and the role of *Sph.* III.10 in the proof (Figure on p. 182, Figures 60, 61).[102] The key is to notice that, in terms of the arcs of the greatest parallel circle that have a significant relationship to **Arc**(QK) of the oblique great circle, that relationship which was handled in *Sph.* III.11 is related to **Arc**(NX) $\sim$ **Arc**(RK), while that which is involved in this proposition is **Arc**(BD) $\sim$ **Arc**(SK). Hence, we can simply consider the relationship between **Arc**(RK) and **Arc**(SK) on **pC**(OK). Then, in the diagram, it is clear that for **Arc**(AQS) to produce S on OK between O and K, where **Arc**(OK) = 2**Arc**(RK), it is only necessary that A lie between **Arc**(EQK) and **Arc**(PQO) in the same region of the sphere as pole L. Moreover, if **pC**(EP) > **pC**(AG), namely if $0° < \varphi < 90° - \varepsilon$, point A must lie between **Arc**(EQK) and **Arc**(PQO) on the side of pole L. Hence, within these bounds on the size of **pC**(EP), it is clear that **Arc**(SK) < 2**Arc**(RK), that is **Arc**(BD) < 2**Arc**(NX). Then, since *Sph.* III.11 provides the least upper bound on **Arc**(MN) : **Arc**(EQ), if this ratio can be related to **Arc**(NX) : **Arc**(QK), then we can extend what we know from *Sph.* III.11 to the arcs at stake in this proposition. It seems clear that *Sph.* III.10 was specially designed to make such a connection, namely **Arc**(MN) : **Arc**(EQ) > **Arc**(NX) : **Arc**(QK). The rest of the argument follows by substitutions from what was already asserted. Furthermore, it should be noted that the argument given in the text follows for both of the orientations considered, so that, within the stated bounds of $0° < \varphi < 90° - \varepsilon$, the claim of this proposition holds for any arc on **gC**(EZ) and that which corresponds to it on **gC**(BZ). Finally, it should be noted that whereas the considerations on the analemma make it clear that *Sph.* III.11 gives the least upper bound on **Arc**(MN) : **Arc**(EQ), the application of *Sph.* III.10, with its non-constructive ratio inequality, implies that *Sph.* III.12 does not assert the least upper bound, but simply *an* upper bound.

This proposition only uses the second expression for ratio inequalities discussed in the Commentary to *Sph.* III.10.

As in *Sph.* III.11, the key arcs, **Arc**(BD) and **Arc**(QK), are introduced for the first time in the *specification*. Indeed, points Q and K are completely unspecified, since they were not mentioned in the *exposition*. This similarity of structure was probably meant to highlight the underlying geometrical similarities binding *Sph.* III.10–III.12, although there are some important differences between the configuration of *Sph.* III.12 and that of III.10 and III.11. In *Sph.* III.12, however, although **Arc**(BD) and **Arc**(QK) are locally

---

[102] Schmidt (1943, 46) gives an explanation of the bound for this proposition that is mathematically satisfying but does not do much to help us understand how it was determined using ancient methods.

unspecified, the letter-names may have been introduced alphabetically. (See the "Textual Comments," below, for a discussion of the issue of *M*.)

The *demonstration* begins by setting up the conditions of the claim that **Arc**(*OR*) = **Arc**(*RK*) using expressions that make it seem like this was a known fact. Although this step does not result directly from a previous proposition in the *Spherics*, it does follow from *Sph.* II.11, as is shown in a marginal gloss. Scholium 456 reads as follows (Czinczenheim 2000, 436–437; Figure 62):[103]

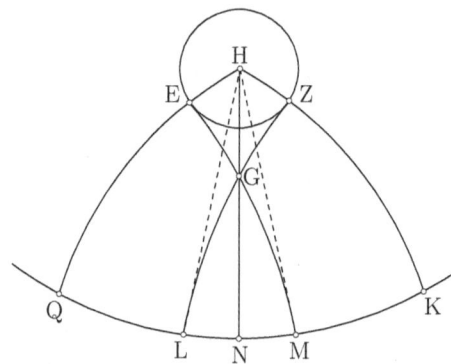

Figure 62: Scholium 456 to *Sph.* III.12. In **A** (36v), the diagram is drawn in a different ink from the other diagrams to the scholia and the dotted lines are missing.

For, let there be parallel circles and great ⟨circles⟩, *EM* and *ZL*, touching *EZ* at points *E* and *Z*, and through the pole, *H*, and ⟨point⟩ *G* a great ⟨circle⟩, *HGN*.

I say that *LN* is equal to *NM*.

For, ⟨let there be drawn⟩ great circles through the pole and the ⟨points of⟩ contact, *HEQ* and *HZK* – and it is obvious that they are upright on circles *EM* and *ZL*. And let *HL* and *HM* be joined.

Then, equal segments of circles, *EH* and *HZ* and continuing, are set up at an upright on the diameters, those from *E* and *Z*, of equal circles, *EM* and *ZL*, and equal circumferences *EH* and *HZ*, less than half of the set-up segments are cut-off, and the straight line joined from *H* to *G* is common to them, so that circumference *EG* is equal to circumference *GZ*. Then, through the same ⟨reasons⟩, since *HL* is equal to *HM*, *EM* is also equal to circumference *ZL*. So that circumference *GL*, as a remainder, is equal to circumference *GM*. And, because circle *HGN* cuts *QNK* through the poles, it is upright to it. Then, an upright segment, *GN* and what continues with it, is set up on the diameter, that from *N*, of a circle, *QNK*, and *GN* is cut off, being less than half of the set-up segment, and circumference *GM* is equal to circumference *GL*, so that, also, the straight line from *L* to *G* is equal to that from *G* to *M*,[104] so that, also, circumference, *LN*, is equal to circumference, *NM*.

Since the text of the scholium is somewhat loose in its expression, it may be helpful to summarize the argument. The initial assumption is that there are two parallel circles, **pC**(*EZ*), **pC**(*QNK*), and two great circles, **gC**(*EGM*), **gC**(*ZGL*) that are tangent to **pC**(*EZ*) at points *E* and *Z*, and intersect at a point *G* between **pC**(*EZ*) and **pC**(*QNK*), while a third great circle, **gC**(*HGN*) passes through the poles of the parallel circles and the point of intersection, *G*.

The *construction* consists of drawing **gC**(*HEQ*) through points *H* and *E*, and **gC**(*HZK*) through *H* and *Z*, *Sph.* I.20. These great circles must be perpendicular to **gC**(*EGM*) and **gC**(*ZGL*) by *Sph.* II.5, I.15. Also, polar radii *HL* and *HM* are joined.[105]

The argument of the lemma consists of setting up the conditions of *Sph.* II.11 three times. First, it shows that, since **Arc**(*EH*) = **Arc**(*HZ*) and *HG* is common, therefore **Arc**(*EG*) = **Arc**(*GZ*), and again, that since **Arc**(*EH*) = **Arc**(*HZ*) and *HL* = *HM*, therefore **Arc**(*EM*) = **Arc**(*ZL*). Hence, by subtraction, **Arc**(*GM*) = **Arc**(*GL*). Finally, the conditions of *Sph.* II.11 are again established, in which the two segments of *Sph.* II.11 are both the same segment considered from different sides in this scholium, so that, since **Arc**(*GN*) is common and *GL* = *GM*, subtending equal arcs in equal circles, then **Arc**(*LN*) = **Arc**(*NM*).

In the course of the *demonstration* of *Sph.* III.12, Theodosios infers from

$$\text{diameter}_{sphere} : \text{diameter}_{\textbf{pC}(EPH)} > \textbf{Arc}(NX) : \textbf{Arc}(QK)$$

that if the antecedents are doubled then the ratio inequality will be maintained, in order to apply the substitution of **Arc**(*BD*) = 2**Arc**(*NX*). This can be regarded as an extension of *Elem.* V.4 to ratio inequalities that follows almost immediately from *Elem.* V.4 itself. *Elem.* V.4 shows that, where

---

[104] Czinczenheim (2000, 37) follows the reading of **A** (36v) over that of **B** (78v), in taking the two individual Γs (*G*s) as *N*s. But whether because it preserves an earlier reading or is a correction, the reading in **B** is better and is mathematically necessary.

[105] Although lines *HL* and *HM* are produced in the *construction* and referred to as such in the *demonstration* they are not drawn into the diagram in the margin of **A** (36v). There is no diagram for this scholium in **B**. Notice that whereas these two lines are constructed and named, the lines from *H* to *G*, *L* to *G*, and *G* to *M*, which serve a similar function in the argument, are neither constructed nor named.

$m, n \in \{1, 2, 3, \dots\}$, $a_1 : a_2 = b_1 : b_2 \Rightarrow ma_1 : na_2 = mb_1 : nb_2$. Hence, using an argument similar to the application of *Sph.* III.10 in III.12, we note that

$$\text{diameter}_{sphere} : \text{diameter}_{\mathbf{pC}(EPH)} = \mathbf{Arc}(NX) : \chi,$$

where $\chi < \mathbf{Arc}(QK)$. That is, by *Elem.* V.4,

$$2\,\text{diameter}_{sphere} : \text{diameter}_{\mathbf{pC}(EPH)} = 2\mathbf{Arc}(NX) : \chi,$$

and since $\chi < \mathbf{Arc}(QK)$,

$$2\,\text{diameter}_{sphere} : \text{diameter}_{\mathbf{pC}(EPH)} > 2\mathbf{Arc}(NX) : \mathbf{Arc}(QK),$$

as stated in the text.

From the perspective of spherical astronomy, we can regard $\mathbf{gpC}(MBZ)$ as the celestial equator, of which L is a pole, $\mathbf{gC}(EQKZ)$ as the ecliptic tangent to the tropic circle $\mathbf{pC}(EPH)$, $\mathbf{gC}(LEMH)$ as the solstitial colure, and $\mathbf{gC}(AQB)$, $\mathbf{gC}(GKD)$ as two configurations of the horizon both tangent to v-circle $\mathbf{pC}(AG)$. Then, $\mathbf{Arc}(QK)$ is any arc of the ecliptic, λ, cut off by the two configurations of the horizon and $\mathbf{Arc}(BD)$ is the corresponding arc of the equator – that is, the arc of the equator that rises in the same time as λ – namely $\varrho(\lambda, \varphi)$. Then, *Sph.* III.12 shows that, where $0° < \varphi < 90° - \varepsilon$, for any λ,

$$2\,\text{diameter}_{sphere} : \text{diameter}_{tropic\ circle} > \varrho(\lambda, \varphi) : \lambda.$$

In this way, Theodosios has derived a mathematical claim about the rising times of all arcs of the ecliptic, including those in $\mathbf{semiC}(\text{Can} \rightarrow \text{Sag})$, for which *Sph.* III.8 does not hold (Schmidt 1943, 44–46). Because $\text{diameter}_{tropic\ circle}$ depends on the magnitude of the obliquity of the ecliptic, ε, this proposition determines a bound on the ratio between an arc of the ecliptic and its rising time that is related to the parameter ε.

Basis: *Elem.* I.post.3 (spherical), III.29 (by implication), V.4 (generalized to ratio inequalities), V.8, *Sph.* I.15 (by implication), I.20, I.21, II.10, II.11 (by implication), II.13, II.15, III.10, III.11.

Usage: Unused in the *Spherics*.

Diagram: The diagram for this proposition uses a solid representation of global objects, using only intrinsic objects. It follows the same conventions as the diagrams for *Sph.* III.7 and III.8, with a minor difference. One of the parallel circles is depicted not as touching the base circle, but as cutting it. This is because the base circle has a different geometrical and astronomical interpretation in this proposition than in the foregoing propositions. The effect of this is that in this diagram, like those of *Sph.* II.17 and II.19, the two hemispheres are encoded as inside and outside of the base circle, following the convention for planar representations. As in the diagrams for *Sph.* III.7 and III.8, the great circles that meet the full parallel are tangent to it, not intersecting. There are no false crossings, so the diagram depicts the topology of the configuration.

Textual Comments: The diagram in **A** (37r) is poorly drawn in two respects. In the first place, Z appears to mark the intersection of the three great circles EZ, MB, and ΛEM (*LEM*), but this is impossible because MB is the greatest of the parallels, not a parallel circle equal to circle EH.

Secondly, point A is depicted on great-circle arc ΛK (*LK*), but there is no reason that this should be the case. Hence, this is a case in which the tendency in the manuscript diagrams towards overspecification has led to the representation of a special case – that is, while it is not necessarily incorrect, it implies more conditions than are required by the text.

A number of the Greek manuscripts avoid the mathematical error of making great circle EZ intersect great circles MB and ΛEM at the same point by drawing EZ crossing MB inside the base circle ΛEM. But they also show point Z placed on ΛEM not MB (**B**, 79v; **D**, 15v; **F**, 45v). Indeed, these manuscripts denote the great circle that **A** calls MBZ with the letter-name MB or MBΞ (*MBX*) in the text (Heiberg 1927, 158; Czinczenheim 2000, 171). They also continue to place point A on arc ΛK. Some later manuscripts have the same incorrect diagram as **A**, for example Par.gr. 2342 (128v). Heiberg's preferred **K** breaks off before this diagram would have appeared.

When great circle MBZ is introduced in the *exposition* as the greatest of the parallels, M is used before H, which follows in the next clause. It may be that the original name of this great circle was BZ and that the M was added later by an editor who was swayed by the significance of circle ΛEM in the diagram. If the circle had originally been called BZ, then M would have been introduced in the standard order following Λ in the *construction*.

Comparison with Arabic: In manuscript **I** (51b) of Thābit's version, the diagram is essentially the same as that in **A**, although rotated 90°. Moreover the oblique great circle is called م ب ز (*MBZ*) in the text (Kunitzsch and Lorch 2010, 300), supporting the reading of **A** and indicating that there was at least one other manuscript in the 9th century that also had this reading.

In many of the diagrams for this proposition in the manuscripts of al-Ṭūsī's *Revision of Theodosios's Spherics*, points ١ (*A*) and ح (*G*) are drawn meeting circle ح١ outside circles ط ل (*LQ*) and ح ل (*LK*) to the right. For example, the diagram in Sipahsalar 4727 (p. 129), a copy that was made from a copy that had been corrected by al-Ṭūsī himself, shows ١ and ح to the right, but also depicts circle ح١ not as a full circle, but as an arc inside of great circle ه م ح (*EMH*). The diagram in our translation has been influenced by those in the al-Ṭūsī tradition.

Comparison with Latin: In Gerard's Latin, *MBZ* is called *MS* (*MX* in our translation). Moreover, the diagram in one of the Latin manuscripts also shows *A* as any point on circle *AG*, in this case inside great circles *LT* (*LQ*) and *LK* (Kunitzsch and Lorch 2010, 301, 340 n. 12).

Comparison of the projections of equal and contiguous arcs in opposite halves of a great circle onto another oblique great circle (*Sph.* III.13)

This proposition deals at once, and in one figure, with the two configurations previously handled separately in *Sph.* III.5, III.6, III.9–III.11, on the one hand, and *Sph.* III.7, III.8, III.12, on the other – namely, great circles that either pass through the pole of the parallel circles or are tangent to a pair of equal parallels.

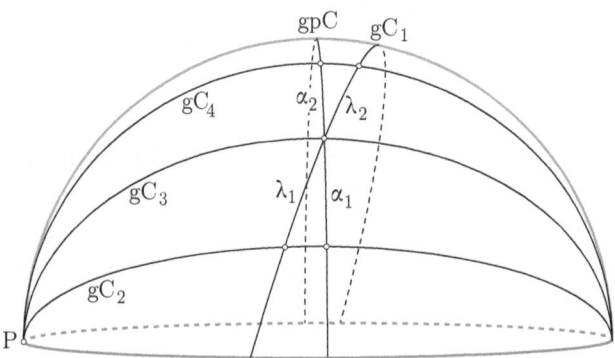

Figure 63: Perspective diagram, *Sph.* III.13, first configuration.

Specifically, if we assume that some great circle, gC_1, is oblique to the greatest of the parallels, gpC, and equal arcs, $\lambda_1 = \lambda_2$, are cut off of gC_1 on each side of its intersection with gpC, and if three other great circles, gC_2, gC_3, gC_4 – either (first configuration) passing through the poles or (second configuration) tangent to the same parallel circle, pC – are passed through the endpoints of arcs λ_1 and λ_2, then they will cut off between themselves equal arcs of gpC, say $\alpha_1 = \alpha_2$ or $\varrho_1 = \varrho_2$. That is, assuming the stated construction,

$$\lambda_1 = \lambda_2 \Rightarrow \alpha_1 = \alpha_2 \text{ or } \varrho_1 = \varrho_2.$$

Astronomically, this proposition is related to the second part of *Phen.* 8, 12, and 13. Namely, if we set gpC as the celestial equator, gC_1 as the ecliptic of which $\lambda_1 = \lambda_2$ are equal arcs about an equinoctial point, and gC_2, gC_3, gC_4 are configurations of the horizon, *Sph.* III.13 explicitly shows that the rising and setting times of arcs of the ecliptic, namely $\alpha_1 = \alpha_2$ or $\varrho_1 = \varrho_2$,

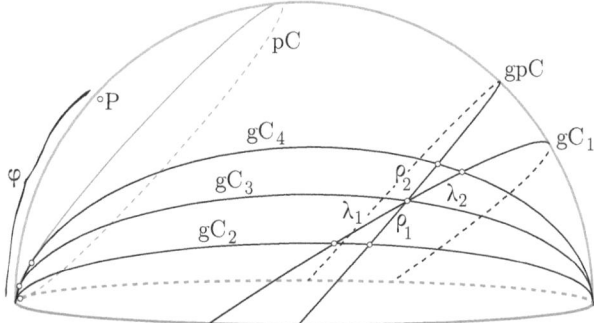

Figure 64: Perspective diagram, *Sph.* III.13, second configuration.

are symmetrical about each of the equinoxes, as stated in the second part of *Phen.* 12, and 13. Furthermore, in the course of the proof it is shown that the arcs of gC_3 and gC_4 cut off between gC_1 and gpC are equal, $\eta_1 = \eta_2$ (not labeled in the Figures), so that equal arcs of the ecliptic on either side of an equinox rise over equal arcs of the horizon, as shown the second part of *Phen.* 8.

In *Phen.* 12, Euclid argues directly that equal arcs of the ecliptic in **Quad**(Can → Vir) set in decreasing times and then uses an argument involving a claim similar to that made in *Sph.* III.13 to extend these considerations to all of **semiC**(Can → Sag). Another two-stage argument is made for the rising times of **semiC**(Cap → Gem) in *Phen.* 13. In the Commentary to *Sph.* III.8 above, however, we argued that the polar symmetry of diagram allows these claims about the setting times of all of **semiC**(Can → Sag) and the rising times of **semiC**(Cap → Gem) to be made directly from the results of that theorem. Moreover, it seems likely that the avoidance of directional terminology, such as was used in the *Phenomena*, was intentional on Theodosios's part, and was meant to highlight the polar symmetry. If, however, one does not believe that Theodosios intended us to read any symmetry into the diagram for *Sph.* III.8, then *Sph.* III.13 could be used to extend the results of III.8 to the other quadrant of the relevant semicircle. Namely, to claim that

$$\bar{\varrho}(\text{Sag}) \quad > \quad \bar{\varrho}(\text{Sco}) \quad > \quad \bar{\varrho}(\text{Lib}),$$

and hence, also,

$$\varrho(\text{Gem}) \quad > \quad \varrho(\text{Tau}) \quad > \quad \varrho(\text{Ari}).$$

Sph. III.13 (Theorem)

Although this proposition relies on a number of theorems of *Spherics* II and *Elements* III, the key to the proof is *Sph.* III.3, dealing with congruence.

Indeed, because *Sph.* III.3 is the latest result from the *Spherics* on which this proposition rests, *Sph.* III.13 could have immediately followed that theorem. We should also consider the possibility that this proposition was not part of the original composition.

As mentioned above, astronomically this proposition is related to *Phen.* 8, 12, and 13, however, the role that the geometrical fact that is shown in *Sph.* III.13 plays in the mathematical development of the two treatises is rather different. In *Phen.* 8, a claim that is shown along the way in *Sph.* III.13 is demonstrated using something equivalent to *Sph.* III.7. The first part of *Phen.* 12 relies on something like what is shown in *Sph.* III.8, while in the second part something equivalent to *Sph.* III.13 is shown and then used to extend the claims of the proposition to another quadrant of the ecliptic. Although the methods of the two treatises are somewhat different – because the *Phenomena* relies on the explicit concept of motion while the *Spherics* is static geometry – it seems clear that a proposition like *Sph.* III.13, shown geometrically in the same way as we now read in the *Spherics*, was not available to Euclid when he wrote his *Phenomena*.

Basis: *Elem.* III.def.1 (by implication), III.def.11, III.26, III.27, III.28, *Sph.* I.def.1(by implication), I.def.2 (by implication), I.6 (by implication), II.10, II.13, II.17, II.18, III.3.

Usage: Unused in the *Spherics*.

Diagram: The diagram represents local objects that are only intrinsic, but, like that of *Sph.* II.10 and III.3, it could be classified as either planar or solid. In fact, it shows only the local interactions of the objects under discussion that are relevant to the proposition. Hence, the only way that we can tell that these objects lie on the surface of the sphere is through the fact that they are represented with the gentle arcs of large circles. This style of diagram would be further developed by Menelaos in his *Spherics*.

Comparison of arcs of parallel circles cut off between two great circles that are both oblique to the parallels (*Sph.* III.14)

In terms of its geometrical expression, this proposition seems to have little bearing on the topics of the rest of this book, but the objects introduced in the proof, as well as the diagram, make it clear that this proposition is related to *Phen.* 4. In terms of its astronomical interpretation, it can be understood to deal with with the relative risings and settings of points of the ecliptic (Schmidt 1943, 21–24).

What *Sph.* III.14 actually states and shows, however, is that if there is some base great circle, gC, tangent to a parallel circle, pC_1, and if there

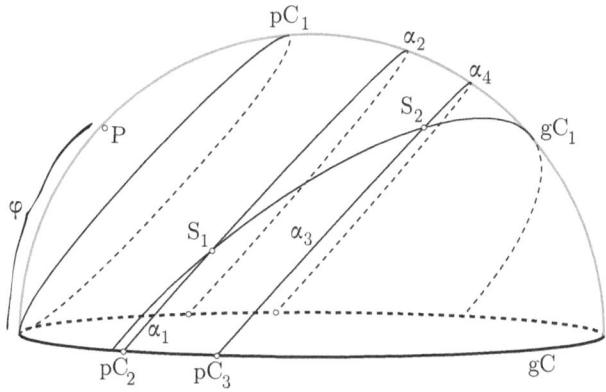

Figure 65: Perspective diagram, *Sph.* III.14, *Phen.* 4.

is another great circle, gC_1, oblique to the parallels and tangent to two parallels greater than pC_1 (not shown), then gC and gC_1 will cut off between themselves arcs of the parallels, α_1, α_2, α_3, α_4, such that, as these arcs are taken farther from the pole on the same side of gC_1 as the arcs, they will be progressively less than similar to the arcs nearer to the same pole (Figure 65). That is,

$$\alpha_2 \succ \alpha_4$$

monotonically decrease as they are taken farther away from P, while

$$\alpha_3 \succ \alpha_1$$

monotonically decrease in the direction towards P.

As mentioned, this theorem is closely related to *Phen.* 4, which shows that, of stars on a great circle oblique to the parallel circles and tangent to a pair of parallel circles greater than the *v*-circle, whichever rises earlier also sets earlier, and whichever sets earlier also rises earlier. That is, if we set gC as a horizon tangent to *v*-circle pC_1 and such that $0° < \varphi < 90° - \varepsilon$, and gC_1 as the ecliptic, being oblique to the parallels by ε, then if S_1 rises earlier than S_2 it will set earlier, and if it sets earlier, it will have risen earlier. In *Sph.* III.14, however, points of the ecliptic, say S_1 and S_2, are not mentioned in the *enunciation* or *exposition* but are introduced in the *construction*, seemingly as auxiliaries to the *demonstration*. Hence, we may understand *Sph.* III.14 to be a proposition about the underlying geometrical facts that give rise to the proposition asserted and demonstrated in *Phen.* 4, and, indeed, the argument in each is similar and hinges on the geometrical fact demonstrated in *Sph.* II.13. The details of the argument given in *Sph.* III.14 also establish the further implication that, within the bounds of $0° < \varphi < 90° - \varepsilon$, if two stars in general rise at the same time, that nearer the visible pole will set later (Schmidt 1943, 22–23), and if two stars set at the same time, that nearer the visible pole will have risen earlier.

Sph. III.14 (Theorem)

This proposition depends on nothing else in *Spherics* III, and hence its connection with the rest of the book is somewhat tenuous. As mentioned, the key to this proposition is *Sph.* II.13, which can be applied following the construction permitted by *Sph.* II.15. A number of factors raise the possibility that *Sph.* III.14 either originated in an earlier treatment of spherics or was later added to the *Spherics* (Czinczenheim 2000, 946): its content is only loosely related to that of the rest of *Spherics* III upon which it does not depend, its claim is ambiguously articulated, its structure is more involved than is necessary, and the letter-names are introduced in a non-standard way.

As with most of the previous propositions in this book, the bounds of $0° < \varphi < 90° - \varepsilon$ are set by the fact that the oblique great circle, $\mathbf{gC}(BGE)$, is assumed to be tangent to a parallel circle that is greater than the parallel, $\mathbf{pC}(AD)$, to which the initial great circle, $\mathbf{gC}(ABG)$, is assumed to be tangent.

As mentioned, there are some ambiguities in the way that this proposition is articulated. In the first place, the statement regarding the distance between the arcs of the parallel circles and "whichever," ὁπότερος, pole is not very precisely expressed, but it must mean the distance between the parallel circle and whichever pole is on the same side of $\mathbf{gC}(BEG)$ as the arc in question, as specified by *Sph.* II.20. Furthermore, in the first *construction*, points E and K are taken arbitrarily on $\mathbf{gC}(BEG)$, but there must be some conditions placed on the location of these points. Indeed, if we assume that E and K really can be anywhere on $\mathbf{gC}(BEG)$, then there are some cases in which the proposition will not hold, and there is also the possibility that E and K lie on the same parallel circle, in which case the proposition is not even expressible. One possibility would be to assume that the configuration is the same as that in *Sph.* III.7 and III.8, such that points B and G represent the points of tangency of $\mathbf{gC}(BEG)$ and the two equal and parallel circles to which it is tangent, and that $\mathbf{gC}(ABG)$ and $\mathbf{gC}(BEG)$ are perpendicular – but this introduces more conditions than are necessary. Hence, the simplest qualification would be to assume that E and K must fall on the same semicircle between the points of tangency. In this case, the proof of *Sph.* III.14 follows even if $\mathbf{gC}(BEG)$ is oblique to $\mathbf{gC}(ABG)$.

The structure of this proposition is somewhat unusual. There are two *specifications* and two *constructions*. The first *specification* is asserted in the same general terms as the *enunciation* and the second uses the letter-names of objects introduced in the first *construction*, and which are necessary for making the claim explicit. The first *construction* introduces arbitrary points, E and K, on $\mathbf{gC}(BEG)$, and the parallel circles through these points. The parallel circles are drawn without first finding the pole, as was done in *Sph.* III.8, and often in the *Phenomena*. The second *construction* introduces the great circles that will be needed to complete the argument.

Basis: *Elem.* I.post.3 (spherical), I.c.n.5, *Sph.* I.21 (by implication), II.13, II.15.

Usage: Unused in the *Spherics*.

Diagram: The diagram is a solid representation of global objects, containing only intrinsic objects. It follows the same conventions as the diagrams for *Sph.* III.7 and III.8. Because there are no false crossings, it accurately depicts the topology of the configuration.

Textual Comments: The letter-names are not introduced in alphabetical order, and there is no way to rectify this situation without some minor rewriting. This fact adds to the impression that this proposition was an addition made before the 9th century.

In the oldest Greek manuscript, **A** (37v), arc ΞNK (*XNK*) is concave about Λ (*L*), just as ΔEM (*DEM*), giving the visual impression that both semicircles are non-intersecting with the semicircle from **A** in the direction of Θ (*Q*), which ΞNK should not be. In a number of the later manuscripts, however, it is concave in the opposite direction, which produces a visual impression in better agreement with the text (for example, **B**, 81v; **D**, 15v; **F**, 45v). We have followed these later manuscripts in our diagram.

Comparison with Arabic: The diagram in **I** (52b) is similar to that in **A**.

Part IV

Paraphrase

Paraphrase

General Introduction

As mentioned in the Introduction to this volume, the Paraphrase provides a more modernized, or idiomatic, version of the treatise, accompanied by perspective diagrams, both of which modern readers may find more quickly intelligible than the Translation and its diagrams. One thing that it does not provide, which would increase intelligibility, is reasons why statements are true. This was not possible without massive redundancy, as virtually every statement in the Translation has some such indication footnoted. This means that, while the paraphrase indicates what the treatise says, mathematically but not linguistically speaking, it does not set out to say more than the treatise nor to explain it. The footnotes to the Translation must be consulted for justifications, as well as the Commentary for explanations both mathematical and astronomical. For each proposition, what is used to prove it and what it is used to prove are listed at the end of the Commentary as "Basis" and "Usage," that is, what it is used for later in *Spherics* and in the astronomical texts *Phenomena* and *Moving Sphere*.

This is the place to indicate such conventions as are used in the Paraphrase alone. The first is the separation of the paraphrase itself from the text like this in which it is embedded. This will be done with the use of a fleuron, ❧. The text after a paraphrase, headed "Comments," has been placed there because, unlike the introductions, it depends on familiarity with the paraphrase. Commentary within the paraphrase is put in parentheses. Each proposition is divided into conventional parts with bold headings, enunciations being introduced by the kind of thing enunciated, such as "Theorem" or "Corollary," and italicized. Cases and parts are identified in these headings.

A single italicized letter is always a point, but more than one point can be a straight line, circular arc, or whole circle. For definiteness, letters representing straight lines are underlined in the paraphrase and figure captions (for consistency), but not in the embedding text.

All two-dimensional diagrams of three-dimensional objects have limitations. One that the paraphrase's diagrams have that the medieval-style diagrams do not is the difficulty of representing the back side of the sphere. This is largely avoidable and sometimes avoided even when it would be more informative to include the back side. The second diagram for *Sph.* II.17 (Figure 47 (b)) illustrates lines joining points on the sphere, one of which ends

behind the great circle bounding the diagram. The first two diagrams for
Sph. II.22 (Figures 55 and 56) illustrate a discrepancy between two diagrams of the same configuration. In the first, point B is shown although it
is on the back side of the sphere, and in the second it has been moved for
clarity.

The diagrams are mostly perspective diagrams of curves on a sphere
looked at from outside and plotted with Mathematica®, which means that
the circles represented really are what they purport to be, parallel or great.
The sphere that is almost always present is best represented in the first
diagram, represented more mutedly thereafter, and not represented at all
when the geometrical objects of interest take the stage within a bounding
great circle. The sphere is not filled in when a great circle bounding the
curves of interest is present, e.g., Figure 15 and following, because it is
a distraction. If one looks at a sphere, one does not see the great circle
perpendicular to one's line of sight because it is on the rear portion of the
surface, obscured by the visible part of the sphere, which is less than a
hemisphere for a viewer that is not infinitely far from the sphere.

Lines referred to in the proofs are often illustrated in the diagrams despite the fact that they are not drawn in the text. This happens in *Sph.* II.19
(*EZ* in Figure 50) and others.

In a two-dimensional diagram of something three-dimensional, some direction is necessarily taken as vertical and the plane perpendicular to it as
horizontal. The adjective "horizontal" throughout the Paraphrase will refer
to this plane, which differs from the plane of the paper. In some theorems
curves on a sphere will appear that represent astronomical horizons. They
can even be horizontal (*Sph.* II.21). But the adjective "horizontal" in the
Paraphrase has no astronomical implication; it refers to the diagrams' base
planes. The noun "horizon," on the other hand, refers to the astronomical
horizon and appears only in introductions and comments with astronomical
reference. This is easy to keep straight because "horizontal" complements
"vertical" and does not refer to any horizon, whereas the astronomical horizon is only accidentally horizontal in that sense.

Introduction to *Spherics* I

What Book I contributes to the treatise as a whole is a stand-alone introduction to spherical geometry with a utilitarian theme but no particular
slant toward the astronomy that motivates Books II and III. Astronomy appears in comments on the definitions below because they have astronomical
relevance, of which it would be pointless to put off mention. The geometry
in the whole treatise is mainly the study of great circles on a sphere and in
later books also small circles, typically being a parallel set covering the surface except for their common poles, about which they can be drawn with a

compass. Great circles too can be drawn with a compass, and it is one of the aims of *Spherics* I to determine the compass setting needed. *Sph.* I.20 (of 21 propositions regarded as authentic) shows how to draw the great circle through a given pair of points, which would have allowed the non-Euclidean geometry of the spherical surface to be pursued at any time, since circles can already be drawn and great circles substitute for straight lines (see p. 10). Instead, the book concludes in the next proposition by showing how to find the pole of a given circle, the point about which parallel circles can be drawn.

The first proposition stands alone in establishing the topic of the treatise as a whole, circles both great and small: the plane section of a solid sphere is a solid circle. While the definition of a sphere requires a center, its location is given in more operational terms in *Sph.* I.2 with converse *Sph.* I.7. *Sph.* I.3–5 concern planes tangent to the sphere, a topic not further developed. *Sph.* I.6 introduces great circles. *Sph.* I.8–10 concern the configuration consisting of a circle and its axis, singling out its center and poles and the sphere's center. The topic of the converse pair *Sph.* I.11 and 12 returns to great circles and they prove a definitive property. *Sph.* I.13–15 concern a single configuration, a great circle through the poles of another circle, and show the equivalence of three ways of describing it. *Sph.* I.16 and I.17 show the definitive size of the polar radius of great circles, but they do not indicate how to produce it for a given sphere. To the end of doing so, *Sph.* I.18 and I.19 show how to find the diameters of a circle on the sphere and then of the sphere itself. With the diameter of the sphere given, the polar radius of great circles can be constructed (in an external plane) and is used in *Sph.* I.20 to draw the great circle through two points. This capacity is used in *Sph.* I.21 to find the pole of a given circle. The constructions of *Sph.* I.20 and I.21 are the goal of the book and are used in Books II and III. There are spurious additions called *Sph.* I.22 and I.23 in the Translation only.

Definitions

1. A *sphere* is a solid figure contained by a single surface, all lines to which from a single point that lies within the figure are equal to one another.

2. The point is the *center of the sphere.*

3. An *axis of the sphere* is a line passing through the center and bounded in each direction by the surface of the sphere, around which immobile line the sphere rotates.

4. The *poles of the sphere* are endpoints of the axis.

5. A *pole of a circle in a sphere* is a point on the surface of the sphere from which all lines to the circumference of the circle are equal to one another.

6. A plane is said to be *similarly inclined* to a plane, when, in each of the planes, lines produced at right angles to the intersection of the planes at the same point contain equal angles.

§◦

Comments. For discussion see Commentary on Definitions (p. 195).

In the definition of "sphere" in *Elem.* XI.def.14, a semicircle is rotated about its diameter, which makes that diameter the axis (*Elem.* XI.def.15) of the sphere, a definition parallel to those of axis of cone (*Elem.* XI.def.19) and cylinder (*Elem.* XI.def.22). This rotation is not the rotation of *Sph.* I.def.3.

It will be convenient to use the term "axis of a circle" for the line joining its poles, but this is not the axis of *Sph.* I.def.3. The term "circle" will be used somewhat ambiguously for the solid circle within the sphere and its circumference, the surface circle. The ambiguity appears from *Sph.* I.3. See Commentary on *Sph.* I.def.5 (p. 197).

It will be necessary later in this book to introduce some thoroughly anachronistic discussion in order better to convey to a twenty-first-century reader what the geometrical situation is before describing how it was written down two thousand years ago. Here at the beginning it may be useful to say something *about* what is going on in the theorems and problems, in particular about agency. For discussion of constructions and how they relate to those in the *Elements*, see the section "Constructive assumptions and procedures" of the Commentary (p. 198). Right from the beginning things have been done by an agency unknown. A sphere is cut by a plane; some agency has cut the sphere with the plane, since the plane cannot do that or anything else on its own. This same sort of *deus ex machina* produces the starting objects in all of the propositions and also performs constructions in the proofs and once – in *Sph.* I.2 – performs the official construction before the proof. What is done is not limited as it is in the early books of the *Elements* to things that we could do ourselves with less exactitude with straightedge and compass but includes things like cutting the sphere with a plane and erecting perpendiculars to planes. However, an important goal of the book is four problems that are to allow us to approximate actions that would be child's play for the kind of agent on display earlier in the book. In problems I.18–I.21 it is shown how things can be done with a compass on the surface of the sphere that are quite surprising. It is to describe those situations that anachronisms will be helpful. The first is to call the starting objects what is "given," since that is a single word that is now customary.

Propositions

Introduction to *Sph.* I.1

The proof is carried out by a thought-experimental construction that has no need of being carried out to make its point. These thought-experimental constructions will continue until *Sph.* I.18 when something will have to be done – actually constructed and available for use. It is sufficient here to imagine the dropping of a perpendicular from the center of the sphere to a cutting plane that does not pass through the center. One likewise can imagine radii to all of the points on the curve – identified by three points on it – that will prove to be a circle. The three points on the curve – on the sphere and on the plane – identify the curve by identifying the plane. Many other curves pass through three points, but only one plane. The other way that Euclid identifies a plane, intersecting lines, would be quite unsuitable here, since they would not lie on the sphere.

Theorem. *If a sphere is cut by a plane, the curve produced on the surface of the sphere is the circumference of a circle.*

Exposition. Let a plane make curve ABG in the surface of the sphere by cutting it.

Specification. It is to be proved that curve ABG is the circumference of a circle, whose center is determined in the course of the proof.

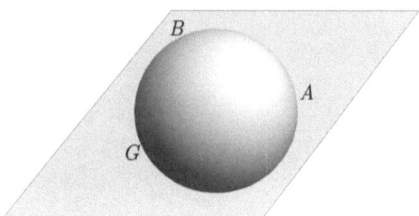

Figure 1: *Sph.* I.1. Points A, B, and G on a plane through center of the sphere.

Demonstration (Case 1: plane through center of sphere (Figure 1)). It is clear that curve ABG is the circumference of a circle, for the lines from the center of the sphere to the surface of the sphere are equal to one another, and curve ABG is in the surface, and so the lines from the center of the sphere to curve ABG are equal to one another. The plane curve ABG is the circumference of a circle, whose center is also that of the sphere.

Construction (Case 2: plane not through center of sphere (Figure 2 (a))). Let the cutting plane not be through the center of the sphere, and let the center of the sphere be point D. Let the perpendicular to the cutting plane be $\underline{DE}$. Join $\underline{EA}$ and $\underline{EB}$, $\underline{DA}$ and $\underline{DB}$ (Figure 2 (b)).

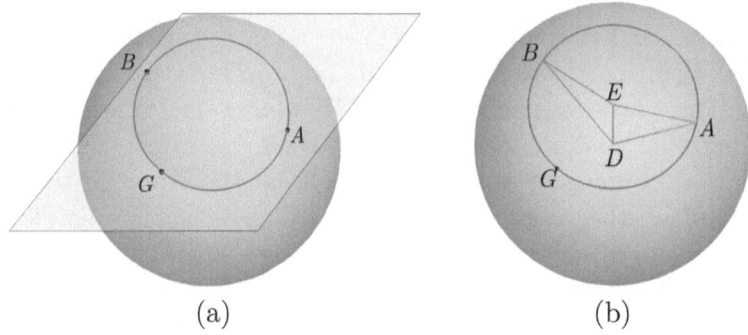

(a)   (b)

Figure 2: *Sph.* I.1. (a) Points A, B, and G on a plane not through center of sphere. (b) Right triangles ADE and BDE with common side from the sphere center D to the circle center E.

Demonstration (Case 2). In right triangles ADE and BDE, $\underline{AD}$ is equal to $\underline{BD}$ and the square on $\underline{AD}$ is equal to the sum of the squares on $\underline{AE}$ and the common side $\underline{ED}$, and the square on $\underline{BD}$ is equal to the sum of the squares on $\underline{BE}$ and $\underline{ED}$. The squares on $\underline{AE}$ and $\underline{ED}$ are equal to the squares on $\underline{BE}$ and $\underline{ED}$. If the common square on $\underline{ED}$ is taken away, the remaining square on $\underline{AE}$ is equal to the square on $\underline{BE}$. Therefore $\underline{AE}$ is equal to $\underline{BE}$. Similarly we will show that all lines running from point E to curve ABG are equal to one another. Therefore, curve ABG is the circumference of a circle whose center is E.

Corollary. *If a circle is in a sphere, the perpendicular produced from the center of the sphere to it will fall at its center.*

Comments. This proof introduces without mentioning it one of the two main concerns of the whole work prior to their definitive characterization at *Sph.* I.6, circles through the center of the sphere, *great circles*. Other circles, necessarily smaller are *small circles*.

Justifications like *Elem.* I.47, which one might expect to be mentioned, are not stated in keeping with the standard expository style. Authorities for most statements throughout the treatise are given as footnotes to the Translation. Since A and B are any two points on the curve, we would now regard the proof as complete, but the Greek style is to say that the completeness follows, i.e., that all pairs of points are equidistant from E, because we could repeat the proof for any such pair.[1] The text says "we

---

[1] This is one interpretation of the unfulfilled promise. Such unfulfilled promises are a standard feature of Greek mathematical texts.

will show," as though we will do so in response to someone else's choice of a pair of points. This shows a slightly greater attention to being convincing without dependence on the logic of the situation, which is nowadays regarded as obvious.

Introduction to *Sph.* I.2

The second proposition answers the question not raised in the first, just where is the center of a sphere. The definition of a sphere is not much help in this, since the center's being the unique point from which the distances to the points on the surface of the sphere is no help in finding it. The proof of *Sph.* I.1 just uses the center; it does not trouble to find it first. All that is needed there is to imagine that it exists.

§◦

Problem. *To find the center of a given sphere.*
Exposition. Let a sphere be given.
Specification. It is required to find its center.
Construction. Cut the given sphere by a plane, which cuts the sphere in a circle. Let ABG be the circle in which the plane cuts the given sphere, and let its center be D (Figure 3 (a)). Set up at D a perpendicular to the plane of the circle ABG, extended to meet the surface of the sphere at E and Z (Figure 3 (b)). Bisect $\underline{EZ}$ at H.
Specification. It is required to show that H is the center of the sphere.

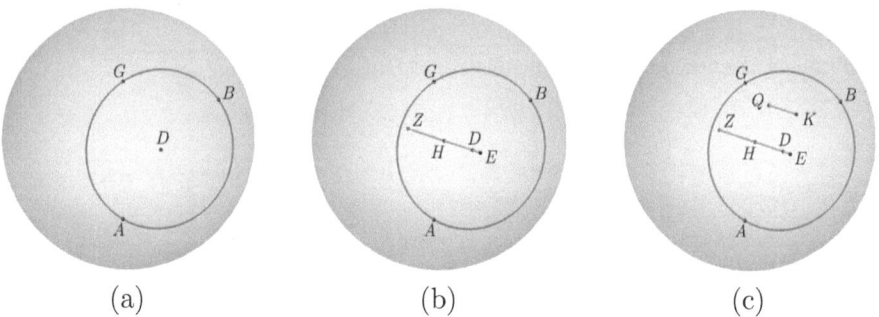

(a)   (b)   (c)

Figure 3: *Sph.* I.2. (a) Circle ABG and center D. (b) Perpendicular (sphere diameter) $\underline{EZ}$ and center H of sphere. (c) False sphere center Q and perpendicular to K.

Construction. Suppose H is not the center of the sphere, but let the center of the sphere instead be Q, and from Q a perpendicular $\underline{QK}$ be dropped to plane ABG (Figure 3 (c)).

Demonstration. Then point K is the center of the circle ABG. But the center is also D; therefore the supposition is impossible. Therefore point Q, different from H, is not the center of the sphere nor is any other such point. Therefore point H is the center of the sphere, as was required to show.

Corollary *If a circle is in a sphere and a perpendicular is erected at its center, the center of the sphere is on it.*

⸎

Comments. The default circle ABG in the sphere is a small circle so that the possibility that the cutting plane passes through the center of the sphere is not separately considered but is included implicitly as a possibility (H could be D).

Introduction to *Sph.* I.3–I.5

Sph. I.3–I.5 concern planes tangent to the sphere, not a matter of subsequent interest in this treatise. They also mark a new beginning, being independent of *Sph.* I.2. Astronomical relevance begins here; see Commentary on *Sph.* I.3–I.5 (p. 206).

⸎

Sph. I.3. Theorem. *A sphere touches a plane that does not cut it in not more than one point.*
Demonstration. Suppose a sphere touches a plane in two points, A and B. Let the center of the sphere be G. Join $\underline{AG}$ and $\underline{BG}$, and produce the plane containing $\underline{AG}$ and $\underline{BG}$. The plane ABG cuts the surface of the sphere in a circle DAB and the given plane in a straight line $\underline{EABZ}$ (Figure 4).

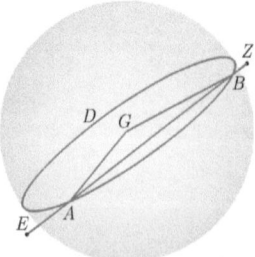

Figure 4: *Sph.* I.3. Intersection $\underline{EABZ}$ of plane ABG with plane tangent at A and B to sphere with center G.

Since the given plane does not cut the sphere, neither does the circle DAB cut the straight line $\underline{EABZ}$, which lies outside the circle DAB. Since the two touching points A, B, lie on the circumference of the circle DAB, the straight line joining A to B falls within the circle DAB. But it is also outside; this absurdity proves the theorem.

$\mathcal{S}\!\!\bullet$

Introduction to *Sph.* I.4

See Introduction to *Sph.* I.3–I.5 (p. 370).

$\mathcal{S}\!\!\bullet$

Theorem. *Let a plane not cut a sphere but touch the sphere at a point. Then the line joining the point of contact to the center is perpendicular to the plane.*

Exposition. Let a plane touch a sphere at point A and not cut it, and let the center of the sphere be point B. Join $\underline{BA}$ (Figure 5 (a)).

Specification. It is to be proved that $\underline{BA}$ is perpendicular to the plane.

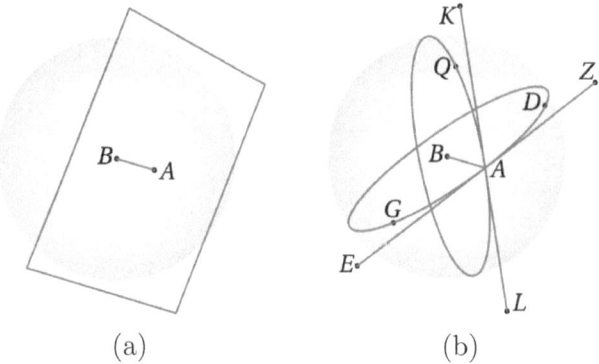

(a)   (b)

Figure 5: *Sph.* I.4. (a) The given sphere, plane, and radius $\underline{BA}$. (b) Construction adding circles and lines.

Construction. Produce a plane through $\underline{BA}$. It cuts the surface of the sphere in the circumference of a circle GAD and the touching plane in a line $\underline{EAZ}$ (Figure 5 (b)). Produce another plane through $\underline{BA}$ cutting the surface of the sphere in a circle AQ and the touching plane in a line $\underline{KAL}$.

Demonstration. Since the plane touches the sphere, the line $\underline{EAZ}$ touches the circle GAD. Since the line $\underline{EAZ}$ touches the circle GAD at A and $\underline{BA}$ joins the center of the sphere to A, $\underline{BA}$ is perpendicular to $\underline{EAZ}$. Similarly $\underline{BA}$ is perpendicular to $\underline{KAL}$. Therefore, since $\underline{BA}$ is perpendicular to the two lines $\underline{EZ}$ and $\underline{KL}$ at their point of intersection A, $\underline{BA}$ is also perpendicular to the plane containing them, as required.

$\mathcal{S}\!\!\bullet$

Introduction to *Sph.* I.5

See Introduction to *Sph.* I.3–I.5 (p. 370). This theorem is what converse there can be to *Sph.* I.4.

Sₒ

Theorem. *If a sphere touches a plane not cutting it, then the center of the sphere will be on a perpendicular erected into the sphere at the point of contact.*

Exposition. Let a plane not cutting a sphere touch it at A, and let the perpendicular to the plane erected at A meet the surface of the sphere at a point B (Figure 6 (a)).

Specification. It is to be proved that the center of the sphere is on $\underline{AB}$.

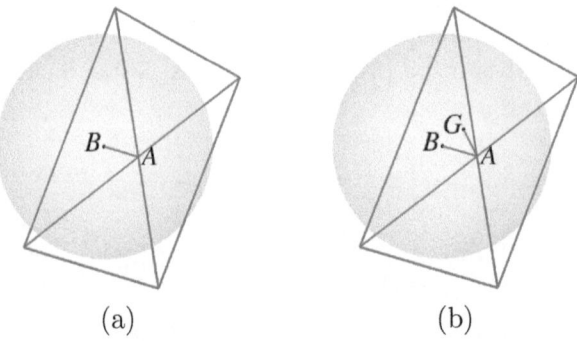

(a)   (b)

Figure 6: *Sph.* I.5. (a) The given configuration of sphere and tangent plane with diameter $\underline{AB}$. (b) The false perpendicular $\underline{GA}$ added.

Demonstration. The center of the sphere is on $\underline{BA}$ because if it were not, suppose it were a point G not on $\underline{BA}$ (Figure 6 (b)). Since $\underline{GA}$ joins the center of the sphere to the point of contact, $\underline{GA}$ is a second distinct perpendicular to the plane at A, which is impossible. Therefore G is not the center of the sphere. As the same can be shown for any point not on $\underline{BA}$, the center of the sphere is on $\underline{BA}$, the perpendicular at the point of contact.

Sₒ

Comments. It is necessary that BA pass into the sphere rather than lie entirely outside it for the proof to work, since it calls upon *Elem.* XI.13 to guarantee that at a point there can be only one perpendicular to a plane *on the same side* of the plane.

Introduction to *Sph.* I.6

Great circles are introduced as unique for each direction of the axis, each having been seen previously in *Sph.* I.1 and I.2.

Theorem. *Circles through the center of a sphere are great circles. Other circles in a sphere are equal to one another if equidistant from the center of the sphere, and the farther away from the center the smaller the circles.*

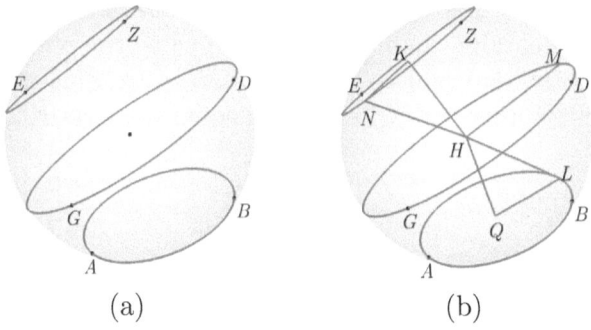

(a) (b)

Figure 7: *Sph.* I.6. (a) The given three circles in the sphere, *EZ* and *AB* equal and *GD* through the center. (b) The construction with equal circles.

Exposition (Parts 1 and 2 of 3). Let *AB*, *GD*, and *EZ* be circles in a sphere, let *GD* be through the center of the sphere, and let *AB* and *EZ* be equidistant from its center (Figure 7 (a)).

Specification. It is to be proved that *GD* is a great (i.e., largest) circle, and *AB* and *EZ* are equal.

Construction (Parts 1 and 2). Let the center of the sphere be *H*, which is also the center of circle *GD*. Drop perpendiculars <u>HQ</u> and <u>HK</u> from *H* to the planes of the circles *AB* and *EZ* at *Q* and *K* respectively. Then *Q* and *K* are the centers of circles *AB* and *EZ* by the corollary to *Sph.* I.1. Construct circle radii from *Q*, *H*, and *K* to the circumference of the circles *AB*, *GD*, and *EZ* respectively, namely <u>QL</u>, <u>HM</u>, and <u>KN</u>. Join <u>HL</u> and <u>HN</u> (Figure 7 (b)).

Demonstration (Part 1). Since <u>HQ</u> is perpendicular to the plane of the circle *AB*, it also makes right angles with all lines that meet it and lie in the plane of the circle *AB* such as <u>QL</u>. Therefore angle *HQL* is a right angle. Similarly angle *HKN* is a right angle.

Further, since angle *HQL* is a right angle, angle *LHQ* is less than a right angle, therefore less than angle *HQL*. Therefore in triangle *HQL* side <u>HL</u> is longer than side <u>QL</u>. And the sphere's radii <u>HL</u> and <u>HM</u> are equal. Therefore <u>HM</u> is longer than <u>QL</u>. But <u>HM</u> is a radius of circle *GD*, and <u>QL</u> is a radius of circle *AB*. Therefore circle *GD* is greater than circle *AB*.

Similarly circle *GD* is also greater than all circles in the sphere not passing through the center of the sphere.

Demonstration (Part 2). Since circles *AB* and *EZ* are equidistant from the center of the sphere, $\underline{HQ}$ is equal to $\underline{HK}$. And since *H* is the center of the sphere, radius $\underline{HL}$ is equal to radius $\underline{HN}$; therefore the square on $\underline{LH}$ is equal to the square on $\underline{HN}$. But in right triangle *LHQ*, the squares on $\underline{QL}$ and $\underline{QH}$ added together are equal to the square on $\underline{LH}$, and in right triangle $\overline{NKH}$, the squares on $\underline{NK}$ and $\underline{KH}$ added together are equal to the square on $\underline{HN}$; the squares on $\underline{QL}$ and $\underline{QH}$ are therefore equal to the squares on $\underline{NK}$ and $\underline{KH}$, of which the square on $\underline{QH}$ is equal to the square on $\underline{KH}$. Therefore the remaining square on $\underline{QL}$ is equal to the remaining square on $\underline{NK}$. Therefore $\underline{QL}$ is equal to $\underline{NK}$. But $\underline{QL}$ is a radius of circle *AB*, and $\underline{NK}$ is a radius of circle *EZ*. Circle *AB* is equal to circle *EZ*.

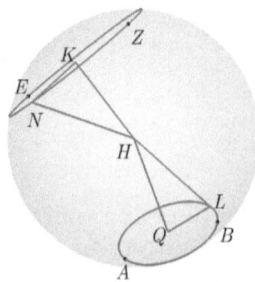

Figure 8: *Sph.* I.6. The construction with unequal circles.

Exposition (Part 3). Now let circle *AB* be farther from the center of the sphere than circle *EZ*.

Specification (Part 3). It is to be proved that *AB* is smaller than *EZ*.

Demonstration (Part 3). With the same construction of triangles *LHQ* and *NKH* (Figure 8), since circle *AB* is farther from the center of the sphere than *EZ*, $\underline{QH}$ is longer than $\underline{KH}$. And since $\underline{HL}$ is equal to $\underline{HN}$, the squares on $\underline{QL}$ and $\underline{QH}$ added together (that is to say, the square on $\underline{LH}$) are therefore equal to the squares on $\underline{NK}$ and $\underline{KH}$ added together, of which the square on $\underline{QH}$ is larger than the square on $\underline{KH}$. Therefore the remaining square on $\underline{QL}$ is smaller than the remaining square on $\underline{NK}$, and so $\underline{QL}$ is shorter than $\underline{NK}$. But $\underline{QL}$ is a radius of circle *AB* and $\underline{NK}$ is a radius of circle *EZ*. Therefore circle $\overline{AB}$ is smaller than circle *EZ*.

Summary. In the triangles made by perpendiculars $\underline{HQ}$ and $\underline{HK}$ to the circles, radii $\underline{QL}$ and $\underline{KN}$ of the circles, and the equal hypotenuse radii $\underline{HL}$ and $\underline{HN}$ of the sphere, the sum of the squares on the perpendiculars and circle radii are constant.

3. Corresponding to the longer perpendicular must be the shorter circle radius.

2. If the perpendiculars are equal, then the circle radii are equal.

1. When the perpendicular has no length at all, then the circle radius is at a maximum.

Introduction to *Sph.* I.7–I.10

The first theorem is as much of a converse of *Sph.* I.2.cor. as there can be. It is here to elucidate the following theorem and is used at *Sph.* I.14.

The axis of a circle in a sphere joins its poles and the sphere and circle centers perpendicularly. The next three theorems all refer to this configuration (Figure 10 (a)). The configuration is shown to be obtainable by extending the perpendicular *DE* in *Sph.* I.8, which, by I.7, is just extending the line *DE*, extending the perpendicular *HE* in I.9, and joining the poles *Z* and *H* in I.10.

See the introductory Commentary to the Translation of *Sph.* I.7–I.10 (p. 210) for more on this configuration.

Sph. I.7. Theorem. *A straight line joining the center of a sphere to the center of a circle in the sphere is perpendicular to the circle.*

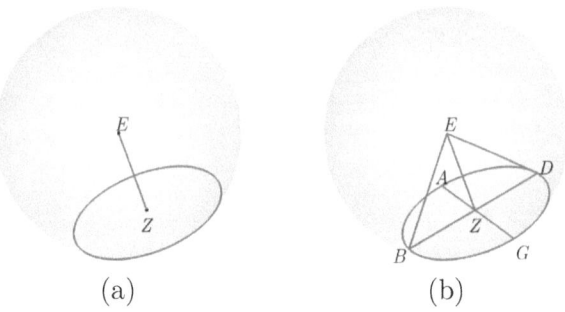

(a)　　　　　　　　(b)

Figure 9: *Sph.* I.7. (a) The given line *EZ* and circle. (b) The circle diameters and sphere radii added.

Exposition. Let the circle in the sphere be *ABGD*, and let *EZ* join *E*, the center of the sphere, to *Z*, the center of the circle (Figure 9 (a)).

Specification. It is to be proved that *EZ* is perpendicular to circle *ABGD*.

Construction. Draw diameters *AZG* and *BZD* in the circle, and join *EB* and *ED* (Figure 9 (b)).

Demonstration. In triangles *BZE* and *DZE*, since *BZ* is equal to *DZ* and *ZE* is common, and the third side *BE* is equal to the third side *DE*, the angle *BZE* is equal to the angle *DZE*. Therefore each of the angles *BZE* and *DZE* is a right angle, and *ZE* is perpendicular to *BD*.

Similarly we will show also that *ZE* is perpendicular to *AG*.

Since line *EZ* is set up perpendicularly to the two straight lines *AG* and *BD* that cut each other at *Z*, it is perpendicular to the plane through *BD*, that is, the circle *ABGD*.

Introduction to *Sph.* I.8

See the Introduction to *Sph.* I.7–I.10 (p. 375).

Theorem. *If a perpendicular is dropped from the center of a sphere to a circle in the sphere and extended in both directions, it meets the sphere at the poles of the circle.*

Exposition. Let *ABG* be a circle in a sphere, let *D* be the center of the sphere. Drop perpendicular *DE* from *D* to *E* in the plane of the circle *ABG*. Then *E* is the center of the circle *ABG*. Then extend *DE* in both directions, meeting the surface of the sphere at points *Z* and *H* (Figure 10 (a)).

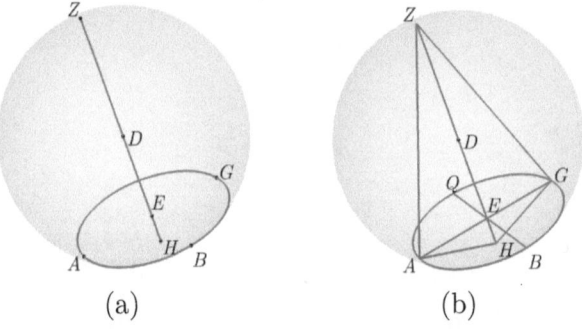

(a)   (b)

Figure 10: *Sph.* I.8. (a) The given line and circle. (b) The circle diameters and polar radii added.

Specification. It is to be proved that *Z* and *H* are poles of the circle *ABG*.
Construction. Draw diameters *AEG* and *BEQ* across the circle, and join *AZ*, *GZ*, *AH*, and *GH* (Figure 10 (b)).
Demonstration. Since line *ZE* is perpendicular to circle *ABG*, it also makes right angles with all straight lines that meet it and lie in the plane of the circle *ABG*; therefore each of the angles *ZEA*, *ZEG*, *ZEB*, and *ZEQ* is a right angle. In triangles *ZEA* and *ZEG*, since *EA* is equal to *EG* and *ZE* is common and perpendicular, the hypotenuse *AZ* is equal to the hypotenuse *GZ*.

Similarly we will show that all straight lines from Z to the circumference of circle ABG are equal to one another. Therefore Z is a pole of circle ABG. Similarly we will show that H is a pole. Therefore Z and H are poles of the circle ABG.

∗

Introduction to *Sph.* I.9

See the Introduction to *Sph.* I.7–I.10 (p. 375).

∗

Theorem. *If a perpendicular is dropped to a circle in a sphere from one of its poles, it will fall on the center of the circle, and extended it meets the surface of the sphere at the other pole of the circle.*

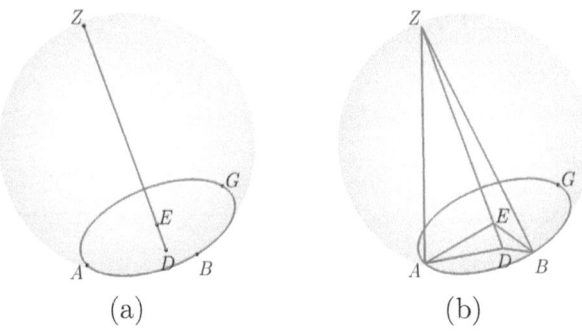

(a)   (b)

Figure 11: *Sph.* I.9. (a) The given line and circle. (b) The circle radii and polar radii added.

Exposition. Let circle ABG be in a sphere. From one of its poles D drop perpendicular DE to the plane of the circle at E and extend it to meet the surface of the sphere at Z (Figure 11 (a)).

Specification. It is to be proved that E is the center of circle ABG (Part 1) and that Z is the other pole of the circle (Part 2).

Construction. Draw EA and EB from E, and join AD, BD, AZ, and BZ (Figure 11 (b)).

Demonstration (Part 1). Since DE is perpendicular to circle ABG, it also makes right angles with all straight lines that meet it and lie in the plane of circle ABG. And each of EA and EB meets it and lies in the plane of ABG; therefore each of the angles DEA and DEB is a right angle.

Since polar radius AD is equal to polar radius BD, the square on AD is equal to the square on BD. But in triangles ADE and BDE, the squares on ED and EA added together are equal to the square on hypotenuse AD, and the squares on ED and EB added together are equal to the square on hypotenuse BD, equal to the square on AD. Subtract what is common from the sums of the squares; then the remaining square on EA is equal to the remaining square on EB. Therefore EA is equal to EB. Similarly we will show that all straight lines from E to curve ABG are equal. Therefore E is the center of circle ABG.

Demonstration (Part 2). In triangles AEZ and BEZ, since circle radius AE is equal to circle radius BE and the common side EZ is perpendicular to AE and BE, the hypotenuse AZ is equal to the hypotenuse BZ.

Similarly we will show that all straight lines from Z to curve ABG are equal.

Therefore Z is a pole of circle ABG.

Therefore E is the center of ABG and Z is the pole opposite D.

§

Introduction to *Sph.* I.10

See the Introduction to *Sph.* I.7–I.10 (p. 375).

§

Theorem. *If a circle is in a sphere, the line joining its poles is perpendicular to the circle and passes through the centers of the circle and of the sphere.*

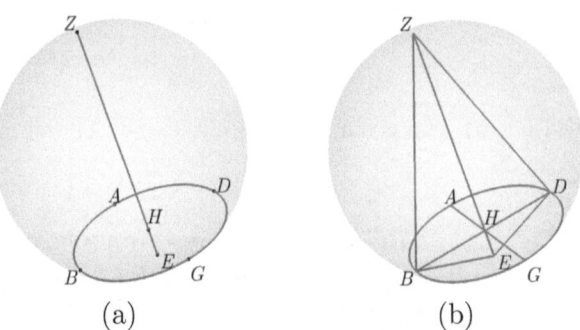

(a) (b)

Figure 12: *Sph.* I.10. (a) The given line and circle. (b) The diameters and polar radii added.

Exposition. Let circle $ABGD$ be in a sphere, and let line EZ join its poles (Figure 12 (a)).

Specification. It is to be proved that EZ is perpendicular to circle $ABGD$ (Part 1), and passes through the centers of $ABGD$ (Part 2) and of the sphere (Part 3).

Construction. Let EZ meet the plane of circle $ABGD$ at H, draw AHG and BHD through H, and join BE, DE, BZ, and DZ (Figure 12 (b)).

Demonstration (Part 1). In triangles *BEZ* and *DEZ* since <u>*BE*</u> is equal to <u>*DE*</u>, side <u>*EZ*</u> is common, and <u>*BZ*</u> is equal to <u>*DZ*</u>, angle *BEZ* is equal to angle *DEZ*.

Next, in triangles *BEH* and *DEH*, <u>*BE*</u> is equal to <u>*DE*</u>, <u>*EH*</u> is common, and angle *BEH* is equal to angle *DEH*, and so side <u>*BH*</u> is equal to side <u>*DH*</u>, triangle *BEH* is congruent to triangle *DEH*, and the remaining angles are equal to the remaining angles that the equal sides subtend. In particular, angle *BHE* is equal to angle *DHE*. Then each of angles *BHE* and *DHE* is a right angle. Therefore <u>*EH*</u> is perpendicular to <u>*BD*</u>.

Similarly we will show that <u>*EH*</u> is perpendicular to <u>*AG*</u>. Therefore <u>*EZ*</u> is perpendicular to circle *ABGD*.

Demonstration (Part 2). Since *ABGD* is a circle in a sphere and <u>*EH*</u> runs perpendicularly from one of its poles to it meeting the plane at *H*, *H* is the center of *ABGD* by *Sph.* I.9.

Demonstration (Part 3). Since *ABGD* is a circle in a sphere and <u>*EHZ*</u> is perpendicular to the plane of the circle at its center *H*, the center of the sphere is on <u>*EHZ*</u> by *Sph.* I.2.cor. Therefore <u>*EZ*</u> is through the center of the sphere.

§☙

Introduction to converses *Sph.* I.11 and I.12

Each of these theorems is the converse of the other, stating an important, as well as definitive, fact about great circles.

§☙

***Sph.* I.11. Theorem.** *In a sphere, two great circles bisect each other.*

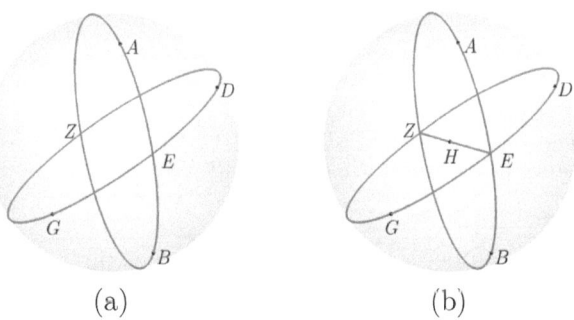

Figure 13: *Sph.* I.11. (a) Two great circles in a sphere. (b) Great circles with diameter.

Exposition. Let the circumferences of two great circles *AB* and *GD* in a sphere cut each other at *E* and *Z* (Figure 13 (a)).

Specification. It is to be proved that circles *AB* and *GD* bisect each other.

Construction. Let their common center be H, the center of the sphere. Join $\underline{HE}$ and $\underline{HZ}$ (Figure 13 (b)).

Demonstration. Since E, H, and Z are on both planes AB and GD, they are in their intersection, which is a straight line. Therefore EHZ is a straight line. And since H is the center of circle $\underline{AB}$, $\underline{EZ}$ is the diameter of it. Therefore EAZ and EBZ are semicircles. Further, since H is similarly the center of circle GD, $\underline{EZ}$ is a diameter of it. Therefore each of EGZ and EDZ is a semicircle, and circles AB and GD bisect each other.

Comments. What is known about great circles from their characterization in *Sph.* I.6 is that they are circles in the sphere and pass through the center of the sphere. That gives the two a common point. It does not give, without further consideration, that they intersect in a line.

Introduction to *Sph.* I.12

See Introduction to *Sph.* I.11 and I.12 (p. 379).

Theorem. *In a sphere, circles that bisect each other are great circles.*

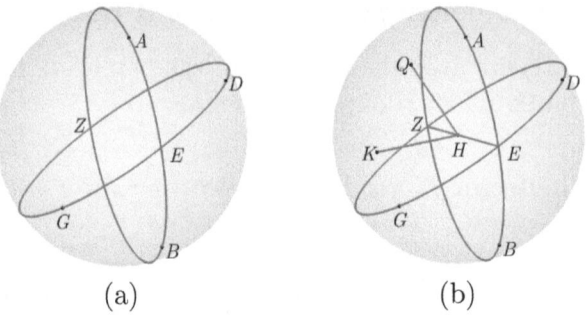

(a)   (b)

Figure 14: *Sph.* I.12. (a) Two circles bisecting each other in a sphere. (b) Great circles with intersection and normals $\underline{HQ}$ and $\underline{HK}$.

Exposition. In a sphere, let the circumferences of two circles AB and GD bisect each other at points E and Z (Figure 14 (a)).

Specification. It is to be proved that circles AB and GD are great circles.

Construction. Join $\underline{EZ}$. Then $\underline{EZ}$ is a diameter of circles AB and GD. Bisect $\underline{EZ}$ at H; then H is the center of the circles. Set up, at point H, $\underline{HQ}$ perpendicular to the plane of circle GD, and $\underline{HK}$ perpendicular to the plane of circle AB (Figure 14 (b)).

Demonstration. Since circle GD is in a sphere and $\underline{HQ}$ is set up perpendicular to the plane of the circle at its center, the center of the sphere is on

$\underline{HQ}$ extended, by the corollary to *Sph.* I.2. Likewise we will show that the sphere center is on $\underline{HK}$ extended. Therefore the intersection H of $\underline{HQ}$ and $\underline{HK}$ is the center of the sphere. But H is also the center of AB and $\overline{GD}$ and the circles, being through the center of the sphere, are great circles.

Comments. Converse of *Sph.* I.11.

Introduction to *Sph.* I.13–I.15

The cycle of theorems I.13–I.15 concerns the same configuration in a sphere, a circle *EBZD* and a great circle *ABGD* through its poles A and G (Figure 15 (a)). These theorems show that three conditions are equivalent by showing that each implies the other two: (1) the great circle *ABGD* bisects *EBZD* at B and D (2) great circle *ABGD* cuts *EBZD* perpendicularly and (3) the great circle passes through the poles A and G of *EBZD*. This configuration turns up again and again.

Sph. **I.13. Theorem.** *If a great circle in a sphere cuts a circle in the sphere at right angles, it will bisect it and pass through its poles.*

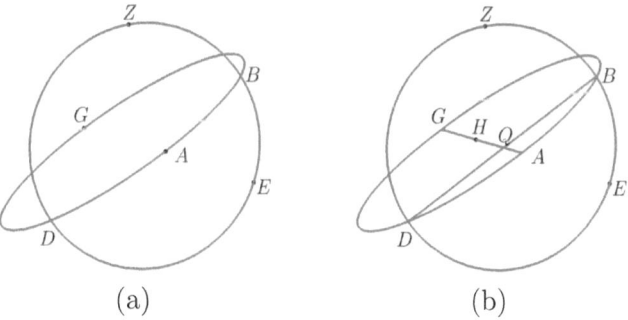

Figure 15: *Sph.* I.13–I.15. (a) The given configuration of circles. (b) The diameter and axis added.

Exposition. Let great circle *ABGD* in a sphere cut circle *EBZD* in the sphere at right angles. And let the circumference of *ABGD* in the sphere cut the circumference of *EBZD* at B and D.

Specification. It is to be proved that *ABGD* will bisect *EBZD* (Part 1) and pass through its poles (Part 2).

Construction. Draw the line of intersection $\underline{BD}$, and find the center of the circle *ABGD* (and also of the sphere) H. Drop a perpendicular $\underline{HQ}$ from H to Q on the line $\underline{BD}$, and let it be extended in both directions, meeting the surface of the sphere at points A and G (Figure 15 (b)).

Demonstration (Part 1). Since the planes of circles *ABGD* and *EBZD* are perpendicular to each other, and *QA* runs perpendicularly to their intersection *BD* in plane *ABGD*, line *AG* is perpendicular to the plane *EBZD*. Then, since *HQ* runs perpendicularly from the center of the sphere to *Q* on circle *EBZD*, *Q* is the center of circle *EBZD* by *Sph.* I.1.cor. Therefore line *BQD* is a diameter of circle *EBZD*, and arcs *BED* and *BZD* are semicircles. That is, circle *ABGD* bisects circle *EBZD*.

Demonstration (Part 2). Points *A* and *G* are the poles of circle *EBZD* by *Sph.* I.8. Therefore great circle *ABGD* passes through the poles of circle *EBZD*.

§◦

Introduction to *Sph.* I.14

See Introduction to *Sph.* I.13–I.15 (p. 381). When circle *EBZD* is great, the circles must bisect each other from *Sph.* I.11 and so bisection cannot imply the other conditions, but then they are equivalent by *Sph.* I.10.

§◦

Theorem. *If a great circle in a sphere bisects a small circle in the sphere, it will cut it at right angles and pass through its poles.*

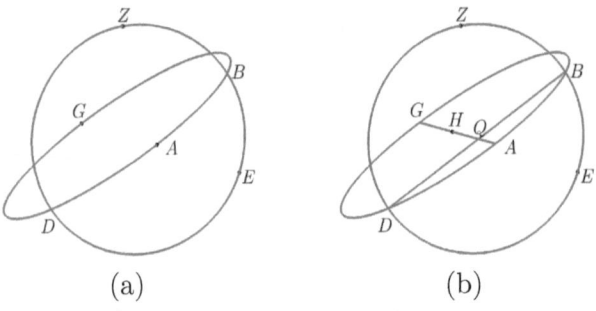

(a) (b)

Figure 16: *Sph.* I.14. (a) The given configuration of circles. (b) The diameter and axis added.

Exposition. In a sphere, let great circle *ABGD* bisect a small circle *EBZD* in the sphere (Figure 16 (a)).

Specification. It is to be proved that *ABGD* cuts *EBZD* at right angles (Part 1) and passes through its poles (Part 2).

Construction. Draw the line of intersection *BD*. Since great circle *ABGD* bisects circle *EBZD*, each arc *BED* and *BZD* is a semicircle, and *BD* is a diameter of circle *EBZD*. Then bisect *BD* at *Q*, which is the center of circle *EBZD*. Let point *H* be the center of great circle *ABGD*, which is also the center of the sphere. Join HQ and extend it in each direction to meet the surface of the sphere at A and G, (Figure 16 (b)) needed for Part 2.

Demonstration (Part 1). Since $EBZD$ is a circle in a sphere and straight line HQ joins its center to the center of the sphere, HQ is perpendicular to $EBZD$ by *Sph.* I.7, and all planes containing HQ are perpendicular to $EBZD$. Since circle $ABGD$ is one of the planes containing HQ, $ABGD$ is perpendicular to $EBZD$.

Demonstration (Part 2). Points A and G are the poles of $EBZD$ by *Sph.* I.8.

Therefore circle $ABGD$ passes through the poles of circle $EBZD$.

§⊙

Unstated corollary. *If a great circle in a sphere bisects a small circle in the sphere, the poles of the small circle are the midpoints of the arcs of the great circle cut off by the small circle.*

Introduction to *Sph.* I.15

See Introduction to *Sph.* I.13–15 (p. 381).

§⊙

Theorem. *If a great circle in a sphere cuts a circle in the sphere through its poles, it will bisect it at right angles.*

Exposition. Let the poles of circle $EBZD$ be points A and G on great circle $ABGD$, both in a sphere (Figure 17 (a)), and join AG (Figure 17 (b)).

Specification. It is to be proved that $ABGD$ bisects $EBZD$ (Part 1) and cuts $EBZD$ at right angles (Part 2).

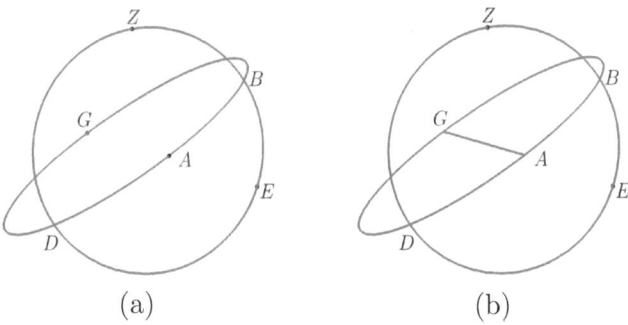

(a)   (b)

Figure 17: *Sph.* I.15. (a) The given configuration of circles. (b) The circles with the upright diameter.

Demonstration (Part 2). Since $EBZD$ is a circle in a sphere and AG runs between its poles, AG is perpendicular to circle $EBZD$ by *Sph.* I.10. Therefore all planes containing AG are perpendicular to $EBZD$. Circle

ABGD is one of the planes containing $\underline{AG}$ and is therefore perpendicular to circle *EBZD*.

Demonstration (Part 1). Therefore *ABGD* bisects *EBZD* (if *EBZD* is a small circle by *Sph.* I.13 Part 1. And all pairs of great circles bisect each other by *Sph.* I.11).

$\mathcal{S}$⟜

Comments. Circle *EBZD* need not be a small circle, unlike *Sph.* I.14.

Introduction to converses *Sph.* I.16 and I.17

None of the characteristic features of a great circle, passing through the center of the sphere, being the greatest circle or the necessary and sufficient condition of bisecting and being bisected by another great circle is any use for drawing one. For that the polar radius is required. Unlike the other properties, that depends on the size of the sphere, which will be determined in *Sph.* I.19. It will be useful to introduce a neologism, the *great square*, for the square inscribed in a great circle. The equal arcs its sides subtend will be called *quadrants* (p. 221).

$\mathcal{S}$⟜

***Sph.* I.16. Theorem.** *The polar radius of a great circle in a sphere is equal to the side of a great square.*

Exposition. Let *ABGD* be a great circle.

Specification. It is to be proved that its polar radius is equal to the side of the great square.

Construction. Let $\underline{AG}$ and $\underline{BD}$ be two diameters of *ABGD* perpendicular to each other at *E*, the center of the sphere (Figure 18). At *E*, set up a perpendicular to the plane of circle *ABGD* and extend it to the surface of the sphere at *Z*. Then *Z* is a pole of circle *ABGD*. Join $\underline{AB}$ and $\underline{AZ}$. Then $\underline{AB}$ is a side of the square inscribed in circle *ABGD*, and $\underline{AZ}$ is a polar radius.

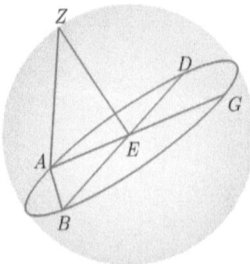

Figure 18: *Sph.* I.16. The sphere, circle *ABGD*, pole *Z*, polar radius $\underline{AZ}$ and square side $\underline{AB}$.

Specification. It is to be proved that $\underline{AZ}$ is equal to $\underline{AB}$.

Demonstration. Since $\underline{ZE}$ is perpendicular to $ABGD$, it makes right angles with all lines in the plane of circle $ABGD$ that meet it; in particular, $\underline{ZE}$ is perpendicular to each of $\underline{AE}$, $\underline{BE}$, $\underline{GE}$, and $\underline{DE}$. Since E is the center of the sphere, in triangles AEB and AEZ, $\underline{BE}$ is equal to $\underline{ZE}$, $\underline{AE}$ is common, and right angles AEB and AEZ are equal. Therefore hypotenuse $\underline{AB}$ is equal to hypotenuse $\underline{AZ}$. But $\underline{AZ}$ is a polar radius of circle $ABGD$, and $\underline{AB}$ is a side of a great square, as required.

§❧

Introduction to *Sph.* I.17

See the Introduction to *Sph.* I.16 and I.17, above.

§❧

Theorem. *If the polar radii of a circle in a sphere are equal to the side of a great square, then the circle is a great circle.*

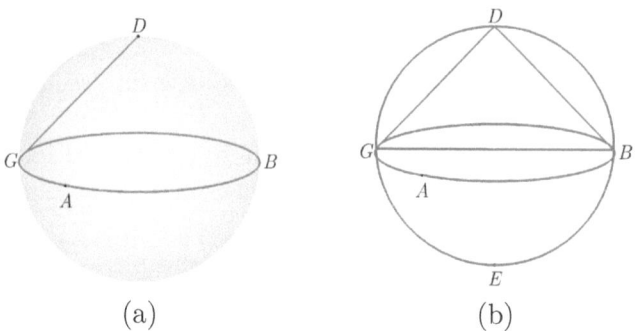

(a) (b)

Figure 19: *Sph.* I.17. (a) The given circle ABG and polar radius $\underline{DG}$. (b) The given circle ABG, its diameter $\underline{BG}$, the great circle $BDGE$, and the polar radii $\underline{DB}$ and $\underline{DG}$.

Exposition. Let circle ABG with pole D be in a sphere and the polar radius $\underline{DG}$ be equal to the side of a great square (Figure 19 (a)).

Specification. It is to be proved that circle ABG is a great circle.

Construction. Let a plane extend through $\underline{DG}$ and the center of the sphere. It will make the curve of intersection in the surface of the sphere be a great circle $BDGE$ (Figure 19 (b)). Let the intersection of circles $BDGE$ and ABG be the line $\underline{BG}$. And join $\underline{DB}$.

Demonstration. The polar radii $\underline{DG}$ and $\underline{DB}$ are equal to two sides of a great square $BDGE$. Therefore arc BDG is a semicircle. Then $\underline{BG}$ is a diameter of great circle $BDGE$ and so of the sphere. Since D is a pole of circle ABG, circle $BDGE$ cuts circle ABG through the poles. By *Sph.* I.15,

BDGE bisects *ABG* at right angles. Since their intersection is line $\underline{BG}$, $\underline{BG}$ is a diameter of circle *ABG*. And it is a diameter of the sphere. Therefore circle *ABG* is a great circle.

§◦

Comments. Converse of *Sph.* I.16.

Introduction to problems *Sph.* I.18–I.21

Some discussion will help the reader to understand the situation of the final four problems. The insight put to work in *Sph.* I.18 is that, while the problem determines the diameter of a circle in the sphere, it is entirely a plane problem. A circle in the sphere is given and three arbitrary points on it. These are three points in the plane of the circle (and in the surface of the sphere). *Sph.* I.19 determines the diameter of the sphere using *Sph.* I.18, and *Sph.* I.20 shows how to draw great circles using the diameter from *Sph.* I.19 and the knowledge of *Sph.* I.17. Finally *Sph.* I.21 finds the poles of a circle in the sphere.

§◦

Sph. I.18. Problem. *To construct in a plane a line equal to the diameter of a given circle in a sphere.*
Exposition. Let *ABG* be the given circle in a sphere (Figure 20 (a)).
Specification. It is required to construct a line in the plane equal to the diameter of circle *ABG*.
Construction. Construct triangle *DEZ* in the plane from three lines so that $\underline{DE}$ is equal to $\underline{AB}$, $\underline{DZ}$ is equal to $\underline{AG}$, and $\underline{EZ}$ is equal to $\underline{BG}$. Also, on the opposite side of $\underline{EZ}$ from *D*, erect perpendiculars to $\underline{ED}$ and $\underline{ZD}$ at *E* and *Z* to intersect at *H* (Figure 20 (b)). Join $\underline{DH}$ (Figure 20 (c)).
Specification. It is required to show that the line $\underline{DH}$ is equal to the diameter of circle *ABG*.
Demonstration. Draw the diameter $\underline{AQ}$ of circle *ABG*, and join $\underline{AB}$, $\underline{BG}$, $\underline{GA}$, and $\underline{GQ}$ (Figure 20 (d)).

Since the three sides of triangle *DEZ* have been constructed equal to the three sides of triangle *ABG*, angle *ABG* is equal to angle *DEZ*. But angle *ABG* is equal to angle *AQG*, subtended by $\underline{AG}$ in triangle *ABG*, and angle *DEZ* is equal to angle *DHZ*, subtended by $\underline{DZ}$ in circumcircle *DEHZ*, and so angle *AQG* is equal to angle *DHZ*. Beyond that, in triangles *AQG* and *DHZ*, right angle *AGQ* is equal to right angle *DZH*, and $\underline{AG}$ is equal to $\underline{DZ}$, and therefore $\underline{AQ}$ is equal to $\underline{DH}$.

§◦

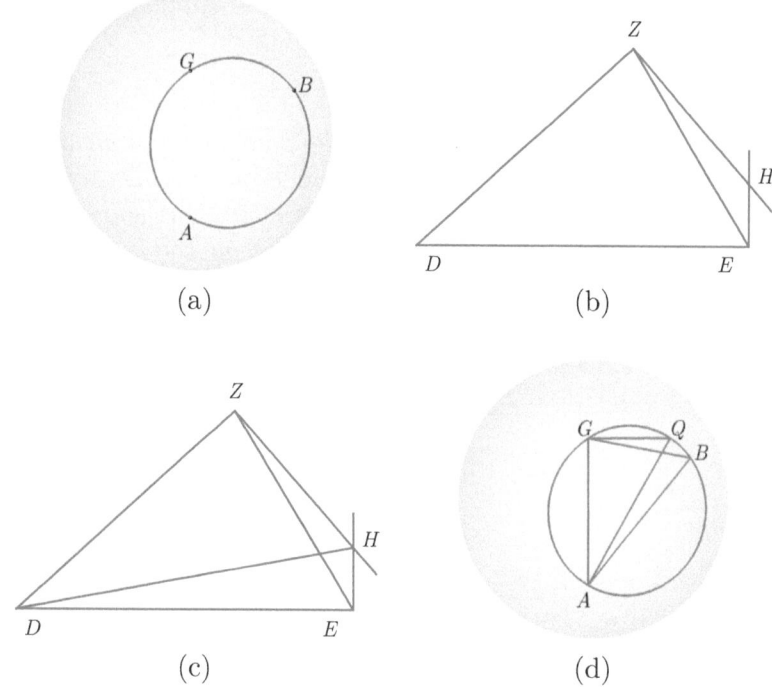

Figure 20: *Sph.* I.18. (a) The given circle *ABG*. (b) The triangle in the plane. (c) The construction of the diameter <u>*DH*</u> in the plane. (d) The construction in the circle.

Comments. The different sorts of construction in this and the following problems are discussed in the Commentary to *Sph.* I.18 (p. 224). It is interesting to note that the locating of *Q* is of the theoretical kind in the text, obtained by postulates and problems of the *Elements* in the plane of the circle internal to the sphere, but it would also have been possible to obtain it with a real construction by moving triangle *EZH* with a compass from the plane to *BGQ* on the sphere as *ABG* was moved to *DEZ*. The theoretically obtained *Q* is sufficient because the point *Q* is only wanted for the proof.

Introduction to *Sph.* I.19

See the Commentary's introduction to *Sph.* I.18–I.21 (p. 223) and the Introduction above (p. 386). The sphere is all that is given. *Sph.* I.18 and the two-circle configuration of *Sph.* I.13–I.15 are combined to find the diameter of great circles in the sphere, the diameter of the sphere. The task is taken in two stages. *Sph.* I.18 is used as written to find the diameter of a small circle. Then its polar radius twice and that diameter are used as the side lengths of a triangle whose vertices are on a great circle. This uses the procedure of *Sph.* I.18, which cannot itself be applied where no circle is given.

§◦

Problem. *To construct a line equal to the diameter of a given sphere.*

Construction. Take any two points A and B on the surface of the sphere. With pole A and polar radius $\underline{AB}$ draw circle BGD (Figure 21 (a)). From *Sph.* I.18, it is possible to construct a line $\underline{ZH}$ in the plane equal to the diameter $\underline{BD}$ of circle BGD. Construct the triangle EZH so that $\underline{EZ}$ and $\underline{EH}$ are equal to the polar radius of circle BGD (Figure 22 (a)). On the opposite side of $\underline{ZH}$ from E, erect perpendiculars to $\underline{EZ}$ at Z and to $\underline{EH}$ at H to intersect at Q. Join $\underline{EQ}$.

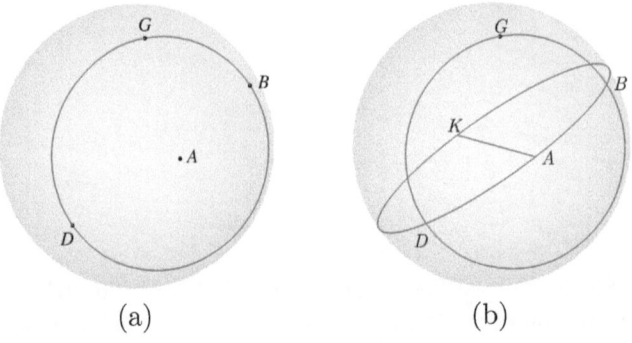

(a)   (b)

Figure 21: *Sph.* I.19. (a) The circle through B with pole A; D is diametrically opposite B. (b) The other pole K of circle BGD and the great circle bisecting it through B and D.

Specification. It is to be proved that $\underline{EQ}$ is equal to the diameter of the sphere.

Construction. Consider the diameter $\underline{AK}$ of the sphere and, containing $\underline{AK}$, a plane intersecting the sphere in the great circle containing B and D (Figure 21 (b)). Join $\underline{AB}$, $\underline{DA}$, $\underline{DB}$, and $\underline{DK}$ (Figure 22 (b)).

Demonstration. In triangles ABD and EZH, since the three sides $\underline{AB}$, $\underline{BD}$, and $\underline{AD}$ are respectively equal to sides $\underline{EZ}$, $\underline{ZH}$, and $\underline{EH}$, angle ABD is equal to angle EZH. But angle ABD is equal to angle AKD (subtended by the same secant in the undrawn circumcircle $EZQH$ equal to great circle $ABKD$). Likewise, angle EZH is equal to angle EQH, and so angle AKD equals angle EQH.

So the two triangles AKD and EQH, having angle AKD equal to angle EQH, right angle ADK equal to right angle EHQ, and side $\underline{AD}$ equal to side $\underline{EH}$ by construction, has the remaining sides respectively equal to the remaining sides. Therefore $\underline{EQ}$ is equal to $\underline{AK}$, the diameter of the sphere.

§◦

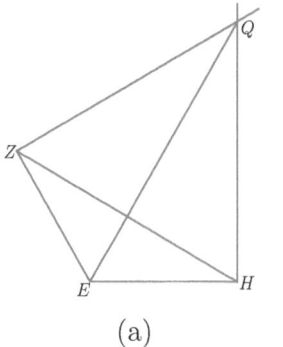 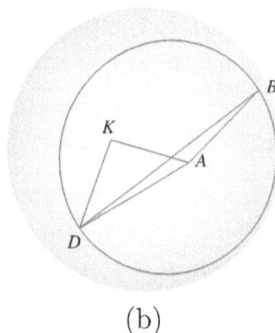

(a) (b)

Figure 22: *Sph.* I.19. (a) The construction in the plane showing EQ equal to the diameter of the sphere. (b) The construction in the sphere for the proof.

Comments. Euclid sets out the diameter of a given sphere several times in the *Elements*, viz. *Elem.* XIII.13–XIII.18 except XIII.17, which uses XIII.15, without any authority. It can be claimed that Euclid's definition of a sphere as the surface of rotation of a semicircle allows him to set out the diameter of that semicircle, but the definition is not enough. Faced with a sphere, the diameter of the semicircles that can produce it by rotation is no more accessible than the diameter of the sphere itself or its radius or center. While it is impossible to know what Euclid had in mind in writing so, it is indisputable that Theodosios had *Elem.* XIII in mind.

There is an interesting contrast between the way in which Theodosios organizes the two problems *Sph.* I.18 and I.19 and the following more concise rewrite, which some readers may find more appealing. *Sph.* I.18 is set up to deal with circle BGD, and so circle $ABKD$ has to be done, albeit with the same technique, as though *Sph.* I.18 were not there. If a problem that we can call *Sph.* I.18* were set out to find the diameter of a circle through any triangle in the sphere with given side lengths, having the same construction and proof as *Sph.* I.18, then it would shorten *Sph.* I.19 because the second invocation of *Sph.* I.18* would give the diameter of the sphere as the diameter of a great circle.

Construction and Demonstration. Take any two points A and B on the surface of the sphere. With pole A and polar radius AB draw circle $BGDE$ (Figure 21 (a), where point is E not shown). (The locations of G and E are immaterial, of D initially undetermined.) Call BD the diameter of circle BGD, which the great circle through B and the pole A bisects perpendicularly, in the configuration of *Sph.* I.13–15. Call AK the diameter from A of the great circle $ABKD$ and of the sphere. It is required to construct in the plane a line QL equal to the diameter AK. Using I.18* and side lengths from triangle BGE, construct a line HZ in the plane equal to the diameter

$\underline{BD}$ of circle BGD. Using I.18* and side lengths from isosceles triangle ABD, construct a line $\underline{QL}$ in the plane equal to the diameter $\underline{AK}$ of the great circle $ABKD$. $\underline{QL}$ equals a diameter of the sphere because $ABKD$ is great.

Introduction to *Sph.* I.20

See the Commentary's introduction to *Sph.* I.18–I.21 (p. 223) and the Introduction above (p. 386). The way to construct a great circle through two non-antipodal points is the standard way one would draw a circle of given polar radius through two given points in a plane. *Sph.* I.17 and I.19 showed how to obtain the radius.

So

Problem. *To draw a great circle through two given points on a spherical surface.*
Exposition. Let the two given points on a spherical surface be A and B.
Specification. It is required to draw a great circle through A and B.
Construction and Demonstration. (If $\underline{AB}$ is a diameter of the sphere, it is plain that arbitrarily many great circles can be drawn through points A and B.)

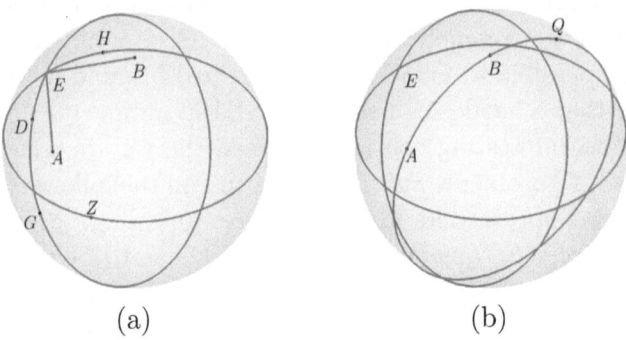

(a)   (b)

Figure 23: *Sph.* I.20. (a) The construction of great circles with poles A and B and lines $\underline{EA}$ and $\underline{EB}$, polar radii of the required great circle. (b) The required circle ABQ.

Let $\underline{AB}$ then not be a diameter of the sphere, and, with poles A and B and appropriate polar radius from *Sph.* I.17, draw great circles GDE and ZEH respectively intersecting at E (Figure 23 (a)). Join $\underline{EA}$ and $\underline{EB}$. Each of $\underline{EA}$ and $\underline{EB}$ is equal to the side of the square inscribed in the great circle. Then the circle BQ drawn with pole E and polar radius $\underline{EB}$ passes through A as well as B (Figure 23 (b)). Call it ABQ. Then circle ABQ is a great circle through A and B, for its polar radius is appropriate from *Sph.* I.17.

So

Comments. The Commentary on the translation discusses why the case of antipodal *A* and *B* is dismissed and how one would tell whether that is the case (see p. 233).

Introduction to *Sph.* I.21

See the Commentary's introduction to *Sph.* I.18–I.21 (p. 223) and the Introduction above (p. 386).

§❧

Problem. *To find the pole of a given circle in a sphere.*
Exposition. Let *ABG* be the given circle in a sphere (location of *G* to be determined) (Figure 24 (a)).
Specification. It is required to find the pole of circle *ABG*.
Construction. Let *A* be any point on the circumference, let two equal arcs *AD* and *AE* be cut off it, and let the remaining arc *DE* be bisected at *Z*.

Now either circle *ABG* is small (Case 1, Figure 24) or great (Case 2, Figure 25).

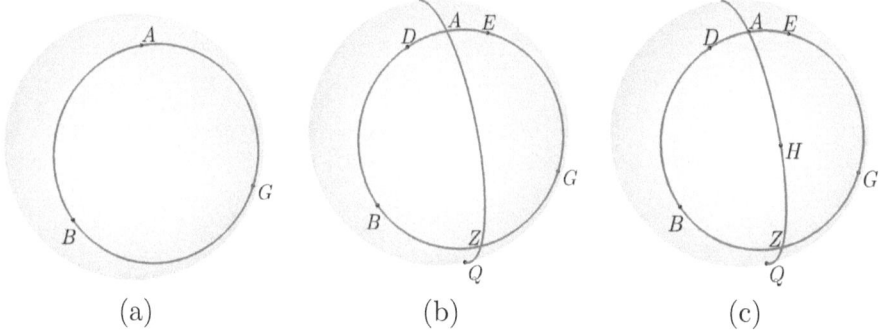

(a)   (b)   (c)

Figure 24: *Sph.* I.21. (a) The given circle *ABG* if small. (b) The points *D*, *E*, and *Z* on the given small circle *ABG* and great circle *AZQ* added. (c) The point *H* added, mid-point of arc *AZ* of great circle *AZQ*.

Construction and Demonstration (Case 1). Let *ABG* be small, and through diametrically opposite points on it *A* and *Z*, which are on the surface, draw great circle *AZQ* (Figure 24 (b)), bisecting circle *ABG*.

By *Sph.* I.14, arc *DA* is equal to arc *AE*, and arc *DZ* is equal to arc *ED*; therefore arc *ADZ* is equal to arc *AEZ*, so that circle *AZQ* bisects circle *ABG*; *AZQ* cuts circle *ABG* at right angles and through its poles. Bisect arc *AZ* of great circle *AZQ* at *H* (Figure 24 (c)), which is the pole of circle *ABG*.

Construction and Demonstration (Case 2). Let circle ABG be great (Figure 25 (a)). As before, arc ADZ is equal to arc AEZ. Bisect either arc AZ at point G. Then each of the arcs AG and GZ is a quadrant. Therefore the circle drawn with pole G and polar radius $\underline{GZ}$ (Figure 25 (b)) will pass through A as well as Z since $\underline{AZ}$ is a diameter.

Then circle AZQ is great by *Sph.* I.17. And since G is the pole of circle AZQ, circle ABG cuts circle AZQ through the poles. Then, by *Sph.* I.15, circle ABG bisects circle AZQ at right angles. Then since great circle AZQ cuts circle ABG at right angles, circle AZQ bisects circle ABG through the poles. Let arc AZ of great circle AZQ be bisected at point H, which is therefore the pole of circle ABG (Figure 25 (c)).

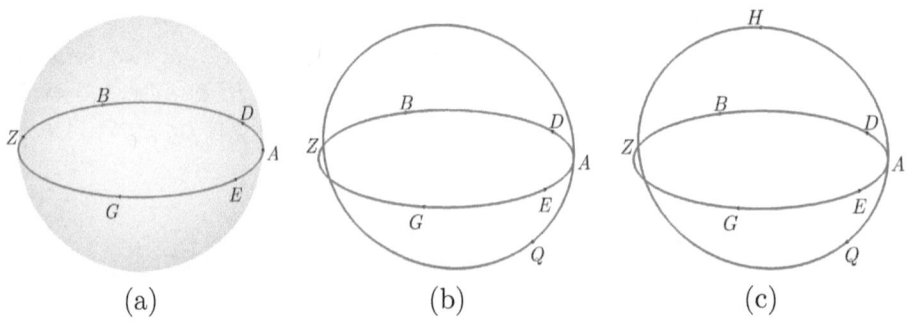

Figure 25: *Sph.* I.21. (a) The points D, E, and Z, on the given great circle ABG. (b) The great circle AZQ added. (c) The point H.

§◦

Introduction to *Spherics* II

Book II covers a lot of ground. First (*Sph.* II.1, II.2) it considers parallel circles with the same poles. It second (*Sph.* II.3–II.5) considers the configuration of two circles that touch and the great circle through their poles and point of contact (Figure 26). Then (*Sph.* II.6–II.8) treats two equal parallel circles touching a great circle. *Sph.* II.8 is where direct astronomical relevance becomes usual. In *Sph.* II.9 arc lengths determined by intersections with great circles are the focus. *Sph.* II.10, II.13, and II.16 are a group concerning great circles intersecting a set of parallel circles. It is interrupted by two geometrical pairs, each containing something needed for the next of the group, *Sph.* II.11 and II.12 for II.13 and *Sph.* II.14 and II.15 for II.16. The two solid-geometrical propositions (*Sph.* II.11, II.12) intervene concerning a configuration that can occur in a sphere but need not and are converses.

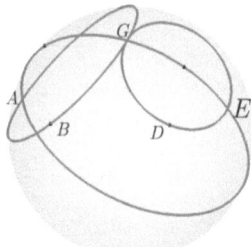

Figure 26: Two circles *ABG* and *GDE* that touch at *G* and the great circle through *A*, *G*, *E*, and their poles (*Sph.* II.3).

Again two quite different propositions (*Sph.* II.14, II.15) intervene because they are used to prove *Sph.* II.16. They are the last problems, to construct a great circle tangent to a small circle through a given point either on the circle or off the circle but unavoidably between it and the circle equal and parallel to it. Taken as a pair, their position is dictated by their depending on *Sph.* II.11 and II.12.

The study of arc lengths begun in *Sph.* II.10, II.13 and II.16 is now continued (*Sph.* II.17–II.20) with parallel circles and great circles that intersect them, the sizes of the parallel circles, and the lengths of the arcs that result.

Sph. II.21 is a geometrical preliminary to the final two propositions.

The final two propositions (*Sph.* II.22, II.23) concern the way a great circle tangent to a small circle is variously inclined to any other great circle if its point of tangency rotates around the small circle.

Starting with *Sph.* II.5, all propositions can be claimed to be used in *Moving Sphere* and *Phenomena*. These uses are explained in the Commentary proposition by proposition and also in groups, *Sph.* II.5–II.8, II.10–II.13, II.16–II.18, II.19, II.20, II.22, II.23, and in the latter part of the Introduction to *Spherics* II. The first half of the *Phenomena* and *Phen.* 12 and 14 are more or less translations of spherical geometrical facts into astronomical facts. Because in the *Spherics* nothing moves and in astronomy nothing stays still, there is a serious matter of translation involved, and that may be one of the reasons both texts remained current in curricula into the Early Modern period.

§

Definition

Two circles in a sphere are said to *touch* each other when the line of intersection of their planes touches both circles.

So

Comments. For discussion of tangency of circles, see Commentary (p. 245).

Propositions

Introduction to *Sph.* II.1 and II.2

These two propositions are converses concerning parallel circles and their poles.

So

***Sph.* II.1. Theorem.** *In a sphere, parallel circles have the same poles.*
Exposition. Let parallel circles *ABG* and *DEZ* be in a sphere (Figure 27 (a)).

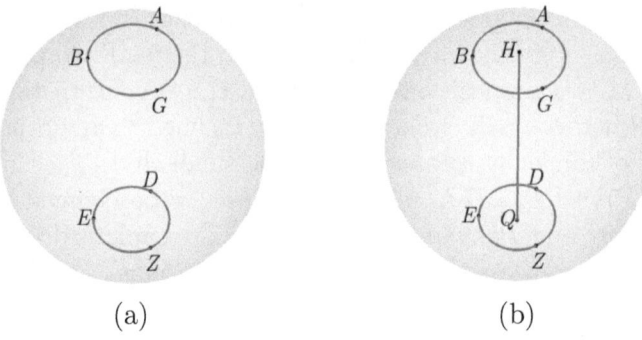

(a)   (b)

Figure 27: *Sph.* II.1. (a) Given parallel circles. (b) Line $\overline{HQ}$ joining the poles of the two circles.

Specification. It is to be proved that circles *ABG* and *DEZ* have the same poles.
Construction. Find the poles of circle *ABG*, *H* and *Q*, and join $\overline{HQ}$ (Figure 27 (b)).
Demonstration. Since, if a circle is in a sphere, the straight line produced through its poles is perpendicular to the circle and through its center and the sphere's, $\overline{HQ}$ is perpendicular to circle *ABG* and is through the center of it and of the sphere. Since $\overline{HQ}$ is perpendicular to circle *ABG* and since *ABG* is parallel to circle *DEZ*, $\overline{HQ}$ is also perpendicular to circle *DEZ*. Then, since circle *DEZ* is in a sphere, the perpendicular to it from the center of

the sphere HQ, extended in both directions, meets the surface of the sphere at its poles H and Q. But they are also the poles of circle ABG.

§◦

Introduction to *Sph.* II.2

See Introduction to *Sph.* II.1 and II.2 (above). Converse of *Sph.* II.1.

§◦

Theorem. *In a sphere, circles that have the same poles are parallel.*
Exposition. Let circles ABG and DEZ in a sphere have the same poles H and Q (Figure 28 (a)).

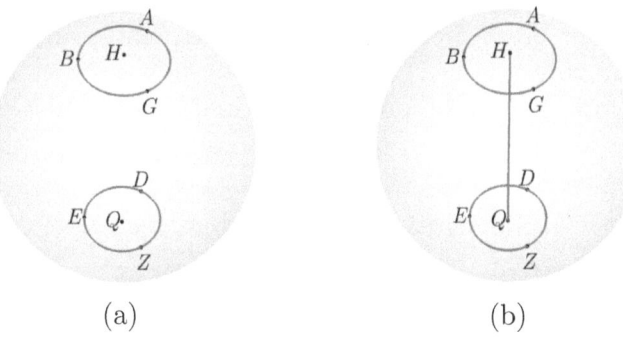

Figure 28: *Sph.* II.2. (a) Given circles with common poles H and Q. (b) Line HQ joining the poles of the two circles.

Construction. Join HQ (Figure 28 (b)).
Demonstration. Since ABG is a circle in a sphere and the straight line HQ runs through its poles, HQ is perpendicular to circle ABG. Similarly HQ is also perpendicular to circle DEZ. But planes, to which the same straight line is perpendicular, when extended do not meet; therefore the planes of circles ABG and DEZ do not meet. Therefore circle ABG is parallel to circle DEZ.

§◦

Comments. These two theorems imply the transitivity of parallelism for circles on the sphere.

Introduction to *Sph.* II.3–II.5

This group concerns the tangency of two circles. First is a condition that makes tangency happen, and then tangency is shown to imply both parts of that condition. These propositions concern what happens when the configuration of *Sph.* I.13–I.15, a great circle through the poles of another circle, is beside another with the bisecting circle common to the two configurations and the other circles, one of which must be small, touching. For more elaborate analysis of the conditions, see the Commentary's introduction to the group (p. 247). The group is analogous to the single *Sph.* II.9 with respect to two circles meeting at two points rather than at one, the class of situations for which the present is a limiting case.

§◦

Sph. II.3. Theorem. *In a sphere, if two circles cut the circumference of a certain great circle at the same point and have their poles on it, then the circles touch each other.*

Exposition. Let two circles *ABG* and *GDE* in a sphere cut the circumference of a certain great circle *AGE* at the same point *G* and have their poles on it (Figure 26).

Specification. It is to be proved that circles *ABG* and *GDE* touch each other.

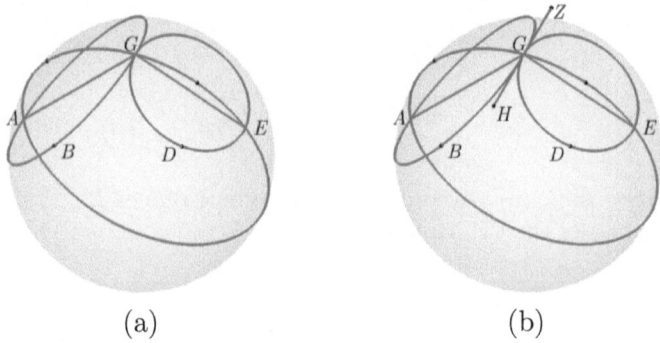

(a)   (b)

Figure 29: *Sph.* II.3. (a) The given circles and their diameters. (b) Line of intersection $\underline{ZGH}$ of the circles' planes added.

Construction. Let the lines of intersection of *AGE* and *ABG* be $\underline{AG}$, of *AGE* and *GDE* be $\underline{GE}$ (Figure 29 (a)), and of *ABG* and *GDE* be $\underline{HGZ}$ (Figure 29 (b)).

Demonstration. Since in a sphere great circle *AGE* cuts circle *ABG* in the sphere through its poles, *AGE* bisects *ABG* perpendicularly; *AG* is a diameter of circle *ABG*. Similarly $\underline{GE}$ is also a diameter of circle *GDE*. Since *AGE* is perpendicular to each of circles *ABG* and *GDE*, each of *ABG* and *GDE* is perpendicular to circle *AGE*, and therefore their line of intersection $\underline{ZGH}$ is perpendicular to circle *AGE*. Their line of intersection $\underline{ZGH}$ is therefore perpendicular to all lines it cuts lying in the plane of circle *AGE*, and it cuts each of $\underline{AG}$ and $\underline{GE}$, in the plane of AGE. Therefore $\underline{ZH}$ is perpendicular to each of $\underline{AG}$ and $\underline{GE}$.

Then since $\underline{ZH}$ runs perpendicularly from the end of the diameter AG of circle ABG in its plane, $\underline{ZH}$ touches circle ABG at point G. Similarly $\underline{ZH}$, the line of intersection of the planes of circles ABG and GDE, touches circle GDE at point G. Therefore circles ABG and GDE touch each other.

§◦

Introduction to *Sph.* II.4

See Introduction to *Sph.* II.3–II.5 (p. 395).

§◦

Theorem. *In a sphere, if two circles touch each other, then the great circle drawn through their poles goes through their point of contact.*

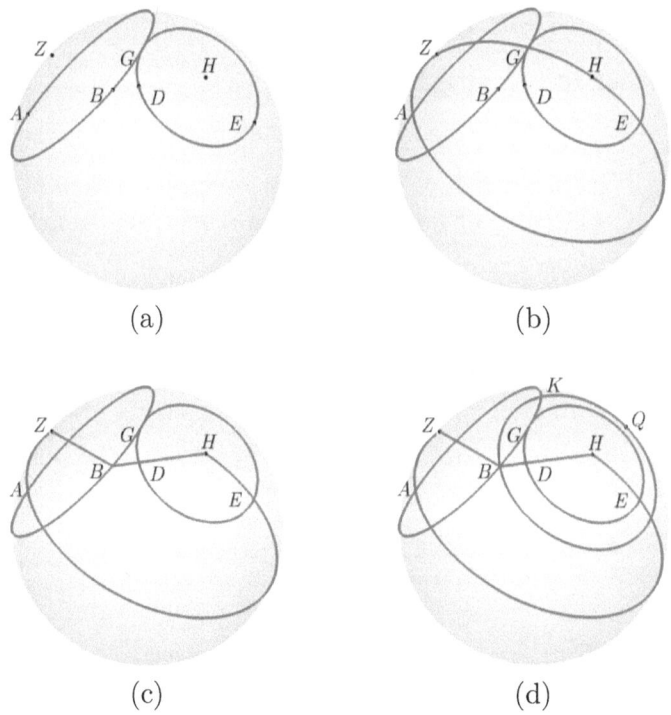

(a)   (b)

(c)   (d)

Figure 30: *Sph.* II.4. (a) The given circles and their poles. (b) The great circle *ZGH* through poles and point of contact. (c) The impossible great circle *ZBDH* added (partly with lines) not through *G*, arcs *ZB* and *BH* drawn straight. (d) The construction circle *BKQ* added.

Exposition. Let two circles in a sphere, AG with pole Z and GE with pole H, touch each other at G (Figure 30 (a)).

Specification. It is to be proved that the great circle drawn through the poles Z and H goes through point G (Figure 30 (b)).

Construction. Suppose not, and if it were possible, draw the great circle $ZBDH$, B being on the circle AG and D on GE, both different from G (Figure 30 (c)), and with pole H and polar radius $\underline{HB}$ draw circle BKQ (Figure 30 (d)).

Demonstration. Then circle GDE is parallel to circle BKQ, for they have the same poles. And since the two circles in a sphere ABG and BKQ cut the circumference of a great circle ZBH (illustrated as a broken line) at point B and have their poles on it, circles ABG and BKQ touch each other. But they also cut each other, which is impossible. Therefore the great circle drawn through the poles of circles ABG and GDE goes through their point of contact.

§

Introduction to *Sph.* II.5

See Introduction to *Sph.* II.3–II.5 (p. 395).

§

Theorem. *In a sphere, if two circles touch each other, then the great circle drawn through the poles of one and the point of contact goes through the poles of the other.*

Exposition. Let two circles in a sphere, ABG with pole Z and GDE with pole H, touch each other at point G (Figure 31 (a)).

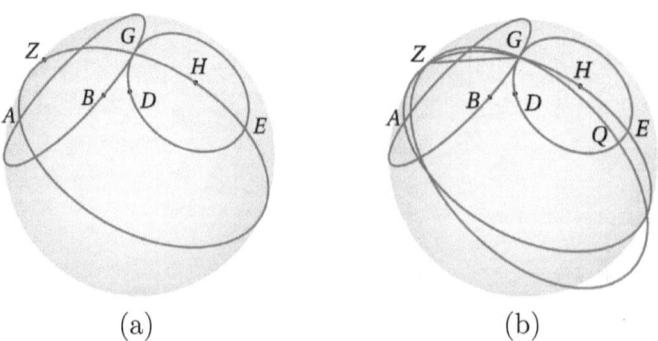

(a)   (b)

Figure 31: *Sph.* II.5. (a) The given circles and the great circle ZGH. (b) The assumed great circle ZGQ and line $\underline{ZG}$ added.

Specification. It is to be proved that the great circle drawn through points Z and G also goes through point H.

Construction. Suppose not and, if it were possible, let it be *ZGQ* not through *H*, and draw through poles *Z* and *H* the great circle *ZGH* (Figure 31 (b)). Of course it goes through *G*. Join <u>*ZG*</u>.

Demonstration. Since each of *ZGH* and *ZGQ* is great, they bisect each other; therefore <u>*ZG*</u> is a diameter of both great circles and of the sphere. But it is the polar radius of circle *ABG*, which is impossible. Therefore the great circle drawn through points *Z* and *G* also goes through point *H*.

Introduction to *Sph.* II.6–II.8

These propositions concern the configuration of a great circle oblique to the parallel circles and touching two that are equal. This configuration is described differently in *Sph.* II.6 and II.7 and differently again in *Sph.* II.8. The various conditions are analyzed in the Commentary's introduction to the group (p. 250).

Sph. **II.6. Theorem.** *In a sphere, if a great circle touches a certain circle in the sphere, then it also touches another circle equal and parallel to it.*

Exposition. Let great circle *AG* in a sphere touch a circle *GD* at point *G* (Figure 32 (a)).

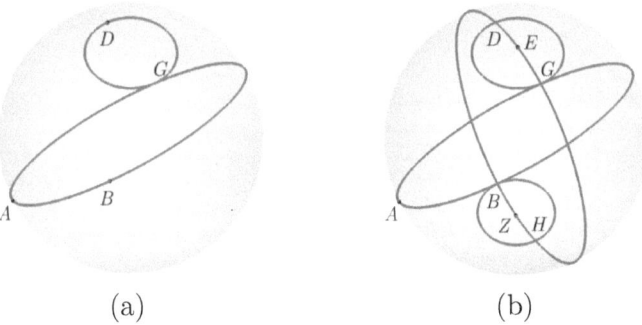

(a)   (b)

Figure 32: *Sph.* II.6. (a) Touching great circle *ABG* and small circle *GD*. (b) Additions: poles *E* and *Z*, great circle *GED*, circle *BH*.

Specification. It is to be proved that circle *AG* also touches another circle equal and parallel to circle *GD*.

Construction. Find the pole of circle *GD*, point *E*, and draw through points *G* and *E* a great circle *GED* intersecting circle *AG* at *B* (Figure 32 (b)). Then cut off arc *BZ* equal to arc *GE*, and with pole *Z* and polar radius <u>*ZB*</u> draw a circle cutting circle *GEDBZ* at *H*.

Demonstration. Since the two circles ABG and GD in the sphere touch each other and the great circle $GEDBZH$ has been drawn through one of the poles E and the point of contact G, $GEDBZH$ goes through the poles of circle ABG. And since the two circles ABG and BH in the sphere cut the circumference $GEDBZH$ of a great circle at the same point B and have their poles on it, circles ABG and BH touch each other. In circle $GEDBZH$, if EB is added to each of the equal arcs GE and BZ, then the whole arc GB is equal to whole arc EZ. But GB is a semicircle, and so EZ is a semicircle. And point E is a pole of circle GD; therefore pole Z of BH is the other pole of circle GD. Likewise, since EZ is a semicircle and Z is a pole of circle BH, E is the other pole of circle BH. Therefore circles GD and BH, having the same poles, are parallel. And since GE is equal to BZ by construction, circle GD is also equal to circle BH. Therefore circle ABG also touches the other circle BH equal and parallel to GD.

Comments. For justification that circle GD is equal to circle BH, see Commentary (p. 252).

Introduction to *Sph.* II.7

See Introduction to *Sph.* II.6–II.8 (p. 399).

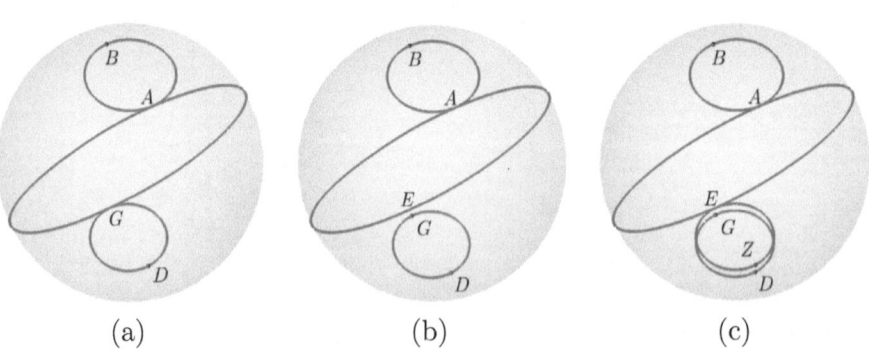

Figure 33: *Sph.* II.7. (a) Equal parallel circles and great circle touching one. (b) Equal parallel circle added touching the great circle. (c) Equal parallel circles and great circle touching them.

Theorem. *If two equal and parallel circles are in a sphere, then a great circle touching one of them also touches the other.*
Exposition. Let AB and GD be two equal parallel circles in a sphere.

Specification. It is to be proved that the great circle touching AB also touches GD (Figure 33 (a)).

Demonstration. If it were possible, let great circle AE touch AB at point A but not touch GD (Figure 33 (b)).

Since great circle AE in a sphere touches a certain circle AB in the sphere, it touches another circle equal and parallel to AB; accordingly let the equal and parallel circle that it touches be EZ (Figure 33 (c)). Since then AB is equal and parallel to EZ, but AB is equal and parallel to GD, GD is also equal and parallel to EZ. Therefore there are in a sphere three equal and parallel circles, which is impossible. The great circle touching AB touches GD.

Introduction to *Sph.* II.8

See Introduction to *Sph.* II.6–II.8 (p. 399). As mentioned above, this is the point in the treatise at which astronomical relevance begins to be consistent with specific geometrical exceptions, *Sph.* II.9, II.11 and II.12, II.14 and II.15, and II.21.

Theorem. *A great circle cutting a circle in the sphere not through its poles touches two equal circles parallel to it.*

Exposition. Let a great circle ABG cut circle BD in the sphere obliquely (i.e., not through the poles, not perpendicularly) at B and D (Figure 34 (a)).

Specification. It is to be proved that ABG touches two equal circles parallel to each other and to BD.

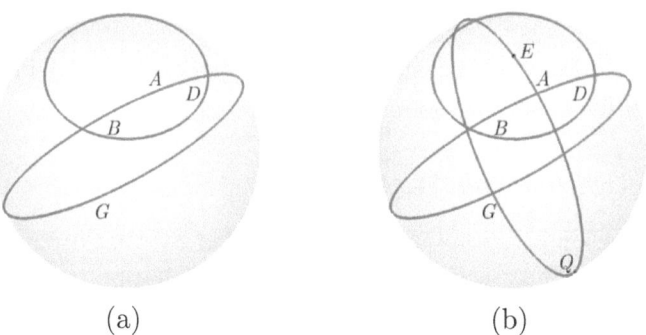

Figure 34: *Sph.* II.8. (a) The given intersecting small circle BD and great circle ABG. (b) Great circle $AEGQ$ added through poles E and Q of BD and ABG.

Construction. Since ABG is oblique to BD, the pole E of BD is not on circle ABG; through point E and one of the poles Q of the great circle draw great circle $AEGQ$ (Figure 34 (b)), determining points A and G on circle ABG. Draw circle AZ with the same pole E as BD determining Z (Figure 35 (a)).

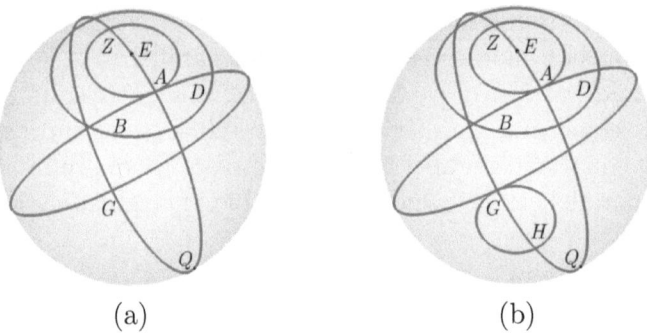

(a)   (b)

Figure 35: *Sph.* II.8. (a) Circle AZ added parallel to BD. (b) Circle GH added parallel to BD and AZ.

Demonstration. Since they have the same poles, circles AZ and BD are parallel. Since the two circles ABG and AZ in the sphere cut the circumference of great circle AEQ at the same point A and have their poles on it, they touch each other; therefore circle ABG touches circle AZ at A. Then since great circle ABG touches circle AZ in the sphere, it also touches another circle equal and parallel to AZ. Let that circle be GH (Figure 35 (b)), where H is the other intersection of parallel circle GH and $AEZGH$. Then since circle AZ is equal and parallel to GH, but AZ is parallel to BD, GH is also parallel to BD. Therefore circle ABG touches the two equal circles AZ and GH, parallel to each other and to BD.

Comments. For the astronomical significance of this group of theorems, see the Commentary (p. 251).

Introduction to *Sph.* II.9

This proposition concerns intersecting circles, stating a condition that follows from their intersecting. It concerns what happens when one configuration of *Sph.* I.13–I.15, a great circle through the poles of another circle, overlaps another with the bisecting circle common to the two configurations and the other circles, one of which must be small, intersecting. It is analogous to *Sph.* II.4 in calling on the common great circle.

For astronomical significance, see Commentary (p. 256).

Theorem. *In a sphere, if two circles cut off arcs of each other and a great circle is drawn through their poles, then it bisects the arcs cut off.*

Exposition. Let two circles in a sphere $ZAEB$ and $ZGED$ cut each other at points Z and E and great circle $AGBD$ pass through their poles (Figure 36 (a)).

Specification. It is to be proved that circle $AGBD$ bisects the arcs of the circles cut off, that is to say, that arc ZA is equal to arc AE, and ZB to BE, and ZG to GE, and ZD to DE.

Construction. Let the line of intersection of circle $AGBD$ and $ZAEB$ be $\underline{AB}$ and of $AGBD$ and $ZGED$ be $\underline{GD}$ intersecting at H (Figure 36 (b)), and join $\underline{ZH}$ and $\underline{HE}$.

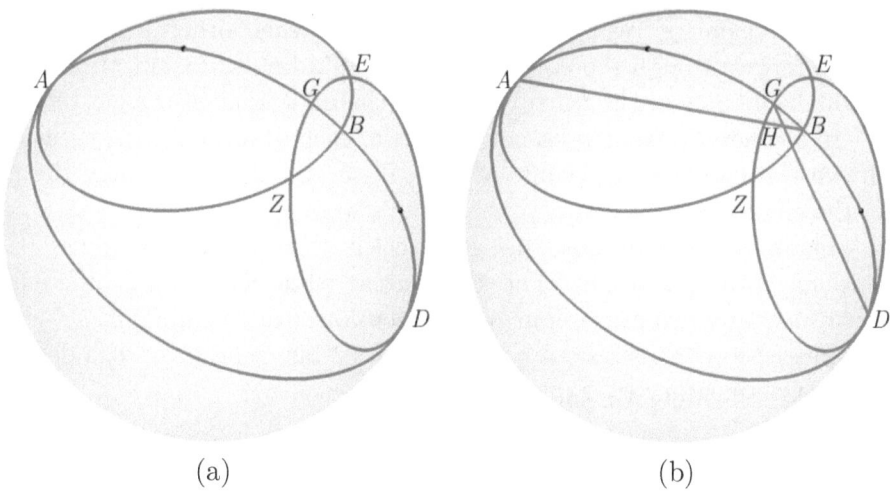

(a)   (b)

Figure 36: *Sph.* II.9. (a) The given configuration of intersecting circles and the great circle through their poles. (b) $\underline{AHB}$ and $\underline{GHD}$ are the intersections of circle $AGBD$ with $ZAEB$ and $ZCED$. $\underline{ZH}$ and $\underline{HE}$ are joined, making line $\underline{ZE}$.

Demonstration. Since points Z, H, and E are in the plane of circle $AEBZ$ and in the plane of circle $ZDEG$, Z, H, and E are on the line of intersection of the two planes – and the line of intersection of every two planes is straight – line $\underline{ZH}$ is collinear with $\underline{HE}$. And since great circle $AGBD$ in a sphere cuts circle $ZAEB$ in the sphere through its poles, it bisects it perpendicularly; therefore $\underline{AB}$ is a diameter of circle $ZAEB$. Similarly we will show that $\underline{GD}$ is a diameter of circle $ZDEG$. And since circle $AGBD$ is perpendicular to each of circles $ZAEB$ and $ZDEG$, each of $ZAEB$ and $ZDEG$ is perpendicular to circle $AGBD$. Therefore the line of intersection $\underline{ZHE}$ of $ZAEB$ and $ZDEG$ is also perpendicular to plane $AGBD$. Therefore $\underline{ZHE}$ is also perpendicular to circle $AGBD$, so that it makes a right angle with every line meeting it and in the plane of circle $AGBD$. But $\underline{ZHE}$ meets each of $\underline{AB}$ and $\underline{GD}$ in the plane of circle *AGBD*; therefore each of $\underline{AB}$ and $\underline{GD}$ is perpendicular to $\underline{ZHE}$.

Then since in circle *ZAEB* diameter $\underline{AB}$ cuts line $\underline{ZHE}$ perpendicularly not through the center of the circle, $\underline{AB}$ bisects $\underline{ZHE}$ (triangle *EAH* is congruent to triangle *ZAH*). Therefore $\underline{ZH}$ is equal to $\underline{HE}$, and $\underline{HA}$ is perpendicular to the line of intersection $\underline{ZE}$. Therefore $\underline{ZA}$ equals $\underline{AE}$ and arc *ZA* is equal to arc *AE*. Similarly we will show also that *ZB* is equal to *BE* and *ZG* to *GE* and *ZD* to *DE*. Therefore circle *AGBD* bisects the cut-off arcs of the circles *ZAE*, *ZBE*, *ZGE*, and *ZDE*.

§●

Comments. In the way that *Sph.* II.9 is similar to *Sph.* II.4, there could also be, similar to *Sph.* II.3 and II.5, the following propositions.

In a sphere, if two circles cut off arcs of each other, then the great circle drawn through the pole of one and a mid-point of a cut-off arc passes through the pole of the other circle and the mid-point of the cut-off arc.

In a sphere, if two circles cut off arcs of each other, then the great circle drawn through the mid-points of the cut-off arcs passes through the poles of the circles.

These are true because the great circle through the two poles passes through both mid-points. The single great circle has several different descriptions, any two of the four points mentioned determining it.

This theorem suggests a construction for bisecting arcs; it is described in the Commentary (p. 255).

Introduction to *Sph.* II.10, II.13, II.16

Sph. II.10 indicates a consequence of great circles cutting parallel circles that is quite different from *Sph.* II.8. It begins a topic that continues, interrupted, in *Sph.* II.13 and II.16, which is roughly a converse of *Sph.* II.10 and II.13 jointly. Rather the way *Sph.* II.9 picks up the matter of *Sph.* II.3–II.5, II.13 picks up the matter of *Sph.* II.10. *Sph.* II.3–II.5 concern a limiting case of *Sph.* II.9, intersecting circles becoming tangential. *Sph.* II.10 concerns a limiting case of *Sph.* II.13, which will be explained there.

The astronomical interpretation of the group *Sph.* II.10, II.13, and II.16 is given in the introduction to the Commentary on the group (p. 258) and specifically of *Sph.* II.10 (p. 259).

§●

***Sph.* II.10. Theorem.** *In a sphere, if great circles are drawn through the poles of parallel circles, then the arcs of the parallel circles between the great circles are similar and the arcs of the great circles between the parallel circles are equal.*

Exposition (Parts 1 and 2). In a sphere, let *ABGD* and *EZHQ* be parallel circles, and let great circles *AEHG* and *BZQD* pass through their poles *K* and *N* (Figure 37).

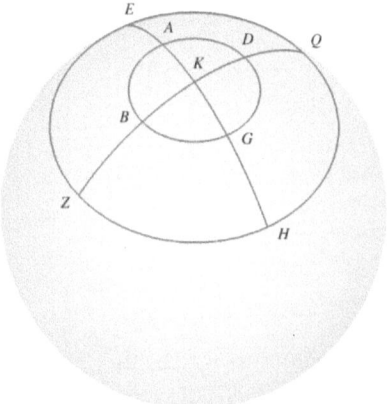

Figure 37: *Sph.* II.10. The given configuration of parallel small circles and two great circles through their poles.

Specification (Parts 1 and 2). It is to be proved (Part 1) that the arcs of the parallel circles between the great circles are similar, that is to say that arc *BG* is similar to arc *ZH*, and *GD* to *HQ*, and *DA* to *QE*, and *AB* to *EZ*, and (Part 2) that the arcs of the great circles between the parallel circles are equal, that is to say that the four arcs *ZB*, *HG*, *QD*, and *EA* are equal to one another.

Demonstration (Part 1). Let the line of intersection of circle *ABGD* and *BZQDK* be *BD*, and the line of intersection of circle *ABGD* and *AEHGK* be *AG*, and the line of intersection of circle *EZHQ* and *BZQDK* be *ZQ*, and the line of intersection of *EZHQ* and *AEHGK* be *EH* (Figure 38 (a)).

  Since great circle *AEHG* cuts circle *ABGD* through its poles, it bisects it perpendicularly; therefore *AG* is a diameter of circle *ABGD*. Similarly we will show also that *BD* is a diameter of circle *ABGD*; therefore point *L*, the point of intersection of *AG* and *BD*, is the center of circle *ABGD* (Figure 38 (b)).

  Likewise, since great circle *AEHG* cuts circle *EZHQ* through its poles, it bisects it perpendicularly; therefore *EH* is a diameter of circle *EZHQ*. Similarly we will show also that *ZQ* is a diameter of circle *EZHQ*. Therefore *M*, the point of intersection of *EH* and *ZQ*, is the center of circle *EZHQ*. Since two parallel planes *ABGD* and *EZKHQ* cut the plane *BZQD*, their lines of intersection *BD* and *ZQ* are parallel. Similarly we will show also that *AG* is parallel to *EH*.

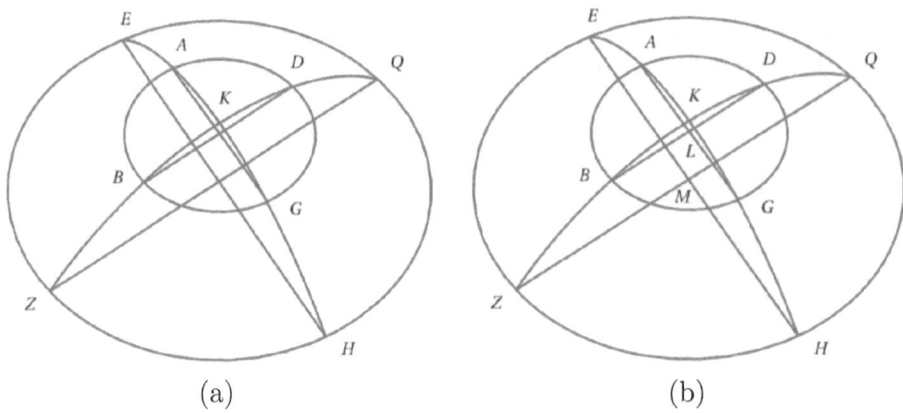

Figure 38: *Sph.* II.10. (a) Diameters <u>AG</u>, <u>BD</u>, <u>EH</u>, and <u>ZQ</u> added. (b) Circle centers L and M labeled.

Then since the two intersecting lines <u>BL</u> and <u>LG</u> correspond to the two intersecting lines <u>ZM</u> and <u>MH</u>, not in the same plane, they contain equal angles; therefore angle ZMH equals angle BLG. And since, at the centers, angle ZMH is subtended by arc ZH and angle BLG is subtended by arc BG, arc ZH is similar to arc BG. Similarly we will show also that arc GD is similar to arc HQ, AD to EQ, and AB to EZ.

Specification (Part 2). It is to be proved that the arcs of the great circles between the parallel circles, EA, ZB, HG, and QD, are equal.

Demonstration (Part 2). Since K is a pole of circle $ABGD$, the four arcs KA, KB, KG, and KD are equal to one another. Likewise, since K is a pole of circle $EZHQ$, the four arcs KE, KZ, KH, and KQ are equal to one another. Therefore the remaining four arcs EA, ZB, HG, and QD are equal to one another.

Introduction to *Sph.* II.11 and II.12

These propositions are converses referring to the same configuration, which is three-dimensional but not spherical. The configuration can, however, be embedded in a sphere. That is how the propositions are used.

Sph. II.11. **Theorem.** *If on diameters in equal circles equal segments of circles are set up perpendicularly, and on them equal arcs from the ends of the segments are cut off less than half of the whole segments, and from the points so determined equal lines are extended to the circumferences of the* *first circles, they cut off equal arcs of the first circles from the ends of the diameters.*

Exposition. On the diameters $\underline{AG}$ and $\underline{DZ}$ in equal, let us say horizontal, circles let equal segments AIG and DOZ of circles with mid-points I and O be set up vertically (Figure 39 (a)), and on the segments equal arcs AH and DQ be cut off from their endpoints A and D less than arcs AI and DO, and from points H and Q equal lines $\underline{HB}$ and $\underline{QE}$ be extended to the circumferences of the circles ABG and DEZ.

Specification. It is to be proved that arc AB is equal to arc DE.

Construction. Find the centers of the horizontal circles, points M and N. Drop perpendiculars $\underline{HK}$ and $\underline{QL}$ from points H and Q to the horizontal planes of circles ABG and $DE\overline{Z}$, K and L falling on the lines of intersection (Figure 39 (b)), that is to say on $\underline{AMG}$ and $\underline{DNZ}$. Now join $\underline{KB}$, $\underline{BM}$, $\underline{LE}$, and $\underline{EN}$.

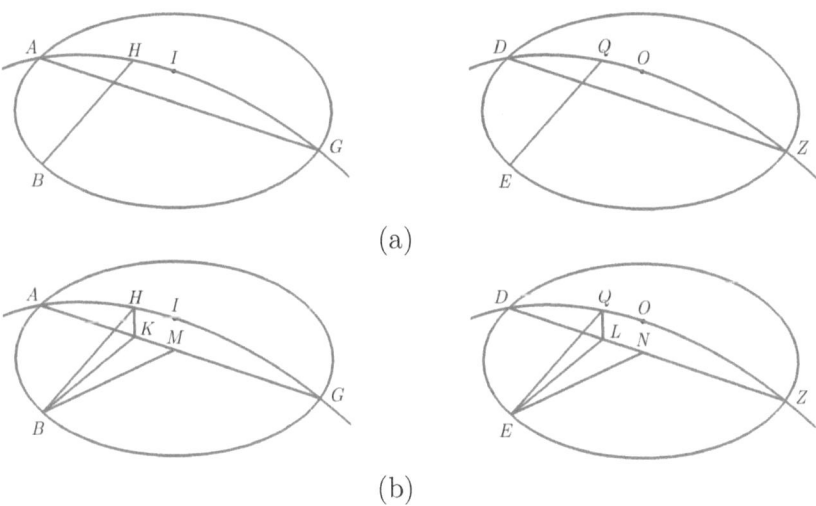

(a)

(b)

Figure 39: *Sph.* II.11 and 12. (a) The two identical initial configurations. (b) The constructions.

Demonstration. Since $\underline{HK}$ is perpendicular to the plane of circle ABG, it makes a right angle with all horizontal lines meeting it, and $\underline{KB}$ meets it; therefore angle HKB is right. Similarly angle QLE is right. And since the two segments AHG and DQZ are equal and the arcs AH and DQ are cut off equal and $\underline{KH}$ and $\underline{QL}$ are produced perpendicular, $\underline{AK}$ is equal to $\underline{DL}$ and $\underline{HK}$ to $\underline{QL}$. And since $\underline{BH}$ is equal to $\underline{EQ}$, the square on $\underline{BH}$ is equal to the square on $\underline{EQ}$. In right triangles BHK and EQL, the equal squares on $\underline{BH}$ and $\underline{EQ}$ less the equal squares on $\underline{HK}$ and $\underline{QL}$ equal the squares on $\underline{KB}$ and $\underline{LE}$ respectively. Therefore $\underline{KB}$ is equal to $\underline{LE}$. And since $\underline{AM}$ is equal to

DN, of which <u>*AK*</u> is equal to <u>*DL*</u>, in triangles *KBM* and *LEN* the remaining <u>*KM*</u> is equal to the remaining <u>*LN*</u>, and <u>*BM*</u> is equal to <u>*EN*</u>, and the third sides <u>*KB*</u> and <u>*LE*</u> are equal; therefore angle *KMB* is equal to angle *LNE*. And in equal circles equal angles are subtended by equal arcs whether at the centers or at the circumferences. Therefore arc *AB* is equal to arc *DE*.

§⟜

Introduction to *Sph.* II.12

See Introduction to *Sph.* II.11 and II.12 (p. 406). The configuration here is the same as that in *Sph.* II.11, which is its converse.

§⟜

Theorem. *If on diameters in equal circles equal segments of circles are set up perpendicularly, and on them equal arcs from the ends of the segments are cut off less than half of the whole segments, and in the same directions equal arcs are cut off from the first circles from the ends of the diameters, then the lines joining the points so determined are equal to each other.*

Exposition. On the diameters <u>*AG*</u> and <u>*DZ*</u> in equal, let us say, horizontal, circles *ABG* and *DEZ*, let equal segments *AIG* and *DOZ* of circles, with *I* and *O* being their midpoints, be set up vertical (Figure 39 (a)), and on the segments equal arcs *AH* and *DQ* be cut off from their endpoints *A* and *D* less than arcs *AI* and *DO*, and from the horizontal circles equal arcs *AB* and *DE* be taken in the same directions. And let <u>*HB*</u> and <u>*QE*</u> be joined.

Specification. It is to be proved that <u>*HB*</u> is equal to <u>*QE*</u>.

Construction. Find the centers of the horizontal circles *ABG* and *DEZ*, *M* and *N*. Drop perpendiculars <u>*HK*</u> and <u>*QL*</u> from points *H* and *Q* to the planes of circles *ABG* and *DEZ* falling on the lines of intersection, that is to say on <u>*AMG*</u> and <u>*DNZ*</u>, and join <u>*KB*</u>, <u>*BM*</u>, <u>*LE*</u>, and <u>*EN*</u> (Figure 39 (b)).

Demonstration. Since arc *AB* is equal to arc *DE*, angle *AMB* is also equal to angle *DNE*. And since the two segments *AHG* and *DQZ* are equal, and arcs *AH* and *DQ* are cut off equal, and <u>*KH*</u> and <u>*QL*</u> have been produced vertically, <u>*AK*</u> is equal to <u>*DL*</u> and <u>*HK*</u> to <u>*QL*</u>. Then in triangles *KBM* and *LEN*, since <u>*AM*</u> is equal to <u>*ND*</u>, of which <u>*AK*</u> is equal to <u>*DL*</u>, the remaining <u>*KM*</u> is equal to <u>*LN*</u>, and <u>*BM*</u> is also equal to <u>*EN*</u>. In triangles *KMB* and *LNE*, the two <u>*KM*</u> and <u>*BM*</u> are respectively equal to the two <u>*LN*</u> and <u>*EN*</u>, and the angle *KMB* is equal to angle *LNE*. Therefore the third side <u>*KB*</u> is equal to <u>*LE*</u>. Since <u>*HK*</u> is perpendicular to the plane of circle *ABG*, it makes a right angle with all horizontal lines meeting it, but <u>*KB*</u> meets it. Therefore angle *HKB* is right. Similarly we will show that angle *QLE* is right. In triangles *HBK* and *QEL*, <u>*HK*</u> is equal to <u>*QL*</u> and <u>*KB*</u> is equal to

LE, and they contain right angles. Therefore the third sides *HB* and *QE* are equal.

§❧

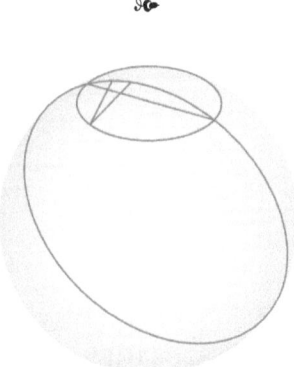

Figure 40: *Sph.* II.12, Comments. Configuration of *Sph.* II.11 and II.12 embedded in a sphere, illustrating the ambiguous (and therefore undiscussed) case in which equal lines and horizontal arcs allow two different vertical arcs (see p. 260).

Comments. "Less than half of the segments" is not needed; see the Commentary's introduction to the group *Sph.* II.11, II.12 (p. 261). The undiscussed case is illustrated in Figure 40.

This topic returns at the beginning of Book III.

Introduction to *Sph.* II.13

See Introduction to *Sph.* II.10, II.13 and II.16 (p. 404). *Sph.* II.13 is about something that has not previously been mentioned, and so it needs it own introduction. This is supplied in the Commentary on it (p. 265).

The astronomical interpretation of this proposition appears in the Commentary on *Sph.* II.13 (see p. 266). In the Greek of the demonstration no line is mentioned by name; but lines are named when convenient in the paraphrase.

§❧

Theorem. *If there are parallel circles in a sphere and two great circles are drawn touching one of them and cutting those parallel to it, then the arcs of each parallel circle between the non-intersecting semicircles of the great circles are similar, and the arcs of the great circles between two parallels are equal.*

Exposition. Let *ABGD*, *EZHQ*, *KL*, and *FU* be parallel circles in a sphere (Figure 41 (a)), and let great circles *BZLQDU* and *AEKHGF* be tangent to *KL* at *K* and *L*, and to *FU* at *U* and *F*.

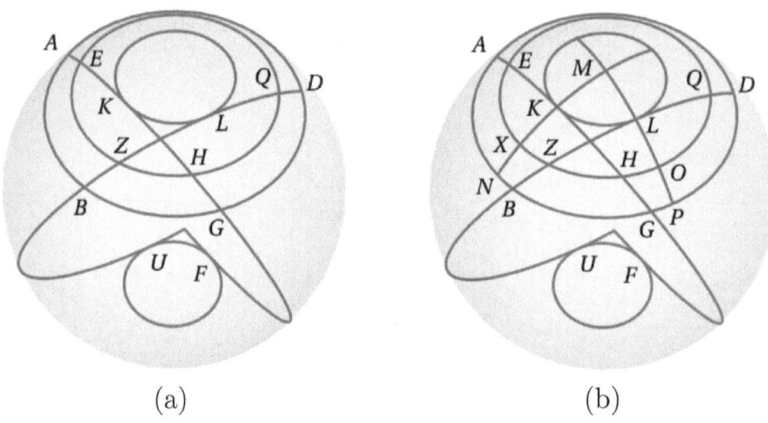

(a)   (b)

Figure 41: *Sph.* II.13. (a) The three initial parallel circles, great circles $DQLZBU$ and $AEKHGF$, and FU antipodal to KL. Semicircle $KEAF$ is non-intersecting with semicircle $LZBU$ and $KHGF$ non-intersecting with $LQDU$. (b) The great circles $MKXN$ and $MLOP$ through the poles of the parallels added.

Specification. It is to be proved that (Part 1) arcs KL, EZ, and AB are similar, and KL, HQ, and GD are similar and (Part 2) the four arcs EK, KH, ZL, and LQ are equal to one another and the four AE, BZ, HG, and QD are equal to one another.

Demonstration (Part 1). Find the pole of the parallel circles, point M, and through M and each of the points K and L draw a great circle $MKXN$ and $MLOP$ intersecting $ABGD$ in N and P and $EZHQ$ in X and O (Figure 41 (b)).

Since in a sphere two circles $AEKHGT$ and KL touch each other at point K, and through the poles of one of them, KL, and the point of contact K a great circle $MKXN$ is drawn, it runs through the poles of $AEKHG$, and it is perpendicular to it. Similarly we will show that $MLOP$ runs through the poles of $BZLQD$ perpendicularly.

Since equal segments of circles beginning with KM and LM are set up perpendicularly on diameters $\underline{KF}$ and $\underline{LY}$ in equal circles $AEKHGF$ and $BZLQDU$, and from them equal arcs KM and LM are cut off less than half of the whole, and the line $\underline{MA}$ is equal to the line $\underline{MD}$ (M being pole of circle $ABGD$), the arcs cut off,

KA and LD, are equal.   (1)

Since M is the pole of $EZHQ$, $\underline{ME}$ equals $\underline{MQ}$; then similarly

KE is also equal to $\overline{LQ}$.   (2)

And since in a sphere two circles $ABGD$ and $AEKHG$ cut each other and great circle $MKXN$ is drawn through their poles, $MKXN$ bisects the segments cut off,

arc KEA is equal to arc KHG.   (3)

and
$$NA \text{ equal to } NG. \qquad (4)$$
Similarly we will show also that
$$LB \text{ is equal to } LD, \qquad (5)$$
and
$$PB \text{ equal to } PD. \qquad (6)$$
Then since (1) is that arc KEA is equal to arc LQD and arc $AEKG$ is double arc AEK by (3) and $DQLB$ is double LQD by (5), $AEKG$ is equal to $DQLB$. And the circles, being great, are equal; therefore the line $\underline{AG}$ is equal to the line $\underline{DB}$. And for this reason arc ABG is equal to arc DGB in the same circle. And AN is half of arc ABG by (4), and BP is half of BPD by (6); therefore AN is equal to BP. Let arc NB, common to ANB and NBP, be added; then the whole ANB is equal to the whole NBP. And they are of the same circle; therefore arc ANB is similar to arc NBP. But NBP is similar to KL; therefore ANB is similar to KL.

Similarly we will also show that EXZ is similar to KL; therefore AB is similar to EZ. Similarly, that is by way of KL, we will also show that GPD is similar to HOQ.

Demonstration (Part 2). Since great circle KN bisects segment EKH,
$$\text{arc } EK \text{ equals } KH. \qquad (7)$$
Similarly, LP bisects segment ZLQ, and so
$$\text{arc } ZL \text{ equals } LQ. \qquad (8)$$
The proof proceeds by chains of equalities.
$$\text{Arc } ZL \text{ by (8) equals } LQ \text{ by (2) equals } EK \text{ by (7) equals } KH. \qquad (9)$$
Also
$$\text{arc } BZL \text{ by (5) equals } LQD \text{ by (1) equals } AEK \text{ by (3) equals } KHG. \qquad (10)$$
By subtraction respectively of the quantities of (9) from the quantities of (10), the second part is proved.
$$\text{Arc } BZ \text{ equals } QD \text{ equals } AE \text{ equals } HG$$
as required.

❧

Introduction to *Sph.* II.14 and II.15

These propositions are the concluding problems, each constructing a great circle tangent to a small circle and through a point. In *Sph.* II.14 the point is the point of tangency on the small circle, and in *Sph.* II.15 the point lies between the small circle and its antipodal circle.

❧

Sph. II.14. Problem. *Given a small circle in a sphere and a point on its circumference, to draw a great circle touching the given circle at the given point.*

Exposition. In a sphere, let the given small circle be AB, and let the given point on its circumference be B (Figure 42 (a)).

Specification. It is required to draw a great circle touching circle AB at B.

Construction. Let the pole of circle AB be G, and through points B and G draw a great circle BG. Cut arc BD off the great circle (Figure 42 (b)) equal to the sphere quadrant (i.e., one quarter of a great circle, known from *Sph.* I.20), and with pole D and polar radius $\underline{DB}$ draw circle EBZ.

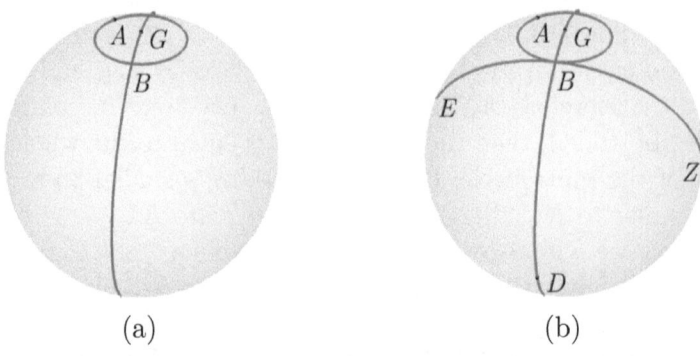

(a)   (b)

Figure 42: *Sph.* II.14. (a) The given circle AB, its pole G, and the great circle through B and G. (b) The pole D and the required great circle EBZ added.

Demonstration. EBZ is a great circle by construction. And since in a sphere the two circles AB and EBZ cut the circumference of a certain great circle BGD at the same point B, having their poles G and D respectively on it, the circles touch each other. Therefore great circle EBZ touches circle AB at B.

Comments. This problem could have been dealt with at any time after *Sph.* II.3, which is needed for the proof. It was presumably kept until this position to be adjacent to *Sph.* II.15, which requires *Sph.* II.12.

Introduction to *Sph.* II.15

See Introduction to *Sph.* II.14 and II.15 (p. 411). Additional cases appear in the translation with reasons in footnotes why they are not here.

Problem. *Given a small circle in a sphere and a certain point on the surface of the sphere between it and the circle equal and parallel to it, to draw a great circle through the given point touching the given circle.*

Exposition. In a sphere let the given small circle be AB (Figure 43 (a)), and let point G be on the surface of the sphere between AB and the circle equal and parallel to it.

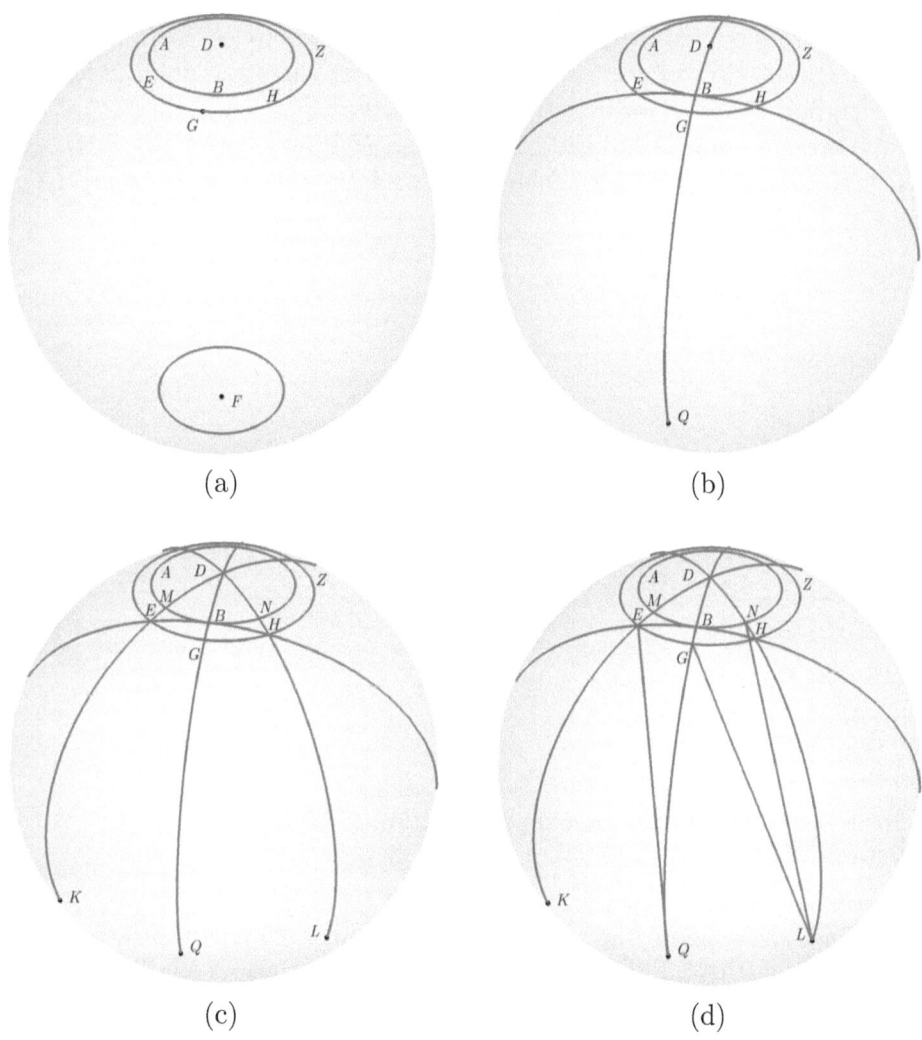

(a) (b)

(c) (d)

Figure 43: *Sph.* II.15. (a) The given circle AB, antipodal circle, and point G between them. Poles D and F and parallel circle EGH added. (b) Great circle $DBGQ$ added as far as Q and EBH added tangent to AB at B. (c) Great circles $DMEK$ and $DNHL$ added. (d) Given circle AB, point G, and construction circles and lines.

Specification. It is required to draw through point G a great circle touching circle AB.

Construction. Find the pole of circle AB, point D. With pole D and polar radius $\underline{DG}$ draw circle $GEZH$. Through points D and G draw great circle DG intersecting circle AB in point B.

Cut off BQ, outward from B, equal to the quadrant, and with pole Q and polar radius $\underline{BQ}$ draw circle EBH (Figure 43 (b)). EBH is a great circle by construction. And it touches AB, for the two circles EBH and AB cut the circumference of the great circle $DBGQ$ at the same point B and have their poles Q and D on it.

Through point D and each point E and H draw great circles $DMEK$ and $DNHL$ (Figure 43 (c)) with each of EK and HL cut off equal to arc GQ.

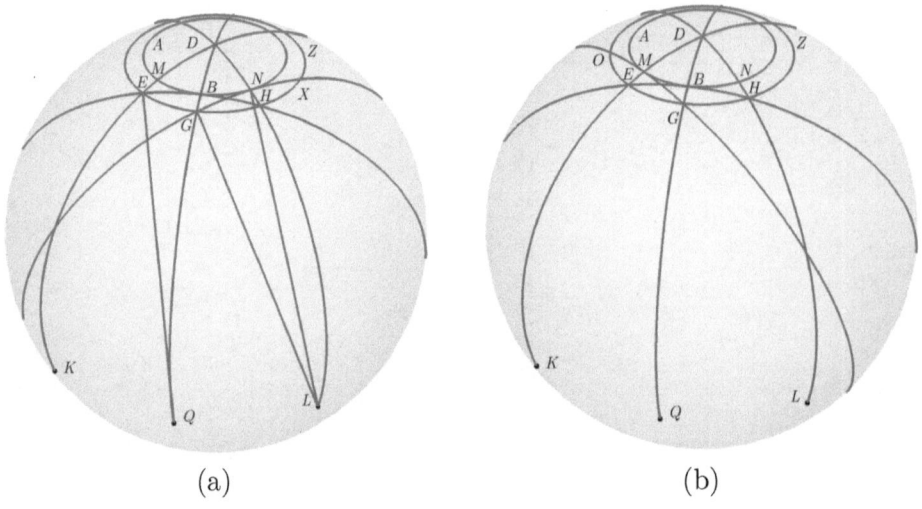

(a)   (b)

Figure 44: *Sph.* II.15. (a) Required circle through N added. (b) Required circle through M added.

Since in a sphere the two circles EBH and $GEZH$ cut each other and great circle $DBGQ$ has been drawn through their poles Q and D, it bisects the segment EGH cut off. Therefore arc EG is equal to GH. Since the three arcs DE, DG, and DH are equal to one another and arcs DM, DB, and DN are equal to one another, the remaining arcs ME, BG, and NH are equal to one another. But arcs EK, GQ, and HL are also equal to one another; therefore the whole arcs MK, BQ, and NL are equal to one another. And BQ is equal to the quadrant; therefore arcs MK and NL are also equal to the quadrant. Since the great circle $DBGQ$ cuts circle $ZEGH$ through the poles, it bisects it perpendicularly. Similarly we will show that $DNHL$ and $DMEK$ are perpendicular to $ZEGH$.

Join $\underline{LN}$, $\underline{LG}$, and $\underline{QE}$ (Figure 43 (d)).

Since equal segments of circles, beginning with equal arcs *GQ* and *HL* cut off less than half of the whole, are set up perpendicularly on diameters of *ZEGH* from points *G* and *H*, and arc *GE* is equal to arc *HG*, line $\underline{QE}$ is equal to $\underline{LG}$. And $\underline{QE}$ is the side of a great square; therefore $\underline{LG}$ is equal to the side of a great square. Therefore a circle drawn with pole *L* and polar radius $\underline{LG}$ runs through point *N* (Figure 44 (a)). Let it be *GNX*.

Then *GNX* is a great circle.

Since in a sphere two circles cut the circumference of a great circle *DNHL* at the same point *N*, having their poles on it, they touch each other. Therefore great circle *GNX* touches circle *AB*. Similarly, the circle *GO* drawn with pole *K* and polar radius $\underline{KG}$ touches circle *AB* at *M* (Figure 44 (b)). Therefore through the given point *G* and touching circle *AB*, two great circles have been drawn, *GNX* and *GMO*.

So

Introduction to *Sph.* II.16

See Introductions to *Sph.* II.10, II.13 and II.16 (p. 404) and to *Sph.* II.13 (p. 409). *Sph.* II.16 is a joint converse to *Sph.* II.10 and II.13.

So

Theorem. *In a sphere, two great circles cutting off similar arcs of parallel circles either pass through the poles of the parallels or touch the same one of the parallels.*

Exposition. In a sphere, let great circles *AEHG* and *BZQD* cut off between them similar arcs of the parallel circles *ABGD* and *EZHQ*, that is to say *AB* is similar to *EZ* (Figure 45 (a)).

Specification. It is to be proved that *AGH* and *BDQ* either are through the poles of the parallels or touch the same parallel circle.

Either circle *AGH* is through the poles of the parallels (Case 1) or not (Case 2).

Exposition (Case 1). Let *AGH* be through the poles of the parallels.

Specification (Case 1). It is to be proved that circle *BDQ* is through the poles of the parallels, that is to say, that the point of intersection *K* of *AGH* and *BDQ* is the pole of the parallel circles *ABGD* and *EZHQ*.

Construction (Case 1). Suppose not, but if it were possible let point *L* be the pole of the parallels, and draw a great circle *LMZ* through points *L* and *Z* intersecting circle *ABGD* at *M* distinct from *B* (Figure 45 (b)).

Demonstration (Case 1). Then arc *ABM* is similar to *EZ*, but *EZ* is similar to *AB*; therefore *MA* is similar to *AB*. And they are of the same circle; therefore arc *AM* is equal to arc *AB*, contradicting the distinctness of *B* and *M*.

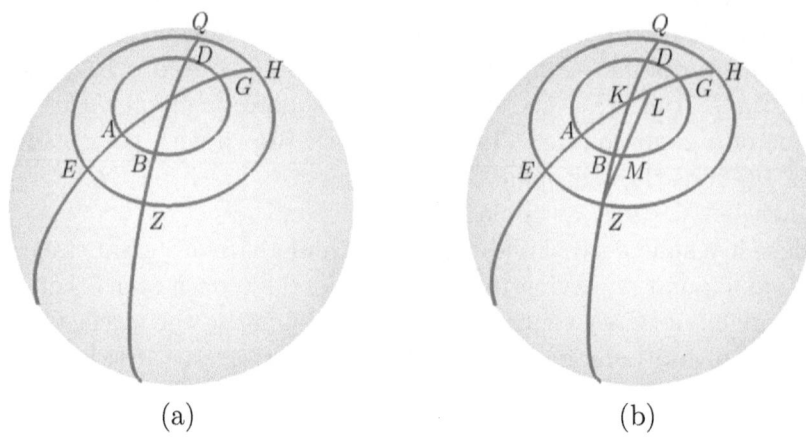

Figure 45: *Sph.* II.16, Case 1. (a) Given. (b) False pole L and arc ZML added.

Therefore L is not the pole of the parallel circles. Similarly any point except K is not the pole. Therefore K is the pole of the parallels, and circles AGH and BQD pass through the poles of the parallels.

Exposition (Case 2). Suppose that AGH is not through the poles of the parallels. Either it touches circle $EZHQ$ (Figure 46 (a)) (Case 2.1) or it is oblique to it (Case 2.2).

Exposition (Case 2.1). Let the great circle through A touch circle EZQ at E.

Specification (Case 2.1). It is to be proved that BZ touches EZQ at Z.

Construction (Case 2.1). If it were possible let ZB not touch EZQ, and let great circle ZG', with G' distinct from B, touch EZQ at point Z, making the semicircle beginning AE non-intersecting with the semicircle beginning $G'Z$ (Figure 46 (b)).

Demonstration (Case 2.1). Then arc $G'A$ is similar to ZE and hence to BA, and they are of the same circle. Therefore arc $G'A$ is equal to arc BA, contradicting the distinctness of B and G'.

Therefore it is not the case that circle BZ does not touch circle EZQ; therefore it touches the same circle as AE.

Exposition (Case 2.2). Now suppose AHG is oblique to the parallels as in the the diagrams. Then it touches two circles equal to each other and parallel to $ABGD$ and $EZHQ$.

Specification (Case 2.2). It is to be proved that $BZQD$ is oblique to the parallels and that it touches the same circles as $AEHG$.

Construction (Case 2.2). Let circle $AEHG$ touch a certain of the parallels MX at point L, and if it were possible let MLX not touch $BZQD$ (Figure 46 (c)). Through point Z between circle ML and the circle equal and

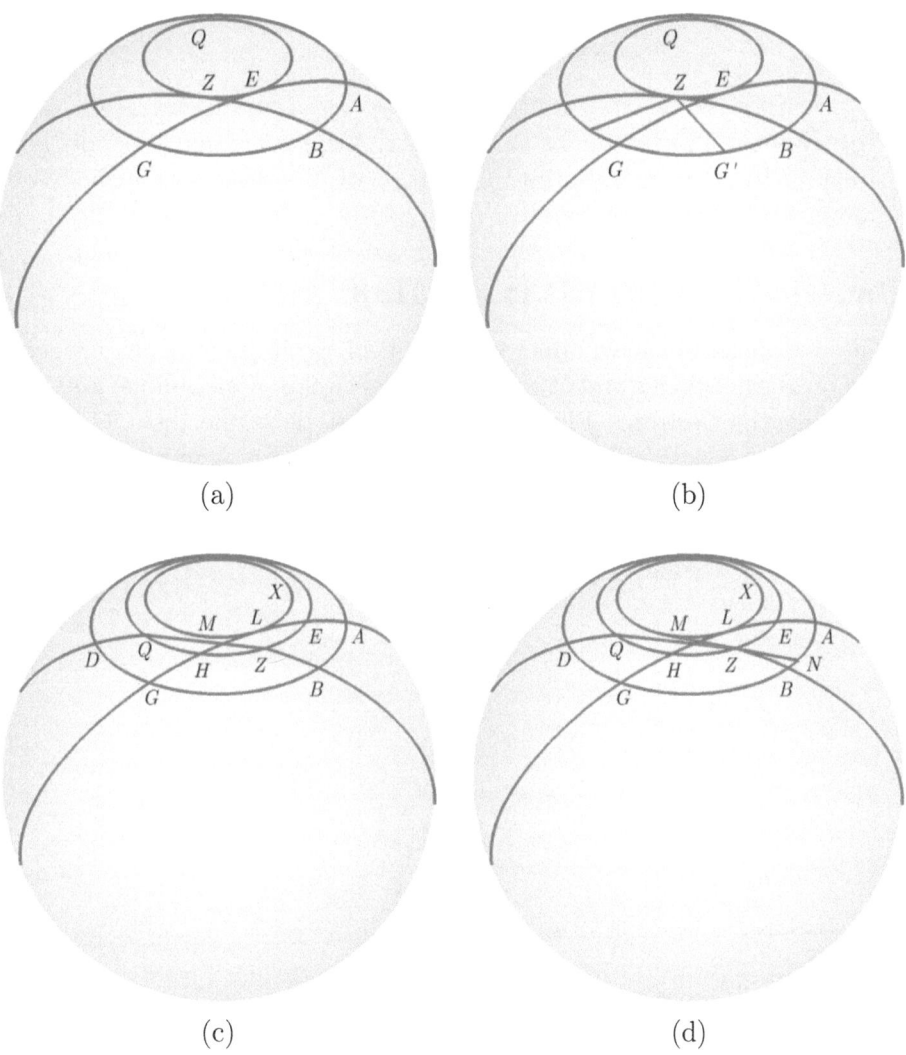

Figure 46: *Sph.* II.16. (a) Case 2.1. Given. (b) Construction of impossible arc beginning *G'Z* as a broken line. (c) Case 2.2. Great circle *AEHG* touching at point *L* circle *MLX*, which is parallel to circles *ABGD* and *EZHQ*. Great circle *BZQD* does not touch *MLX*. (d) The impossible arc *NZM* added as a straight line.

parallel to *MLX*, draw a great circle *MZN*, with *N* distinct from *B*, touching *LM* at point *M* and making the semicircle beginning *LEA* non-intersecting with semicircle *MZN* (Figure 46 (d)).

Demonstration (Case 2.2). Then arc *NA* is similar to arc *ZE*, but arc *ZE* is similar to *BA*. Therefore *NA* is similar to *BA*, and they are of the same circle. Therefore arc NA is equal to arc BA, contradicting the distinctness of B.

❧

Comments. The reason for naming G' in Case 2.1 is that it is not the G of circle $AEHG$. The manuscript duplication of G is discussed in the "Textual Comments" of the Commentary (see p. 276).

Introduction to *Sph.* II.17 and II.18

These propositions have direct astronomical significance. They concern the size of parallel circles and of arcs of parallel circles or an oblique great circle, typically the horizon. First, parallel circles the same distance from the equator are equal and diminish with distance. This is proved in *Sph.* II.17 and the converse in II.18. They are logically equivalent to *Sph.* I.6, as is discussed in the Commentary's introduction to the group together with the astronomical significance (p. 278).

❧

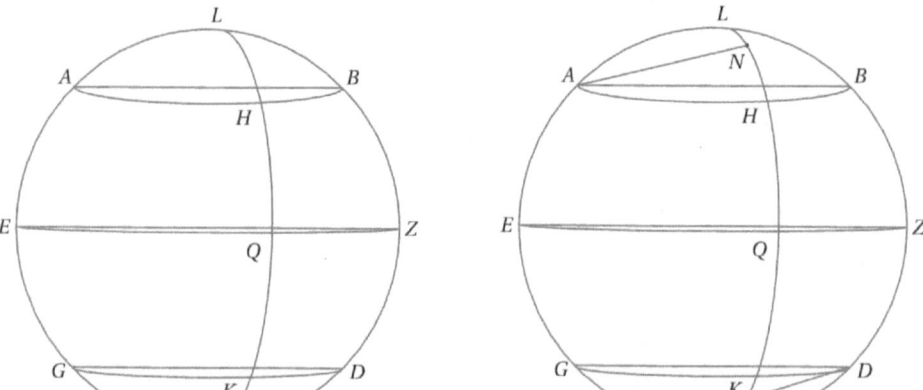

(a)   (b)

Figure 47: *Sph.* II.17. (a) The front hemisphere bounded by the given great circle, the given parallel circles, their intersections, and the great semicircle $LHQKM$ constructed in Part 1, Case 2. (b) The extension of the semicircle and the polar radii added.

***Sph.* II.17. Theorem.** *In a sphere, if a great circle has equal arcs cut off it between each of two parallel circles and the parallel great circle, then the two parallel circles are equal, and the longer the arcs the smaller both circles.*

Exposition (Part 1). In a sphere, let parallel circles AB and GD cut off in the first place equal arcs DZ and ZB of great circle $ABDG$ beginning at the parallel great circle EZ (Figure 47 (a)).

Specification (Part 1). It is to be proved that circle AB is equal to GD.

Construction (Part 1). Let the lines of intersection of $ABDG$ and parallel circles AB, EZ, and GD be $\underline{AB}$, $\underline{EZ}$, and $\underline{GD}$.

Demonstration (Part 1). Since the two parallel planes EQZ and GKD are cut by the plane of the circle $ABDG$, their lines of intersection are parallel. Therefore line $\underline{EZ}$ is parallel to line $\underline{GD}$. Similarly we will show also that $\underline{AB}$ is parallel to $\underline{EZ}$. And since in the circle $ABDG$ the two lines $\underline{EZ}$ and $\underline{GD}$ are parallel, arc DZ is equal to arc EG, for if we join $\underline{ED}$ alternate angles will be equal, and in equal circles equal angles are subtended by equal arcs. Therefore arc EG is equal to arc DZ. Similarly we will show also that BZ is equal to AE. So BZ, ZD, AE, and EG are all equal, and therefore AE and BZ together are equal to the sum EG and ZD. Since the whole arc $EALBZ$ is equal to the whole arc $EGMDZ$ – for EQZ and $ABDG$, being great, bisect each other, the remaining ALB is equal to the remaining GMD, and they are of the same circle. Therefore line $\underline{AB}$ is equal to line $\underline{GD}$.

Of course either circle $ABDG$ cuts AHB and GKD through the poles (Case 1) or not (Case 2).

Demonstration (Part 1, Case 1). Let $ABDG$ cut AHB and GKD through the poles. Then it also bisects them. Therefore $\underline{AB}$ is a diameter of circle AHB, and $\underline{GD}$ of circle GKD, and $\underline{AB}$ is equal to $\underline{GD}$. Therefore circle AHB is equal to circle GKD.

Demonstration (Part 1, Case 2). Let circle $ABDG$ not cut circles AHB and GKD through the poles, and let one pole of the parallels be N. Through point N draw a great circle $LNHQKM$ through the poles of circle $ABDG$, and cut off MX equal to arc LN. (Figure 47 (b)).

Then since arc LN is equal to arc MX, let the same arc NKM be added to both so that the whole arc $LNKM$ is equal to the whole arc $NKMX$. Since $LNKM$ is a semicircle, $NKMX$ is a semicircle. Therefore point N is diametrically opposite to point X, and X is the other pole of the parallel circles. And since in the sphere two circles $ABDG$ and GKD cut each other, and $LQKX$ is a great circle drawn through their poles, $LQKX$ bisects arc GMD. Therefore GM is equal to arc MD, and arc GMD is twice arc DM. Similarly we will show also that arc ALB is twice arc AL, and arc GMD is equal to arc ALB. Therefore arc MD is equal to arc AL.

It is now the case that segment $LQMX$ from L to M is set up perpendicular on a diameter of circle $ABDG$ (Figure 47 (b)) and from it equal arcs LN and MX are cut off less than half the whole, and from the first circle equal arcs LA and MD are cut off. Therefore the line $\underline{NA}$ is equal to the line $\underline{XD}$. The line $\underline{NA}$ is the polar radius of circle AHB, and the equal line

$\underline{XD}$ is the polar radius of circle *GKD*; therefore circle *AHB* is equal to circle
GKD.

Exposition (Part 2). Suppose arc *DZ* is longer than arc *ZB* (Figure 48 (a)).

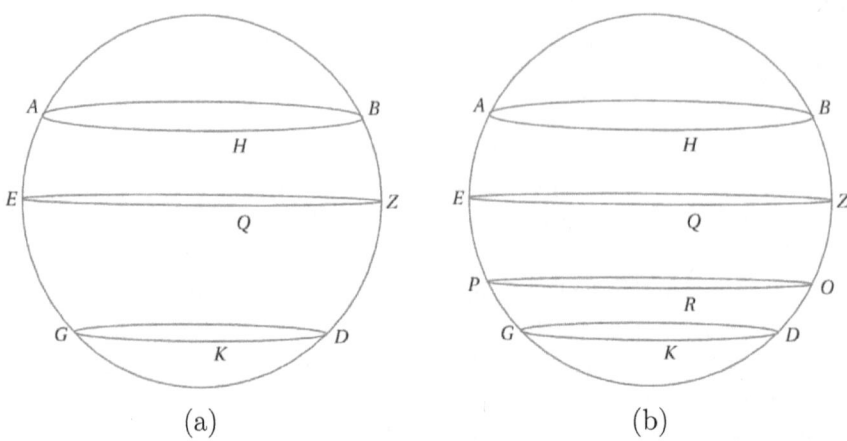

Figure 48: *Sph.* II.17. (a) The given circles of Part 2. (b) The constructed circle added.

Specification (Part 2). It is to be proved that circle *GKD* is smaller than circle *AHB*.

Construction (Part 2). Since arc *DZ* is longer than arc *ZB*, cut off *ZO* from *DZ* equal to *ZB*, and through *O* draw circle *ORP* parallel to circle *EQZ* (Figure 48 (b)).

Demonstration (Part 2). By Part 1, circle *PRO* is equal to circle *AHB*, and circle *PRO* is larger than circle *GKD*: for circle *PRO* is closer to the center of the sphere than circle *GKD*. Therefore circle *GKD* is smaller than circle *AHB*.

❧

Introduction to *Sph.* II.18

See the Introduction to *Sph.* II.17 and I.18 (p. 418). This proposition is the converse of *Sph.* II.17.

❧

Theorem. *In a sphere, equal parallel circles cut off, between them and the greatest of the parallels, equal arcs of a great circle, and the larger the circles the shorter the arcs.*

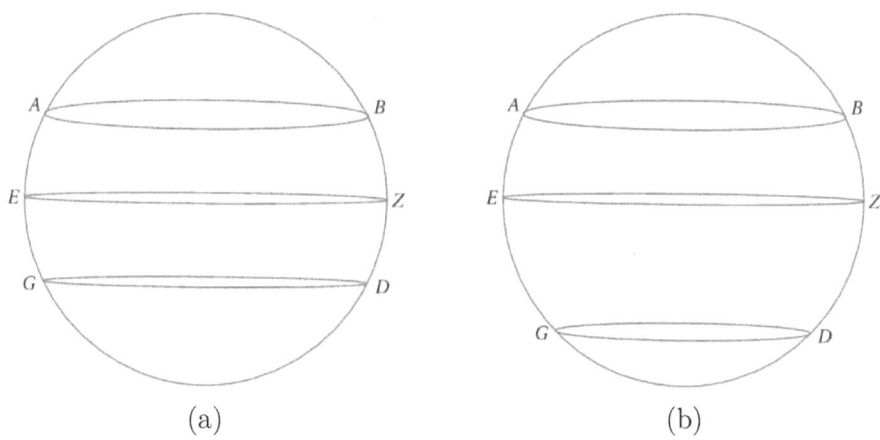

(a)   (b)

Figure 49: *Sph.* II.18. (a) The given circles of Part 1. (b) The given circles of Part 2.

Exposition (Part 1). In a sphere, let equal parallel circles AB and GD cut off a great circle $ABDG$ arcs ZB and ZD beginning at the parallel great circle EZ (Figure 49 (a)).

Specification (Part 1). It is to be proved that arc BZ is equal to arc DZ.

Demonstration (Part 1). If arc BZ is not equal to DZ, then circle AB is not equal to circle GD, and so BZ is equal to DZ.

Exposition (Part 2). Let circle AB be larger than circle GD (Figure 49 (b)).

Specification (Part 2). It is to be proved that BZ is shorter than arc DZ.

Demonstration (Part 2). If arc BZ is not shorter than arc DZ, then circle AB is not larger than circle GD, and so BZ is shorter than arc DZ.

Introduction to *Sph.* II.19 and II.20

The equality of the parallel circles at equal distances on opposite sides of the equator extends to symmetrical longer and shorter arcs cut off by the oblique great circle. This and the astronomical significance is discussed in the Commentary's introduction to the group (see p. 282).

***Sph.* II.19.** **Theorem.** *In a sphere, if a great circle cuts some parallel circles in the sphere not through their poles, it cuts them into unequal segments except for the parallel great circle. Cut-off segments between the parallel great circle and their pole in one hemisphere are larger than semicircles and cut-off segments on the same side of the cutting circle between the parallel great circle and the other pole are smaller than semicircles. And*

the segments of equal parallel circles on opposite sides of the cutting circle are equal to each other.

Exposition (Part 1). In a sphere, let great circle *ABGD* cut parallel circles *AD*, *EZ*, and *BG* not through their poles (Figure 50 (a)), and let *EZ* be the parallel great circle, *AB*, *EZ*, and *BG* being the sections of the parallel circles by *ABGD*.

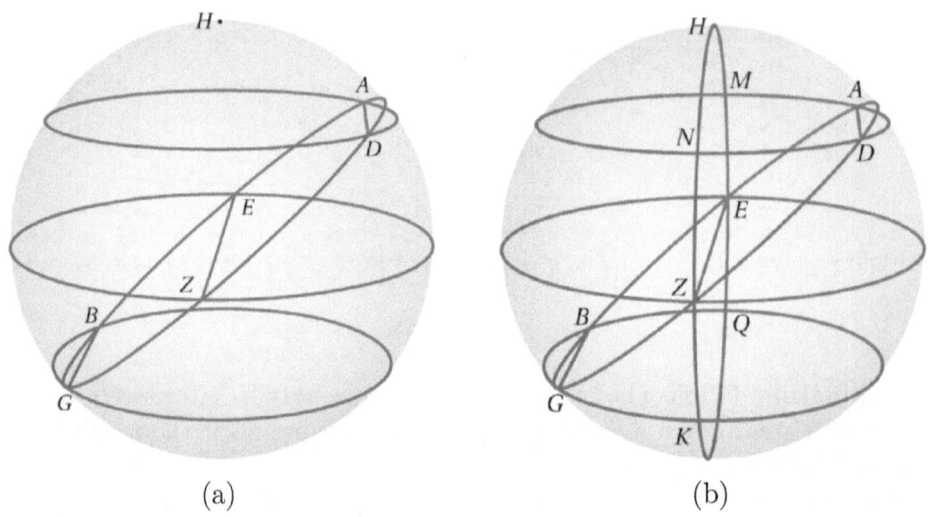

(a)   (b)

Figure 50: *Sph.* II.19. (a) The given configuration of circles. (b) The great circle *HMEQKZN* added.

Specification (Part 1). It is to be proved that *ABGD* cuts them in unequal segments except for *EZ*, the parallel great circle. And the segments in one hemisphere cut off by *ABGD* are larger than semicircles between *EZ* and its pole *H* and smaller on the other side of *EZ* in that hemisphere.

Construction (Part 1). Draw great circle *HMEQKZN* through *E* and *H* (Figure 50 (b)). Then it also runs through point *Z*, for *EZ* is the common diameter of great circles *EZ* and *ABGD*.

Demonstration (Part 1). Since in a sphere great circle *HMEQKZN* cuts through their poles the circles in the sphere *AMND*, *EZ*, and *QBGK*, it bisects them perpendicularly. Therefore the arcs *MN*, *EZ*, and *QBGK* (on one side of *ABGD*) are semicircles. Then since *MN* is a semicircle, the segment *AMND* containing it is larger than a semicircle. Similarly we will show also that all such segments between circle *EZ* and the pole *H* are larger than a semicircle.

Again, since *QBGK* is a semicircle, the segment *BG* within it is smaller than a semicircle. Similarly we will show also that all segments between circle *EZ* and the other pole, also in the same hemisphere with respect to great circle *AEBGZD* are smaller than semicircles.

Exposition (Part 2). Now let us suppose that circle *AD* is equal and parallel to circle *BG*.

Specification (Part 2). It is to be proved that segments of the circles *AD* and *BG* on opposite sides of *ABGD* are equal to each other.

Demonstration (Part 2). Since circle *AD* is equal and parallel to circle *BG*, arc *AE* is equal to arc *EB*, and arc *DZ* is equal to arc *ZG*. Therefore *AE* and *DZ* together are equal to *EB* and *ZG* together. And the whole semicircle *EADZ* is equal to the whole semicircle *EBGZ*; therefore the remaining arc *AD* is equal to the remaining arc *BG*. And *AD* and *BG* are arcs of the same circle *ABGD*. Therefore the line *AD*, subtended by the arc *AD*, is equal to the line *BG*, subtended by the arc *BG*. And in equal circles equal lines cut off equal arcs, the longer equal to the longer and the shorter equal to the shorter. Therefore the longer arc of circle *AD*, longer than a semicircle, is equal to the longer arc of circle *BG*, and the shorter arc of circle *AD* is equal to the shorter arc of circle *BG*, smaller than a semicircle. Therefore the segments of circles *AD* and *BG* on opposite sides of *ABGD* are equal to each other.

Introduction to *Sph.* II.20

See the Introduction to *Sph.* II.19 and II.20 (p. 421).

This theorem is more readily understood by reading its exposition and specification, in conjunction with the first diagram, before the enunciation.

Theorem. *In a sphere, if a great circle cuts parallel circles not through their poles, then, in a hemisphere defined by the great circle, arcs cut off by the great circle that are closer to that hemisphere's pole of the parallel circles are longer than similar to those farther away.*

Exposition. In a sphere, let great circle *ABDG* cut parallel circles *AB*, *GD*, and *EZ* not through their pole *H* (Figure 51 (a)).

Specification. It is to be proved that, of the arcs cut off circles *AB*, *GD*, and *EZ* in the hemisphere containing *H*, that nearer to *H* is always longer than similar to that farther off, that is to say, arc *AB* is longer than similar to arc *GD* and arc *GD* is longer than similar to arc *EZ*.

Construction. Through *H* and each of *G* and *D* draw a great circle *HQG* and *HKD* with *Q* and *K* on arc *AB* (Figure 51 (b)).

Demonstration. Then *HQG* and *HKD* cut off similar arcs between themselves; therefore either arc *QK* is similar to the corresponding arc *GD*. And so *AQKB* is longer than similar to *GD*. Similarly we will show also that

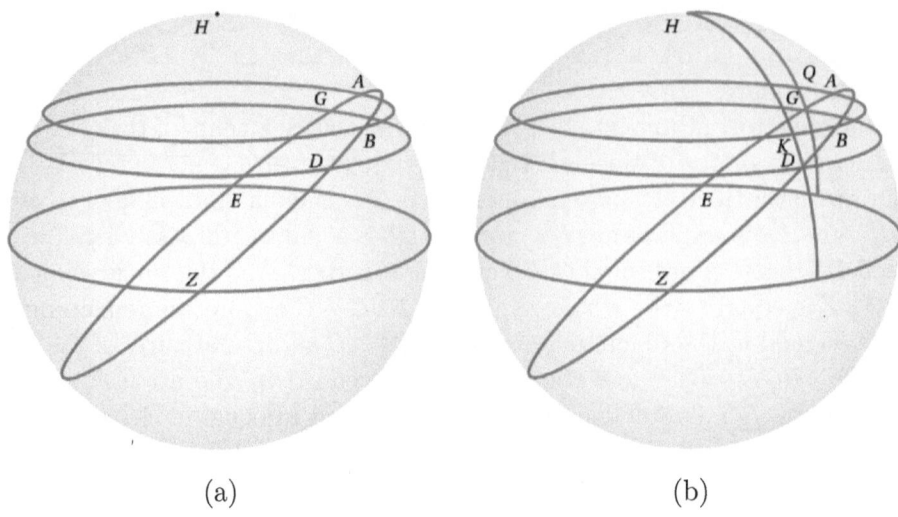

(a)   (b)

Figure 51: *Sph.* II.20. (a) The given configuration of circles and the pole *H*. (b) Construction added.

GD is longer than similar to *EZ*, having drawn great circles through *H* and each of *E* and *Z*.

Introduction to *Sph.* II.21

This proposition gives a condition for different and equal inclinations of one great circle with respect to a second, taken to be horizontal for definiteness. In the situations of interest, it will be the horizon. Moreover, the basic position of the horizon, as indicated in *Sph.* II.10, is that of observers on the terrestrial equator. Following the terminology in the Greek text, whatever has a greater inclination is called *more inclined.*

Theorem. *In equal spheres, if great circles are inclined to horizontal great circles, that is more inclined whose pole is raised up higher, and those are similarly inclined whose poles are equally distant from the horizontal plane.*
Exposition (Part 1). In equal spheres, let great circles *BKD* and *ZLQ* with poles *M* and *N* be inclined to horizontal great circles *ABGD* and *EZHQ* with *M* raised up higher than *N* (Figure 52).
Specification (Part 1). It is to be proved that circle *BKD* is inclined to circle *ABG* more than circle *ZLQ* is to circle *EZHQ*.

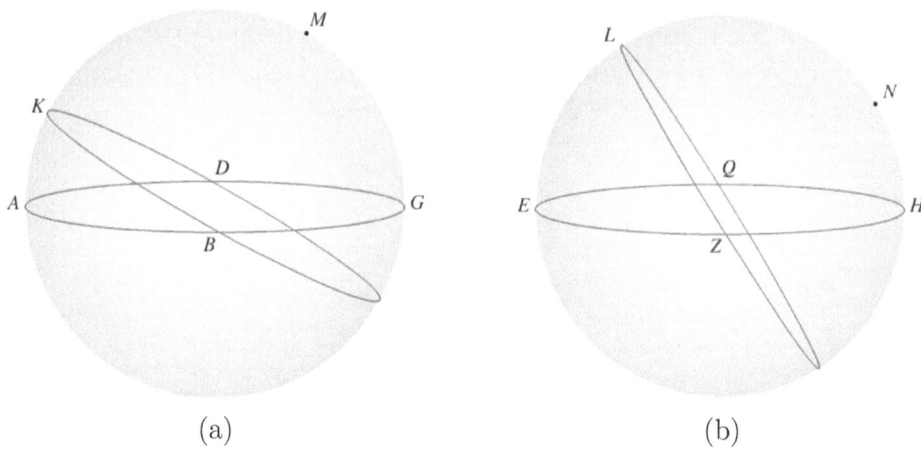

(a)   (b)

Figure 52: *Sph.* II.21. Inclined great circles in equal spheres.

Construction (Part 1). Draw great circle *AKMG* through *M* and one of the poles of the circle *ABGD* (Figure 53 (a)) and great circle *ELNH* through *N* and one of the poles of circle *EZHQ*. Let

- *BD* be the line of intersection of circles *ABGD* and *BKD*,
- *AG* be the line of intersection of circles *ABGD* and *AKMG*,
- *KX* be part of the line of intersection of circles *BKD* and *AKMG*,
- *ZQ* be the line of intersection of circles *EZHQ* and *ZLQ*,
- *EH* be the line of intersection of circles *EZHQ* and *ELNH*, and
- *LO* be part of the line of intersection of circles *ZLQ* and *ELNH*.

Demonstration (Part 1). Since great circle *AKMG* cuts circles *ABGD* and *BKD* through their poles, it bisects them perpendicularly; each of *ABGD* and *BKD* is perpendicular to circle *AKMG*. Therefore the line of intersection *BD* of *ABGD* and *BKD* is perpendicular to circle *AKMG*, so that it makes right angles with every line meeting it and in the plane of circle *AKMG*, in particular with *AX* and *KX*. The angle *KXA* is the inclination with which plane *BKD* is inclined to plane *ABGD*. Similarly we will show also that angle *LOE* is the inclination with which plane *ZLQ* is inclined to plane *EZHQ*.

Specification (Part 1). It must be shown that angle *KXA* is smaller than angle *LOE*.

Demonstration (Part 1). Since *M* is raised up higher than *N*, the perpendicular (Figure 53 (b)) from *M* to *AG* is longer than the perpendicular from *N* to *EH*, and so arc *MG* is longer than arc *NH*. And arc *MK* is equal to arc *NL*, for each of them is equal to a quadrant (Figure 53 (b)). Therefore the whole arc *KMG* is longer than arc *LNH*. Then since the semicircle *AKMG* is equal to the semicircle *ELNH*, in which *KMG* is longer than *LNH*, the remaining arc AK is shorter than the remaining arc EL. But AK subtends the angle KXA, and LE subtends angle LOE. Therefore the inclination of circle BKD to circle $ABGD$ is greater than the inclination of circle ZLQ to circle $EZHQ$.

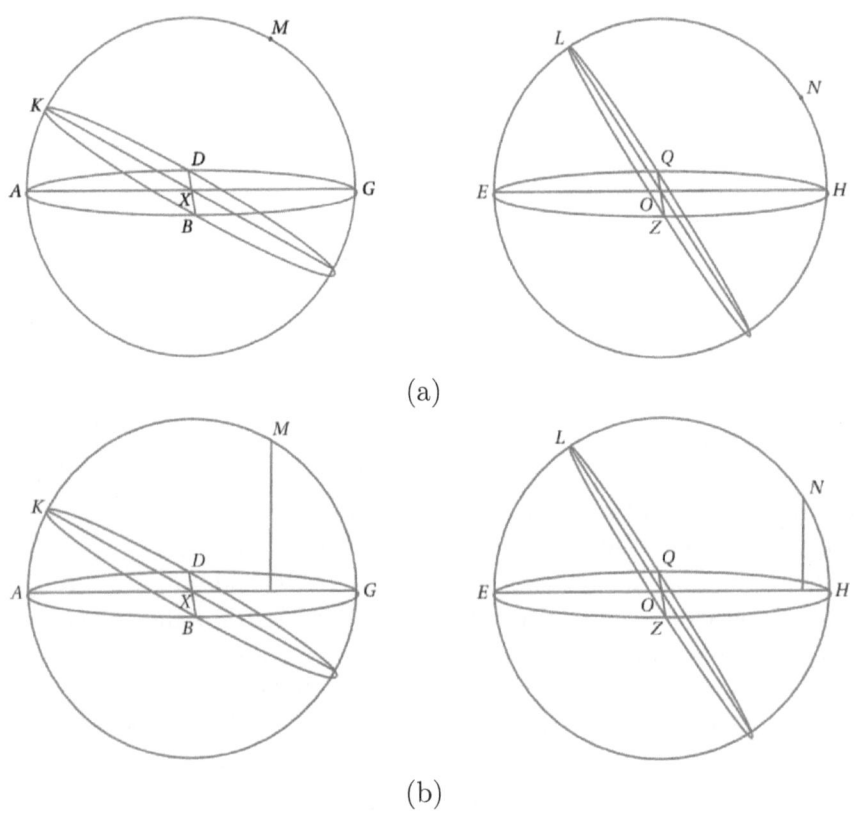

(a)

(b)

Figure 53: *Sph.* II.21. (a) Great circles added through the poles of the inclined circles and the lines of intersection. (b) Perpendiculars dropped to the horizontal planes from the poles of inclined circles.

Exposition (Part 2). Suppose that the poles of circles BKD and ZLQ are equally distant from the planes, that is to say, let the perpendicular produced from M to $\underline{AG}$ in the plane of circle $ABGD$ be equal to the perpendicular produced from N to $\underline{EH}$ in the plane of circle $EZHQ$.

Specification (Part 2). It is to be proved that circles BKD and ZLQ are similarly inclined to $ABGD$ and $EZHQ$, that is to say that angle KXA is equal to angle LOE.

Demonstration (Part 2). Since the perpendicular from M to $\underline{AG}$ is equal to the perpendicular from N to $\underline{EH}$, arc MG is equal to arc NH. And MK is also equal to LN for each is equal to the quadrant; therefore the whole arc *KMG* is equal to the whole arc *LNH*. Then since the semicircle *AKMG* is equal to the semicircle *ELNH*, the remaining arc *AK* is equal to the remaining arc *EL*. But *AK* subtends the angle *KXA*, and *LE* subtends angle *LOE*. Therefore the inclination of circle *BKD* to circle *ABGD* is equal to the inclination of circle *ZLQ* to circle *EZHQ*.

So

Comments. A lemma required for the demonstrations appears in the Commentary (p. 288).

Introduction to *Sph.* II.22 and II.23

The common configuration of these two theorems involves the astronomical arrangement of a great circle (the horizon), the visible parallel circle (*v*-circle) tangent to it, and various positions of the ecliptic great circle tangent to the upper tropic parallel circle (by the tropic definition), which is below the pole of the horizon (as a matter of fact for the range of terrestrial latitudes of interest, see p. 31). The sole concern of *Sph.* II.23 and the main concern of II.22 is the angles between various positions of the ecliptic and the horizon. In addition, II.22 says something about the poles of the various positions of the ecliptic as its Part 6. That part requires some explanation.

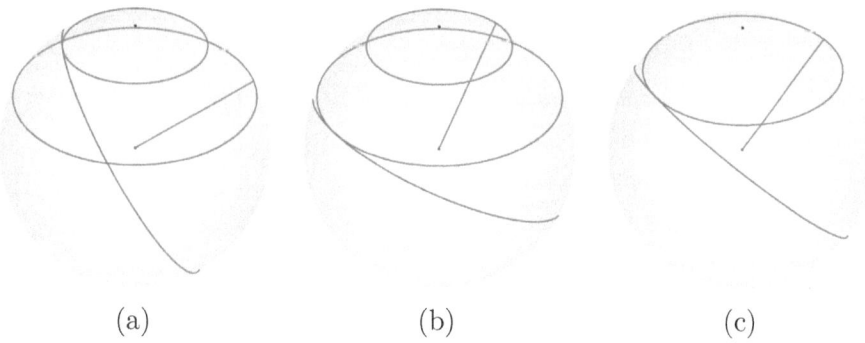

(a)   (b)   (c)

Figure 54: *Sph.* II.22 and II.23, Introduction. (a) Small circle, single tangent great circle with axis and pole, and the circle such poles compose. (b) A larger parallel circle with tangent circle, axis, pole, and circle of poles. (c) The parallel circle whose circle of the poles is itself.

The poles of great circles tangent to a circle compose a circle parallel to the tangent circle (Figure 54 (a) and (b)), which can be called the circle of the poles. There is a mapping from the tangent circle to the circle of the poles. The geodesic distance over the pole of the tangent circle from the point of tangency to the pole of the great circle is a quadrant. Because of the constancy of the over-the-pole distance and the fact that the two small circles share their pole, this circle-to-circle mapping is reciprocal in the sense that the image of an image is the original. If one begins with the circle of the poles and goes over the pole that distance, one gets back the first circle. If the small circle is exactly half way from its pole to the great circle parallel to it, then the circle of the poles is itself, fixed under the reciprocal mapping (Figure 54 (c)). I will refer to it as the fixed circle. The theorem's demand that the pole of the horizon be between the *v*-circle, *AD*, and the tropic, *EZHQ*, amounts to the independent demands that the circle of the poles of the horizon, reciprocal to *AD*, be larger than *AD* and smaller than *EZHQ*. The circle that Part 6 of the theorem is about is the circle of the poles of the ecliptic, *EZHQ*. Because *EZHQ* is larger than the circle reciprocal to *AD*, its reciprocal circle is smaller than *AD*, as the theorem shows. See the Commentary for a discussion of these details (p. 290).

Sph. II.22 is more readily understood by reading the exposition, construction, and specification, in conjunction with the first diagram, before the enunciation.

Sph. II.22. Theorem. *In a sphere, if a horizontal great circle touches a small circle and it cuts a larger small circle both parallel to the first and between the center of the sphere and the pole of the horizontal great circle, and also great circles are drawn touching the larger of the parallels, then*

1. *they are inclined to the horizontal great circle,*

2. *the most upright is that having its point of contact at the bisector of the larger segment of the larger parallel,*

3. *the least upright is that having its point of contact at the bisector of the smaller segment,*

4. *of the others, those having their point of contact equally distant from either of the bisectors are similarly inclined, and*

5. *that having its point of contact farther from the bisector of the larger segment is more inclined.*

6. *And the poles of the great circles touching the larger parallel are on another parallel circle smaller than the smaller parallel.*

Exposition. In a sphere, let great circle *ABG* touch a certain circle *AD* of those in the sphere at *A* (Figure 55), and cut at *E* and *H* another circle *EZHQ* parallel to *AD* and between the center of the sphere and *K*, the pole of *ABG* (that is, *K* between circles *EH* and *AD*). On *EH* let *Z* be the bisector of the larger segment, *Q* be the bisector of the smaller segment, *N* and *P* be equally distant from either bisector on *EZHQ*, and *T* be an arbitrary point in the arc *NQP*.

Construction. Let there be drawn great circles *BZG*, *UQ*, *MNX*, *OPR*, and *TS* touching the larger of the parallels *EZHQ* at *Z*, *Q*, *N*, *P*, and *T*.

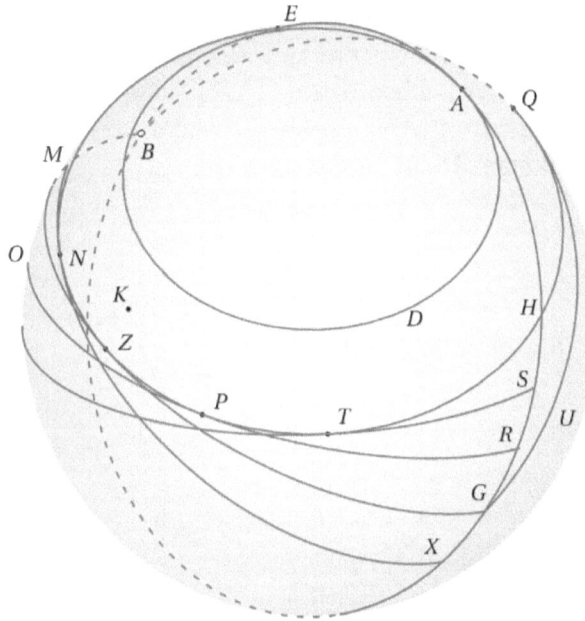

Figure 55: *Sph.* II.22. The given circles and the pole *K* of circle *ABG*. Dashed arcs are on the back side of the sphere running to *B*, which is antipodal to *G*.

Specification. It is to be proved

1. that great circles touching the larger of the parallels *EZHQ* at *Z*, *Q*, *N*, *P*, and *T* are inclined to circle *ABG*, and
2. that the most upright of them is *BZG*,
3. the least upright *UQ* (*BQG*),
4. *MNX* and *OPR* are similarly inclined, and
5. *ST* is more inclined to *ABG* than *OPR*;
6. and the poles of those at *Z*, *Q*, *N*, *P*, and *T* are on one parallel circle smaller than *AD*.

Construction. Let the pole of the parallels *AD* and *EZHQ* be *L*, and through points *A* and *L* draw great circle *AL*. Since in the sphere the two circles *ABG* and *AD* touch each other and great circle *AL* is drawn through the poles of *AD* and the point of contact, great circle *AL* goes through the poles of circle *ABG* perpendicularly, and a pole of circle *ABG* is *K*; therefore *AL* extended goes through *K* making arc *ALK*.

Since the two circles *ABG* and *EZHQ* cut each other and great circle *ALK* is drawn through their poles, great circle *ALK* bisects the segments of the circles cut off, but Z is the bisector of segment EZH and Q is bisector of EQH; therefore ALK goes through points Z and Q making arc $QALKZ$. Since point K is a pole of great circle ABG, arc AK is a quadrant; therefore arc AKZ is longer than a quadrant. And since circle $EZHQ$ is small and is between the center of the sphere and AD, and its pole is point L, LZ is shorter than a quadrant. Then, since AKZ is longer than a quadrant and LZ is shorter, if we cut off AKZ an arc ZF equal to a quadrant from point Z, F is between points A and L (Figure 56).

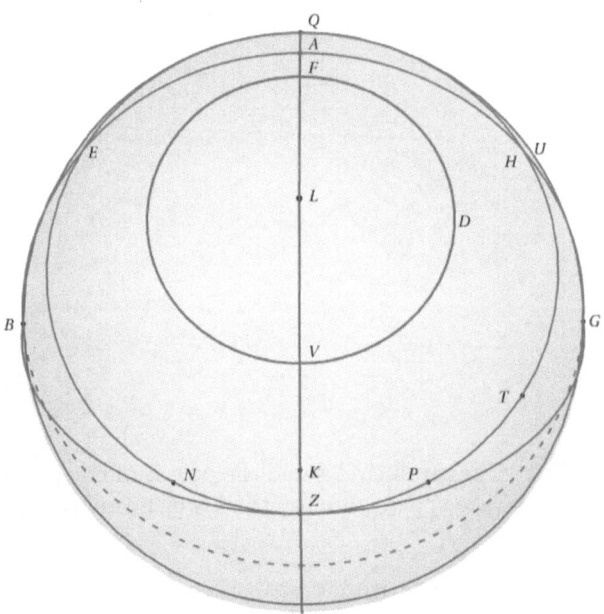

Figure 56: *Sph.* II.22. Additions: the pole L of the parallel circles, the great circle $QALKZ$, the parallel circle (pole L) through the pole F of ZGB. Deletions: the tangent great circles at N, P, and T.

With pole L and polar radius $\underline{LF}$ let a circle $FWVYC$ be drawn. Circle $FWVYC$, smaller than AD since F is between A and L, is parallel to AD and $EZHQ$. Through point L and each of points P, N, and T let a great circle NLW, PLC, and TLY be drawn with W, C, and Y respectively beyond L all on the circle $FWVYC$ (Figure 57).

Demonstration (Part 6, Translation Part 1). Since NL is equal to LZ because L is pole of circle $EZHQ$ and LW is equal to LF because L is pole of circle FWC, the whole NLW is equal to the whole of ZLF, and ZLF is equal to a quadrant by construction. Similarly we will show also that each of CLP, YLT, and VLQ is equal to a quadrant.

Since the great circle NLW is drawn through the poles of $EZHQ$ and the point of contact of $EZHQ$ and the great circle MNX touching it at N, NLW

goes through the poles of *MNX* perpendicularly. And *NLW* is equal to a quadrant; therefore <u>*WN*</u> is the polar radius of the great circle *MNX*. And so *W* is the pole of *MNX*.

Similarly we will show also that point *F* is the pole of *BZG* and *C* of *OPR*, *Y* of *ST*, and *V* of *UQ*. Therefore the great circles at *Z*, *Q*, *N*, *P*, and *T* have their poles on one circle *FCW* parallel to and smaller than *AD*.

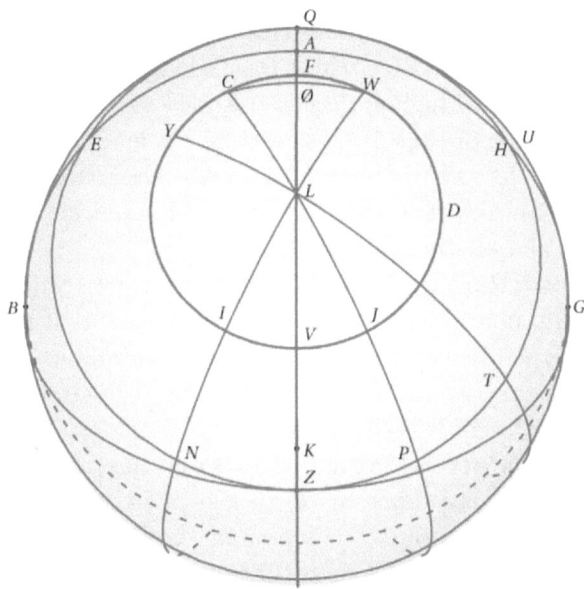

Figure 57: *Sph.* II.22. The great circles through *L* and *N*, *P*, and *T* and the arc parallel to great circle *ABG* (pole *K*) through the poles *C* and *W* intersecting *LF* in Ø are added.

Demonstration (Part 2, Translation Part 2.1). Since equal arcs *NZ* and *ZP* are of the same circle, they are similar. But *NZ* is similar to *IV* and arc *ZP* is similar to arc *JV*, and *IV* and *JV* are of the same circle (therefore equal), but arc *IV* is equal to arc *FW*, for they are vertically opposite, and *JV* to *FC* for the same reason; therefore arc *FW* is equal to *FC*.

Since the segment of circle *VKZF*, beginning with the arc *VK* and cut off less than half of the whole, is set up perpendicularly on diameter <u>*VF*</u> of circle *VWFC* and equal arcs *FC* and *FW* (equivalent to *VC* equal to *VW*) are cut off in opposite directions from circle *VWFC*, the line <u>*KC*</u> is equal to <u>*KW*</u>. Therefore the circle drawn with pole *K* and polar radius <u>*KC*</u> goes also through *W*; let it be *CW* (Figure 57). Then *CW* is parallel to *BAHG*, for they have the same pole *K*.

Since the circle *CW* is parallel to circle *ABG*, the perpendiculars from *C*, Ø, and *W* to the plane of circle *ABG* are equal. And the perpendicular from *C* to the plane of circle *ABG* is longer than the perpendicular from

F to the plane. Therefore the perpendicular from *W* to the plane is longer than the perpendicular from *F*. Therefore the pole *W* of circle *MNX* is raised up higher than the pole *F* of circle *BZG*. Circle *MNX* is therefore more inclined than circle *BZG*, and so circle *BZG* is more upright than circle *MNX*. Similarly we will show also that, of all the circles touching *EZHQ*, *BZG* is the most upright because *F* is the point on circle *FCYW* closest to *ABG*. Therefore *BZG* is the most upright.

Demonstration (Part 3, Translation Part 2.2). Since the perpendicular from *V* to the plane of circle *ABG* is longer than the perpendicular from *Y*, *V* is raised up higher than *Y*. And *V* is the pole of the circle at *Q*, and *Y* is the pole of that at *T*; therefore circle at *Q* is more inclined to circle *ABG* than that at *T*. Similarly we will show also that that at *Q* is more inclined than all other great circles touching circle *EZHQ*. Therefore the great circle at *Q* is the least upright.

Demonstration (Part 4, Translation Part 2.3). Since the perpendicular from *W*, the pole of the circle at *N*, to the plane of the circle *ABG* is equal to the perpendicular from *C*, pole of the circle at *P*, points *W* and *C* are equally distant from the plane. Therefore the circles at *N* and *P* are similarly inclined to circle *ABG*.

Demonstration (Part 5, Translation Part 2.4). Since the perpendicular from *Y*, pole of the circle at *T*, to the plane of circle *ABG* is longer than the perpendicular from *C*, the pole of the circle at *P*, the circle at *T* is more inclined to circle *ABG* than that at *P*. (Part 1 is proved implicitly by Parts 2–5.)

So

Comments. Part 1 is really proved only by showing that specific circles have various inclinations to circle *ABG*, a general statement proved by arbitrary examples and so not complete until Part 5 has been proved.

The proof in the text deals with the case of *AD* smaller than the fixed circle in the reciprocal mapping discussed in the Introduction to *Sph.* II.22 and II.23. The proof needs little modification if *AD* is the fixed circle or between the fixed circle and *EZHQ*, itself always larger than it. The bigger *AD* is, the easier it is for circle *FV* to be inside it. The most delicate case is the one proved. See the discussion of the bounds for this theorem in the Commentary (pp. 292–295).

Introduction to *Sph.* II.23

See the Introduction to *Sph.* II.22 and II.23 (p. 427). This proposition offers an alternative hypothesis for conclusion 4 of *Sph.* II.22 concerning the same configuration and derived by using *Sph.* II.22.

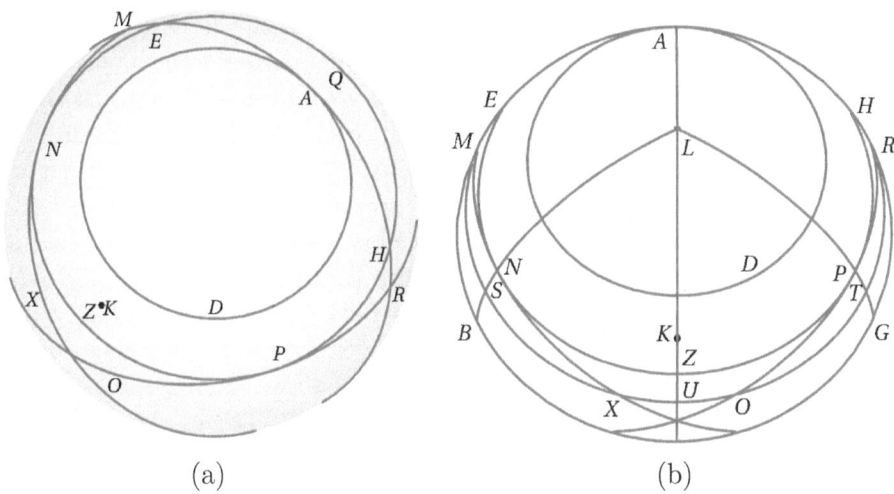

Figure 58: *Sph.* II.23. (a) Initial configuration. (b) Removed: *Q.* Added: pole *L*, great circles *QALKZ*, *LNB*, and *LPG*, parallel circle *MXOR*.

Sph. II.23 is more readily understood by reading the exposition and specification, in conjunction with the first diagram, before the enunciation.

Theorem. *In a sphere, if a horizontal great circle touches a small circle and it cuts another small circle both parallel to the first and between the center of the sphere and the pole of the horizontal great circle, and also two great circles are drawn touching the larger of the parallels, and if the arcs running from the points of intersection of the two great circles and the horizontal great circle to their points of contact are equal, then they are similarly inclined to the horizontal.*

Exposition. The configuration part way through the proof of *Sph.* II.22, without point *T* or the circles tangent at *Q*, *T*, and *Z* and with different *B*, *G*, *S*, *T*, and *U*, has parallel circles *EQHZ* and *AD*, with *AD*'s pole *K* between them, tangent great circles *BEAHG*, *MNX*, and *OPR* at *A*, *N*, and *P* as before. Let the arcs *NM* and *PR* running from the points of contact *N* and *P* to the same circle *ABG* be equal (Figure 58 (a)).

Specification. It is to be proved that circles *MNX* and *OPR* are similarly inclined to circle *ABG*.

Construction. Find the pole of the parallel circles *AD* and *EZHQ*, point *L*, and through points *A* and *L* draw great circle *QALKZ* (Figure 58 (b)), also through *K*, pole of circle *ABG*. Through *L* and each of *N* and *P* draw a great circles *LNB* and *LPG*.

Demonstration. Since in the sphere the two circles *EZHQ* and *MNX* touch each other and great circle *BLN* is drawn through the poles of one and their point of contact, circle *BLN* passes through the poles of *MNX* and is perpendicular to it. Similarly circle *GLP* passes through the poles of *OPR* perpendicularly. Since, on the (undrawn) diameters from points *N* and *P* in the equal circles *MNX* and *OPR*, equal and perpendicular segments of circles beginning with equal cut-off arcs *NL* and *PL*, less than half the whole, are set up and from them and equal arcs *NM* and *PR* are cut off from the beginning of the full circles, the (undrawn) lines <u>LM</u> and <u>LR</u> are equal. Therefore the circle drawn with pole *L* and polar radius <u>LM</u> also goes through *R*. Let it be circle *MXOR* (Figure 58 (b)), cutting *LBN* and *LGP* at *S* and *T*. It is parallel to *AD* and *EZHQ*, having the same poles. Since in the sphere the two circles *ABG* and *MXOR* cut each other, the great circle *QALKZ* drawn through their poles cuts *MXOR* at *U* such that the segments of the circles cut off, *MU* and *UR* are equal.

Again, since in a sphere the two circles *MNX* and *MSXUOTR* cut each other, the great circle *BLN* drawn through their poles bisects the segment of the circle cut off; therefore arc *MN* is equal to arc *NX* and *MS* to *SX*. Similarly we will show also that *OP* is equal to *PR* and *OT* to *TR*.

Then since arc *MN* is equal to arc *PR* and *MNX* is double *MN* and *OPR* double *PR*, arc *MNX* is also equal to *OPR*: and they are equal (great) circles. Therefore the line <u>MX</u> subtending arc *MNX* is equal to the line <u>OR</u> subtending *OPR*. But the equal lines <u>MX</u> and <u>OR</u> also subtend arcs *MSX* and *OTR* of the same circle. Therefore arc *MSX* is equal to arc *OTR*. And *MS* is half of arc *MSX* and *TR* half of *OTR*, and also the whole *MSXU* is equal to the whole *UOTR*. With the equal halves removed, the remaining *SXU* is equal to the remaining *UOT*, and they are of the same circle; therefore arc *SXU* is similar to arc *UOT*. But arc *SXU* is similar to *NZ*, and arc *UOT* is similar to arc *PZ*; therefore *NZ* is also similar to *ZP*, and they are of the same circle; therefore arc *NZ* is equal to arc *ZP*. Since the points of contact of circles *MNX* and *OPR* are equally distant from either midpoint *Z* or *Q* and circles equally distant from either midpoint are similarly inclined circles, *MNX* and *OPR* are similarly inclined to circle *ABG*.

So

Comments. Points *N* and *P* need not lie on arcs *EZ* and *ZH* but can lie instead on arcs *AE* and *AH*. The proof still works.

Introduction to *Spherics* III

Book III begins with preliminary theorems – two solid and not necessarily spherical and two spherical – which are used in later proofs, and then returns to the main subject of projecting arcs of one great circle onto another. *Sph.* III.5 and III.7 begin this as well as preparing for *Sph.* III.6 and III.8 respectively. *Sph.* III.9 shows how *Sph.* III.6 can be generalized from adjacent arcs of the ecliptic to non-adjacent arcs by a method that could be applied to other propositions like *Sph.* III.5, III.7, and III.8. *Sph.* III.11 and III.12 (*Sph.* III.10 being preparatory to them) have some similarity to *Sph.* III.6 and III.8. *Sph.* III.13 can be interpreted to extend the astronomical reach of *Sph.* III.6, III.8, III.11, and III.12, but that is not indicated in the text nor in the *Phenomena*, suggesting (as does its verging on trigonometry) that it is later than the *Phenomena*.

While the propositions of this book after the four preliminaries have astronomical relevance, little of it is through the *Phenomena*. *Sph.* III.3, and something like *Sph.* III.7, and III.8 are used in *Phen.* 8 and 12, but each of *Sph.* III.5–III.14 is of direct astronomical relevance, as is explained in the Commentary proposition by proposition and also in two groups, *Sph.* III.5–III.8 and *Sph.* III.10–III.12. This absence from the *Phenomena* suggests that it is not only *Sph.* III.13 that is later than the *Phenomena* but also the whole book. The apparent use of *Sph.* III.7 and III.8 in the *Phenomena* by no means requires *Spherics* III as a document to predate it (see p. 48).

Each theorem of this book is more readily understood by reading its exposition and specification in conjunction with the first diagram before the enunciation.

Introduction to *Sph.* III.1 and III.2

The first propositions are not spherical but are extensions into three dimensions of a fact that is taken as obvious in two dimensions, namely that, from a fixed point in the plane other than the center of a circle, the distance to the circumference of the circle varies in a simple way. If the point is in or on the circle, a unique diameter lies through the point. If the point is outside the circle, it lies on the extension of a unique diameter. The distance from the point is minimized at the end of the diameter near it, maximized at the diameter's other end, and increases strictly monotonically from one end to the other, symmetrically on account of the reflective symmetry about the diameter. Aspects of this complex fact are called upon several times in the multi-part proof.

As befits so obvious a fact, it is easy to prove. Let the circle be the unit circle $x^2 + y^2 = 1$ with center at $(0,0)$. Let the fixed point be $(0, a)$ with $a > 0$. Representing a point on the circle as $(\cos \vartheta, \sin \vartheta)$, the square of the distance from the point to the circumference is $\cos^2 \vartheta + \sin^2 \vartheta - 2a \sin \vartheta + a^2$, with the derivative with respect to ϑ equal to $-2a \cos \vartheta = -2ax$. Accordingly, the distance decreases with increasing ϑ from its maximum at the far end of the diameter $(0, -1)$ when x is positive and, increases from its minimum at the near end of the diameter $(0, 1)$ when x is negative. We may be inclined to think of this as the square of the length of a vector from the fixed point to a point moving on the circle.

The above proof has a simple description of how it is overly powerful. It determines whether one distance is bigger than another by determining precisely the *rate* at which the distance is changing from the one to the other. The whole notion of rate, indeed of motion, is absent from the treatise. There are statements of considerable sophistication about difference between quantities here and there but no thought of change and none therefore of rate of change.

Introduction to *Sph.* III.1

Taking a circle to be horizontal for exposition, the proposition is about points above the circle. From any such point not over the center, a perpendicular can be dropped to a point in the interior of the circle, which is on a unique diameter. The square of the distance to the circumference, which is the hypotenuse of a right triangle with vertical and horizontal sides, is equal to the sum of the squares on those two sides. The square on the perpendicular is constant, and so the square on the hypotenuse increases from the near end of the diameter to the far end of the diameter with the increase in the length of the horizontal side. Therefore the length of the hypotenuse likewise increases from minimum to maximum.

The geometrical situation is discussed at length in the Commentary's introduction to *Sph.* III.1 and III.2 (p. 304).

More is accomplished in the proof than the three comparisons specified in the enunciation of the theorem. See the summary following the paraphrase or pp. 304–306.

So

***Sph.* III.1. Theorem.** *If a secant is drawn in a circle cutting off a segment, and on it a segment of a circle not greater than a semicircle is set up perpendicularly, and the circumference of the segment erected is divided by a point unequally, then the lines from the point to the*

   1. *The line to the nearer end of the secant if not a diameter is the shortest of those to the longer arc of the circle.*

   2. *If the secant is a diameter, then the line to its nearer end is shortest of all and*

   3. *that to its farther end is longest of all.*

Exposition (Case 1, Part 1). Let a certain line $\underline{BD}$ be drawn in the circle $ABGD$ cutting the circle in unequal parts and let arc BGD, where A and G are points to be chosen later, be longer than arc BAD (Figure 59 (a)).

And let segment *BED* of a circle not greater than a semicircle (cf. *Sph.* II.11, II.12), with *E* closer to *B* than to *D*, be set up perpendicularly on *BD*. And let *EB* be joined.

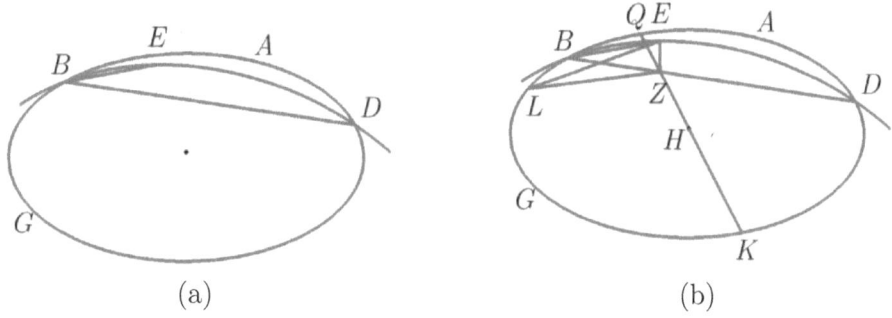

Figure 59: *Sph.* III.1. (a) The given configuration for Case 1. (b) First comparison.

Specification (Case 1, Part 1). It is required to prove that *BE* is shortest of all lines from *E* to arc *BGD*. (False if *E* is closer to *D* than to *B*.)

Construction (Case 1, Part 1). Drop perpendicular *EZ* from *E* to the plane of circle *BGD*; clearly it will fall on the intersection of planes *ABGD* and *BED* – on the line *BD* – because segment *BED* is perpendicular to circle *ABGD*. Let the center of circle *ABGD* be *H*, join *ZH* and extend it to *K* on arc *BGD* and *Q* on arc *BAD*. Join *E* and *Z* to *L* on arc *BGD* (Figure 59 (b)).

Demonstration (Case 1, Part 1). Since *EZ* is perpendicular to the plane of circle *ABGD*, it will make right angles with all lines passing through *Z* and lying in the plane of circle *ABGD*. *EZ* meets each of *ZB* and *ZL*, both in the plane of circle *ABGD*. Therefore each of the triangles *BZE* and *LZE* is right. Since *ZB* is shorter than *ZL* (because *B* is closer to *Q* than *L* is), the square on *ZB* is smaller than the square on *ZL*. Add the common square on *EZ*; then the sum of the squares on *EZ* and *ZB*, which equals the square on *EB*, is smaller than the sum of the squares on *EZ* and *ZL*, which equals the square on *EL*. Therefore *EB* is shorter than *EL*.

Likewise *EB* is shortest of any line from *E* to arc *BGD*.

Therefore *EB* is shortest of all the lines from *E* to arc *BGD*.

Specification (Case 1, Part 2). It is to be proved that, of the lines drawn from *E* to arc *BK*, that nearer to *EB* is shorter than that farther away.

Construction (Case 1, Parts 2 and 3). Draw another line *EG* to *LK*, and join *ZG* (Figure 60 (a)).

Demonstration (Case 1, Part 2). Since *ZL* is shorter than *ZG*, the square on *ZL* is smaller than the square on *ZG*. Add the common square on *EZ*. Then the sum of the squares on *EZ* and *ZL*, equal to the square on *EL* is less than the sum of the squares on *EZ* and *ZG*, equal to the square on

ZEG. The square on *EL* is therefore less than the square on *EG*; therefore *EL* is shorter than *EG*.

Similarly, of the lines drawn from *E* to arc *BK*, that nearer to *EB* is shorter than that farther away.

Construction (Case 1, Part 3). Join *EK* and *ED* (Figure 60 (b)).

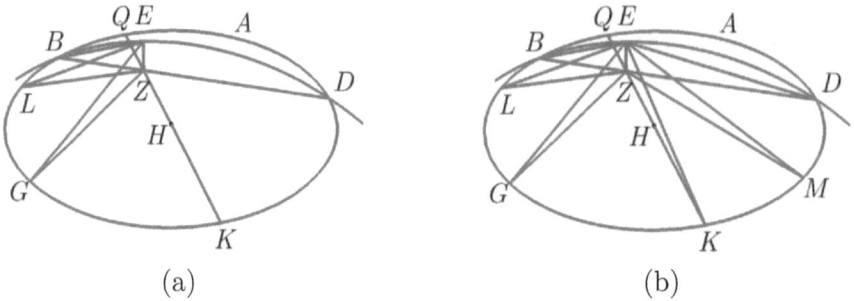

Figure 60: *Sph.* III.1. (a) Case 1. Adding *G*. (b) Adding *M*.

Specification (Case 1, Part 3). It is to be proved that *EK* is longest of all the lines from *E* to arc *KD* and *ED* the shortest of all lines from *E* to *KD*.

Demonstration (Case 1, Part 3). Since *ZK* is longer than *ZG* (because *K* is the far end of the diameter *QK*), the square on *ZK* is also larger than the square on *GZ*. Add the common square on *EZ*; then the sum of the squares on *ZK* and *EZ*, equal to the square on *EK*, is larger than the sum of the squares of *ZG* and *ZE*, equal to the square on *EG*. Therefore *EK* is longer than *EG*.

Similarly, of all lines from *E* to arc *KD*, *EK* is the longest.

Therefore *EK* is the longest of all lines to arc *BGD*.

Specification (Case 1, Part 4). It is to be proved that *ED* is the shortest of all lines from *E* to arc *KD*.

Construction (Case 1, Part 4). Draw another line *EM* to arc *KD*, and join *MZ*.

Demonstration (Case 1, Part 4). Since *ZD* is shorter than *ZM*, the square on *DZ* is less than the square on *MZ*. Add the common square on *EZ*; then the sum of the squares on *EZ* and *ZD*, equal to the square on *ED*, is less than the sum of the squares on *EZ* and *ZM*, equal to the square on *EM*. Therefore *ED* is shorter than *EM*.

Similarly, of all lines from *E* to arc *KD*, *ED* is shortest.

Therefore *ED* is the shortest of all the lines from *E* to arc *KD* and, (unproved Part 5 of Case 1) of the lines to arc *KD*, that nearer to *ED* is shorter than that farther away.

Exposition (Case 2). Now let the dividing line $\underline{BD}$ be instead a diameter of circle $ABGD$ and the rest be assumed the same (Figure 61).

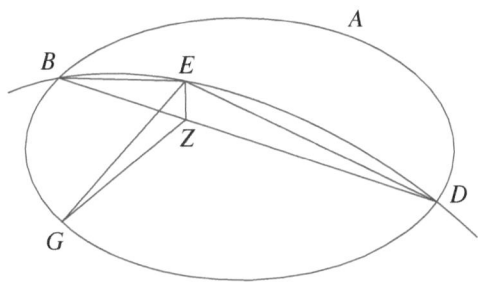

Figure 61: *Sph.* III.1. Case 2.

Specification (Case 2). It is to be proved that $\underline{EB}$ is shortest of all the lines from E to the circumference of circle $ABGD$ and that $\underline{ED}$ is longest.

Demonstration (Case 2). With the same construction, since arc $\underline{ED}$ is longer than arc $\underline{EB}$ and $\underline{EZ}$ is perpendicular, $\underline{ZD}$ is longer than $\underline{BZ}$. And $\underline{BD}$ is a diameter of circle $ABGD$; therefore the center of the circle is on $\underline{ZD}$. Therefore $\underline{ZD}$ is longer than $\underline{ZG}$ and $\underline{ZG}$ than $\underline{ZB}$, so that the square on $\underline{ZD}$ is larger than the square on $\underline{ZG}$ and the square on $\underline{ZG}$ than the square on $\underline{ZB}$. Add the common square on $\underline{ZE}$; then the squares on $\underline{ZD}$ and $\underline{ZE}$, equal to the square on $\underline{ED}$ is larger than the squares on $\underline{GZ}$ and $\underline{ZE}$, equal to the square on $\underline{GE}$, and the squares on $\underline{GZ}$ and $\underline{ZE}$, equal to the square on $\underline{GE}$, is larger than the squares on $\underline{BZ}$ and $\underline{ZE}$, equal to the square on $\underline{BE}$. Therefore $\underline{ED}$ is longer than $\underline{EG}$ and $\underline{EG}$ than $\underline{EB}$.

Similarly we will show also that of all the lines from point E to the circumference of circle $ABGD$, $\underline{ED}$ is longest and $\underline{EB}$ shortest.

Summary. Case 1 (BD not diameter):

1. EB is the shortest line from E to BGD.

2. Line lengths strictly increase along BK from B.

3. EK is the longest line to BGD.

4. ED is the shortest line to KD.

5. Line lengths strictly increase along KD from D (not proved).

Case 2 (BD diameter):

1. EB is the shortest line to the circumference.

2. ED is the longest.

Introduction to *Sph.* III.2

See also Introduction to *Sph.* III.1 and III.2 (p. 435). *Sph.* III.2 concerns arcs similar to *Sph.* III.1 but differently arranged. The secant can be a diameter or not as before, but the segment set up on it is no longer a vertical circle but is inclined toward the shorter (or equal) part of it. All that matters for the proof is that a perpendicular can be dropped to the plane of the circle from the arc and not hit the center of the circle.

The configuration, while not itself spherical, can be and will be embedded in a sphere as in Figure 62.

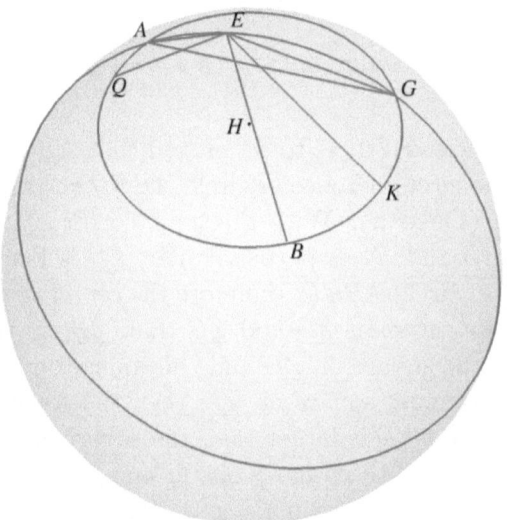

Figure 62: *Sph.* III.2. Embedding in sphere.

§◦

Theorem. *If a secant is drawn in a circle cutting off a segment not less than a semicircle, and on it a segment of a circle not greater than a semicircle is set up inclined toward the segment not greater than a semicircle, and the circumference of the segment erected is divided by a point unequally, then the line from the point to the nearer end of the secant is the shortest of all lines from the same point to the longer arc of the circle.*

Exposition. Let a line $\underline{AG}$ be drawn in the circle $ABGD$ (with diameter $\underline{BD}$ to be specified later) cutting off segment ABG not less than a semicircle, and on $\underline{AG}$ let a segment of a circle AEG not greater than a semicircle be set up inclined toward ADG and divided unequally by E (Figure 63 (a)). Let arc EG be greater than arc EA. Join $\underline{EA}$.

Construction (Part 1). Drop a perpendicular $\underline{EZ}$ from E to the plane of circle $ABGD$; of course it falls between line $\underline{AG}$ and arc ADG on account

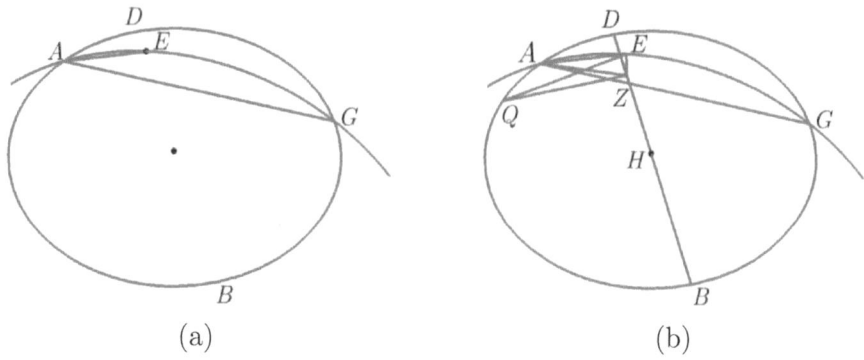

(a)   (b)

Figure 63: *Sph.* III.2. (a) The given configuration for Part 1. (b) Diameter $\underline{BD}$ added and secant removed for clarity.

of the inclination of segment AEG toward segment ADG (Figure 63 (b)).[2] Let H be the center of $ABGD$.

Demonstration (Part 1). H is either on $\underline{AG}$ (Case 2) or between line $\underline{AG}$ and arc ABG (Case 1) because segment ABG is assumed to be a semicircle or more than a semicircle.

Construction (Part 1, Case 1). Let H be between line $\underline{AG}$ and arc ABG (Parts 1–5), extend $\underline{ZH}$ in both directions to join D and B, from E draw line $\underline{EQ}$ to an arbitrary point Q on arc ABG, and join $\underline{AZ}$ and $\underline{QZ}$ (Figure 63 (b)).

Demonstration (Part 1, Case 1). Since $\underline{EZ}$ is perpendcular to the plane of circle $ABGD$, $\underline{EZ}$ also makes right angles with all lines meeting it and in the plane of circle $ABGD$. But $\underline{EZ}$ meets each of $\underline{AZ}$ and $\underline{ZQ}$, both in the plane of circle $ABGD$. Therefore each of angles AZE and QZE is right. And since $\underline{AZ}$ is shorter than $\underline{ZQ}$, the square on $\underline{AZ}$ is less than the square on $\underline{QZ}$. Add the common square on $\underline{ZE}$; then the sum of the squares on $\underline{AZ}$ and $\underline{ZE}$, equal to the square on $\underline{EA}$, is less than the sum of the squares on $\underline{EZ}$ and $\underline{ZQ}$, equal to the square on $\underline{QE}$. Therefore $\underline{EA}$ is shorter than $\underline{EQ}$. Similarly we will show also that $\underline{EA}$ is shortest of all the lines from E to arc AB.

Statement (Part 2). Similarly, of the lines drawn from E to arc AB between points A and B, that nearer A is shorter than that farther off.

Construction (Part 3). Join $\underline{EB}$ (Figure 64 (a)).

Specification (Part 3). It is to be proved that $\underline{EB}$ is the longest of all the lines from E to arc ABG.

---

[2] Z falls on the ADG side of the line $\underline{AG}$ but not necessarily on the $\underline{AG}$ side of arc ADG without restriction not present in the enunciation. This does not matter to the validity of the proof. For discussion of this see the Commentary on this theorem (p. 307).

Demonstration (Part 3). Since BZ is longer than ZQ, the square on BZ is greater than the square on ZQ. Add the common square on ZE. Then the sum of the squares on EZ and ZB, equal to the square on EB, is greater than the sum of the squares on EZ and ZQ, equal to the square on EQ. Therefore BE is longer than EQ. Similarly, of all the lines from E to arc ABG, EB is the longest.

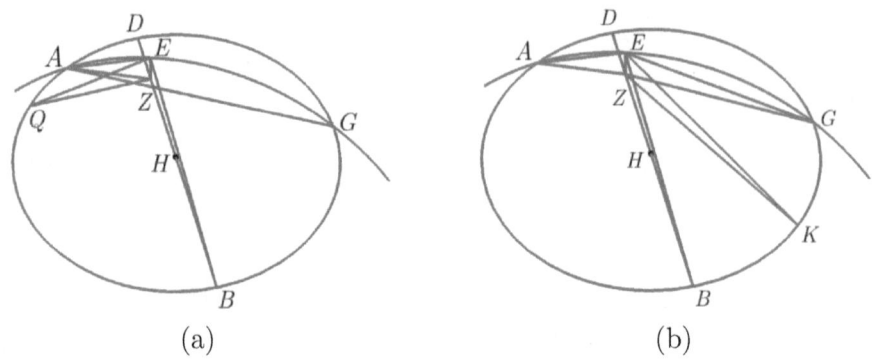

Figure 64: *Sph.* III.2. (a) Part 3. EB added. (b) Part 4. K added.

Construction (Part 4). Join EG (Figure 64 (b)). For arbitrary K on arc BG, draw EK, and join EG, ZG, and ZK.

Specification (Part 4). It is to be proved that EG is the shortest of all lines from E to arc BG.

Demonstration (Part 4). Since GZ is shorter than KZ, the square on GZ is less than the square on ZK. Let the common square on ZE be added. Then sum of the squares on GZ and ZE, equal to the square on GE, is less than the sum of the squares on KZ and ZE, equal to the square on KE. Therefore GE is shorter than KE.

Similarly, of all the lines from E to arc BKG, EG is the shortest.

Statement (Part 5). And again we will show that, of the lines from point E to arc BG, that nearer to EG is shorter than that farther off.

Statement (Part 1, Case 2). Similarly, it will be shown if ABG is a semicircle that AE is shorter than all lines from E to arc ABG.

❧

Summary.

1. EA is the shortest line from E to ABG.

2. Line lengths strictly increase along AB from A.

3. EB is the longest line to ABG.

4. EG is the shortest line to BG.

5. Line lengths strictly increase along *BG* from *G* (not proved).

6. If *ABG* is a semicircle, *EA* is the shortest line from *E* to *ABG* (not proved).

Introduction to *Sph.* III.3

This theorem is used in *Phen.* 12 to show that arcs of two equal circles have equal secants and so are themselves equal arcs. It is used to prove *Sph.* III.6, III.8, and III.13.

§◆

Theorem. *If two great circles in a sphere cut each other, and equal arcs are cut off from each of them contiguously in both directions from a point at which they meet, then the lines joining the endpoints of the arcs in corresponding directions are equal to each other.*

Exposition. In a sphere, let two great circles *AB* and *GD* cut each other at point *E*, let equal contiguous arcs be cut off each of them in both directions from *E*, *AE* equal to *EB* and *GE* equal to *ED*, and let *AG* and *BD* be joined (Figure 65 (a)).

Specification. It is to be proved that line *AG* is equal to line *BD*.

Distinction of cases. The circle drawn with pole *E* and polar radius *EA* goes through *B*. Either it will also go through *G* (Case 1) or not (Case 2).

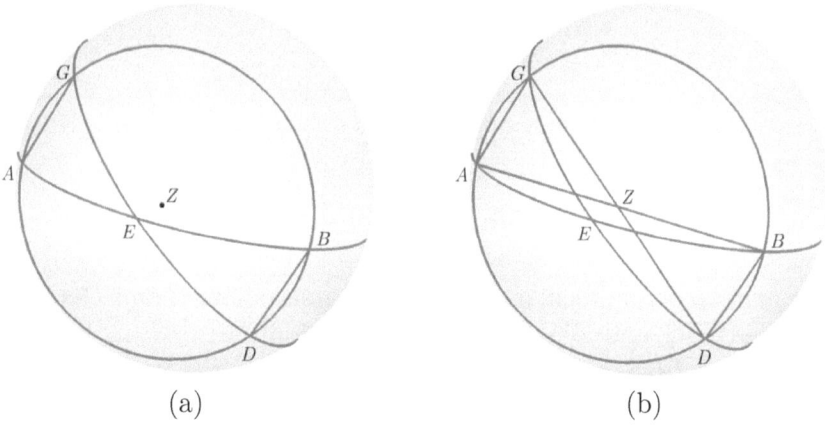

(a) (b)

Figure 65: *Sph.* III.3. (a) Case 1. (b) Case 1 construction.

Demonstration (Case 1). In the first place, let the circle also go through *G*. Then it will also go through *D*, for arc *GE* is equal to arc *ED*. Let it be circle *AGBD* with center *Z*, and let *AB* be the intersection of *AGBD* and great circle *AEB* and *GD* be the intersection of *AGBD* and great circle *GED* (Figure 65 (b)).

Since in a sphere great circle AEB cuts circle $AGBD$ in the sphere through its poles, it bisects it perpendicularly. Therefore line $\underline{AB}$ is a diameter of circle $AGBD$. Similarly line $\underline{GD}$ is a diameter of circle $AGBD$. Therefore the four lines $\underline{ZA}$, $\underline{ZG}$, $\underline{ZB}$, and $\underline{ZD}$ are equal. Then since in triangles AZG and BZD the two $\underline{ZA}$ and $\underline{ZG}$ are equal to $\underline{ZD}$ and $\underline{ZB}$, and angle AZG is equal to angle DZB, the third side $\underline{AG}$ is equal to the third side $\underline{BD}$ as required.

Demonstration (Case 2). Now let the circle drawn with pole E and polar radius $\underline{EA}$ not go through G but let it fall beyond it (Figure 66 (a)). Then it will go through B but fall beyond D. Let it be circle $AHBQ$ with center Z, and H and Q be on great circle GED. Let $\underline{AB}$ be the intersection of circle $AHBQ$ and great circle AEB, and let $\underline{HQ}$ be the intersection of circle $AHBQ$ and HEQ (great circle GED). Similarly to Case 1, each of AEB and $HGEDQ$ is perpendicular to circle $AHBQ$.

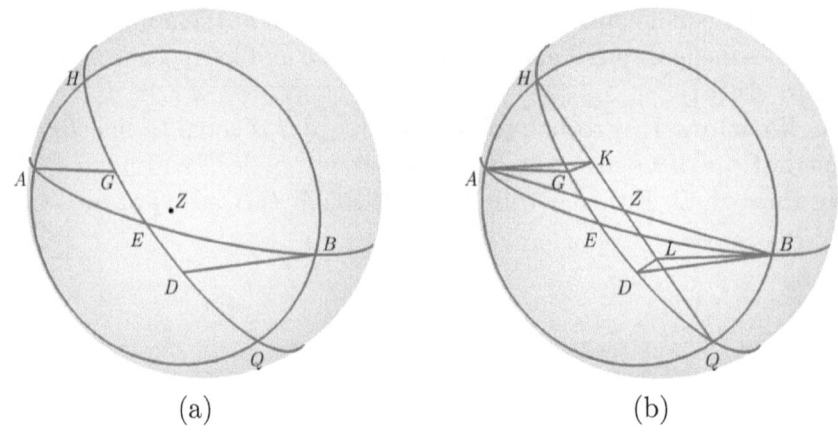

(a)   (b)

Figure 66: *Sph.* III.3. (a) Case 2. (b) Case 2 construction.

Now drop perpendiculars $\underline{GK}$ and $\underline{DL}$ from points G and D to the plane of circle $AHBQ$, and join $\underline{AK}$ and $\underline{BL}$ (Figure 66 (b)).

Since arc EH is equal to arc EQ and in them EG is equal to ED, the remainder GH is equal to the remainder DQ. Then, since HEQ is a perpendicular segment of a circle and the equal arcs GH and DQ are cut off and $\underline{GK}$ and $\underline{DL}$ are produced perpendicular, $\underline{GK}$ is equal to $\underline{DL}$ and $\underline{HK}$ to $\underline{QL}$. But the whole $\underline{HZ}$ is equal to the whole $\underline{ZQ}$; therefore the remainder $\underline{KZ}$ is equal to the remainder $\underline{ZL}$. And $\underline{AZ}$ is equal to $\underline{BZ}$, and so $\underline{AK}$ and $\underline{BL}$ are equal (since triangles AZK and BZL are congruent).

In triangles AKG and BLD, $\underline{AK}$ is equal to $\underline{BL}$ and $\underline{GK}$ to $\underline{DL}$. And angle GKA is equal to angle DLB, for each of them is right.

Therefore the third side $\underline{AG}$ is equal to the third side $\underline{DB}$, which was to be proved.

§⋗

Comment. One may think of a third case when the circle drawn with pole E and polar radius $\underline{EA}$ does not go through G but rather falls short of it. When ED and EG are the longer arcs, renaming A and B as G and D and vice versa turns the situation into Case 2. Then EA and EB are the longer of the arcs and EG and ED the shorter of the arcs. So the circle with pole E and polar radius $\underline{EA}$ must pass beyond G because the arc EG is known to be shorter than EA. There are really only two cases.

Introduction to *Sph.* III.4

This proposition allows *Sph.* III.5 and III.7 to be proved.

§⋗

Theorem. *If two great circles in a sphere cut each other, and equal contiguous arcs are cut off one of them in both directions from a point at which they cut each other, and through the points determined parallel planes are produced, of which one intersects the line of intersection of the planes of the circles outside the surface of the sphere beyond the aforementioned point, and each of the equal arcs is longer than each of the two arcs from the same point cut off by the produced planes, then the arc between the point and the non-intersecting plane is longer than that of the same circle between the point and the intersecting plane.*

Exposition. Let great circles in a sphere cut each other along a diameter ending at point E, and from one of them AEB let equal arcs AE and EB be cut off contiguously in both directions from E, and through A and B let parallel planes AD and BG be drawn, of which AD meets the line of intersection of the great circles AEB and GED at X outside the surface of the sphere beyond E, and each equal arc AE and EB is longer than each of arcs GE and ED (Figure 67 (a)).

It is to be proved that arc GE is longer than arc ED.

Construction. The circle $AHBZ$ (Figure 67 (b)), drawn with pole E and polar radius $\underline{EA}$ goes through B and falls beyond points G (at H on circle GED) and D (at Z on circle GED) because each of AE and EB is longer than GE and ED. Let L be the center of circle $AHBZ$ (Figure 68 (a)); line $\underline{LE}$ is part of the line of intersection of circles AEB and $HGDZ$. Let circle AD, the intersection of plane AD and the surface of the sphere, meet circle $AHBZ$ at point Q; and circle BG, the intersection of plane BG and the surface of the sphere, meet circle $AHBZ$ at point K. Let the lines of intersection of $AHBZ$ and circles AEB, HEZ, ADQ, and KGB be $\underline{AB}$, $\underline{HZ}$, $\underline{AQ}$, and $\underline{KB}$

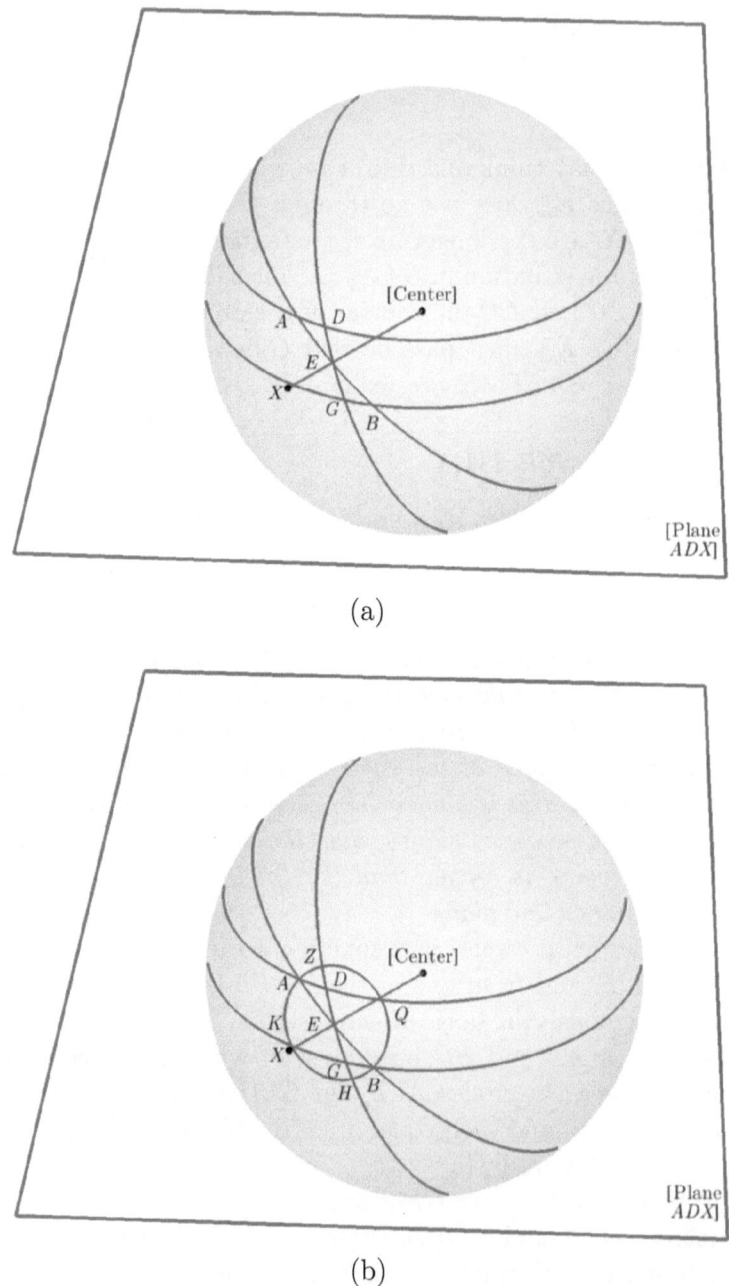

(a)

(b)

Figure 67: *Sph.* III.4. (a) The great circles *AEB* and *DEG*, the parallel small circles *AD* and *GB*, an indication beyond circle *AD* of its plane *ADX*, *X* being where the radius to *E* extended meets the plane. (b) Circle *AZQBHK* added.

respectively, and the lines of intersection of circle *HEZ* and circles *ADQ* and *KGB* be <u>*DM*</u> and <u>*GN*</u>.

Demonstration. Since plane ADQ intersects $\underline{LE}$ extended outside the sphere, call the point of intersection X. Then X is in planes ADQ and HEZ, but so are M and D. Therefore $\underline{MD}$ extended meets $\underline{LEX}$ at X (Figure 68 (b)).

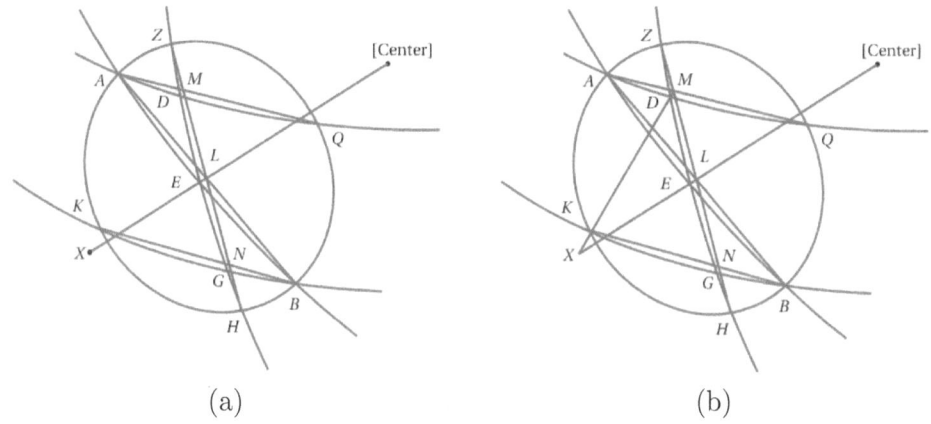

Figure 68: *Sph.* III.4. (a) Enlarged view of the neighborhood of circle $AZQBHK$ showing its intersections $\underline{AQ}$, $\underline{KB}$, $\underline{AB}$, and $\underline{ZH}$ with the planes of the four other circles and their intersections M, N, and L, the center of circle $AZQBHK$. (b) Enlarged view with line $\underline{MDX}$ added.

Since the great circle AEB in a sphere cuts the circle $AHBZ$ through the poles, it will bisect it perpendicularly. Therefore $\underline{AB}$ is a diameter of circle $AHBZ$. Similarly, $\underline{HZ}$ is a diameter of circle $AHBZ$ and L is the center of circle $AHBZ$. Since the parallel planes KGB and ADQ cut the plane of $AHBZ$, the lines of intersection $\underline{KB}$ and $\underline{AQ}$ are parallel. Again, since the parallel planes KGB and ADQ cut the plane of HEZ, their lines of intersection

$$\underline{GN} \text{ and } \underline{DM} \text{ are parallel.} \tag{1}$$

Since each of the planes AEB and HEZ is perpendicular to the plane $AHBZ$, their line of intersection $\underline{EL}$ is perpendicular to plane $AHBZ$, so that it makes right angles with all lines meeting it and in the plane of circle $AHBZ$. Therefore $\underline{EL}$ is perpendicular to each of $\underline{AB}$ and $\underline{HZ}$. Since angle XLN is outside triangle XLM, it is greater than the opposite angle XML inside triangle XML. But angle XLN is right,

$$\text{angle } XML \text{ is acute, and angle } XMZ \text{ is obtuse.} \tag{2}$$

And since $\underline{GN}$ is parallel to $\underline{DM}$ and $\underline{HZ}$ meets these, angle GNH is also acute.

In triangles AML and BNL, $\underline{AM}$ is parallel to $\underline{NB}$, and $\underline{AL}$ is equal to $\underline{LB}$; therefore $\underline{NL}$ is equal to $\underline{LM}$. But the whole $\underline{HL}$ is equal to the whole $\underline{LZ}$; therefore the remaining

<u>HN</u> is equal to the remaining <u>MZ</u>.   (3)

Since *HEZ* is a section of a circle and the ends of secant <u>HNLMZ</u>, <u>HN</u> and <u>MZ</u> are equal (from (3)), and parallels <u>GN</u> and <u>DM</u> are drawn through *N* and *M* making angles *GNH* and *DML* equal (from (1)) and acute (from (2)), arc *HG* is shorter than arc *DZ*.

Then since the whole arc *HE* is equal to the whole arc *ZE* (by construction), of which *HG* is shorter than *DZ*, the remaining arc *GE* is longer than the remaining arc *ED*.

⇛

Comment. The reasoning of the penultimate paragraph can be thought of in terms of vectors from symmetrically arranged points like *N* and *M* to the circular arc *HEZ*. The arc lengths from the respective ends of the secant and arc to the ends of the vectors are equal if the vectors move with mirror symmetry either toward the ends or away from them. The proof of the proposition depends on the fact that, if they move differently, then the arc lengths will differ. One can think of the vectors beginning vertical. The vector from *M* is made to move away from its end *Z* (to *D*, making angle *DML* acute), increasing arc length, and the vector from *N* is made to move toward its end *H* (to *G*, making angle *GNH* acute), decreasing arc length. The result is that the arc *ZD* is longer than arc *HG*. A lemma proving what is needed is in the Commentary (see p. 312).

Introduction to *Sph.* III.5–III.8

The interest in *Sph.* III.5–III.8 lies in their application to astronomy. This is exhaustively discussed in the Commentary (see p. 314); here the briefest indication of what the propositions are about will be given. In *Sph.* III.7 and III.8, the smallest parallel circle mentioned (the *v*-circle, relative to location, within which stars do not rise and set) is tangent to all configurations of the horizon of a given site in the northern or southern terrestrial hemisphere. The great circles tangent to it are configurations of the horizon at various times for such an observer off the equator. In *Sph.* III.5 and III.6, on the other hand, the horizon passes through the pole of the parallels, the situation of an observer on the terrestrial equator where all stars rise and set. As with the group *Sph.* II.10, II.13, II.16, the simpler case, not involving non-intersecting semicircles, is proved first.

Even with several configurations of the horizon, which would occur at different times, the configuration is static but conveys dynamic information.

Introduction to *Sph.* III.5

See the Introduction to *Sph.* III.5–III.8 and for astronomy the Commentary (p. 319). This is an important enabling result for later propositions. Its astronomical determination is of the projection of adjacent arcs of the great circle oblique to the parallels, the ecliptic, onto a colure, that is, a great semicircle from pole to pole. Such curves are also the horizons of locations on the terrestrial equator.

§◦

Theorem. *If two great circles, one of the parallels and the other oblique to the parallels, cut a great circle through the poles of the parallels perpendicularly, and equal contiguous arcs are cut off the oblique great circle on the same side of the greatest of the parallels, and through the points produced parallel circles are drawn, then they cut off contiguously between them unequal arcs of the third great circle, and that nearer to the greatest of the parallels is progressively longer than that farther off.*

Exposition. Let the pole of the parallels be point A on the circumference of great circle ABG (Figure 69), and let two great circles BZG and DZE cut it perpendicularly, of which BZG is one of the parallels and DZE is oblique to the parallels. From the oblique circle DZE let equal contiguous arcs KQ and QH be cut off on the same side of the parallel great circle BZG. And through points K, Q, and H let parallel circles OKP, NQX, and LHM be drawn.

Specification. It is to be proved that circles OKP, NQX, and LHM cut off unequal arcs of the first great circle ABG, and they are progressively longer than those farther off the closer they are to BZG. In particular, it is to be proved that arc ON is longer than arc NL.

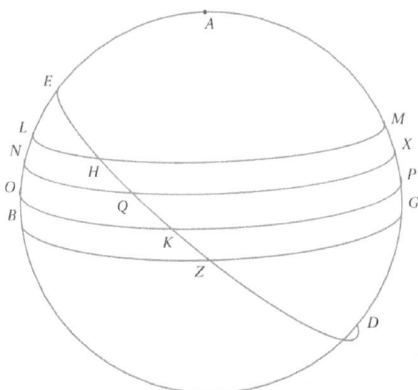

Figure 69: *Sph.* III.5. The great circles ABG, BZG, and EZD, with the equal arcs HQ and QK, and the circles LHM, NQX, and OKP parallel to BZG through the endpoints H, Q, and K of the arcs.

Construction. Draw great circle $ASQR$ through points A and Q, where S and R are the intersections of its circumference with arcs LHM and OKP (Figure 70).

Demonstration. Since A is the pole of circle OKP, arc ANO is equal to arc AQR; and, since A is the pole of circle NQX, arc ALN is equal to arc ASQ. Therefore the remainder NO is equal to the remainder QR. Similarly we will show also that arc LN is equal to SQ.

Since great circle AQR cuts circle OKP through the poles, it will bisect it perpendicularly. Therefore circle AQR is perpendicular to circle OKP. Now a segment of circle RQA beginning with arc RQ is set up perpendicularly on a diameter of circle OKP, and arc RQ is cut off less than half the segment set up; therefore the line joining Q to R is the shortest of all the lines from Q to the circumference of circle OKP. Therefore the secant joining Q to R is shorter than the secant joining Q to K; and they are secants of equal great circles. Therefore arc QR is shorter than arc QK. Similarly QS is shorter than QH. And KQ is equal to QH; therefore each of KQ and QH is longer than each of RQ and QS.

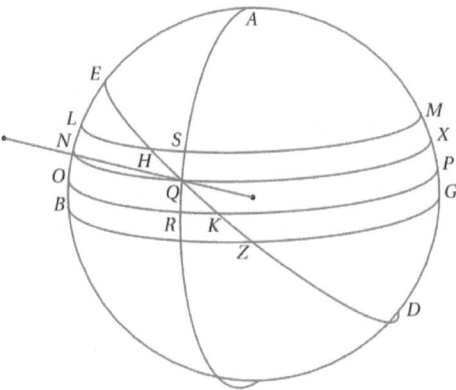

Figure 70: *Sph.* III.5. The great circle $ASQR$ added through the point Q. The line from the center of the sphere to the point where the radius to Q produced intersects the plane of circle LSM.

Sph. III.4 will be used when one further condition is established. Since circle BZG is parallel to LHM and BZG meets the line of intersection of circles HQK and AQR at the center of the sphere, the plane of circle LHM meets the line of intersection of circles HQK and AQR outside the surface of the sphere beyond Q (Figure 70). Then, since the two great circles HQK and SQR cut each other, and equal arcs KQ and QH are cut off HQK contiguously on each side of point Q, and through points H and K parallel planes LHM and OKP have been extended, of which LHM meets the line of intersection of planes HQK and SQR outside the surface of the sphere beyond point Q, and each of the equal arcs KQ and QH is longer than each of RQ and QS, arc RQ is longer than QS.

But *RQ* is equal to *ON*, and *QS* to *NL*; therefore arc *ON* is longer than arc *NL*.

§⬥

Comment. The configuration of *Sph.* III.5 will recur in *Sph.* III.6, III.9–III.11.

The word rendered "progressively" in the enunciation of the proposition alludes to the fact that there can be several arcs like *LN* and *MN*, and they are successively longer than the next without need for further proof. This occurs several more times (*Sph.* III.6–III.9, III.14).

Introduction to *Sph.* III.6

See the Introduction to *Sph.* III.5–III.8 (p. 448) and for astronomy the Commentary (p. 320). Here equal arcs of the oblique great circle (ecliptic) are projected by colures onto the greatest of the parallels.

§⬥

Theorem. *If two great circles, one of the parallels and the other oblique to the parallels, cut a great circle through the poles of the parallels perpendicularly, and equal contiguous arcs are cut off the oblique great circle on the same side of the greatest of the parallels, and through the points produced and the poles great circles are drawn, then they cut off between them unequal arcs of the greatest of the parallels, and that nearer to the third great circle (through the poles) is progressively longer than that farther off.*

Exposition. Let the pole of the parallels be point *A* on the circumference of great circle *ABG*, and let two great circles *BZG* and *DZE* cut it perpendicularly, of which *BZG* is one of the parallels and *DZE* is oblique to the parallels (Figure 71 (a)). From the oblique circle *DZE* let equal contiguous arcs *KQ* and *QH* be cut off on the same side of the greatest of the parallels *BZG*. Through *A* and each of the points *H*, *Q*, and *K* let great circles *AHL*, *AQM*, and *AKN* be drawn, where *L*, *M*, and *N* lie on *BZG*.

Specification. It is to be proved that arc *LM* is longer than arc *MN*.

Construction. Draw parallel circles *XHFO*, *PQR*, and *SUKT* through points *H*, *Q*, and *K* with *X*, *P*, and *S* on one side of *ABG* and *O*, *R*, and *T* on the other side of *ABG* (Figure 71 (b)). Arc *SP* is longer than arc *PX*. But *SP* is equal to *UQ* and *PX* to *QF*; therefore *UQ* is longer than *QF*. Then let *C* be on arc *QU* such that arc *QC* is equal to arc *QF*; and *HQ* is equal to *QK*; therefore the line joining *H* to *F* is equal to the line joining *C* to *K*. Now draw circle *ICYW* through *C* parallel to the earlier circles with *I* and *W* on *ABG* and *Y* on *AKN*.

Demonstration. Since great circle *AYKN* cuts circle *CYW* through the poles, it bisects it perpendicularly. Therefore circle *AYKN* and circle *CYW* are perpendicular. Since the two parallel planes *BZG* and *CYW* cut plane

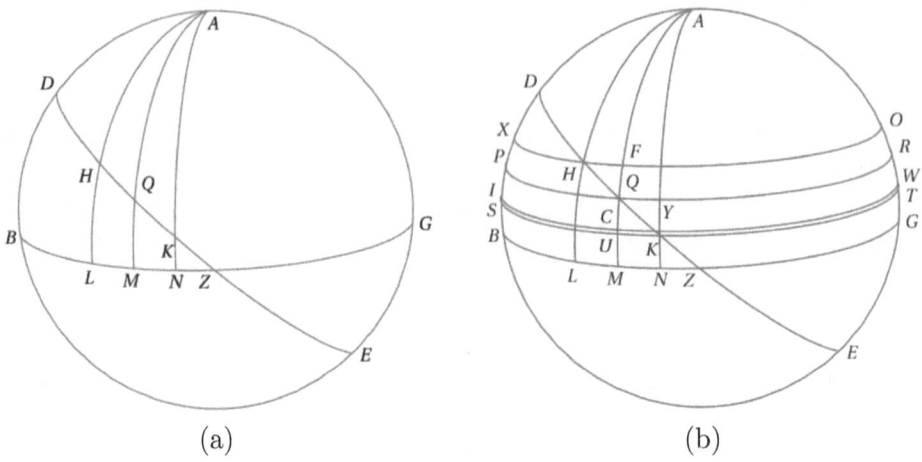

(a)   (b)

Figure 71: *Sph.* III.6. (a) Three evenly spaced points K, Q, and H on an oblique great circle and great circles from the pole of the parallels A through K, Q, and H to N, M, and L on the greatest of the parallel circles. (b) Parallel circles through K, Q, H, and Y added.

$AYKN$, their intersections are parallel. Therefore the intersection of $AYKN$ and BZG, which is the diameter from point N of the circle $AYKN$, and the intersection of $AYKN$ and CYW are parallel (Figure 72).

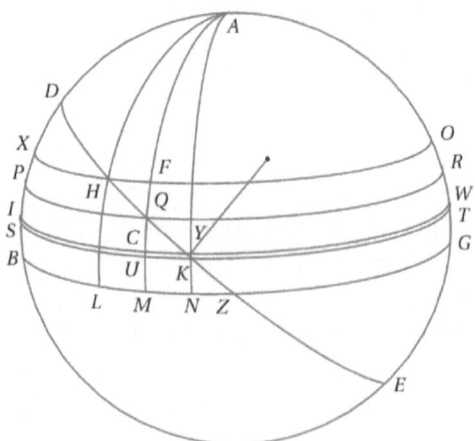

Figure 72: *Sph.* III.6. The diameter of the parallel circle $ICYW$, its intersection with circle $AYKN$, added.

Now the line of intersection of circles $AYKN$ and $ICYW$ cuts the circle $AYKN$ not through the center, and on it a segment of circle YCI beginning with arc YC is set up perpendicularly, and the circumference of the set-up

segment is divided unequally at *C*, and the arc *YC* is shorter than half of
the set-up segment; therefore the line joining *C* to *Y* is shortest of all lines
from point *C* to arc *YKN*. Therefore the line *CK* is longer than line *CY*.
And since line *CK* is equal to line *HF*, line *HF* is longer than line *CY*. And
since circle *CYW* is closer to the center of the sphere than circle *XHFO*,
circle *ICYW* is larger than circle *XFHO*. Then since *XHO* is shorter than
CYW and line *HF* in *XHFO* longer than *CY* in *ICYW*, arc *HF* is longer
than similar to arc *CY*. But arc *HF* is similar to arc *LM* and arc *CY* is
similar to arc *MN*; therefore arc *LM* is longer than similar to arc *MN*, and
they are of the same circle. Therefore arc *LM* is longer than arc *MN*.

§⟐

Introduction to *Sph.* III.7

See the Introduction to *Sph.* III.5–III.8 (p. 448) and for astronomy the Commentary (p. 322). The interest in this and the next proposition lies in their
application to astronomy for observers not on the terrestrial equator. The
smallest parallel circle is tangent to all the configurations of the horizon of
a site off the equator. The tangent circles (only one here) are positions of
the horizon at different times.[3]

§⟐

Theorem. *If two great circles in a sphere touch parallel circles in the
sphere, the first touching smaller circles than the second and the points of
contact of the second circle being on the first circle, and equal arcs are cut
off contiguously from the second circle on the same side of the greatest of
the parallels, and parallel circles are drawn through the ends of the arcs;
then they cut off between them unequal arcs of the first great circle, and that
nearer to the greatest of the parallels is progressively longer than that farther
off.*

Exposition. Let great circles *ABG* and *EZH* in a sphere touch parallel circles through *A* and *E* respectively, the parallel circle through *E* being larger,
points *B*, *Z*, and *G* being on the greatest of the parallels (Figure 73 (a)).
And let equal arcs *LK* and *KQ* be cut off contiguously from the second circle
EZH on the same side of the greatest of the parallels *BZG*, with *Q* farther
from *BZG* than *L*. Through points *Q*, *K*, and *L* let parallel circles *MQN*,
XKO, and *PLR* be drawn with *M*, *X* and *P* on one side of *ABG* and *N*, *O*,
and *R* on the other side.

---

[3] In the discussion of mathematical presuppositions in the translation of the *Phenomena* by Berggren and Thomas (1996, 26–27), the equal arcs of the ecliptic are said
to begin at its most northerly point (tangent to the summer tropic). This configuration,
which is not the general case, is all that Euclid needs, as is illustrated in Figure 7 of that
work.

Specification. It is to be proved that arc *PX* is longer than arc *XM*.

Construction. Draw great circle *DYKS* through point *K* touching the parallel circle through *A* at *D* so that the semicircle from *D* in the direction of *K* is non-intersecting with the semicircle from *A* in the direction of *B* (Figure 73 (a)). Let the pole of the parallels be point *U*, and through points *U* and *K* draw great circle *UTKF* with *T* on *MQN* and *F* on *PLR*.

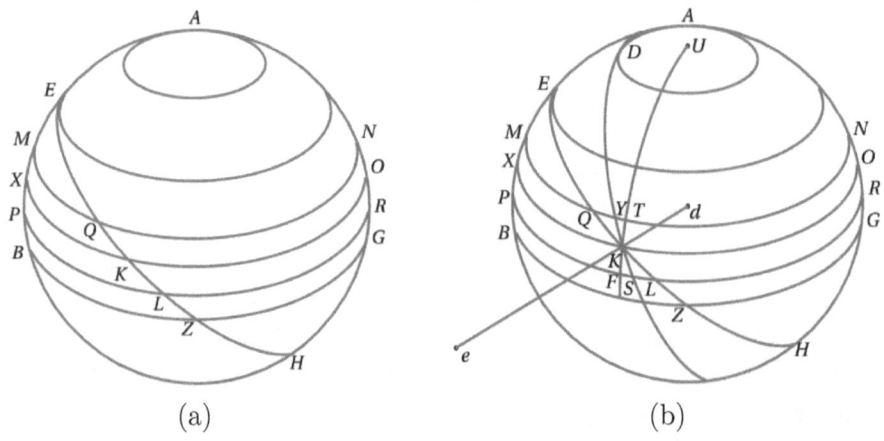

(a)   (b)

Figure 73: *Sph.* III.7. (a) The given configuration of parallel circles and oblique great circles. (b) The semicircle *DYKS*, non-intersecting with *AEB*, the great-circle segment *UTKF*, and the line from the center of the sphere through *K* to the plane *MQN* added.

Demonstration. Since great circle *UTKF* in a sphere cuts circle *PLR* through the poles, it bisects it perpendicularly. We have the configuration of *Sph.* III.1: A segment of circle *FKTU* is set up perpendicularly on a diameter from *F* of circle *PLR*, and the circumference of the set-up segment is divided unequally at point *K*, and *FK* is less than half. Therefore the line *KF* is shortest of all lines from point *K* to the arc *PLR*, and lines nearer to *F* are progressively shorter than those farther away. Therefore line *KS* is shorter than line *KL*. And circles *DS* and *ELH* are equal, for they are great. Therefore arc *KL* is longer than arc *KS*. Now similarly we will show also that arc *KQ* is longer than arc *KY*. (The similarity of this argument is not quite as routine as such similarities usually are. A great circle *aQUb* is required, where *a* and *b* are on *XKO*, *a* near *Q* and *b* opposite. A great circle tangent to circle *AD* and through *Q* is also required, cutting circle *XKO* at *c*. Because the first circle passes through the poles of circle *XaKO*, the line *ab* is its diameter, and a configuration of *Sph.* III.1 again occurs and with the same result, namely that *QK* is longer than *Qc*, which is longer than *Qa*. But *Qc* is equal to *YK*, a step unnecessary above.) And *QK* is equal to *KL*. Therefore each of *QK* and *KL* is longer than each of *YK* and (from above) *KS*.

Since great circle *BZG* is parallel to circle *MQN* and circle *BZG* meets the intersection of *QKL* (*EZH*) and *YKS* within the surface of the sphere at the center of the sphere *d*, the plane of circle *MQN* extended meets the intersection of *QKL* and *YKS* outside the surface of the sphere beyond *K* at, say, *e* (Figure 73 (b)).

We have the configuration of *Sph.* III.4: Since great circles *QKL* and *YKS* in the sphere cut each other at *K* and from one of them *QKL* equal arcs *QK* and *KL* are cut off contiguously on each side of the point at which they cut each other, and through points *Q* and *L* parallel planes *PLR* and *MQN* are produced, of which *MQN* meets the common section of circles *QKL* and *YKS* outside the surface of the sphere beyond *K* at *e*, and each of the equal arcs *QK* and *KL* is longer than each of *SK* and *KY*, *SK* is longer than *KY*. But arc *KF* equals arc *PX* and arc *KT* equals arc *KM*. Therefore *PX* is longer than *XM*.

§⟶

Introduction to *Sph.* III.8

See the Introduction to *Sph.* III.5–III.8 (p. 448) and for astronomy the Commentary (p. 326). This proposition is more general than *Sph.* III.6 in projecting by non-intersecting semicircles tangent to the same small parallel circle onto the equator.[4]

§⟶

Theorem. *Let two great circles in a sphere touch parallel circles in the sphere, the first touching smaller circles than the second, the points of contact of the second circle being on the first circle. Let equal arcs be cut off contiguously from the second circle on the same side of the greatest of the parallels, and through the ends of the arcs great circles also touching the parallel circle the first circle touches be drawn (cutting off between them similar arcs of parallel circles) making the semicircles from the points of contact toward the points through which they were drawn be non-intersecting with the semicircle of the first great circle containing the point of contact of the second great circle on the same side of the greatest of the parallels; then the non-intersecting semicircles cut off between them unequal arcs of the greatest of the parallels, and that nearer to the first great circle is progressively longer than that farther off.*

---

[4] In the discussion of mathematical presuppositions in the translation of Euclid's *Phenomena* by Berggren and Thomas (1996, 26–27), the equal arcs of the ecliptic are said to begin at its most northerly point (tangent to the summer tropic). Again, this configuration, is all that Euclid needs and is illustrated in Figure 8 of that work.

Exposition. Consider a small circle *AD* and the parallel great circle *BZ* in a sphere. Let the bounding great circle *ABG* be tangent to the small circle at the top of both, point *A* (Figure 74). Let another great circle *EZG* less oblique to the parallels have its points of contact *E* and *G* with parallel circles on the bounding circle (upper left and lower right). Let two equal and contiguous arcs *HQ* and *QK* be cut off circle *EG* between its point of contact *E* with the upper parallel circle and its point of intersection *Z* with the parallel great circle. Through *H*, *Q*, and *K* let great circles *DHL*, *MQN*, and *XKO* touch the small circle at *D*, *M*, and *X*, with *L*, *N*, and *O* on *BZ* and and with semicircles non-intersecting with both the bounding circle's left half and one another, each pair *DHL* and *MQN*, *MQN* and *XKO* cutting off similar arcs of parallel circles.

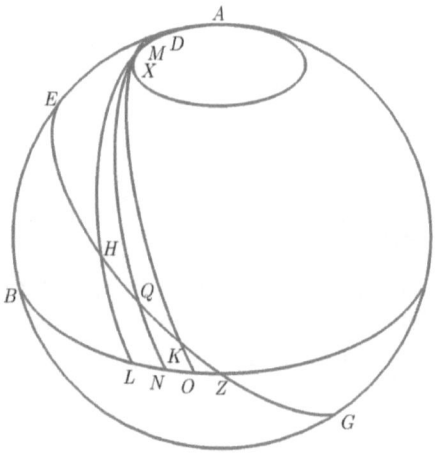

Figure 74: *Sph.* III.8. The configuration of non-intersecting semicircles *DHL*, *MQN*, and *XKO* projecting equal arcs *HQ* and *QK* of the oblique great circle *EZG* onto similar arcs *LN* and *NO* on the greatest of the parallel circles *BZ*.

Specification. It is to be proved that the arcs *LN* and *NO* cut off the parallel great circle are unequal and that *LN*, nearer the left side of the bounding circle, is the longer.

Construction. Through points *H*, *Q*, and *K* draw parallel circles *PHR*, *SQ*, and *TUK* with *P*, *S*, and *T* on arc *AB* and *R* and *U* on arc *MQN* (Figure 75 (a)). By *Sph.* III.7, arc *ST* is longer than arc *SP*. But *ST* is equal to *QU*, and *SP* is equal to *QR*. Therefore *UQ* is longer than *QR*. Then let *F* be the point on *UQ* such that *QF* is equal to *QR*. But *HQ* is equal to *QK*. Therefore, by *Sph.* III.3, the line *HR* is equal to the line *FK*. Now draw through *F* the parallel circle *CFY* with *Y* on arc *XKO* (Figure 75 (b)) and through the pole *W* of the parallel circles draw great circle *WO*.

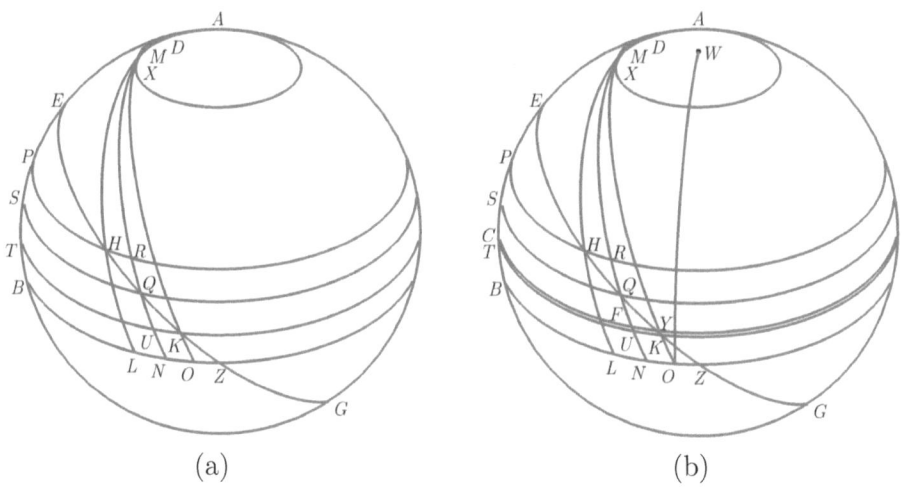

(a)   (b)

Figure 75: *Sph.* III.8. (a) Parallel circles *PHR*, *SQ*, and *TUK* added through the ends of the equal arcs *HQ* and *QK*. (b) Fourth parallel circle *CFY* added and a great circle *WO* through the pole *W* of the parallels.

Demonstration. Since great circle *WO* cuts circle *BZ* through the poles, it bisects it perpendicularly. Therefore *XKO* is inclined to *BZ* in the direction of *E* and *B*. That is, *BZ* is inclined to *XKO* in the direction of *E* and *X*. And *CFY* is parallel to *BZ*. Therefore *CFY* is inclined to *XKO* in the direction of *E* and *X*. Since two parallel planes *BZ* and *CFY* cut the plane *XKO*, the intersection of *XKO* and *CFY* is parallel to the intersection of *XKO* and *BZ*. But the intersection of *XKO* and *BZ* is the diameter of great circle *XKO* from point *O*. Therefore the intersection of *XKO* and *CFY* is parallel to the diameter of circle *XKO* from *O*.

This is the configuration of *Sph.* III.2: The intersection of *XKO* and *CFY* cuts the circle *XKO* into unequal parts, and a segment beginning with the arc *YFC* and not greater than a semicircle is set up on *XKO*, inclined toward the segment *YX*, and the circumference of the set-up segment is divided into unequal parts at *F*, and arc *YF* is shorter than half of the set-up segment. Therefore the line <u>*FY*</u> is shorter than all of the lines from *F* to the circumference beginning with arc *YO*, not less than a semicircle (being the complement of the segment *YX*, not greater than a semicircle). Therefore the line <u>*FY*</u> is shorter than the line <u>*FK*</u>.

But the line <u>*FK*</u> is equal to the line <u>*HR*</u> by *Sph.* III.3, so that the line <u>*HR*</u> is longer than the line <u>*FY*</u>. And since circle *CFY* is nearer to the center of the sphere than parallel circle *PHR*, circle *CFY* is larger than circle *PHR*. Then since *CFY* and *PHR* are unequal circles, and circle *PHR* is the smaller, and lines are drawn in them, <u>*HR*</u> in *PHR* and <u>*FY*</u> in *CFY*, and that in the smaller circle is longer than that in the larger circle, arc *HR* is longer than similar to arc *FY*. But arc *HR* is similar to arc *LN*, and *FY* is similar to *NO*; therefore *LN* is longer than similar to *NO*. And they are of the same circle. Therefore arc *LN* is longer than arc *NO*.

Sc

Introduction to *Sph.* III.9

This proposition generalizes *Sph.* III.6 to non-adjacent equal arcs of the oblique great circle. Of interest is the mode of proof with its division into cases depending on incommensurability of arcs. It is clear enough, under the assumption of continuity appropriate to these circumstances, that if the adjacent arcs covered by *Sph.* III.6 are moved apart slightly, the conclusion about their projections onto the greatest of the parallels ought not to change much, if at all. Since how far they are separated (within bounds) does not change the conclusion of that theorem at all it is worth proving and interesting how it is done.

Sc

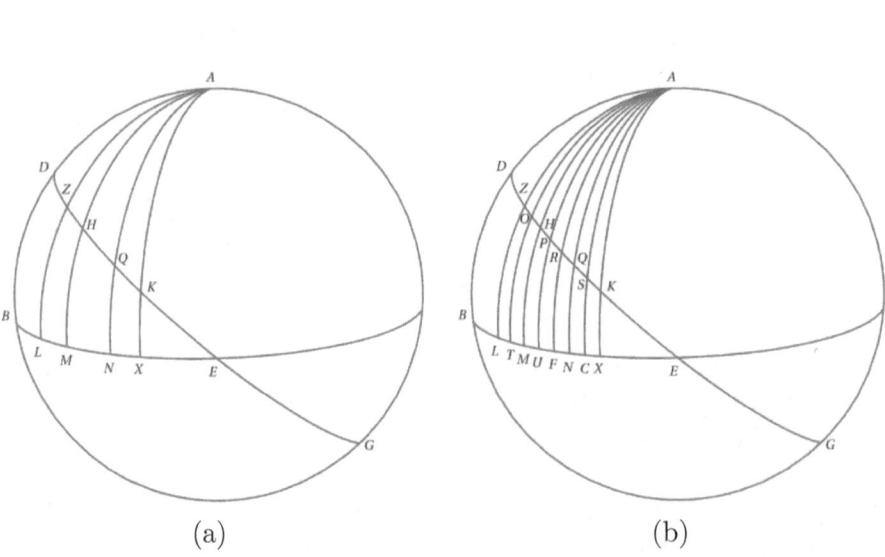

(a)   (b)

Figure 76: *Sph.* III.9. (a) Noncontiguous equal arcs *ZH* and *QK* of the oblique great circle *DEG* projected by great circles though the poles of the parallels onto arcs *LM* and *NX* on the greatest of the parallels *BE*. (b) Case 1. Arcs *ZH* and *QK* divided by great circles though the poles of the parallels.

Theorem. *If the poles of the parallels are on the circumference of a great circle, and two great circles cut it perpendicularly, one of them being the* *greatest of the parallels and the other oblique to the parallels, and arcs equal but not contiguous are cut off the oblique circle on the same side of both the greatest of the parallels and the great circle through the poles, then great circles through the ends of the arcs and the poles cut off between them unequal arcs of the greatest of the parallels, and that nearer to the initial great circle through the poles is progressively longer than that farther off.*

Exposition. Let the pole of the parallels be point *A* on the circumference of great circle *ABG* (Figure 76 (a)), and let oblique great circle *DEG* and parallel great circle *BE* cut circle *ABG* perpendicularly. Let equal but not contiguous arcs *ZH* and *QK* be cut off arc *DE*, and through the pole *A* and points *Z*, *H*, *Q*, and *K* let great circles *AZL*, *AHM*, *AQN*, and *AKX* meet arc *BE* at *L*, *M*, *N*, and *X*.

Distinction of cases. *HQ* is either commensurable with *ZH* and *QK* (Case 1) or not (Case 2).

Construction (Case 1). Let *HQ* be commensurable with *ZH* and *QK*, and let *ZH*, *HQ*, and *QK* be divided into their common measures, for example, at points *O*, *P*, *R*, and *S* (Figure 76 (b)). Through the pole *A* and points *O*, *P*, *R*, and *S* draw great circles *OT*, *PU*, *RF*, and *SC*, with *T*, *U*, *F*, and *C* on *BE*.

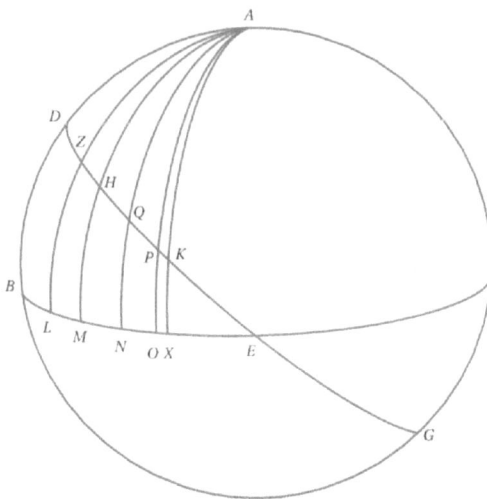

Figure 77: *Sph.* III.9. Case 2.1. Arc *HQ* not commensurable with arcs *ZH* and *QK*.

Demonstration (Case 1). Since contiguous arcs *ZO*, *OH*, *HP*, *PR*, *RQ*, *QS* and *SK* are equal to one another, the contiguous arcs *LT*, *TM*, *MU*, *UF*, *FN*, *NC*, and *CX* decline in length from the longest *LT*. Then since *LT* is longer than *NC* (Greek has *NX*) and *TM* than *CX*, the whole *LM* is longer than the whole *NX*.

Exposition (Case 2). Let *HQ* not be commensurable with *ZH* and *QK*.

Specification (Case 2). It is to be proved that arc *LM* is longer than arc *NX*.

Distinction of Subcases. If *LM* is not longer than *NX*, it is either shorter than it (Case 2.1) or equal to it (Case 2.2).

Construction (Case 2.1). Assume *LM* is shorter than *NX* as it is in Figure 77. Between *N* and *X*, choose *O* so that *NO* is equal to *LM*, and draw an arc of the great circle *PO* through the pole *A* and *O* with *P* on *DE* (Figure 78). Then on *DE* let *R* between *P* and *K* make *QR* commensurable with *HQ* and let *S* between *Z* and *H* make *SH* equal to *QR*. Draw arcs *ST* and *RU* of great circles through the pole *A* and points *S* and *R*, with *T* and *U* on *BE*.

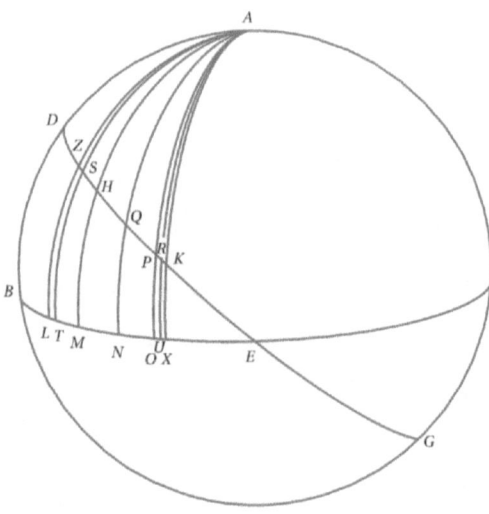

Figure 78: *Sph.* III.9. Constructions.

Demonstration (Case 2.1). Since *SH* is equal to *QR* and *HQ* is commensurable with each of *SH* and *QR*, *TM* is longer than *NU* by Case 1, and *LM* is longer than *TM* and *NU* longer than *NO*. Therefore *LM* is *a fortiori* longer than *NO*. But they are also equal, which is absurd.

Therefore *LM* is not shorter than *NX*.

Specification (Case 2.2). It is to be proved that *LM* is not equal to *NX*.

Construction (Case 2.2). Assume *LM* equal to *NX*. Bisect *ZH* and *QK* at points *O* and *P* as in Figure 79, and draw arcs *OR* and *PS* of great circles through the pole *A* and points *O* and *P*, with *R* and *S* on *BE*.

Demonstration (Case 2.2). Since *ZO* and *OH* are contiguous and equal to each other, projections *LR* and *RM* are contiguous and *LR* longer than *RM*. Therefore *LM* is longer than double *MR*. Again, since *QP* and *PK* are contiguous and equal to each other, *NS* and *SX* are contiguous and *NS* longer than *SX*. Therefore *NX* is shorter than double *NS*. Then since *LM*

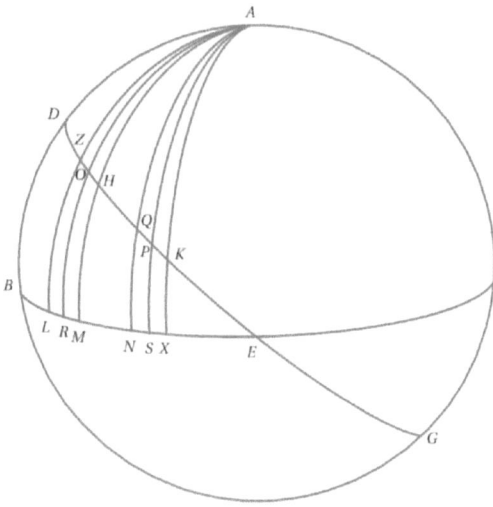

Figure 79: *Sph.* III.9. Case 2.2.

is equal to *NX*, *LM* being longer than double *MR*, and *XN* is shorter than double *NS*, *RM* is shorter than *NS*, which is absurd by Case 1, *OH* and *QP* being assumed equal.

Therefore arc *LM* is not equal to arc *NX*.

It has also been shown that arc *LM* is not shorter than arc *NX*. Therefore arc *LM* is longer than arc *NX*.

Comment. The treatise drops this topic at this point. This may be thought odd, because more can be said in the same vein. *Sph.* III.8 could be generalized, as *Sph.* III.6 is here, with hardly any change in the proof from that above. In a different direction, *Sph.* III.7 and its limiting case *Sph.* III.5 can also be generalized to non-contiguous arcs, again with hardly any change to this proof. On the other hand, it is so easy that its ease can be thought a reason not to bother.

Bisection of arcs of the great circle *DE* is called for in the construction for Case 2.2. This can be done with the general technique for bisection of a circular arc that is an unstated corollary of *Sph.* II.9. There are two other things assumed possible that need justification.

It is assumed that the common measure of the three arcs *HQ*, *ZH*, and *QK*, all arcs of the same great circle, can be found. The procedure for doing this is described in *Elem.* VII.3 for integers and X.4 for magnitudes. It depends on the simpler so-called Eudlidean algorithm for finding the common measure of two integers in *Elem.* VII.2. This is discussed in the

Commentary to this theorem under its ancient name, *anthyphairesis* (see p. 331).

It can reasonably be wondered how, between P and K on DE, to let R make QR commensurable with HQ in the construction for Case 2.1. This is not difficult with the same general technique for bisection. Being able to do arbitrary bisections allows a modification of the method of false position, which is much older than Greek spherics, to determine such an R. A method for general magnitudes appears in a scholium very much as one might write it now. A translation of the scholium appears in the Commentary (see p. 333). A paraphrase follows.

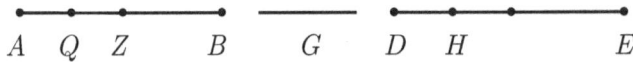

$$A \quad Q \quad Z \qquad B \qquad G \qquad D \quad H \qquad\qquad E$$

Figure 80: *Sph.* III.9. Scholium 416.

Given three magnitudes of the same kind, AB, G, and DE, with $AB > G$, and DE arbitrary, to find Q on BA such that BQ is commensurable with DE (Figure 80). Let Z on BA make $BZ = G$. Bisect DE as often as necessary toward D until H is reached with $DH = DE/2^m < AZ$ for some positive integer m.

Either $DH = BZ/n$ for some positive integer n or not. First, suppose so, and let Q on ZA be such that $ZQ = DH$; then $BQ = (n+1)BZ/n$ is commensurable with DH and so with DE.

Second, suppose $DH \neq BZ/n$ for all n. Then lay off DH from B as often as necessary to reach Q beyond Z by less than $DH < AZ$, therefore before A; then BQ is a multiple of DH and so commensurable with DE.

Introduction to *Sph.* III.10–III.12

Sph. III.10–III.12 concern a different class of inequality; for astronomy see the Commentary (p. 337). Here it is ratios of arc lengths and of lines. Both *Sph.* III.10 and III.11 are used in the proof of *Sph.* III.12, but *Sph.* III.11 is of independent astronomical interest. This is discussed in detail in the Commentary; here the briefest indication of what the propositions are about will be given. *Sph.* III.10 and III.12 are also of interest on account of involving incommensurability like *Sph.* III.9.

Sph. III.10. Theorem. *If the poles of the parallels are on the circumference of a great circle, and two great circles cut it perpendicularly, one of the parallels and the other oblique to the parallels, and two arbitrary points are taken on the oblique circle between the great circle through the poles and the greatest of the parallels, and through each of those points and the poles great*

circles are drawn, then the ratio of the arc of the greatest of the parallels between the first great circle and the next through the poles to the arc of the oblique circle between the same circles is greater than the ratio of the next arc of the greatest of the parallels – that between the great circles through the pole and the taken points – to the arc of the oblique circle between the two taken points.

Exposition. Let point A, the pole of the parallels, be on the circumference of great circle ABG, and let oblique great circle DEG and the greatest of the parallels BE cut circle ABG perpendicularly (Figure 81 (a)). And let two arbitrary points Z and H be taken on the oblique circle DEG between ABG and BE. And through points Z and H and the pole A let great circles AZQ and AHK be drawn, with Q and K on BE.

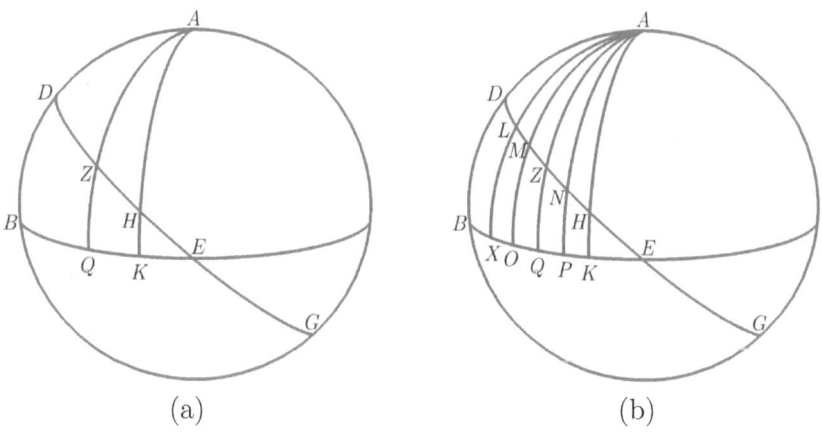

Figure 81: *Sph.* III.10. (a) Given. (b) Construction for Case 1.

Specification. It is to be proved that the ratio of arc BQ to arc DZ is greater than the ratio of arc QK to arc ZH.

Distinction of Cases. Either ZH is commensurable with DZ (Case 1) or not (Case 2).

Construction (Case 1). Let ZH be commensurable with DZ, and divide DZ and HZ into their common measures, for example, at points L and M in DZ and N in HZ (Figure 81 (b)). Through points L, M, and N and the pole A draw great circles LX, MO, and NP, with X, O, and P on BK.

Demonstration (Case 1). Since DL, LM, MZ, ZN, and NH are contiguous and equal to one another, BX, XO, OQ, QP, and PK are contiguous and each longer than the next beginning from the longest BX. Letting arcs represent their lengths in symbols, we have,

$$BX > XO > OQ > QP > PK \text{ and } DL = LM = MZ = ZN = NH \qquad (1)$$

while there are the same number of arcs in

$$\{BX, XO, OQ\} \text{ as in } \{DL, LM, MZ\}$$
$$\text{and in } \{QP, PK\} \text{ as in } \{ZN, NH\},$$

(2)

which together imply $\frac{BQ}{DZ} > \frac{QK}{ZH}$. This concludes Case 1.[5]

Demonstration (Case 2). Let ZH not be commensurable with DZ.

Specification (Case 2). It is to be shown that the ratio of BQ to DZ is greater than the ratio of QK to ZH.

Distinction of Subcases. We assume the contrary, that the ratio of arc BQ to arc DZ is less than the ratio of arc QK to ZH (Case 2.1) or equal to it (Case 2.2).

Construction (Case 2.1). Let $\frac{BQ}{DZ} = \frac{QK}{ZL}$ with $ZL > ZH$ as in Figure 82 (a). Let an arc ZM be shorter than ZL, longer than ZH, that is $ZL > ZM > ZH$, and commensurable with DZ. And through M and the pole A draw great circle MN, with N on BE.

Demonstration (Case 2.1). Since ZM is commensurable with DZ, $\frac{BQ}{DZ} > \frac{QN}{ZM}$. But $\frac{BQ}{DZ} = \frac{QK}{ZL}$. Therefore $\frac{QK}{ZL} > \frac{QN}{ZM}$. And alternately: $\frac{QK}{QN} > \frac{ZL}{ZM}$. But QK is shorter than QN. Therefore ZL is also shorter than ZM. But it is also longer, which is impossible. Therefore it is not the case that the ratio of BQ to DZ is less than the ratio of QK to ZH.

Specification (Case 2.2). It is to be shown that the ratio of BQ to DZ is not equal to the ratio of QK to ZH.

Construction (Case 2.2). Assume BQ is to DZ as QK is to ZH as in Figure 82 (b), and bisect DZ and ZH at L and M. Through each of L and M and the pole A draw a great circle LN and MX, with N and X on BK.

Demonstration (Case 2.2). Since DL and LZ are contiguous and equal to each other, BN and NQ are contiguous and BN longer than NQ. Therefore

---

[5] The desired conclusion is $\frac{BQ}{DZ} > \frac{QK}{ZH}$. The argument in the Commentary (p. 339) is as follows.

Condition (1) implies (3) $\frac{BX}{DL} > \frac{XO}{LM} > \frac{OQ}{MZ} > \frac{QP}{ZN} > \frac{PK}{NH}$. From such a chain of inequalities

$$\frac{a_1}{a_2} > \cdots > \frac{z_1}{z_2} \text{ it follows that } \frac{a_1}{a_2} > \frac{a_1 + \cdots + z_1}{a_2 + \cdots + z_2} > \frac{z_1}{z_2},$$

where $+$ means however the magnitudes involved are appropriately combined.

Now (3) allows the right-hand inequality to show that

$$\frac{BQ}{DZ} = \frac{BX + XO + OQ}{DL + LM + MZ} > \frac{OQ}{MZ}$$

(4)

and the left-hand inequality to show that

$$\frac{QP}{ZN} > \frac{QP + PK}{ZN + NH} = \frac{QK}{ZH},$$

(5)

where $+$ means concatenation. So (4), (3), and (5) show that

$$\frac{BQ}{DZ} > \frac{OQ}{MZ} > \frac{QP}{ZN} > \frac{QK}{ZH}$$

as stated.

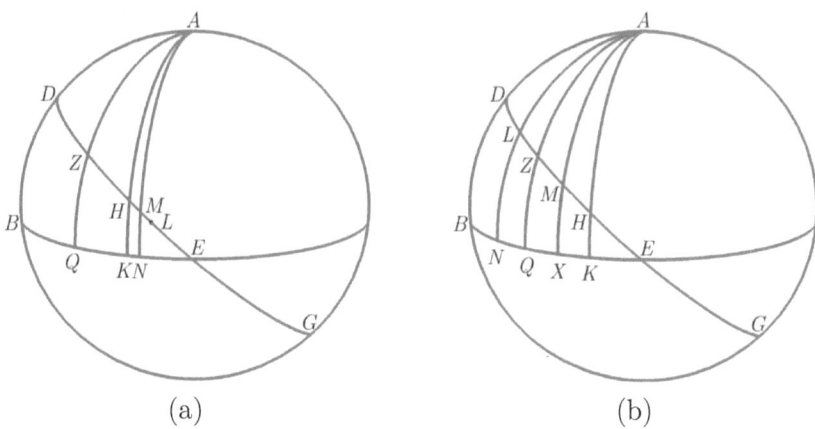

Figure 82: *Sph.* III.10. Constructions for Cases 2.1 and 2.2.

BQ is longer than double NQ. Now we will show similarly that also QK is shorter than double QX. Then, since BQ is longer than double NQ and KQ is shorter than double QX, $\frac{BQ}{NQ} > \frac{QK}{QX}$. And alternately $\frac{BQ}{QK} > \frac{NQ}{QX}$. But it is being supposed that $\frac{BQ}{QK} = \frac{DZ}{ZH}$. Therefore $\frac{NQ}{QX} < \frac{DZ}{ZH}$. But $\frac{DZ}{ZH} = \frac{LZ}{ZM}$. Therefore $\frac{NQ}{QX} < \frac{LZ}{ZM}$. And alternately: $\frac{NQ}{LZ} < \frac{QX}{ZM}$, shown impossible in Case 1. Therefore it is not the case that $\frac{BQ}{DZ} = \frac{QK}{ZH}$.

And it is not the case that $\frac{BQ}{DZ} < \frac{QK}{ZH}$. Therefore $\frac{BQ}{DZ} > \frac{QK}{ZH}$.

֍

Introduction to *Sph.* III.11

See Introduction to *Sph.* III.10–III.12 (p. 462) and for astronomy the Commentary (p. 346).

֍

Theorem. *If the poles of the parallels are on the circumference of a great circle, and two great circles cut it perpendicularly, one of the parallels and the other oblique to the parallels, and another great circle through the poles of the parallels cuts the oblique circle between the greatest of the parallels and the parallel that the oblique circle touches, then the ratio of the diameter of the sphere to the diameter of the circle that the oblique circle touches is greater than the ratio of the arc of the greatest of the parallels between the first great circle and the next great circle through the pole to the arc of the oblique circle between the same great circles.*

Exposition. Let points A and K, the poles of the parallels, be on the circumference of great circle $ABKG$ (Figure 83). Let the greatest parallel circle *BEG* and oblique great circle *DEZ* cut circle *ABG* perpendicularly. Let *DM* be the parallel that *DEZ* touches. Let another great circle *AHQK* through the poles of the parallels cut *DM* at *L*, *DEZ* at *H* between *D* and *E* and *BE* at *Q*.

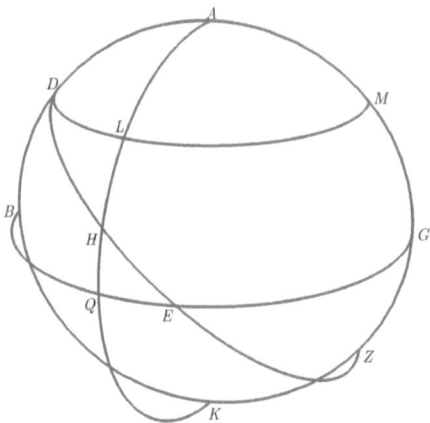

Figure 83: *Sph.* III.11. The given configuration of circles.

Specification. It is to be proved that the ratio of the diameter of the sphere $\underline{DZ}$ to the diameter of circle *DLM*, $\underline{DM}$, is greater than that of arc *BQ* to arc *DH*.

Construction. Draw the parallel circle *NHX* through *H* (Figure 84 (a)), and let the lines of intersection of *ABG* with planes *AHQK*, *DHEZ*, *DLM*, *NHX*, and *BEG* be $\underline{AK}$, $\underline{DZ}$, $\underline{DM}$, $\underline{NX}$, and $\underline{BG}$, and of *AHQK* with planes *NHX*, *DHEZ*, and *BEG* be $\underline{HP}$, $\underline{HO}$, and $\underline{QO}$, with *P* on line $\underline{NX}$ and *O* the center of the sphere, and of *DHEZ* with $\overline{NHX}$ be $\underline{HR}$, with *R* on line $\underline{NX}$. Let the point of intersection of $\underline{AK}$ and $\underline{DM}$ be *S*.

Demonstration. Since the great circle *ABG* in a sphere cuts certain circles *DLM*, *NHX*, and *BEG* in the sphere through the poles, it also bisects them perpendicularly. Therefore $\underline{DM}$, $\underline{NX}$, and $\underline{BG}$ are diameters of the circles *DLM*, *NHX*, and *BEG*, and circle *ABG* is perpendicular to each of circles *DLM*, *NHX*, and *BEG*. Then, since *DLM*, *NHX*, and *BEG* are parallel circles in a sphere and the line $\underline{AK}$ joins the poles, $\underline{AK}$ is perpendicular to each of circles *DLM*, *NHX*, and *BEG* and through their centers and the sphere's. Therefore points *S*, *P*, and *O* are the centers of the circles *DLM*, *NHX*, and *BEG* and *O* of the sphere.

And since parallel planes *DLM*, *NHX*, and *BEG* cut the plane *ABG*, their lines of intersection $\underline{DM}$, $\underline{NX}$, and $\underline{BG}$ are parallel. Again, since the parallel planes *NHX* and *BEG* cut the plane *AHK*, their lines of intersection $\underline{HP}$ and $\underline{QO}$ are parallel. Then, since the intersecting lines $\underline{PN}$ and $\underline{PH}$ correspond to the intersecting lines $\underline{OB}$ and $\underline{OQ}$, in different planes, they contain equal

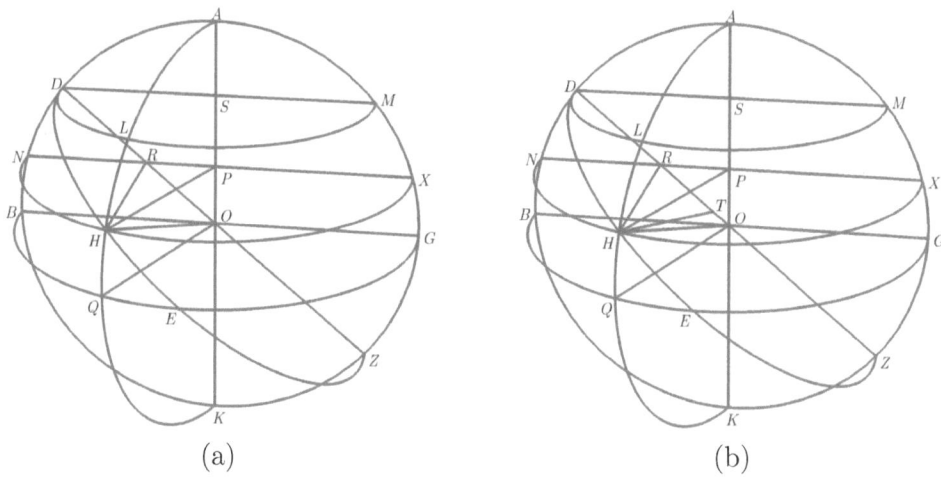

(a)   (b)

Figure 84: *Sph.* III.11. (a) Construction of circle *NHX*. (b) Point *T* constructed and connected to *H*.

angles *NPH* and *BOQ*. And since *NHX* and *DEZ* are perpendicular to circle *ABG*, their line of intersection <u>*HR*</u> is perpendicular to circle *ABG* and makes right angles with every line through *R* in the plane of circle *ABG*. But <u>*HR*</u> meets each of <u>*RP*</u> and <u>*RO*</u> in the plane of circle *ABG*; therefore each of angles *HRP* and *HRO* is right. And since <u>*AK*</u> is perpendicular to <u>*NX*</u>, angle *RPO* is right. Then, since angle *RPO* is right, angle *POR* is acute. Therefore <u>*OR*</u> is longer than <u>*RP*</u>. Therefore let *T* be the point on <u>*RO*</u> such that <u>*RT*</u> is equal to <u>*PR*</u> (Figure 84 (b)), and join <u>*HT*</u>.

In triangles *PRH* and *TRH*, because side <u>*PR*</u> equals side <u>*RT*</u> by choice of *T*, side <u>*HR*</u> is common, and in the previous paragraph it was shown that right angle *HRP* equals angle *HRT*, for it is angle *HRO*, Angle *HTR* is equal to corresponding angle *HPR*, that is angle *HPN*, shown equal to angle *QOB* in the previous paragraph.

Attention now turns to the ratios the theorem is about, relying on a single fact, namely, that since the right triangle *HOR* with right angle at *R* has <u>*HT*</u> drawn across it to *T* on <u>*OR*</u>, the ratio of the side <u>*OR*</u> to the part cut off

$$\frac{OR}{RT} > \frac{\angle HTR}{\angle HOR}. \tag{1}$$

Considering the length ratio of (1), we see that <u>*TR*</u> equals <u>*PR*</u> by choice of *T* so that $\frac{OR}{RT} = \frac{OR}{RP}$. These are sides of triangle *ORP*, similar to triangle

ODS. and so $\frac{OR}{RP} = \frac{OD}{DS}$ first, and then, doubling those lengths,

$$\frac{OR}{RT} = \frac{DZ}{DM},\tag{2}$$

the ratio of diameters sought. Now considering the angle ratio of (1), we recall that the previous paragraph proved that $\angle HTR = \angle QOB$, while $\angle HOR$ is $\angle HOD$. Accordingly

$$\frac{\angle HTR}{\angle HOR} = \frac{\angle QOB}{\angle HOD},\tag{3}$$

which is the ratio of arc BQ to arc DH because O is the center of the sphere. The conditions (2) and (3) transform the inequality (1) into that required.

Therefore the ratio of the diameter of the sphere DZ to diameter DM of circle DLM is greater than that of arc BQ to arc DH.

Comment. The fact labeled (1) above is discussed and proved in the Commentary (see p. 341).

Introduction to *Sph.* III.12

See Introductions to *Sph.* III.10–12 (p. 462) and III.11 (p. 465), and for astronomy the Commentary (p. 352).

Theorem. *If two great circles in a sphere touch the same one of the parallel circles in a sphere cutting off between them similar arcs of parallel circles, and a third oblique great circle touches larger parallel circles and cuts those touching the first parallel circle between the first parallel circle and the greatest of the parallels, then twice the ratio of the diameter of the sphere to the diameter of the second parallel circle is greater than the ratio of the arc of the greatest of the parallels between those touching the first parallel circle to the arc of the oblique circle between the same circles.*

Exposition. In a sphere, let great circles AB and GD touch the same one of the parallels AG at A and G, cutting off between them similar arcs of the parallel circles, including BD on the greatest of the parallels MBZ (Figure 85). And let another oblique great circle EZ touch at E a larger parallel than those AB and GD touch, EH between circles AG and MBZ, the greatest of the parallels. Let the points of intersection of the arcs AB and GD with EZ be Q and K.

Specification. It is to be proved that the ratio of arc BD to arc QK is less than twice the ratio of the diameter of the sphere to the diameter of the circle EH.

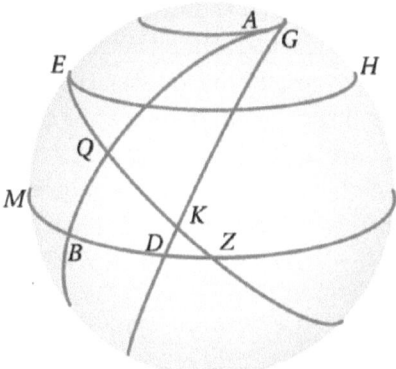

Figure 85: *Sph.* III.12. The given configuration of circles.

Construction. Through L and each of points Q and K draw a great circle LQN and LKX (Figure 86 (a)). Through Q draw great circle QP touching circle EH at P (Figure 86 (b)). And through K draw parallel circle SRK meeting QP, AB, and LQN at O, S, and R.

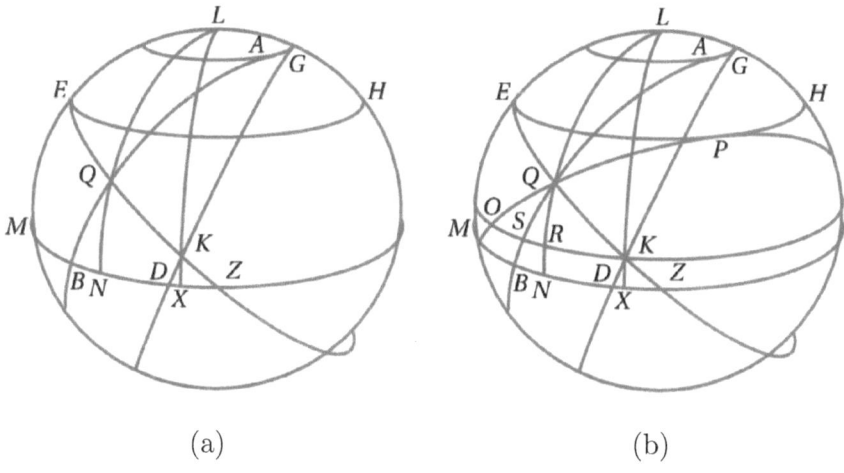

(a)   (b)

Figure 86: *Sph.* III.12. (a) Addition of great circles LEM, LQN, and LKX. (b) Addition of great circle OQP and small circle $OSRK$.

Demonstration. Since OK and EPH are two parallel circles in a sphere, and two great circles $EQKZ$ and OQP are drawn touching EPH at E and P, and through the point Q and the pole L great circle LQR is drawn, arc OR is equal to RK. Therefore SR is shorter than RK. Therefore SK is shorter than twice RK. But SK is similar to BD and RK to NX. Therefore BD is shorter than twice *NX*. And since the ratio of the diameter of the sphere to the diameter of circle *EH* is greater than the ratio of arc *MN* to arc *EQ*, and also the ratio of arc *MN* to arc *EQ* is greater than arc *NX* to *QK*, the ratio of the diameter of the sphere to the diameter of circle *EPH* is greater than the ratio of arc *NX* to arc *QK*. And doubling the antecedents, twice the ratio of the diameter of the sphere to the diameter of circle *EPH* is greater than twice the ratio of arc *NX* to arc *QK*. And twice the ratio of arc *NX* to arc *QK* is greater than arc *BD* to arc *QK*, for twice arc *NX* is longer than *BD*.

Therefore the ratio of arc *BD* to arc *QK* is less than twice ratio of the diameter of the sphere to the diameter of circle *EPH*.

§⟶

Comment. The first sentence of the Demonstration, obvious by symmetry, is justified in the Commentary (see p. 350).

Introduction to *Sph.* III.13

This proposition allows arc lengths to be transferred across the greatest of the parallels as discussed in the Commentary to *Sph.* III.5–III.8 (see p. 314). For astronomy see the Commentary (p. 354).

§⟶

Theorem. *If parallel circles in a sphere cut off equal arcs of a great circle on opposite sides of the greatest of the parallels, and through the points produced great circles are drawn either through the poles of the parallels or touching the same parallels, then they cut off between them equal arcs from the greatest of the parallels.*

Specification. It is to be proved that arc *ZE* is equal to arc *EH*.

Exposition. In a sphere let parallel circles through *A* and *D* cut off equal arcs *AE* and *ED* of great circle *AED* on opposite sides of the greatest of the parallels. And through points *A*, *E*, and *D* let great circles *AZG*, *QEK*, and *BHD* be drawn, either through the poles of the parallels (Figure 87 (a)) or touching the same one of the parallels (Figure 87 (b)), with *Z*, *E*, and *H* being on the greatest of the parallels, *Q* and *B* being on the parallel through *A*, and *G* and *K* on the parallel through *D*.

Demonstration. Since in a sphere, parallel circles *AB* and *GD* cut off equal arcs *AE* and *ED* of great circle *AD* at the greatest of the parallels *ZH*, circle *AB* is equal to circle *GD*. Then, since in a sphere equal and parallel circles *AB* and *GD* cut off arcs *QE* and *EK* of great circle *QK* at the greatest of the parallel circles *ZH*, arc *QE* is equal to *EK*. And *AE* is equal to *ED*. Therefore the line *AQ* is equal to the line *KD*. Therefore arc *AQ* is equal to arc *KD*. And the circles are equal. Therefore arc *AQ* is similar to arc *KD*.

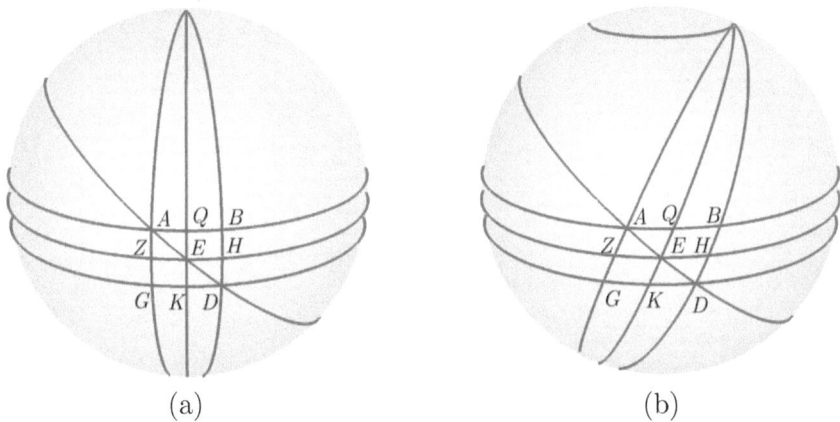

Figure 87: *Sph.* III.13. (a) The great circles *AZG*, *QEK*, *BHD* pass through the poles of the parallels. (b) The great circles *AZG*, *QEK*, *BHD* are tangent to the same small parallel circle.

But *AQ* is similar to *ZE*, and *KD* is similar to *EH*. Therefore *ZE* is similar to arc *EH*, and they are of the same circle.

Therefore arc *ZE* is equal to arc *EH*.

Introduction to *Sph.* III.14

This simple proposition seems to be here for its direct astronomical use, which is discussed in the introduction to the proposition in the Commentary (see p. 357).

Theorem. *If a great circle in a sphere touches a circle among parallel circles in the sphere, and another oblique great circle touches larger parallels than those the first circle touches, they will cut off between them dissimilar arcs of parallel circles, and those nearer to whichever pole are progressively longer than similar to those farther away.*

Exposition. In a sphere let great circle *ABG*, bounding in Figure 88 (a), touch parallel circle *ADX* at *A*. And let another oblique great circle *BG* touch larger parallels than those *ABG* touches.

Specification. It is to be proved that *BG* and *ABG* cut off dissimilar arcs of parallels and that those closer to whichever pole are progressively greater than similar to those farther away.

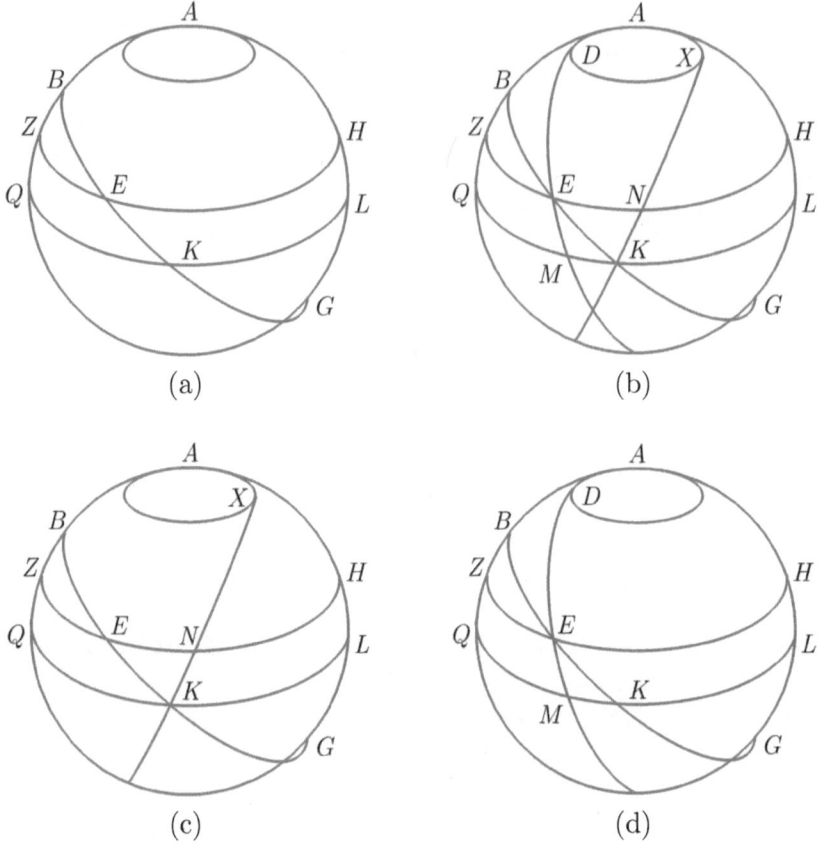

Figure 88: *Sph.* III.14. (a) Given. (b) Construction. (c) First pair of arcs. (d) Second pair of arcs.

Construction. Let two arbitrary points E and K be taken on the oblique circle BG, and through points E and K let parallel circles ZEH and QKL be drawn with Z and Q, H and L on opposite sides of the bounding circle.

Specification. It is to be proved that arc EH is longer than similar to arc KL, and arc QK is longer than similar to ZE.

Construction. Through points E and K draw great circles DEM and XNK touching at D and X the parallel circle touching ABG at A (Figure 88 (b)) and with M on QKL and N on ZEH; so the semicircle from D in the direction of M does not intersect the semicircle from A in the direction of Q and the semicircle from X in the direction of K does not intersect the semicircle from A in the direction of L.

Demonstration. Since the semicircles AL and XK are non-intersecting (Figure 88 (c)) and the arcs they cut off parallel circles are NH and KL, arc

NH is similar to arc *KL*. For the same reason *ZE* is similar to *QM*. And since arc *NH* is similar to arc *KL*, arc *ENH* is longer than similar to arc *KL*.

For the same reasons (Figure 88 (d)) arc *QMK* is also longer than similar to arc *ZE*.

§☛

Comment. This seems a strange way to end a book and the treatise. It is suspected of being an interpolation replacing what previously came after *Sph.* III.13. It could follow *Sph.* II.15.

Bibliography

Primary Sources

Classical Greek and Latin sources are often cited by the abbreviations of Hornblower, S., Spawforth, A., Eidinow, E., 2012. *The Oxford Classical Dictionary*, 4th edition, Oxford University Press, Oxford.

Papyri are cited by the abbreviations of Oates, J.F., Bagnall, R.S., Clackson, S.J., O'Brien, A.A., Sosin, J.D., Wilfong, T.W., and Worp, K.A., 2013. *Check-list of Greek, Latin, Demotic and Coptic Papyri, Ostraca and Tablets*, http://scriptorium.lib.duke.edu/papyrus/texts/clist.html.

Ancient and Medieval Works Referenced by Title or Proposition

Almagest (properly *Mathematical Treatise*, Ptolemy): Greek text edited by Heiberg (1898–1903), German translation by Manitius (1912), English translation and corrections by Toomer (1984).

Analemma (Ptolemy): Greek and Latin texts edited by Heiberg (1907, 189–223), Latin text reedited with English translation and commentary by Edwards (1984).

Architecture (*Arch.*, Vitruvius): Latin text edited with English translation by Granger (1931–1934), Latin text edited with French translation by Soubiran (1969).

Ascensions (Hypsikles): Greek and Arabic texts edited with German translations and commentaries by De Falco, Krause, and Neugebauer (1966), English translation by Montelle (2016).

Collection (*Coll.*, Pappos): Greek text edited with Latin translation by Hultsch (1876–1878). Greek text of Book VII reedited with English translation and commentary by Jones (1986), Greek text of Book IV reprinted with English translation and commentary by Sefrin-Weis (2010).

Commentary on Aristotle's Heavens (Simplikios): Greek text edited by Heiberg (1894). Partially translated into English by Mueller (2005), Bowen (2013), and others.

Commentary on Euclid's Elements I (Proklos): Greek text edited by Friedlein (1873), English translation by Morrow (1970).

Commentary on Ptolemy's Almagest V (Pappos): Greek text edited by Rome (1931–1943, I.1–169).

Commentary on the Phenomena of Aratos and of Eudoxos (Hipparchos): Greek text edited with German translation by Manitius (1894).

Days and Nights (Theodosios): Greek text edited with Latin translation by Fecht (1927), Arabic translation partially edited with English translation by Kunitzsch and Lorch (2011b).

Definitions (*Def.*, attributed to Heron): Greek text with German translation by Heiberg (1912).

Elements (*Elem.*, Euclid): Greek text edited with French translation by Peyrard (1814–1818), Greek text edited by Heiberg and Stamatis (1969–1977), English translation with commentary by Heath (1908), French translation and commentary by Vitrac (1990–2001), Greek text reprinted with Italian translation by Acerbi (2007).

Equilibrium of Planes (*Equi. Plan.*, Archimedes): Greek text edited with Latin translation by Heiberg and Stamatis (1972, II.124–123), revisions to the Greek text with French translation by Mugler (1970–1972, II.80–125).

Geography (Ptolemy): Greek text edited with German translation by Strückelberger and Graßhoff (2006), English translation of the theoretical material by Berggren and Jones (2001).

Geography (Strabon): Greek text edited with German translation and commentary by Radt (2002–2010), English translation by Roller (2014).

Habitations (Theodosios): Greek text edited with Latin translation by Fecht (1927), Arabic and Latin translations edited with English translation by Kunitzsch and Lorch (2011a).

Introduction to Arithmetic (*Int. arith.*, Nikomachos): Greek text edited by Hoche (1866), English translation by D'Ooge (1926).

Introduction to Astrological Effects (Paul of Alexandria): Greek text edited by Boer (1958), English translation by Schmidt (1993).

Introduction to the Phenomena (*Int.*, Geminos): Greek text edited with German translation by Manitius (1898), Greek text reedited with French translation by Aujac (1975), English translation by Evans and Berggren (2006).

Marriage of Philology and Mercury (Martianus Capella): Latin text edited by Willis (1983), English translation by Stahl et al. (1977).

Measurements (*Meas.*, Heron): Greek text edited with German translation by Schöne (1903), facsimile of *Seragliensis* GI 1 (formerly *Contantinopolitani palatii veteris* No.1) along with a transcription of the Greek and an English translation by Bruins (1964), Greek text reedited with French translation by Acerbi and Vitrac (2014).

Method (Archimedes): Greek text edited with Latin translation by Heiberg and Stamatis (1972, 2.426–507). New edition of the Greek in Netz, Noel, Tchernetska, and Wilson (2011, 2.68–127). English translation and summary by Heath (1912, supp. 5–51)

Moving Sphere (*Mov. Sph.*, Autolykos): Greek text edited by Mogenet (1950), English translation by Bruin and Vondjidis (1971), Greek text reedited with French translation by Aujac et al. (2002).

Optics (Euclid): Greek text edited with Latin translation by Heiberg (1895).[1]

Phenomena (*Phen.*, Euclid): Greek text edited with Latin translation by Menge (1916), English translation by Berggren and Thomas (1996).

Planisphere (*Planis.*, Ptolemy): Gerhard of Cremona's Medieval Latin translation edited by Heiberg (1907, 227–259), German translation by Drecker (1927), edition of the medieval Arabic translation with English translation by Sidoli and Berggren (2007).

Restoration of the Spherics (Thābit ibn Qurra): Edition of the Arabic text of Ahmet III 3464 with English translation by Martin (1975), reedition of the same Arabic version, based on all the currently known manuscripts, along with an edition of Gerhard of Cremona's Latin translation by Kunitzsch and Lorch (2010).

Revision of the Spherics (Naṣīr al-Dīn al-Ṭūsī): A printing of the Arabic text is included in the Hyderabad (1939/40) series.

Sand Reckoner (Archimedes): Greek text edited with Latin translation by Heiberg and Stamatis (1972, 2.216–259).

Sizes and Distances of the Sun and Moon (Aristarchos): Greek text edited with English translation by Heath (1913, 352–411).

Spherics (*Sph.*, Menelaos): Greek fragments edited with English translation by Acerbi (2015). Arabic version of Abū Naṣr Manṣūr ibn ʿIrāq

---

[1] It should be noted that both Jones (1994) and Knorr (1994) have argued that the version of the treatise that Heiberg attributed to Theon of Alexandria is, in fact, the less edited version of the text.

edited with German translation by Krause (1936). Arabic version of Abū
ʿAbdallāh Muḥammad ibn ʿĪsā ibn Aḥmad al-Māhānī and Aḥmad ibn
Abī Saʿd al-Harawī edited with English translation by Rashed and Papadopoulos (2017).

Spherics (Sph., Theodosios): Greek text edited with Latin translation by
Heiberg (1927), French translation by Ver Eecke (1927), Greek text edited
with French translation and study by Czinczenheim (2000).

Medieval Manuscripts

Arabic

Beirut, St. Joseph University 223. 16th century. (The original manuscript
was lost during the Lebanese Civil War, but a microfilm is preserved.)

I: Istanbul, Seray, Ahmet III 3464. Late 1220s CE (620s AH), copied by
Muḥammad ibn Abī Bakr ibn Muḥammad.

Leiden, University Library, or. 98.

Tehran, Sipahsalar (now, Kitābkhāna-yi Madrasa-yi ʿAlī Shahīd Muṭahharī)
4727, 1272 CE (671 AH).

Greek

A: Vatican, Biblioteca Apostolica Vaticana, gr. 204. 9th century.

B: Vatican, Biblioteca Apostolica Vaticana, gr. 202. 13th century.

D: Vatican, Biblioteca Apostolica Vaticana, gr. 203. 13th century.

F: Vatican, Biblioteca Apostolica Vaticana, gr. 191. 12th–13th century.

Vatican, Biblioteca Apostolica Vaticana, gr. 180. 10th century.

Vatican, Biblioteca Apostolica Vaticana, gr. 218. 10th century.

K: Paris, Bibliothéque National de France, gr. 2448. 1332 CE.

Paris, Bibliothéque National de France, gr. 2342. Mid-14th century.

Hebrew

New York, The Jewish Theological Seminary of America, Ms. 8182. 1332
CE.

Oxford, Bodleian Library, Hunt. 96. 15th century.

St. Petersburg, Institute of Oriental Manuscripts, C(S) 12. 1499 CE.

St. Petersburg, National Library of Russia, Ms. EVR II A 13, 14th–15th century.

Latin

Milan, Biblioteca Ambrosiana, C 241 inf. 1401 CE.

Paris, Bibliothéque National de France, lat. 9335. Late 12th century.[2]

Vatican, Biblioteca Apostolica Vaticana, lat. 3380.

Modern Scholarship

Acerbi, F., 2007. *Euclide, Tutte le opere*, Bompiani, Milano.

———— 2011. *La sintassi logica della matematica greca*, Archives-ouvertes.fr, Sciences de l'Homme et de la Société, Histoire, Philosophie et Sociologie des Sciences. https://hal.archives-ouvertes.fr/hal-00727063.

———— 2014. "Types, function, and organization of the collections of scholia to the Greek mathematical treatises," *Trends in Classics* 6, 115–169.

———— 2015. "Traces of Menelaus' *Sphaerica* in Greek scholia to the *Almagest*," *SCIAMVS* 16, 91–124.

———— 2016. "Byzantine recensions of Greek mathematical and astrological texts: A survey," *Estudios bizantinos* 4, 133–213.

———— 2017. "The mathematical *scholia vetera* to the *Almagest* with a critical edition of the diagrams and an explanation of their symmetry properties," *SCIAMVS* 18, 133–259.

———— 2020a. "Mathematical generality, letter-labels, and all that," *Pronesis* 65, 27–75.

———— 2020b. "Logistic, arithmetic, harmonic theory, geometry, metrology, optics and mechanics," in Lazaris, S., *A Companion to Byzantine Science*, Brill, Leiden, 105–159.

Acerbi, F., Vinel, N., Vitrac, B., 2010. "Les Prolégomènes à l'Almageste, une édition à partir des manuscrits les plus anciens : Introduction générale – Parties I–III," *SCIAMVS* 11, 53–210.

Acerbi, F., Vitrac, B., 2014. *Metrica: Héron d'Alexandrie*, Fabrizio Serra, Pisa.

---

[2] This manuscript was dated by B. Bischoff (Lorch and Folkerts 1992, 71).

Andersen, K., 2007. *The Geometry of an Art: The History of the Mathematical Theory of Perspective from Alberti to Monge*, Springer, New York.

Arnaud, P., 1984. "L'image du globe dans le monde romain : science, iconographie, symbolique," *Mélanges de l'École française de Rome, Antiquité* 96, 53–116.

Aujac, G., 1970. "La sphéropée, ou la mécanique au service de la découverte du monde," *Revue d'histoire des sciences* 23, 93–107.

———— 1975. *Géminos, Introduction aux phénomènes*, Les Belles Lettres, Paris.

———— 1979. "Regards sur l'astronomie grecque," in Aujac, G., Soubiran, J., eds., *L'astronomie dans l'antiquité classique*, Les Belles Lettres, Paris, 35–54.

———— 1984. "Le langage formulaire dans la géométrie grecque," *Revue d'histoire des sciences* 37, 97–109.

———— 1993. *La sphère, instrument au service de la découverte du monde*, PARADIGME, Caen.

Aujac, G., Brunet, J.-P., Nadal, R., 2002. *Autolycos de Pitane, La sphère en mouvement, Levers et couchers héliaques*, Les Belles Lettres, Paris.

Becker, O., 1933. "Exdoxos-Studien I: Eine voreudoxische Proportionenlehre und ihre Spuren bei Aristoteles und Euklid," *Quellen und Studien zur Geschichte der Mathematik, Astronomie und Physik* B.II, 311–330.

Berggren, J.L., 1976. "Spurious theorems in Archimedes' *Equilibrium of Planes*: Book I," *Archive for History of Exact Sciences* 16, 87–103.

———— 1991. "The relation of Greek spherics to early Greek astronomy," in Bowen, A.C., ed., *Science & Philosophy in Classical Greece*, Garland, New York, 227–248.

Berggren, J.L., Jones, A., 2001. *Ptolemy's Geography*, Princeton University Press, Princeton.

Berggren, J.L., Sidoli, N., 2007. "Aristarchus's *On the Sizes and Distances of the Sun and the Moon*: Greek and Arabic texts," *Archive for History of Exact Sciences* 61, 213–254.

Berggren, J.L., Thomas, R.S.D., 1996. *Euclid's* Phaenomena*: A Translation and Study of a Hellenistic Treatise on Spherical Astronomy*, Garland Publishing, New York.

Bernard, A., Proust, C., Ross, M., 2014. "Mathematics education in antiquity," in Karp, A., Schubring, G., eds., *Handbook on the History of Mathematics Education*, Springer, New York, 27–53.

Besthorn, R.O., Heiberg, J.L., 1897–1905. *Codex leidensis, 399,1: Euclidis Elementa ex interpretatione Al-Hadschdschadschii cum commentariis Al-Narizii*, Libraria Gyldendaliana, Copenhagen (Hauniae).

Bjørnbo, A.A., 1902. *Studien über Menelaos' Sphärik. Beiträge zur Geschichte der Sphärik und Trigonometrie der Griechen, Abhandlungen zur Geschichte der mathematischen Wissenschaften*, Heft 14, Teubner, Leipzig.

———— 1903. "Ein Lehrgang der Mathematik und Astrologie im Mittelalter," *Bibliotheca Mathematica* 4, 288–290.

Blass, F., 1887. *Eudoxi ars astronomica qualis in charta aegyptica superest denuo edita*, Program of the University of Kiel, Kiel.

Boer, E., 1958. *Pauli Alexandrini Elementa apotelesmatica*, Teubner, Leipzig.

Bowen, A.C., 2002, "Simplicius and the early history of Greek planetary theory," *Perspectives in Science* 10, 155–167.

———— 2007. "The demarcation of physical theory and astronomy by Geminus and Ptolemy," *Perspectives in Science* 15, 327–357.

———— 2013. *Simplicius on the Planets and Their Motions*, Brill, Leiden.

Brentjes, S., 2018. *Teaching and Learning the Sciences in Islamicate Societies (800–1700)*, Brepols, Turnhout.

Bruin, F., Vondjidis, A., 1971. *The Books of Autolykos, On a Moving Sphere and On Risings and Settings*, American University of Beirut, Beirut.

Bruins, E.M., 1964. *Codex Constantinopolitanus Palatii Veteris No. 1*, 3 parts, Brill, Leiden.

Bulmer-Thomas, I., 1970. "Theodosius of Bithynia," in Gillispie, C.G., ed., *Dictionary of Scientific Biography*, vol.13, Charles Scribner's Sons, New York, 319–321.

Burford, A., 1972. *Craftsmen in Greek and Roman Society*, Cornell University Press, Ithaca, New York.

Burnett, C., 1999. "Dialectic and mathematics according to Aḥmad ibn Yūsuf: A model for Gerard of Gremona's programme of translation and teaching?" in Biard, J., ed., *Langage, sciences, philosophie au XIIe siècle*, J. Vrin, Paris, 83–92.

——— 2001. "The coherence of the Arabic-Latin translation program in Toledo in the Twelfth Century," *Science in Context* 14, 249–288.

Carman, C.C., 2018. "Accounting for overspecification and indifference to visual accuracy in manuscript diagrams: A tentative explanation based on transmission," *Historia Mathematica* 45, 217–236.

——— 2020. "Vestiges of the emergence of overspecification and indifference to visual accuracy in the mathematical diagrams of medieval manuscripts," *Centaurus* 62, 141–157.

Christianidis, J., Megremi, A., 2019. "Tracing the early history of algebra: Testimonies on Diophantus in the Greek-speaking world (4th–7th century CE)," *Historia Mathematica* 47, 16–38.

Clavius, C., 1586. *Theodosii Tripolitae Sphaericorum libri III ...*, Domenicus Basa, Rome.

Cornford, F.M., 1935. *Plato's Cosmology: The* Timaeus *of Plato*, Routledge, London.

Cribiore, R., 2001, *Gymnastics of the Mind*, Princeton University Press, Princeton.

——— 2009. "Education in the papyri," in Bagnall, R.S., ed., *The Oxford Handbook of Papyrology*, Oxford University Press, Oxford, 330–357.

Cuomo, S., 2000. *Pappus of Alexandria and the Mathematics of Late Antiquity*, Cambridge University Press, Cambridge.

Cuvigny, H., 2004. "Une sphère céleste antique en argent ciselé," in Harrauer, H., Rosario Pintaudi, R., eds., *Gedenkschrift Ulrike Horak (P. Horak)*, Edizioni Gonnelli, Florence, 345–380.

Czinczenheim, C., 2000. *Édition, traduction et commentaire des* Sphériques *de Théodose*, These de docteur de l'Universite Paris IV, Atelier national de reproduction des thèses, Lille.

De Falco, V., Krause, M., Neugebauer, O., 1966. *Hypsikles: Die Aufgangszeiten der Gestirne*, Abhandlungen der Gesellschaft der Wissenschaften zu Göttingen, Philosophisch-Historische Klasse, Dritte Folge, Nr. 62, Göttingen.

Decorps-Foulquier, M., 2018. "'Parts of Text' in the Mathematical Literature of Ancient Greece: From the Author to His Commentator. The Example of *Conics* by Apollonius of Perga," in Bretelle-Establet, F., Schmitt, S., eds., *Pieces and Parts in Scientific Texts*, Springer, Cham, 135–157.

Dekker, E., 2013. *Illustrating the Phaenomena*, Oxford University Press, Oxford.

Dijksterhuis, E.J., 1987. *Archimedes*, Princeton University Press, Princeton.

D'Ooge, M.L., 1926. *Nicomachus of Gerasa: Introduction to Arithmetic*, Macmillan, London.

Drecker, J., 1927. "Das Planisphaerium des Claudius Ptolemaeus," *Isis* 9, 255–278.

Dunbabin, K.M.D., 1999. *Mosaics of the Greek and Roman World*, Cambridge University Press, Cambridge.

Edwards, D.R., 1984. *Ptolemy's* Περὶ ἀναλήμματος: *An Annotated Transcription of Moebeke's Latin Translation and of the Surviving Greek Fragments, with an English Version and Commentary*, PhD Thesis, Brown University, Department of Classics.

Evans, J., 1998. *The History and Practice of Ancient Astronomy*, Oxford University Press, Oxford.

―――― 2016. "Images of Time and Cosmic Connection," in Jones, A., ed., *Time and Cosmos in Greco-Roman Antiquity*, Princeton University Press, Princeton, 143–169.e

Evans, J., Berggren, J.L., 2006. *Geminos's* Introduction to the Phenomena, Princeton University Press, Princeton and Oxford.

Evans, J., Carman, C.C., 2014. "Mechanical astronomy: A route to the ancient discovery of epicycles and eccentrics," in Sidoli, N., Van Brummelen, G., eds., *From Alexandria, Through Baghdad: Surveys and Studies in the Ancient Greek and Medieval Islamic Mathematical Science in Honor of J.L. Berggren*, 145–174.

Fecht, R., 1927. *Theodosii, De habitationibus liber, De deibus et noctibus libri duo*, Abhandlungen der Akademie der Wissenschaften in Göttingen. Philologisch-Historische Klasse; n.F., Bd. 19, Nr. 4, Weidmannsche, Berlin.

Federspiel, M., 1995. "Sur l'opposition défini/indéfini dans la langue des mathématiques grecques," *Les études classiques* 63, 249–293.

—— 2010. "Sur l'élocution de l'ecthèse dans la géométrie grecque classique," *L'Antiquité Classique* T. 79, 95–116.

Feke, J., 2013. "Meta-mathematical rhetoric: Hero and Ptolemy against the philosophers," *Historia Mathematica* 41, 261–276.

—— 2018. *Ptolemy's Philosophy: Mathematics as a Way of Life*, Princeton University Press, Princeton.

Fleck, L., 1935/1979. *Genesis and Development of a Scientific Fact*, translated by Bradley, F., Trenn, T.J., University of Chicago Press, Chicago.

Fowler, D.H., 1999. *The Mathematics of Plato's Academy*, 2nd edition. Oxford University Press, Oxford.

Fowler, D.H., Taisbak, C.M., 1999. "Did Euclid have two kinds of radius?" *Historia Mathematica* 26, 361–364.

Friedlein, G., 1873. *Procli Diadochi in primum Euclidis Elementorum librum commentarii*, Teubner, Leipzig.

Gee, E., 2000. *Ovid, Aratus and Augustus: Astronomy in Ovid's* Fasti, Cambridge University Press, Cambridge.

Goldstein, B.R., Bowen, A.C., 1983. "A new view of early Greek astronomy," *Isis* 74, 330–340.

Granger, F., 1931–1934. *Vitruvius, On Architecture*, 2 vols., Harvard University Press, Cambridge, MA.

Gundel, H.G., 1992. *Zodiakos: Tierkreisbilder im Altertum*, Verlag Philipp von Zabern, Mainz am Rhein.

Hayashi, E., Saito, K., 2009. *Tenbin no Majutsushi: Arukimedesu no Sugaku (Sorcerer of the Scales: Archimedes' Mathematics)*, Kyōritsu Shuppan, Tokyo.

Heath, T.L., 1908. *The Thirteen Books of Euclid's Elements*, 3 vols., Cambridge, Cambridge University Press. (Reprinted: Dover, New York, 1956.)

—— 1912. *The Works of Archimedes, with a Supplement* The Method of Archimedes *Recently Discovered by Heiberg*, Cambridge, Cambridge University Press. (Reprinted: Dover, New York, 2002.)

—— 1913. *Aristarchus of Samos*, Clarendon Press, Oxford.

—— 1921. *A History of Greek Mathematics*, 2 vols., Oxford University Press, Oxford. (Reprinted: Dover, New York, 1981.)

Heglmeier, F., 1988. *Die homozentrischen sphären des Eudoxos und des Kallipos und der itttum des Aristoteles*, Inaugural-Dissertation, Friedrich-Alexander-Universität, Erlagen-Nürnberg.

——— 1996. "Die griechische astronomie zur zeit des Aristoteles: ein neuer anzatz zu den sphärenmodellen des Eudoxos und des Kallipos, *Antike Naturwissenschaft und ihre Rezeption* 6, 51–72.

Heiberg, J.L., 1882. *Litterargeschichtliche Studien Über Euklid*, Teubner, Leipzig.

——— 1883–1888. *Euclidis Elementa, Euclidis opera omnia* vols. 1–5, Teubner, Leipzig.

——— 1894. *Simplicii in Aristotelis de caelo commentaria*, Commentaria in Aristotelem Graeca VII, George Reimer, Berlin.

——— 1895. *Euclidis Optica, Opticorum recensio Theonis, Catoptrica, cum scholiis antiquis, Euclidis opera omnia* vol. 7, Teubner, Leipzig.

——— 1898–1903. *Claudii Ptolemaei Syntaxis mathematica*, Teubner, Leipzig.

——— 1903. "Paralipomena zu Euklid," *Hermes* 38, 46–74, 161–201, 321–356.

——— 1907. *Claudii Ptolemaei opera astronomica minora*, Teubner, Leipzig.

——— 1912. *Heronis definitiones cum variis collectionibus, Heronis quae feruntur geometrica*, Opera quae supersunt omnia, vol. 4, Teubner, Leipzig.

——— 1927. *Theodosius [Tripolites]* Sphaerica,[3] Abhandlungen der Akademie der Wissenschaften in Göttingen. Philologisch-Historische Klasse; n.F., Bd. 19, Nr. 3, Weidmannsche, Berlin.

Heiberg, J.L., Stamatis, E.S., 1969–1977. *Euclidis Elementa, Euclidis opera omnia* vols. 1–5, Teubner, Leipzig.

——— 1972. *Archimedes opera omnia*, Teubner, Stuttgart.

Hoche, R., 1866. *Nicomachi Geraseni Pythagorei introductionis arithmeticae libri ii*, Teubner, Leipzig.

Hogendijk, J., 1986. "Discovery of an 11th-century geometrical compilation: The Istikmāl of Yūsuf al-Mu'taman ibn Hūd, King of Saragossa," *Historia Mathematica* 13, 43–52.

---

[3] Although "Tripolites" appears in the title, it is removed by a corrigendum, p. 3.

—— 1991. "The geometrical parts of the *Istikmāl* of Yūsuf al-Mu'taman ibn Hūd (11th century)," *Archives internationales d'histoire des sciences* 41, 209–281.

—— 2001. "The geometrical works of Abū Saʿīd al-Ḍarīr al-Jurjānī," *SCIAMVS* 2, 47–74.

Høyrup, J., 2014. "Mathematics education in the European middle ages," in Karp, A., Schubring, G., eds., *Handbook on the History of Mathematics Education*, Springer, New York, 109–124.

Huffman, C.A., 2005. *Archytas of Terentum: Pythagorean, Philosopher and Mathematician King*, Cambridge University Press, Cambridge.

Hultsch, F., 1876–1878. *Pappi Alexandrini collectionis quae supersunt*, Weidmann, Berlin.

—— 1886. "Eine Sammlung von Scholien zur Sphärik des Theodosios," *Berichte über die Verhandlungen der Königlich Sächsischen Gesellschaft der Wissenschaften, Philologisch-Historische Klasse* 38/39, 119–128.

—— 1886. "Autolykos und Euklid," *Berichte über die Verhandlungen der Königlich Sächsischen Gesellschaft der Wissenschaften, Philologisch-Historische Klasse* 38/39, 128–155.

Hyderabad 1939/40 (1358 AH) = Naṣīr al-Dīn al-Ṭūsī, "Taḥrīr Kitab al-Ukar li-Thā'ūdhūsiyūs," in *Majmūʿ al-Rasā'il*, Osmania Oriental Publications Bureau, Hyderabad.

ʿId, Y., 1969. "An analemma construction for right and oblique ascensions," *Mathematics Teacher* 62, 669–672.

Johnson, W.A., 2004. *Bookrolls and Scribes in Oxyrhynchus*, University of Toronto Press, Toronto.

Jones, A., 1986. *Pappus of Alexandria, Book γ of the Collection*, Springer, New York.

—— 1991. "The Adaptation of Babylonian Methods in Greek Numerical Astronomy," *Isis* 82, 440–453.

—— 1994. "Peripatetic and Euclidean theories of the visual ray," *Physis* NS 31, 47–76.

—— 2000. "Pappus' notes to Euclid's *Optics*," in Suppes, P., Moravcsik, J. M., Mendell, H., eds., *Ancient & Medieval Traditions in the Exact Sciences: Essays in Memory of Wilbur Knorr*, CSLI Publications, Stanford, 49–58.

—— 2009. "Mathematics, science, and medicine in the papyri," in Bagnall, R.S., ed., *The Oxford Handbook of Papyrology*, Oxford University Press, Oxford, 338–357.

—— 2017. *A Portable Cosmos*, Oxford University Press, New York.

Jones, H.L., 1949. *The Geography of Strabo*, Harvard University Press, Cambridge, MA.

Keyser, P.T., 1998. "Orreiries, the date of [Plato] *Letter* ii, and Eudoros of Alexandria," *Archiv für Geschichte der Philosophie* 80, 241–267.

Knorr, W., 1975. *The Evolution of the Euclidean Elements*, Reidel, Dordrecht.

—— 1978. "Archimedes and the pre-Euclidean proportion theory," *Archives internationales d'histoire des sciences* 28, 183–244.

—— 1985. "Ancient versions of two trigonometric lemmas," *Classical Quarterly* 35, 362–391.

—— 1986a. *The Ancient Tradition of Geometric Problems*, Birkhäuser, Boston. (Reprinted: Dover, New York, 1993.)

—— 1986b. "The medieval tradition of a Greek mathematical lemma," *Zeitschrift für Geschichte der Arabisch-Islamishen Wissenschaften* 3, 230–261.

—— 1992. "When circles don't look like circles: An optical theorem in Euclid and Pappus," *Archive for History of Exact Sciences* 44, 287–329.

—— 1994. "Pseudo Euclidean reflection in ancient optics: A re-examination of textural Issues pertaining to the Euclidean *Optica* and *Catoptrica*," *Physis* 31 Nuova Serie, 1–45.

Krause, M., 1936, *Die* Sphärik *von Menelaos aus Alexandrien in der Verbesserung von Abū Naṣr Manṣūr b. ʿAlī b. ʿIrāq mit Untersuchungen zur Geschichte des Texte bei den islamischen Mathematikern*, Weidmannsche Buchhandlung, Berlin.

Kunitzsch, P., Lorch, R., 2010. *Theodosius* Spherica, *Arabic and Medieval Latin*, Franz Steiner, Stuttgart.

—— 2011a. *Theodosius,* De habitationibus, *Arabic and Medieval Latin Translations*, Bayerischen Akademie der Wissenshaften, Philologisch-Historische Klasse, Heft 1, München, Bayerischen Akademie der Wissenshaften.

——— 2011b. "Theodosius, *De diebus et noctibus*," *Suhayl* 10, 9–46.

Künzl, E., 1997/98. "Der Globus im Römisch-Germanischen Zentralmuseum Mainz: der bisher einzige komplette Himmelsglobus aus dem griechisch – römischen Altertum," *Der Globusfreund* 45/46, 7–153. [German followed by English translation.]

Künzl, E., Fecht, M., Greiff, S., 2000. "Ein römischer Himmels globus der mittleren Kaiserzeit. Studien zur römischen Astralikonographie," *Jahrbuch des Römisch-Germanischen Zentralmuseums Mainz* 47, 495–581.

Lacey, R.A., 1993. *Philoponus: On Aristotle's* Physics 2, Cornell University Press, Ithaca, New York.

Lamoreaux, J.C., 2016. *Ḥunayn ibn Isḥāq on his Galen Translations*, Brigham Young University Press, Provo.

Lasserre, F., 1964. *The Birth of Mathematics in the Age of Plato*, Mortimer, H., trans., American Research Council, Larchmont, New York.

——— 1966. *Die Fregmente des Eudoxos von Knidos*, Walter de Gruyter, Berlin.

Lattis, J.M., 1994. *Between Copernicus and Galileo: Christoph Clavius and the Collapse of Ptolemaic Cosmology*, University of Chicago Press, Chicago.

Le Meur, G., 2012. "Le rôle des diagrams dans quelques traités de la « petite astronomie »," *Revue d'histoire des mathématiques* 18, 157–221.

Lemay, R., 1976. "The teaching of astronomy in medieval universities, principally at Paris in the fourteenth century," *Manuscripta* 20, 197–217.

Leone, P.A.M., 1991. *Maximi monachi Planudis epistulae*, Hakkert, Amsterdam.

Lévy, T., 1997. "The establishment of the mathematical bookshelf of the medieval Hebrew scholar: Translations and translators," *Science in Context* 10, 431–451.

Lo Bello, A., 2009. *The Commentary of al-Nayrizi on Books II–IV of Euclid's* Elements of Geometry, Brill, Leiden.

Lorch, R., 1996. "The transmission of Theodosius' *Spherics*," in Folkerts, M., ed., *Mathematische Probleme im Mittelalter Der lateinische und arabische Sprachbereich*, Harrassowitz, Wiesbaden, 159–183.

———— 2014. "The 'Second' Arabic Translation of Theodosius' *Sphaerica*," in Sidoli, N., Van Brummelen, G., eds., *From Alexandria, Through Baghdad: Surveys and Studies in the Ancient Greek and Medieval Islamic Mathematical Sciences in Honor of J.L. Berggren*, Berlin, Springer, 255–257.

Lorch, R., Folkerts, M., 1992. "Some geometrical theorems attributed to Archimedes and their appearance in the West," in Dollo, C., ed., *Archimede — Mito, Tradizione, Scienza, Nuncius, Studi e Testi*, IV, Biblioteca di Nuncius, Firenze, 61–79. (Reprinted in Lorch, R., 1995, *Arabic Mathematical Sciences*, Variorum, Altershot, article II.)

Luckey, P., 1927. "Das Analemma von Ptolemäus," *Astronomische Nachrichten* 230, 17–46.

Malpangotto, M., 2003. "Sul commento di Pappo d'Alessandria alle *Sferiche* di Teodosio," *Bollettino di Storia delle Scienze Matematiche* 23, 121–148.

———— 2010. "Graphical choices and geometrical thought in the transmission of Theodosius's *Spherics* from Antiquity to the Renaissance," *Archive for History of Exact Sciences* 64, 75–112.

Manitius, K., 1894. *Hipparchi in Arati et Eudoxi Phaenomena commentariorum libri tres*, Teubner, Leipzig.

———— 1898. *Gemini Elementa astronomiae*, Tuebner, Leipzig.

———— 1912. *Ptolemäus, Handbuch der Astronomie*, Tuebner, Leipzig. (Second edition revised by O. Neugebauer, 1963.)

Mansfeld, J., 1998. *Prolegomena Mathematica: From Apollonius of Perga to Late Neoplatonism*, Brill, Leiden.

Marrou, H.-I., 1956, *A History of Education in Antiquity*, Lamb, G., trans., New American Library, New York.

Martin, T.J., 1975. *The Arabic Translation of Theodosius's* Sphaerica, PhD Thesis, University of St. Andrews.

Marx, C., 2020. "The determination of Hipparchus' phenomena and their consistency with the *Almagest* star catalog," *Astronomische Nachrichten* 341, 1043–1053.

Mastorakou, S., 2019. "Aratus' *Phaenomena*, beyond its sources," *Interpretatio* A 3, 1–15.

———— 2020. "Aratus and the popularization of Hellenistic astronomy," in Bowen, A.C., Rochberg, R., eds., *Hellenistic Astronomy: The Science in its Context*, Brill, Leiden, 383–397.

Maurolico, F., 1558. *Theodosii sphaericorum elementorum libri III ...*, Pietro Spira, Messina.

Mendell, H., 1998. "Reflections on Eudoxus, Callippus and their Curves; Hippopedes and Callippopedes," *Centaurus* 40, 177–275.

———— 2000. "The trouble with Eudoxus," in Suppes, P., Moravcsik, J. M., Mendell, H., eds., *Ancient & Medieval Traditions in the Exact Sciences: Essays in Memory of Wilbur Knorr*, CSLI Publications, Stanford, 59–138.

———— 2007. "Two traces of two-step Eudoxan proportion theory in Aristotle: A tale of definitions in Aristotle, with a moral," *Archive for History of Exact Sciences* 61, 3–37.

Menge, H., 1916. *Euclidis Phaenomena ed scripta musica, Euclidis opera omnia, 8*, Teubner, Leipzig, v–223.

Mitchell, A.G., 2012. *Greek Vase-Painting and the Origins of Visual Humour*, Cambridge University Press, Cambridge.

Mogenet, J., 1947. "Les définitions dans l'Ancienne Sphérique," *Annales de la Société scientifique de Bruxelles*, Series 1, 61, 235–241.

———— 1950. *Autolycus de Pitane, histoire du text suivie de l'édition critique des traités De la sphére en mouvement et Des levers et couchers*, Publications universitaires de Louvain, Louvain.

Montelle, C., 2016. "The *Anaphoricus* of Hypsicles of Alexandria," in Steele, J., ed., *The Circulation of Astronomical Knowledge in the Ancient World*, Brill, Leiden, 287–315.

Morrow, G.R., 1970. *Proclus, Commentary on the First Book of Euclid's Elements*, Princeton University Press, Princeton.

Mourelatos, A.P.D., 1980. "Plato's "real astronomy": *Republic* 527D–531D," in Anton, J., ed., *Science and the Sciences in Plato*, Caravan Books, Delmar, NY, 33–73.

Mueller, I., 1980. "Ascending to problems: Astronomy and harmonics in *Republic* VII," in Anton, J., ed., *Science and the Sciences in Plato*, Caravan Books, Delmar, NY, 103–122.

———— 1981. *Philosophy of Mathematics and Deductive Structure in Euclid's Elements*, MIT Press, Cambridge, MA.

———— 2004. "Remarks on Physics and Mathematical Astronomy and Optics in Epicurus, Sextus Empiricus, and Some Stoics," *Aperion* 37, 57–87.

—— 2005. *Simplicius:* On Aristotle's "On the Heavens 2.10–14," Cornell University Press, Ithaca, New York.

Mugler, C., 1959. *Dictionnaire historique de la terminologie géomérique des grecs*, C. Klincksieck, Paris.

—— 1970–1972. *Archimède*, 4 tomes, Les Belles Lettres, Paris.

Nadal, R., Brunet, J.-P., 1984. "Le 'Commentaire' d'Hipparque, I. La sphère mobile," *Archive for History of Exact Sciences* 26, 201–236.

Netz, R., 1997. "Classical mathematics in the classical Mediterranean," *Mediterranean Historical Review* 12: 1–24.

—— 1999a. *The Shaping of Deduction in Greek Mathematics*, Cambridge University Press, Cambridge.

—— 1999b. "Proclus' division of the mathematical proposition into parts: How and why was it formulated?" *Classical Quarterly* (N. S.) 49, 282–303.

—— 2002. "Greek mathematicians: A group picture," in Tuplin, C.J., Rihill, T.E., eds., *Science and Mathematics in Ancient Greek Culture*, Oxford University Press, New York, 196–216.

—— 2009. "Imagination and layered ontology in Greek mathematics," *Configurations* 17, 19–50.

—— 2020. "Why were Greek mathematical diagrams schematic?" *Nuncius* 35, 506–535.

Netz, R., Noel, W., Tchernetska, N., Wilson, N., 2011. *The Archimedes Palimpsest, II, Images and Transcripts*, Cambridge University Press, New York.

Neugebauer, O., 1975. *A History of Ancient Mathematical Astronomy*, Springer, New York.

Nikolantonakis, K., 2016. "Le contenu mathématique et les structures déductives des trois livres du traité *Les Spériques* de Théodose de Tripoli," *Bollettino di Storia delle Scienze Matematiche* 36, 133–171.

Nizze, E., 1852. *Theodosii Tripolitae, Sphaericorum libros tres*, Reimer, Berlin.

Noble, J.V., 1965. *The Techniques of Painted Attic Pottery*, Watson Guptill Publications, New York.

Nokk, A., 1847, *Ueber die Sphärik des Theodosius*, Beilage zum Program des Gymnasiums im Jahr 1847, Carlsruhe.

———— 1850, *Euclid's Phaenomene, übersetzt und erläutert*, Beilage zu dem Freiburger Lyceums-Programme von 1854, Freiburg

Oakley, J.H., 2014. *Athenian Potters and Painters*, Oxbow Books, Oxford.

Ossendrijver, M., 2012. *Babylonian Mathematical Astronomy: Procedure Texts*, Springer, New York.

Pedersen, O., 1975. "The corpus astronomicum and the tradition of medieval Latin astronomy," in Gingerich, O., Dobrzycki, J., eds., *Studia Copernicana XIII*, Polish Academy of Sciences, Wroclaw, 57–97.

Pedersen, O., with Jones, A., 2010. *A Survey of the* Almagest, *with Annotation and a New Commentary, edited by A. Jones*, Springer, New York. (Reprint with additions of the 1974 edition.)

Pena, J., 1558. *ΘΕΟΔΟΣΙΟΥ ΤΡΙΠΟΛΙΤΟΥ ΣΦΑΙΡΙΚΩΝ ΒΙΒΛΙΑ Γ΄, Theodosii Tripolitae Sphaericorum libri tres ...*, André Wechel, Paris.

Pérez Martín, I., Manolova, D., 2020. "Science teaching and learning methods in Byzantium," in Lazaris, S., *A Companion to Byzantine Science*, Brill, Leiden, 53–104.

Peyrard, F., 1814–1818. *Les oevres d'Euclide, en Grec, en Latin et en Français*, C.F. Patris, Paris.

Pingree, D., 1994. "The teaching of the *Almagest* in late antiquity," in Barns, T.D., ed., *The Sciences in Greco Roman Antiquity*, Apeiron 27, 75–98.

Radt, S., 2002–2010. *Strabons Geographika*, Vandenhoeck & Ruprecht, Göttingen.

Rashed, R., al-Houjairi, M., 2010. "Sur un théorème de géométrie sphérique : Théodose, Ménélaüs, Ibn ʿIrāq et Ibn Hūd," *Arabic Sciences and Philosophy* 20, 208–253.

Rashed, R., Papadopoulos, A., 2017. *Menelaus' Spherics: Early Translation and al-Māhānī/al-Harawī's Version*, Walter de Gruyter, Berlin.

Raynaud, D., 2014. "Building the stemma codicum from geometric diagrams: A treatise on optics by Ibn al-Haytham as a test case," *Archive for History of Exact Sciences* 68, 207–239.

Regiomontanus 1496 = *Epytoma Ioannis de Monte Regio in Almagestum Ptolomei*, Johannes Hamman, Venice.

Richter, G.M.A., 1971. *Engraved Gems of the Romans*, Phaidon Press, London.

Riddell, R.C., 1979. "Eudoxan Mathematics and the Eudoxan Spheres," *Archive for History of Exact Sciences* 20, 1–19.

Robins, I., 1995. "Mathematics and the conversion of the mind, *Republic* vii 522c1–531e3," *Ancient Philosophy* 15, 359–391.

Roller, D.W., 2014. *The* Geography *of Strabo*, Cambridge University Press, Cambridge.

Rome, A., 1931–1943. *Commentaires de Pappus et de Théon d'Alexandrie sur l'Almageste*, vols. I-III. Biblioteca Apostolica Vaticana, Vatican.

———— 1933. "Les explications de Théon d'Alexandrie sur le théorème de Ménélas," *Annales de la Société Scientifique de Bruxelles* Série A 53, 39–50.

Rose, P.L., 1975. *The Italian Renaissance of Mathematics*, Librairie Droz, Genève.

Rosenthal, F., 1956. "Al-Kindī and Ptolemy," *Studi Orientalistici in onore di Giorgio Levi Della Vida*, 2 vols., Istituto per l'Oriente, Roma, 436–456.

Saito, K., 2003. "Phantom theories of pre-Eudoxean proportion," *Science in Context* 16, 331–347.

———— 2006. "A preliminary study in the critical assessment of diagrams in Greek mathematical works," *SCIAMVS* 7, 81–144.

———— 2008. "The Diagrams of Book II and III of the *Elements* in Greek Manuscripts," in Saito, K., ed., *Diagrams in Greek Mathematical Texts: Report of Grants-in-Aid for Scientific Research by Japan Society for Promotion of Science (id. number 17300287)*, Osaka Prefecture University, Saikai, 39–80.

———— 2012. "One diagram for multiple cases in Euclid," *Hvmanistica* 7, 17–25.

———— 2014. "The Greek manuscript diagrams of the *Elements* – *Elements*, Book XI," in Saito, K., ed., *Reproduced Diagrams from Greek and Arabic Manuscripts*, Research report, JSPS KAKENHI, "Databasing the manuscript diagrams of sources in ancient and medieval mathematics," Osaka Prefecture University, Sakai.

———— 2018. "Diagrams and traces of oral teaching in Euclid's Elements: labels and references," *ZDM – Mathematics Education* 50, 921–936.

Saito, K., Sidoli, N., 2012. "Diagrams and arguments in ancient Greek mathematics: Lessons drawn from comparisons of the manuscript diagrams with those in modern critical editions," in Chemla, K., ed., *The History of Mathematical Proof in Ancient Traditions*, Cambridge University Press, Cambridge, 135–162.

Schädler, U., 2001. "Griechische Geometrie im Artemision von Ephesos," in Muss, U., ed., *Der Kosmos der Artemis von Ephesos*, Österreichisches Archäologisches Institut, Vienna (Wien), 279–287.

Schiaparelli, G., 1877. "La sfere omocentriche di Eudosso, di Callippo e di Aristotele," (Memoria letta nell'adunanza del 26 novembre 1874.) *Memorie del Reale Istituto Lombardo. Classe di scienze matematiche e naturali* 13, 3–112.

Schmidt, O.H., 1943. *On the Relation between Ancient Mathematics and Spherical Astronomy*, PhD Thesis, Brown University, Department of Mathematics.

Schmidt, R., 1993. *Paulus Alexandrinus: Introductory Matters, Translated by Robert Schmidt*, Golden Hind, Berkeley Springs.

Schöne, H., 1903. *Herons von Alexandria, Vermessungslehre und Dioptra*, Opera quae supersunt omnia, vol. 3, Teubner, Leipzig.

Sedley, D., 1976. "Epicurus and the mathematicians of Cyzicus," *Cronache Ercolanesi* 6, 23–54.

Sefrin-Weis, H., 2010. *Pappus of Alexandria: Book 4 of the* Collection, Springer, London.

Sesiano, J., 2020. "Greek Multiplication Tables," *SCIAMVS* 21, 83–140.

Sezgin, F., 1978. *Geschichte des arabischen Schrifttums, Band VI, Astronomie*, Brill, Leiden.

Sidoli, N., 2004a. "On the use of the term diastēma in ancient Greek constructions," *Historia Mathematica* 31, 2–10.

———— 2004b. "Review of *The Works of Archimedes: Volume I. The Two Books On the Sphere and the Cylinder* by Reviel Netz," *Aestimatio* 1, 148–162.

———— 2006. "The sector theorem attributed to Menelaus," *SCIAMVS* 7, 43–79.

———— 2009. "Drawing diagrams and making arguments in Greek geometry," *Reports of the Tsuda College Institute for Mathematical and Computer Science: Proceedings of the 19th Symposium in the History of Mathematics* 30, 133–150.

———— 2014. "Mathematical tables in Ptolemy's *Almagest*," *Historia Mathematica* 41, 13–37.

———— 2015. "Mathematics education," in Bloomer, W.M., ed., *A Companion to Ancient Education*, Wiley Blackwell, Chichester, 387–400.

———— 2018a. "The concept of *given* in Greek mathematics," *Archive for History of Exact Sciences* 72, 353–402.

———— 2018b. "Uses of construction in problems and theorems in Euclid's *Elements* I–VI," *Archive for History of Exact Sciences* 72, 403–452.

———— 2020a. "Greek mathematics," in Taub, L., ed., *The Cambridge Companion Ancient Greek and Roman Science*, Cambridge University Press, Cambridge, 185–207.

———— 2020b. "Mathematical methods in Ptolemy's *Analemma*," in Juste, D., van Dalen, B., Hasse, D.N., Burnett, C., eds., *Ptolemy's Science of the Stars in the Middle Ages*, Brepols, Turnhout, 35–77.

Sidoli, N., Berggren, J.L., 2007. "The Arabic version of Ptolemy's *Planisphere* or *Flattening the Surface of the Sphere*: Text, translation, commentary," *SCIAMVS* 8, 37–139.

Sidoli, N., Isahaya, Y., 2018. *Thābit ibn Qurra's* Restoration of Euclid's *Data (Kitāb Uqlīdis fī al-Muʿṭaiyāt): Text, Translation, Commentaries*, Springer, New York.

Sidoli, N., Kusuba, T., 2008. "Naṣīr al-Dīn al-Ṭūsī's revision of Theodosius's *Spherics*," *Suhayl* 8, 9–46.

———— 2014. "Al-Harawī's version of Menelaus' *Spherics*," *Suhayl* 13, 149–212.

Sidoli, N., Saito, K., 2009. "The role of geometrical construction in Theodosius's *Spherics*," *Archive for History of Exact Sciences* 63, 581–609.

Soubiran, Jean, *Vitruve. De l'architecture, Livre IX*, Paris: Les Belles Lettres, 1969.

Stahl, W.H., Johnson, R., Burge, E.L., 1977. *Martianus Capella and the Seven Liberal Arts: Volume II, The Marriage of Philology and Mercury*, Columbia University Press, New York.

Stone, E., 1721. *Clavius's Commentary on the Spheriks of Theodosius Tripolitae*, J. Senex, London.

Strückelberger, A., Graßhoff, G., 2006. *Klaudios Ptolemaios, Handbuch der Geographie*, Schwabe, Basel.

Tannery, P., 1893. *Recherches sur l'histoire de l'astronomie ancienne*, Gauthier-Villars & Fils, Paris.

Tarán, L., 1975. *Academica: Plato, Philip of Opus, and the Pseudo-Platonic Epinomis*, American Philosophical Society, Philadelphia.

Taub, L., 2013. "On the variety of 'genres' of Greek mathematical writing: Thinking about mathematical texts and modes of mathematical discourse," in Asper, M., Kanthak, A.-M., eds., *Writing Science*, Walter de Gruyter, Berlin, 333–365.

Thomas, R.S.D., 2018a. "An appreciation of the first book of Spherics," *Mathematics Magazine* 91, 3–15.

——— 2018b. "The definitions and theorems of the *Spherics* of Theodosios," in Zack, M., Schlimm, D., eds., *Research in History and Philosophy of Mathematics*, Birkhäuser, Cham, 1–21.

Thorndike, L., 1949. *The Sphere of Sacrobosco and its Commentators*, University of Chicago Press, Chicago.

Toomer, G.J., 1984. *Ptolemy's Almagest*, Duckworth, London. (Reprinted: Princeton University Press, Princeton, 1998.)

——— 1985. "Galen on the astronomers and the astrologers," *Archive for History of Exact Sciences* 32, 193–206.

Turner, E., Fowler, D.H., Koenen, L., Youtie, L.C., 1985. "Euclid *Elements* I, Definitions, 1–10 (P. Mich. III 143)," *Yale Classical Studies* 28, 13–24.

Ulrich, R.B., 2007. *Roman Woodworking*, Yale University Press, New Haven.

Valleriani, M., 2020. "Prolegomena to the study of early modern commentators on Johannes de Sacrobosco's *Tractatus de sphaera*," in Valleriani, M., ed., De sphaera *of Johannes de Sacrobosco in the Early Modern Period: The Authors of the Commentaires*, Springer, Cham, 1–23.

Various 1488 = *Sphaerae mundi compendium foeliciter inchoat*. Ottavianus Scoto, Venice.

Various 1518a = *Sphera mundi nouiter recognita cum commentarijs & authoribus in hoc volumine contentis videlicet.* Lucantonio Giunta, Venice.

Various 1518b = *Sphera cum commentis in hoc volumine contentis videlicet.* Heirs of Ottavianus Scoto, Venice.

Ver Eecke, P., 1927. *Les* Sphériques *de Théodose de Tripoli,* Desclée De Brouwer, Bruges. (Réimprimé : Albert Blanchard, Paris, 1959.)

Vitrac, B., 1990–2001. *Euclide d'Alexandrie, Les Éléments,* vols. 4, Presses Universitaires de France, Paris.

——— 2012. "The Euclidean ideal of proof in *The Elements* and philological uncertainties of Heiberg's edition of the text," in Chemla, K., ed., *The History of Mathematical Proof in Ancient Traditions,* Cambridge University Press, Cambridge, 69–134.

Vitteli, H., 1887–1888. *Ioannis Philoponi in Aristotelis Physicorum libros tres priores commentaria, Commnetaria in Aristotelem Graeca,* vols. 16 and 17, Berlin, Reimer.

Voegelin, J., 1529. *Theodosii de Sphaericis libri tres ...,* Vienna, Officina Joannis Singrenii.

Wallies, M., 1909. *Ioannis Philoponi in Aristotelis Analytica Posteriora commentaria, Commentaria in Aristotelem Graeca,* Vol. 13, Reimer, Berlin.

Watts, E.J., 2006. *City and School in Late Antique Athens and Alexandria,* University of California Press, Berkeley.

Wendel, C., 1940. "Planudea," *Byzantinische Zeitschrift* 40, 406–445.

Westerlink, L.G., 1967. *Pseudo-Elias (Pseudo-David): Lectures on Porphyry's Isagoge,* North-Holland, Amsterdam.

——— 1971. "Ein astrologisches Kolleg aus dem Jahre 564," *Byzantinische Zeitschrift* 61, 6–21.

Willis, J., 1983. *Martianvs Capella,* Teubner, Leipzig.

Woepcke, F., 1855. "Analyse et extraits d'un recueil de constructions géométriques par Aboûl Wafâ," *Journal asiatique,* cinquième série, 5, 218–256, 309–359.

Wright, M.T., 2019. "The sphere of Archimedes: A precursor to the Antikythera mechanism," in Paipetis, S.A., Kostopoulos, V., eds., *Exact Sciences in Greek Antiquity,* Cambridge Scholars, Newcastle, 254–266.

Yavetz, I., 1998. "On the homocentric spheres of Eudoxus," *Archive for History of Exact Sciences* 53, 221–278.

—— 2001. "A new role for the hippopede of Eudoxus," *Archive for History of Exact Sciences* 56, 69–93.

—— 2003. "On Simplicius' testimony regarding Eudoxan lunar theory," *Science in Context* 16, 319–329.

Zhmud, L., Kouprianov, A., 2018. "Ancient Greek *mathēmata* from a sociological perspective: A quantitative analysis," *Isis* 109, 445–472.

Index of Personal Names

Abū al-ʿAbbās Aḥmad ibn
    al-Muʿtaṣim, 17
Abū, *see* next part of the name
Acerbi, F., 13–15, 19, 45, 60, 77,
    79, 80, 196–198, 202,
    212, 217, 275, 342, 343
Aḥmad ibn Yūsuf, 19
al-, *see* alphabetically by name
Alberti, Leon Battista, 22
Alexander of Aphrodisias, 36
Anaximander, 73
Andersen, K., 22
Andrias, 4
Apollonios, 6, 8, 10, 11, 33, 202,
    266
Aratos, 67
Archimedes, 4, 6, 7, 35, 39, 46,
    68, 73, 299, 303, 331,
    339, 342
Archytas, 5, 35, 38
Aristarchos, 46, 68, 303, 339,
    342
Aristotle, 13, 36–38, 64
Arnaud, P., xviii, 63, 74
Attalos, 7
Aujac, G., 23, 25, 36, 39, 40,
    44–46, 68
Autolykos, 5, 14, 34, 39, 40,
    42–44, 47, 54, 67, 68,
    245, 265, 284
Avery, V., xviii

Bānū Mūsā, 18
Becker, O., 331
Berggren, J.L., 9, 23–25, 34, 44,
    47, 49, 56, 73, 189, 245,
265, 285, 299, 321, 331,
    339, 342
Bernard, A., 7
Besthorn, R.O., 249
Bjørnbo, A.A., 4, 20, 44, 56, 271
Blass, F., 301
Bowen, A.C., 13, 33, 35, 37, 38
Brentjes, S., 15, 18
Bruin, F., 285
Bruins, E.M., 202
Brunet, J.-P., 56
Bryennios, Manuel, 19
al-Būzjānī, *see* Abū al-Wafāʾ
Bulmer-Thomas, I., 5, 43
Burford, A., 71
Burnett, C., 19

Campanus of Novara, 20, 22
    edition of a text, *see Edition*
        in Sources Index
Capella, Martianus, 70, 72, 73
Carman, C.C., 64, 79
Christianidis, J., 26
Cicero, 72, 73
Clavius, Christoph, 22–23, 256
Cornford, F.M., 35
Cribiore, R., 8, 67
Cuomo, S., 73
Cuvigny, H., 73
Czinczenheim, C., xvii, 3, 5, 13,
    16, 19, 45, 75, 77, 87,
    120, 132, 166, 189,
    196–198, 204, 213, 215,
    216, 236, 240, 250, 259,
    260, 262, 264, 268, 273,
    275, 276, 288, 290, 298,
    299, 301, 306, 308, 310,

Index of Sources, Titles, and Propositions

Index of Subjects and Terminology

Index of Greek Terminology